Sustainable Food

myths and solutions

Robert Kourik

Metamorphic Press
634 Scotland Drive, Santa Rosa, CA 95409
www.robertkourik.com

ISBN: 978-0-9615848-8-7

Library of Congress Control Number: 2021906520

Distributed in North America by:
Chelsea Green Publishing, Inc.
85 N. Main St., Suite 120
White River Jct., VT 05001
Orders: 800-639-4099, Phone: 802-295-6300, Fax: 802-295-6444
www.chelseagreen.com
All wholesale and retail sales by Chelsea Green
I do my web-site's sales: www.robertkourik.com

MIX
Paper from responsible sources
FSC® C005010
FSC www.fsc.org

Cover: An overview of the edible landscape that I designed for Lou and Susan Preston at the Preston Farm and Winery, Healdsburg, CA.
Inside title page (i): A vegetable bed in the edible landscape at the Preston Farm and Winery.
Front and back cover photos © Robert Kourik 2021.

Printed in the United States of America, not China. Cruelty-free, not tested on animals.
Vegan paper. Recyclable. Totally groovy.

10 9 8 7 6 5 4 3 2 1

Dedicated to:

Paula Blaydes,
who has always given such good advice
about publishing (and *much, much* more.)

and

The Covid-19 Pandemic,
that gave me *so* much time to complete this book.

Acknowledgments

Paula Blaydes (Santa Rosa, CA) has stood by my side through the good and bad times. A wonderful woman who often makes me laugh, the special woman in my life.

Amie Hill (Sebastopol, CA) has been my steadfast editor over many years. She has added wit, style, and correct grammar to this book and many others.

Lori Almedia (Jenner, CA) The *very* talented artist who did much of the artwork, charts, and graphs. Her lettering is so professional several people asked what typeface I used.

Robert Kourik (Santa Rosa, CA) did all the layout. Any mistakes or odd layouts are my responsibility.

Marty Roberts (Sebastopol, CA), thanks for helping me organize, build, and manage my Web site. (www.robertkourik.com)

Betsy Timm and Don Ketman (Santa Rosa, CA), who have been dear friends over the years (decades!).

Patti Buttitta (Santa Rosa, CA), provided the gorgeous and striking design of the front and back covers.

Michael Weaver of Chelsea Green (White River Junction, VT) who gave me much advice and guided me through the process of getting into the Chelesea Green catalog and Amazon.

Ellen Sherron (Sebastopol, CA) is my trusted indexer. She came out of retirement to do this book. Thanks.

Elvin Bishop and Cara (Marin County, CA) provided much support and encouragment over the years. Also, he's the best winter vegetable grower I know.

Jamie Jobb (Martinez, CA) has been a shrubbery buddy since 1974. Besides turning me on to Monty Python, he's been a great source of information and frivolity. We commiserate about writing and publishing.

Deb Martin (Allentown, PA), was the very professional proofreader. If you find a typo it's due to me missing her notation(s).

Thanks to all my friends and associates who provided kind support. (I swear, your free books are in the mail.): Janet Kourik, Jeff Schalau, Burt Cregg, Cathy Barash, Charlotte at Fulcrum Books, Deep Snow Press (Russia), Kate Burroughs, Linda Chalker-Scott, Ros Creasey, Wendy Krupnick, Robert (Bob) Bugg, Miguel Altieri, Marty Roberts, Joy Larkcom, Frederique Laviopierre, Jenny Blaker, Jono Neiger, Eric Toensmeier, Philip Van Solen, Gladis Zinati (Rodale Inst.), Fred Hoffman, Rachael Freeman Long, Dana Visalli, Keith Ables, Damien McAnany, Daniel Herms, Jude Hobbs, Olivia Yeates, Dan Botkin, Jan Mettler, Tom Christopher, Saxon Holt, Jesse Frohilec, Rose Madrone, Matt Power, Sam Benowitz, Margo Baldwin, Darwen Hennings, Annie Novak, Ben Watson, Mark Morrison, Ned Hearn, Linda Parker, James Owen, Emma Marris, Mimi Luebbermann, Mark Keating, Rawson Gulick, and Sandi Wilson.

Sustainable Food Gardens

Table of Contents

Foreword

If you want your personal landscape to function in harmony with Nature, to flourish without unduly taxing natural resources, Robert Kourik has written your go-to guide. But even if you aren't interested in sustainable gardening, if you are a gardener of any stripe you will find this book invaluable.

It's invaluable because it serves as a corrective to one of the major downfalls of contemporary gardening. It really is a fine book and I foresee a long life for it.

American gardeners are addicted to anecdotal information. Someone notices, for example, that the summer when they planted basil near their roses, the roses flourished, and so they circulate the assertion that roses "like" basil. So much of our current garden literature is based on speculation such as this. That, however, is not Robert Kourik's way.

In contrast, Robert has been haunting agricultural libraries, and more recently, academic sites on the internet, to mine the results of properly constructed scientific studies published in peer-reviewed journals. For decades now he has been collecting facts with the enthusiasm that many other gardeners collect rare plants. What's more, by including in this book countless references about where he found his facts, he invites you to do the same and test his conclusions. This book will not only inform you, it will change the way you look at your craft.

Mind you, I don't mean to minimize the intended message of this book, which is that you can have a beautiful and bountiful personal landscape without depleting the environment around you. This is crucial information, and can even provide a template for addressing the larger environmental challenges that confront us so critically today.

Robert accomplishes this in large part by applying the fruits of his studies to ordinary gardening tasks. If, for example, you want to irrigate efficiently, you'll find Robert's surprising insights about how roots actually grow invaluable. Indeed, in most American climates, his information will relieve you of the need to irrigate at all. Read this book and keep the water in the aquifer or river.

It's easy to follow Robert's directives because the results are so agreeable. Examples of his "edible landscaping" in this book seduce rather than admonish. Likewise, Robert's easy and entertaining writing style make using this book a pleasure. It's typical that after chapters dealing with serious matters such as the persistence of organic matter in the soil in different climates and the pluses and minuses of framing raised beds with recycled plastic lumber, he closes with a chapter about "Sustainable Garden Play."

Gardening, as Robert knows, should be not only sustainable, it must be fun. Take this book as your guide, and it will be.

Tom Christopher, Co-author of *Garden Revolution: How Our Landscapes Can Be a Source of Environmental Change* and the "Growing Greener" Podcast

Introduction

A Tale of Two Titles

My first book (often referred to "permaculture before there was permaculture") is the now-classic *Designing and Maintaining Your Edible Landscape—Naturally*, written in 1986, and still in print. It's 370 pages long, thoroughly indexed, and filled with lots of information, advice, and references to the scientific literature as it existed in the 1970s and 1980s.

How is this book different from that one?" Well, it's miles ahead in its scope, and incorporates many additional years (35) of research and experience. It's pretty much a summary of my work since 1986, with 478 pages, 453 full-color photos, charts, graphs, and black and white illustrations, and a complete 22-page Index with 4,841 entries.

Do you need to buy and read the Edible Landscape Book before reading this one? No. This is a stand-alone book; no need to buy and read the earlier book first (except to support the author).

Why This Book?

Over a decade ago, I decided to put together a sequel to my 1986 book *Designing and Maintaining Your Edible Landscape—Naturally*, a classic (he said modestly) that many people have repeatedly asked me to update. Since its publication, however, my files of scientific papers on a wide range of gardening topics had grown to an impressive size—36 lineal inches of paper-file-cabinet space, 200 books, and at least 1000 digital research documents.

I decided that before I began writing this new book, I really needed to bring some order to my massive collection of other people's work. Most were peer-reviewed papers, often double-blind studies in which two approaches were used to investigate a problem or theory, with a "control" to provide a reference point. The other papers primarily consisted of field research, with both methods and summary reviewed by peers, and suggestions offered as to the written wording of a given study's outcome.

I have to confess that I diddled around for years, getting to this task very randomly between other writing projects. Then the Covid-19 pandemic hit, and suddenly I had a lot of time on my hands. After doggedly imposing some order onto the files, I began to spend three to five hours a day, five or more days per week, writing and laying out each chapter of this extensive and exhaustive tome. (As with most of my books, the writing was the easy part; the better part of the 14 months was spent on editing, graphics, and wrangling software.)

As the book evolved, I came to realize how many gardening "facts" were actually assumptions or myth, with little science to back them up, and that much of the scientific research that I studied seemed to refute popular horticultural trends. With this in mind, I decided to investigate, among other topics, the science behind "permaculture," an ever-increasing trend in horticulture that's becoming rooted in the gardening world. Originally an agricultural term (a combination of "permanent" and "agriculture"), permaculture is promoted as an approach to using nature as a model for organic or natural farming.

Although many of its tropics-based designs in popular permaculture books don't always transfer well to temperate-zone gardening, numerous permaculture concepts continue to be recommended for home gardens by its enthusiasts—the first skid down the slippery slope of gardening myth. Much (but definitely not all) of what I discuss in this book is a science-based review of what can or can't be reduced from a farming scale to home-garden scale, as well as the debunking of a number of myths that need to be addressed for future gardening success.

Almost all food gardens work to some extent. The goal of this book is to make them reach their maximum potential or yields and really flourish—using sustainable methods.

So let's dig in...

Chapter 1
The Basics of Sustainable Gardening

Wikipedia defines a garden as "a planned space, usually outdoors, set aside for the display, cultivation, and enjoyment of plants and other forms of nature."

But what makes a garden sustainable? For years now, gardeners have been searching for an approach to gardening that respects nature, is sensitive to seasonal cycles, works with environmental dynamics, utilizes renewable energy sources, uses compost and home-grown natural fertilizers; and grows healthy plants and food without detrimental chemicals, fertilizers, cultivation methods, and pest controls.

There are a number of gardening approaches and systems that embrace many of the underlying concepts of sustainable gardening, but under different names: regenerative gardening, organic gardening, edible landscaping, agroecology, permaculture, environmentally sound gardening, and appropriate horticulture, to name a few.

A Range of Definitions for Sustainable

One very simple definition of sustainability is "to meet the needs of the present without compromising the ability of future generations to meet their own needs." (Brundtland Commission of the United Nations) A statement from the Sustainable Food Laboratory of White River Junction, Vermont, presents two key concepts of sustainability.

The first is the concept of "needs:" "...in particular the essential needs of the world's poor, to which overriding priority should be given; and the idea of limitations imposed by the state of technology and social organization on the environment's ability to meet present and future needs."

The second concept defines a sustainable food system as "one in which resources (including natural resources such as soil and water, as well as human resources such as labor) are used at their rate of recovery. We define a sustainable food system as one in which the fertility of our soil is maintained and improved; the availability and quality of water are protected and enhanced; our biodiversity...and the flow of energy and the discharge of waste, including greenhouse gas emissions, are within the capacity of the earth to absorb forever."

Here's Wikipedia's summary of permaculture advocate/author Bill Mollison's definition of the practice: "Permaculture is sustainable land-use design. This is based on ecological and biological principles, often using patterns that occur in nature to maximize effect and minimize work. Permaculture aims to create stable, productive systems that provide for human needs, harmoniously integrating the land with its inhabitants. Permaculture principles can be applied to any environment, at any scale, from dense urban settlements to individual homes, from farms to entire regions."

Borage (*Borago officinalis*), above, is a edible blue flower that is attractive to bees and beneficial insects. In moderate climates it blooms nearly all year.

Landscape architect, educator, author, and photographer Owen Dell's basic definition of permaculture is short and sweet: "The development of living systems that are beneficial to individuals, communities, society, and the natural world." He continues, "I see [permaculture] as creating and managing a synthetic ecosystem, one that is as good for everybody (not just humans, of course) as possible." As the author of: *Sustainable Landscaping for Dummies*; *Green Your Home for Dummies*, and *How to Start a Home-Based Landscaping Business,* Owen has been walking the talk for over 40 years and is one of the most knowledgeable, articulate, and prolific landscape designers in the sustainable world.

From Toby Hemenway's *Gaia's Garden: A Guide to Home-Scale Permaculture*: "Ecological [that is, sustainable—ed.] gardens meld the best features of wildlife gardens, edible landscapes, and conventional flower and vegetable gardens. They are based on relatively new concepts such as permaculture and ecological design, yet use time-tested techniques honed to perfection by indigenous people, restoration biologists, organic farmers, and cutting-edge landscape designers. These gardens combine low environmental impact, low maintenance once established, and high yields with elegant aesthetics." (This is part of a longer introduction to permaculture in his book.)

Nature invented fuzzy logic long ago; modern science recently "discovered" it and gave it a cute name. Simply put, fuzzy logic is, as I define it, a concept where there are lots of gray areas—no hard boundaries between right and wrong, yes or no, black or white. This means realizing that there is, for example, no ordained list of "safe" organic pesticides; no simple prescribed time to start summer pruning; no easy way of telling when clay soil can be cultivated without damage; no arbitrarily named "weeds;" no clever jingles or cute sayings with universal applications.

Fuzzy logic can even mean things like rethinking opposition to a preordained list of *verboten* chemical sprays and nutrients, it means that just maybe—if the choice is an intelligent one—one of these potions is appropriate for use in an exceptional situation. Like nature, fuzzy logic is not convenient. The more we garden, the more we realize how much more we *don't* know. Accepting fuzzy logic is a humbling experience.

Egos and Gardening

Unfortunately (although most food gardeners wouldn't agree), even a life-affirming practice like gardening can be wrapped up in ego, rather than sustainability. Consider this scenario:

"I love working with the earth," he croons. He's thin, muscular and very sure of himself. This tanned, supple masseur has spent years handling other bodies for hedonistic and therapeutic reasons. Now, as I watch, he's happily planting bare-root 'Globe' artichoke plants—upside down. He really does want to "work with the soil," and he's confident that his actions will make the earth a better place. Yet, he's murdering innocent artichokes because it's all about him and "working with the earth," and he can't be bothered to learn how to plant them correctly.

While ego and pride are important to human survival, they're not part of the evolution of the natural ecosystem. Beliefs and practices that nurture the fragile human psyche (in the interests of being "right," or being revered as an "expert") can easily maim or kill plants and animals. In the garden, psychological needs may constitute one of our biggest environmental hazards.

We still have direct access to traditions that predated, or just ignored, chemical-farming "progress." In my opinion, the most appropriate sources of organic wisdom are the peasant and non-chemical agrarian cultures that never got attached to synthetic farming and gardening methods. One source of these basic practices may live just down the block—the wise, older gardeners in your neighborhood who learned by doing, not by following fads.

Here's an example: "Ben" was about 75 when we met, and I was astounded to find out that he was growing five types of organic avocadoes on a nearby ridge. Ben was organic by chance, not intent; he just didn't think there was any need to spray his avocado trees.

I learned plenty from him about the types of "avos" which can grow in my marginally subtropical climate. He taught me how to determine the best microclimate for a plant; where to plant a sheltered tree with all the chosen varieties grafted to it as a genetic bank against periodic freezes; and how to execute the complicated grafting that avocados require.

However, there was one big misstep: during our meetings, he talked about working with the famous and respected evergreen pomologist William Chandler, of the College of Agriculture in Los Angeles. I had collected Chandler's rare and out-of-print classic *Evergreen Pomology,* and was impressed by Ben's association with the Master.

Soon Ben began to talk about how he worked with Chandler to develop the "Hass" avocado. I checked Chandler's book; no mention of Ben. Later, when I published an article that implied that Ben had helped to develop our most famous avocado, he flew into a rage. No, he said, he had never had any association with the Hass avocado.

While Ben's story was partially one of deception, self-deception, or misplaced ego, I still learned an immense amount from him, and most of his horticultural facts were accurate and useful. It was a rough lesson for me to learn, though, that one shouldn't assume that everything even a wise gardener says or does is true and correct. A healthy amount of skepticism and a bit of double-checking is usually prudent—and might save a friendship.
There are many valid ways to practice organic, environmentally sound, or sustainable gardening. Unfortunately there's also a tendency in the organic-gardening community to turn originally dynamic and open-ended horticultural approaches into factions based on dogma as inflexible as that of any religious discipline or diet plan. This is especially ironic because of the extent to which nature embodies fluctuation and change.

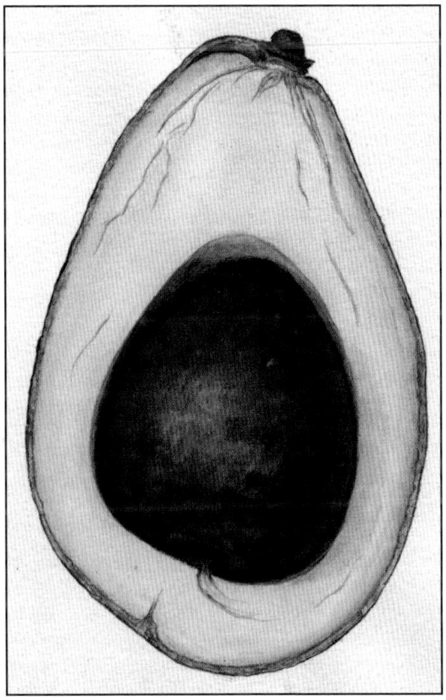

U.S. Department of Agriculture Pomologicatl Watercolor Collection. Rare and Special Collections, National Agricultural Library, Beltsville, MD 20705, c. 4-17-1917.

Dogma

I hate dogma. The reverse spelling of dogma is "Am God," and if there's anything I'm dogmatic about it's dogmatism. This is why most of my books lay out many options for gardening/landscaping, and include the pluses and minuses of most techniques, hardware, and plants.

The best compliment I've ever gotten was "...you don't really tell me which one (technique) to use." That's right you have to make the decision as to what mixture of plants, techniques, and hardware is best for you. Although this book will be full of information on all of the above, it's up to you to use it to figure out your own best practices for sustaining a landscape or garden (but I'll definitely give you lots of options).

As a prime example of dogma, consider French-Intensive Biodynamic gardeners who are often fixated on the practice of "double-digging" to a level of unreasonable passion. Take, for example, one garden near where I live. It's maintained by a non-profit organization, and had been double-dug for over 15 years with countless loads of manure and an endless supply of volunteers and interns, until you could literally dig the soil with your bare hands.

One day a new head gardener proposed the idea of no longer furiously double-digging, but instead just planting according to the no-till methods of Ruth Stout (See page 97 for more details about her). He was nearly thrown out of the organization, and one board member actually slugged him. He stayed, persisted and prevailed; now many of the beds are no longer arbitrarily double-dug, and no-till methods continue to be used successfully in many parts of the garden.

I recommend a two-pronged approach when encountering rigid environmental/organic-horticulture doctrine: first, review and discuss with other gardeners what you perceive to be the facts—taking into consideration both your own and others' actual horticultural experience, and using the tenets of western science as a foundation. Secondly, be prepared to change your views and methods constantly in the light of new insights and scientific discoveries.

I constantly research the scientific literature to substantiate or refute my own gardening assumptions, which could otherwise turn into my own personal myth and dogma. As an example, in my 1986 book, *Designing and Maintaining Your Edible Landscape – Naturally*, I wrote about so-called "dynamic accumulators." That list is now the basis, in some way, for every Internet article and list on the topic.

By 2015, however, I realized that the information and concept I'd published was based on *way* too much anecdotal information, so that fall I contributed a paper to *Permaculture Design* that explained why I'd changed my mind, and provided the new data to replace the outdated concept. (For more see pages 241 to 245.)

Gardening is always an act of faith, because we have so little grasp of how things really work. Nonetheless, I still prefer to apply the rigors of scientific inquiry as a fascinating double-check against my intuitive presumptions.

My Bias

It goes without saying (but I will anyway) that I use organic and environmentally sound gardening techniques. There have been a lot of words and phrases over the years to describe this approach. The oldest (from the sixties and seventies when this concept meant you were, to the "normal" gardening world, a looney outcast heretic) may be "organic."

My introduction to this style of gardening was Bargyla and Gylver Rateaver's 1973 book *The Organic Primer*. (This is where I first, in 1975, encountered the concept of "accumulators." I added "dynamic" in 1986.) This self-published organic gardening book was the first one on the subject that I bought. At the time, such writings were rare indeed, although one could, with a bit of effort, find the early Rodale Press issues of *Organic Gardening* magazine and the brown-papered first issues of *Mother Earth News*.

Next came "appropriate horticulture," a play on the phrase and concept of "appropriate technology." Appropriate technology referred to solar and wind power and other alternative energy technologies. Then, eventually, came the term "sustainable." I like this word better that the more currently fashionable word "permaculture," because it just seems more accurate.

Technically "permaculture," as originally defined by the late Bill Mollison and David Holmgren, is a word synthesized from "permanent" and "agriculture." But I don't farm; I garden. Picky maybe, but that's why I use "sustainable" to describe my design approach. "Regenerative" is the up-and-coming word now. But again, often referring to large-scale agriculture. So, "sustainable" it is, for me, and for this book.

"Import-Export" Economies/Ecologies

Outside inputs to one's garden or landscape—manure, bagged compost, alfalfa meal, liquid kelp, etc.— may be labeled "organic," but all require additional land, water, and resources to produce and move; this is hardly sustainable gardening. An ideal organic garden is, as much as possible, a closed system and a self-sustaining ecology, its only "outputs" nutritious food for the table.

This means that the would-be organic gardener has two choices: (1) import nutrients to force an intensive yield in a smaller space, or (2) use wider spacing and fewer imports when improving the soil and planting.

The thoughtful gardener makes a careful choice between vertical (intensive) versus horizontal (wider spacing) cultivation. This, of course, depends on the size of your garden; vertical cultivation may be your only option, and you may need to import additional resources. By "vertical," I mean that the vegetables are planted so closely together that all vertical space is used. "Horizontal" plantings tend to feature crops grown in rows and spaced farther apart.

Time, patience, and space, versus speed (little time), impatience, and limited space, are the two main garden management choices to consider.

For any vertical garden, the best model, in my opinion, is the French Intensive/Biodynamic method. (The phrase is also written as Biodynamic/French Intensive.) Alan Chadwick brought the method to America from England in 1966. While creating gardens at the University of California in Santa Cruz, he claimed yields of four to six times that of the conventional commercial production of fruits, vegetables, and grains. He also claimed this could be achieved using a quarter of the fertilizer, one-eighth of the water and one-hundredth of the energy per pound of food harvested.

Though those following his tenets have found decent yields, few are recorded as being as tremendous as Chadwick claimed. In a study over three years in a single raised bed at Dawes Arboretum near Newark, Ohio, an average yield of 1.24 pounds per square foot was recorded, as compared to 0.6 pounds per square foot with traditional row-gardening. This is still impressive for vertical gardening. (As reported in *Raised Bed Revolution: Build It, Fill It, Plant It ... Garden Anywhere!*)

John Jeavons has registered the phrase GROW BIOINTENSIVE as a way to distinguish his method from the simplistic approaches that many gardeners use. He has been collecting data for 39 years, and claims that:

"GROW BIOINTENSIVE mini-farming techniques make it possible to grow food using:

- 67% to 88% less water.
- 50% to 100% less fertilizer.
- 99% less energy than commercial agriculture, while using a fraction of the resources.

These techniques can also:

- Produce two to six times more food.
- Build the soil up to 60 times faster than in nature, if properly used.
- Reduce by half or more the amount of land needed."

Edible Landscaping

In the mid-90s someone asked me "What ever became of edible landscaping?" I realized that I wasn't sure. In the '90s, I realized that the phrase had entered the mainstream when I saw a cartoon of two women talking over a wooden fence. In the background their husbands (boyfriends?) are on their hands and knees "grazing" the lawn. The text said, "Must be one of those edible landscapes."

So, if the term has been mainstreamed, where are all the edible-ornamental front yards? It seems that it takes a long time to "turn the boat around" when it comes to changing the direction of a gardening style. It took over 20 years for Dr. Alex Shigo to train us (or at least most of us) not to prune flush with the trunk of a tree. How long will it take before beautiful, edible yards are in fashion? There's a growing trend of planting food gardens in front yards, but many have no pleasing aesthetic, especially in the winter.

Some say permaculture has nothing to do with edible landscaping. They must never have read my book *Designing and Maintaining Your Edible Landscape—Naturally*. Other permaculturists use my book as a reference, and sometimes as required reading. Some permaculture gardeners refer to their work as edible landscaping. These two differently named gardening methods can certainly find common ground and "sleep in the same bed," as it were. But although most edible-landscape books have always promoted or embraced organic, non-petrochemical methods, they certainly don't follow the dogmatic rules required for a permaculture design certificate.

From *Permaculture: Design for Cultivating Ecosystems*:

"*Introduction to Permaculture* and Ross and Jenny Mars' *Getting Started in Permaculture* give many ways to make better use of limited space and recycled materials to increase food production, but Robert Kourik's *Designing and Maintaining Your Edible Landscape—Naturally* is perhaps the most complete text for integrating intensive managed-food production into modern living environments." (End of shameless self-promotion.)

My edible-landscaping book came out in 1986, about four years after Rosalind Creasey's ground breaking book *The Complete Book of Edible Landscaping: Home Landscaping with Food-Bearing Plants and Resource-Saving Techniques*. Though the timing was largely unintentional, Ros's main concept, that of opening the entire yard to ornamental edibles, was followed by my more technical, how-to book. They still make perfect companion books.

A Review of Edible Landscaping

Long before the Berlin wall came down, the arbitrary barriers between vegetable gardens and flower borders, between fruit trees and shade trees; and between herbs and ornamental flowers had begun to crumble. This trend in gardening and landscaping began in the mid-1970s, embraced by those who felt impelled to bring food gardening into its rightful place—around the home.

A beautiful edible landscape or garden can combine a cornucopia of edible plants—a giant step beyond regimented rows of regular old corn and beans. There's the sculptural, lime-green, and tasty 'Romanesco,' a member of the broccoli and cauliflower family the vivid orange and yellow edible petals of the colorful bedding plant calendula, the sweet fragrance of lavender wafting throughout the garden (its foliage can be used in outdoor grilling and its flowers add a wonderful flavor to ice cream and lemoncello), and the spicy taste and colorful highlights of nasturtium's spring blossoms, leaves, and young seedpods

Instead of ordinary shade trees, edible-landscapers can choose to grow the sweetly tart 'Pink Pearl' apple with its marbled pink-and-white flesh. Rather than picking flowers only for ornamental reasons, they can sprinkle their salads with a rainbow of color—the spectacularly edible blossoms of pansies, nasturtiums, and roses, or the sweetly complex anise hyssop.

Edible landscaping means combining the hedonism of great flavors and wonderful fragrances with a rich tapestry of seasonal color. But this bounty-with-beauty doesn't happen without study, tools, neighborly tips, and hard work.

Figure 1. The entry to Ros Creasy's front-yard edible landscape, c. 2014.

Avoiding the Tyranny of Edibility

Some people assume that an edible landscape involves replacing all their ornamental plants with edibles. Heavens, no! A landscape composed entirely of edible plants would be an enormous burden to all but the most compulsive, wealthy—and probably retired—gardeners. The goal of edible landscaping is to create a productive landscape that appears ornamental in its overall design. In my original 1986 book, I recommended: "Those with yards less than 2400 square feet in size should have no more than 50% of the area planted to edibles. And these figures may be way too ambitious."

I now think this is *far* too much area, more than anyone but the most compulsive gardener can nurture, given how our busy lives have ratcheted up so much since the 1980s. My recommended edible quotient is down to more like 100-600 square feet out of 2400 (not including some tree crops and edible ornamental perennials). According to the Cornell Lab of Ornithology, the average area of non-commercial vegetable production is 392 square feet. A survey done by the National Gardening Association says it's an average of 600 square feet for a family of four, with a median (midpoint) of 96 square feet.
My advice has always been to start *very* small and grow only what you can reasonably harvest without your edible landscape turning into a burdensome job.

(Begin tirade.) In regard to the increasing number of books saying you should replace your lawn with edibles (and/or native plants), I say they must have never been to St. Louis County where I grew up. There are miles and miles of homes with huge lawns. Our lawn, at one-half acre, was one of the smallest around. The lawns I mowed to make money in high school ranged from one-half to two acres.

According to the Cornell Lab of Ornithology, the average area of a lot in America is 9,801 square feet (.225 acre; another source places it at 9,100 square feet, or 20% of an acre.) Of that space, the lawn takes up 56 percent, or 5,489 square feet. Try converting that percentage to edibles-only, or even just ornamentals, and you'll see your gardening time soar in comparison with the time needed to just mow your lawn. While the average vegetable garden takes up only 392 square feet, or 4%, of the average yard, vegetable gardens often take as much as 20 hours of tending per season per 100 square feet, while lawns only need an hour of labor per 100 square feet per season.

To look at it another way, the average homeowner spends 150 hours per year mowing; or .4 hours per day (2.8 hours per week). The average gardener spends 4.5 hours per week tending an average 600 square-foot food garden.

I have the same opinion of those who would have us replace all the lawns in America with grass meadows: give me a break. Granted, housing is so expensive now that many houses are built on lots that are so small that the lawn seems to be an afterthought. (End of tirade.)

The Edible Future

I'm happy to report that edible landscaping is far from dead. In fact, there has been a resurrection of the concept, one that has embraced many of the sustainable organic methods of past decades and some of the tenets of permaculture.

Rosalind Creasy's most recent book on the subject, *Edible Landscaping, Now You Can Have Your Gorgeous Garden and Eat it Too!* (Sierra Club Books, 2010), is a tour-de-force, depicting edible landscapes as art *and* food, with brilliant photos and wonderful accompanying text. It also contains a very thorough encyclopedia of dozens of edible plants. In my opinion, no other book on the subject will ever surpass it.

Tropics versus Temperate Zones

In all the reading on "ecology" I was doing in the 1970s, one running thread fascinated me. Much of the advice being giving seemed to apply mainly to semi-tropical or downright tropical ecologies. As I attempted to follow it, I was not very clear on the real differences between tropical and temperate-zone ecologies, and found myself often trying too hard to push the local microclimate's envelope—attempting to grow crops like chayotes, Chilean guavas, cherimoyas, and avocados in not-very-tropical Northern California coastal Zone 9b-10a.

Now that I know considerably more, I'm going to write at some length about the difference between tropical and temperate ecologies, because there's *so* much confusion in the gardening literature. This is especially the case with permaculture books, articles, and blogs, where many recommendations concerning permacultural gardens and forest farms fail to explain that the original ecological modeling was actually based on tropical and subtropical ecosystems.

While tropical forests often contain lots of vertically integrated plants, the temperate American landscape, with its hardwood forests, meadows and prairies, is much less vertically complex. Temperate forests are often dominated by about a half-dozen tree species or so per acre, while a tropical rainforest can contain more than *480* tree species in just 2.5 acres (one hectare). (Another intriguing example of the great discrepancy is that one single bush in the Amazon may host 12 more species of ants than exist in *all* of the United Kingdom—Hölldobler & Wilson (1990), p. 40; from Wikipedia.)

Almost any attempt to take a tropical model of mixed plantings and transfer it to a temperate climate is fraught with problems. In North America's temperate zone, forests and fields store most of their nutrients in the soil, while tropical forests and farms store most nutrients aboveground in the foliage of plants. Because the nutrient mass in a tropical ecosystem exists primarily above the ground, the soil is rather shallow—sometimes less than two inches thick. In P.K. Ramachandran Nair's *An Introduction to Agroforestry,* we read: "Most of the [tropical food] systems are dominated by perennial rather than annual crops, resulting in a relatively high ratio of nutrients stored in the vegetation to that stored in the soil." [*An Introduction to Agroforestry* by P.K. Ramachandran Nair, 1993, Kluwer Academic Publishers, 101 Philip Drive, Norwell, MA 02061, U.S.A.]

In *The Importance of Soil Organic Matter, Key to Drought-Resistant Soil and Sustained Food and Production,* Alexandra Bot and José Benites write: "The relatively faster rate of decomposition induced by the continuous warmth in the tropics implies that high equilibrium levels of organic matter are difficult to achieve in tropical agro-ecosystems. Hence, large annual rates of organic inputs are needed to maintain an adequate soil organic matter pool in cultivated soils. Soils in cooler [temperate] climates commonly have more organic matter because of slower mineralization (decomposition) rates."

From another study: "Studies [of tropical and temperate soils] have shown a rapid loss of added carbon during the first 3 to 6 months in tropical soils (of Nigeria) and that the time scale (for carbon oxidation) was about four times longer in temperate soils (of England). Under intermediate climatic conditions in South Australia, the decomposition rate was half that of Nigeria. This shows the effect of climate on organic matter decomposition and how organic matter is oxidized or lost at a much faster rate in the tropics as compared to subtropical and temperate zones.

While it is true that rates of organic-matter decomposition are higher and therefore it is more difficult to maintain organic matter levels in the tropical as compared to temperate soils, there is little difference in the quality and effectiveness of humus in tropical and temperate soils."

Nitrogen

While some plants in the tropics actually fix nitrogen nodules on their leaves, in a temperate ecosystem the primary reservoir of nitrogen and other nutrients is the deeper soil.

Experiments in a Costa Rican rainforest revealed that fixed nitrogen [from blue-green algae growing on the foliage] is also directly transferred into the leaves of the tropical plants. From the Mongabay.com site, Rhett Butler writes: "The colonial settlers [of tropical South America] did not realize that they were dealing with an entirely different ecosystem from their temperate forests where most of the nutrients exist in the soil. In the rainforest, most of the carbon and essential nutrients are locked up in the living vegetation, dead wood and decaying leaves. As organic material decays, it is recycled so quickly that few nutrients ever reach the soil, leaving it nearly sterile."

From *Nitrogen Fixation in the Canopy of Temperate Forest Trees: A Re-examination*, December 1982. K. Jones, Department of Biological Sciences, University of Lancaster:

"Nitrogen studies...of leaves and shoots of Douglas fir and other (temperate) forest trees do not confirm previous reports that extensive nitrogen fixation occurs on leaf surfaces, and it is concluded that the importance of nitrogen fixation in the canopy of (temperate) forest trees has been exaggerated."

"In temperate climates, nitrogen-fixing bacteria—primarily bacteria found on the roots of the bean family (*Fabaceae* or *Leguminosae*), usually *Rhizobium* spp. Bacteria—grow mostly on roots beneath the soil surface. Generally, rates of nitrogen-fixing for deciduous forests are similar to rates proposed for coniferous forests. High rates for tropical rain forests are mainly the result of a large, nitrogen-fixing canopy lichen population." (Edmisten, 1970). *Nitrogen Fixation In A Deciduous Forest In The South Eastern United States* R. L.. Todd, R. D. Meyer & J. B. Waide)

The amounts and rates of production of nitrogen available to soil microbes and plants are much greater in many tropical-forest soils than in their temperate counterparts; this is due to high rates of nitrogen fixation by free-living microbes and rhizobia associated with legumes, which are especially abundant in many tropical ecosystems. One study found that biological nitrogen fixation rates in tropical forests (15 to 36 kg Nitrogen ha yr) [shorthand for kilograms of nitrogen per hectare per year] are similar to or higher than estimates for their temperate counterparts (7 to 27 kg Nitrogen ha yr). Another, using the same index, found that tropical ecosystems held 88 kg Nitrogen ha yr of fixed nitrogen compared to 12-14 in deciduous hardwood forests.

Several field studies have shown that temperature is a key factor controlling the rate of decomposition of plant residues. This indicates that decomposition normally occurs more rapidly in the tropics than in temperate climates. In a 1985 study, Ladd and Amato reported that: "... despite the importance of soil organic matter differences in plant material and climate patterns, the decomposition of leguminous materials in southern Australian sites followed the same pattern as that of ryegrass for sites in Nigeria and the United Kingdom."

"Reaction rates, however, doubled for each increase of 8–9 °C [46-48°F] in the mean annual air temperature. The relatively faster rate of decomposition induced by the continuous warmth in the tropics implies that high equilibrium levels of organic matter are difficult to achieve in tropical agro-ecosystems. Hence, large annual rates of organic inputs are needed to maintain an adequate available soil organic matter pool in cultivated soils. Soils in cooler climates commonly have more organic matter because of slower mineralization (decomposition) rate." (From: *The Importance of Soil Organic Matter Key to Drought-Resistant Soil and Sustained Food and Production*, 2005; Alexandra Bot and José Benites; Food and Agriculture Organization of the United Nations, Rome.)

The first six to eight inches of soil in the wet tropics is composed of decaying leaves, wood, and other organic matter—the richest sources of nutrients are on the ground. Thus, canopy trees are shallow-rooted, whereas many temperate tree roots extend more than 5 feet deep. Many tropical trees have roots that actually grow out of the ground to form a mat in order to more efficiently collect nutrients. These tiny roots, along with the *mycorrhizae*, or fungal associations, rapidly absorb nutrients.

The shallow roots and great height of tropical trees cause a great deal of instability, especially with wet soils and strong winds. To counter this, many tree species grow extensive root systems that in some cases may extend to over 325 feet, much further than those of trees in temperate-zone climates.

Practical Solutions

Since I have no practical rainforest experience, this book will concentrate on the temperate zones of the US mainland. Here are some quotes taken from the earlier parts of this chapter, followed by useful facts and suggested substitutions and solutions for temperate-zone planting.

- "A tropical rainforest can contain more than 480 tree species in just 2.5 acres (one hectare)."

The temperate zone can't sustainably support this many trees per acre. While a deciduous woodland can contain as many as a half- dozen tree species, it's not known how many the average forest can actually sustain—and of course the numbers would vary considerably from zone to zone and ecotone to ecotone (an ecotone is a region of transition between two biological communities—ecotones between two habitats are often richer in species than either community).

The ideal is to limit the density of your planting, as much as possible, to the minimum number of fruit and nut trees you need.

So many illustrations of forest-farm plantings show a cornucopia of trees, from large to small. (See Figure 2, based on Bill Mollison's *A Permaculture Primer*, 1988). Mollison loved to travel to tropical regions, and many of his illustrations are actually based on tropical ecosystems. Thus, unbeknownst to many readers, the illustrations often portray more plants, or plants closer together, than a temperate climate can ultimately sustain.

Many new food forests/gardens/edible landscapes look great initially when planted with lots of tree crops but, as time goes on, these plantings tend to need more and more resources (such as manure, fertilizers, and compost), to sustain the same level of growth. Be sure to research how many trees your ecosystem tends to support, and adjust the number of tree crops in your yard to this number as best as you can. In Mollison's 1988 *Permaculture, A Designer Manual*, 78 illustrations are based on tropical models, and only 48 for temperate zones (nine of the illustrations are for forest farms).

Figure 2. Bill Mollison traveled to the tropics a lot. In this illustration, based on his book *Permaculture, A Designers' Manual* (Tagara Press, 1988), we see more tropical trees planted in one area than I think would be appropriate in temperate zones. The tropics can sustain 100-750 species of trees per acre, as opposed to only 69 species per acre in a temperate forest. In warmer San Bernardino, CA, it's 40-200 trees per acre, while in New England it's 40-60. Tropical forests usually consist of 200 or more trees per acre.

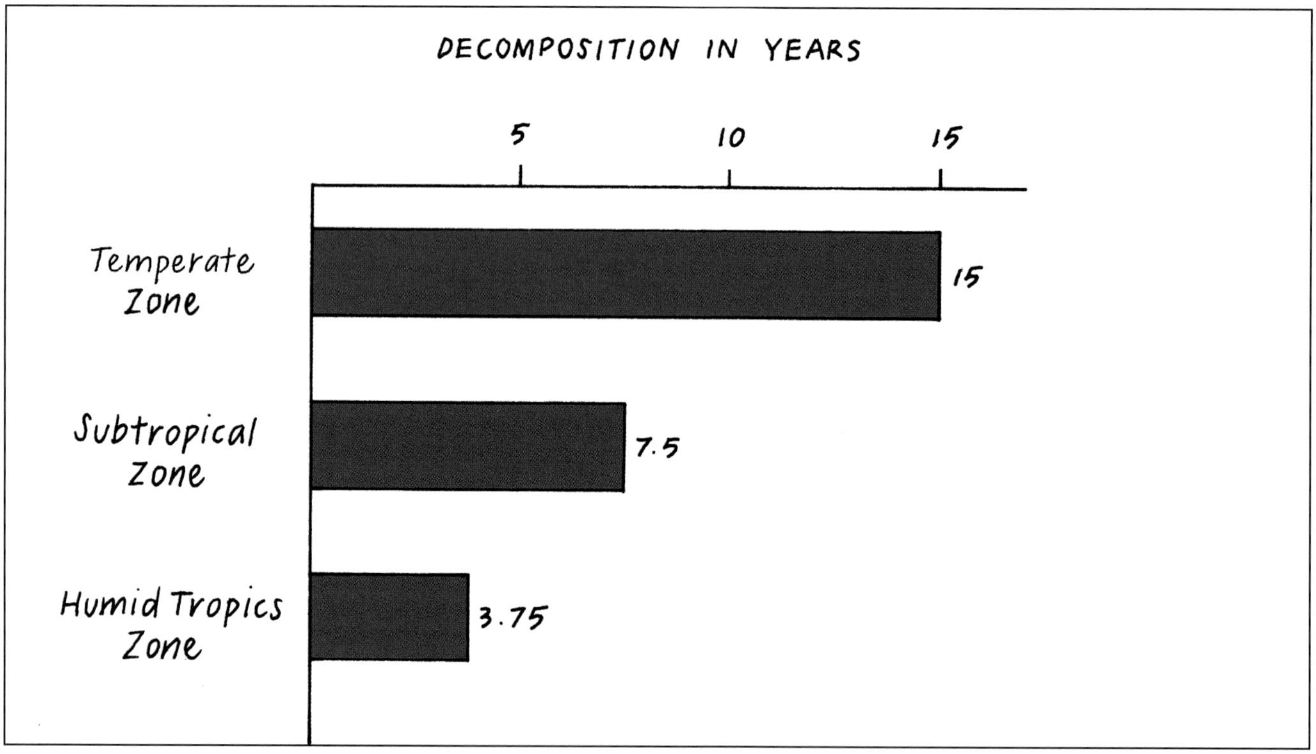

Figure 3. Another limiting condition in the soil is its organic matter, which decomposes into humus (the dark organic material produced in soils by the decomposition of vegetable or animal matter), which is essential to the fertility of the earth. This very important illustration shows how quickly organic matter decomposes in the tropics. The humid tropical climate "burns" through organic matter about five times faster than in more temperate zones.

Some gardening books, like the *Western Garden Book* by Sunset Publishing, which covers the region where I live, suggest the "ideal" distance between various plants. While this can be adjusted, based on soil type, sunlight, pruning, and irrigation, etc., such recommendations, especially those based on regional conditions, are a good place to start.

When putting in fruit trees, knowing the particulars of a given rootstock is vital to knowing the minimum planting interval for it. See my first book (*Designing and Maintaining Your Edible Landscape – Naturally*) for an extensive list of rootstocks and their attributes and limitations.

- "(In the tropics) large annual rates of organic inputs are needed to maintain an adequate soil organic matter pool in cultivated soils. Soils in cooler [temperate] climates commonly contain more organic matter because of slower mineralization (decomposition) rates."

What's the practical aspect? Increase the organic matter. Not too complicated; see the chapter on *Nurturing Soils* for details.

- "This shows the effect of climate on organic-matter decomposition and how organic matter is oxidized or lost at a much faster rate in the tropics as compared to subtropical and temperate zones [four times faster than temperate climates]."

This is good news, as it indicates that our soils can hold onto organic matter better than soils in the tropics can. As seen in Figure 4, a deciduous forest holds on to nitrogen 10 times longer than in a tropical soil, and to carbon, which is essential to forming

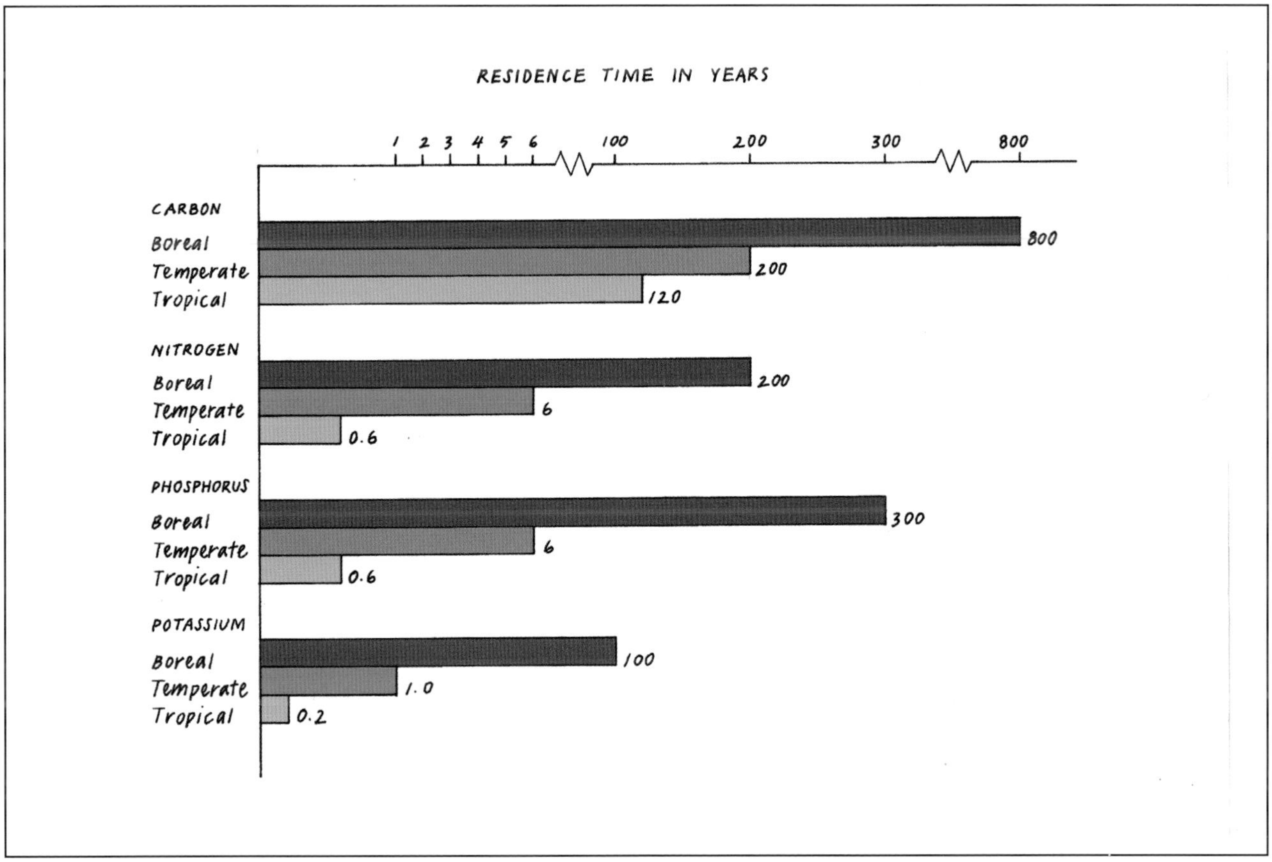

Figure 4. Measured in years, temperate forests (green) hold on to nutrients much longer than tropical forests (blue), but boreal forests—ecosystems in the sub-Arctic Northern Hemisphere, characterized by coniferous trees—(red) are even better at retaining nutrients for long periods of time. Because coniferous forests in colder climates are resistant to decay, nirogen in plant tissues remains 33 times longer (200 years) in boreal forests than in the temperate zones (six years).

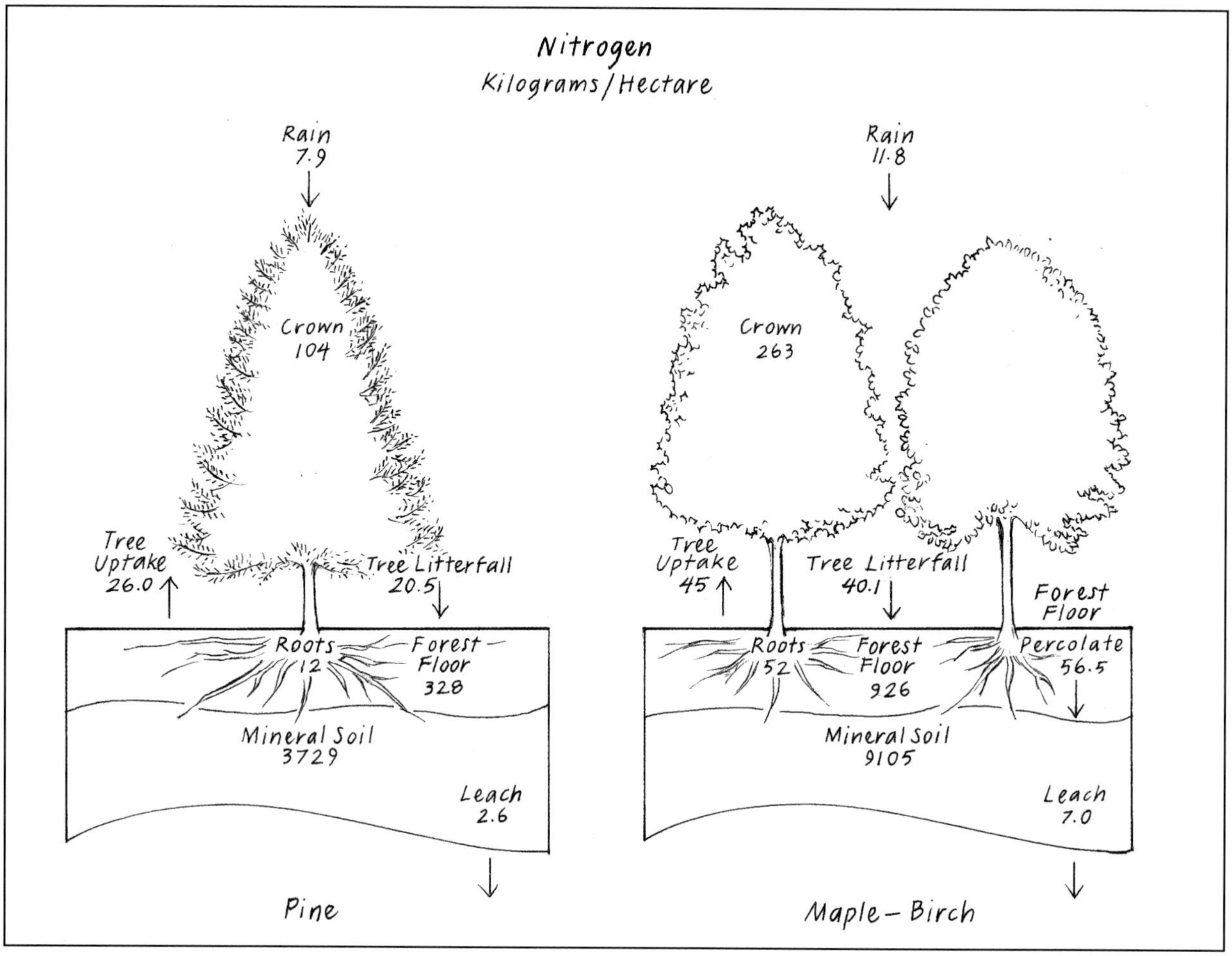

Figure 5. In this chart, the amounts of nitrogen in a pine forest are compared to those in a deciduous forest of maples and birches. The nitrogen in the pines' foliage amounts to about three times more (104 kg) than in their forest-floor soil. Deciduous trees hold 23 times more nitrogen in the forest-floor soil than in the litter/mulch (926 kg divided by 40 kg equals 23), and 10 times more than the lower-mineral soil. The pine-forest-floor soil contains 328 kg of nitrogen and its litter-fall contains 20.5 kg—a factor of 16. The mineral soil under the maple/birch drawing is shown to hold 10 times more nitrogen than the forest floor where most of the roots are found. (From: *The Canadian Journal of Soil Science*, 1986, Vol. 66, No. 4)

humus, eighty years long.

It's important to maintain a certain level of organic matter (and therefore humus) in your garden. If necessary, use leguminous green manures to boost soil nitrogen (See the Chapter *Free Fertilizer*.)

- "In temperate climates, nitrogen-fixing bacteria grow mostly on roots beneath the soil's surface—primarily bacteria found on the roots of the bean family (*Fabaceae* or *Leguminosae*), usually *Rhizobium* spp. bacteria."

While forests in the tropics produce blue-green algae and lichens that pump nitrogen directly into the plants' leaves, nitrogen fixation and accumulation for soils in temperate climates are dependent on a number of various pathways.

The nodules that indicate nitrogen fixation are found on the upper root system and appear as small white lumps. If you don't see such nodules on the roots of a legume, the proper bacteria aren't present, and you'll need to inoculate the seed before planting it. Some roots need specific species of *Rhizobium* to form nodules, so be sure to buy the mix that's best for your crop. Many plant nurseries carry the proper inoculant, and will give you guidance in using it.

If you do (carefully) cut open existing root nodules

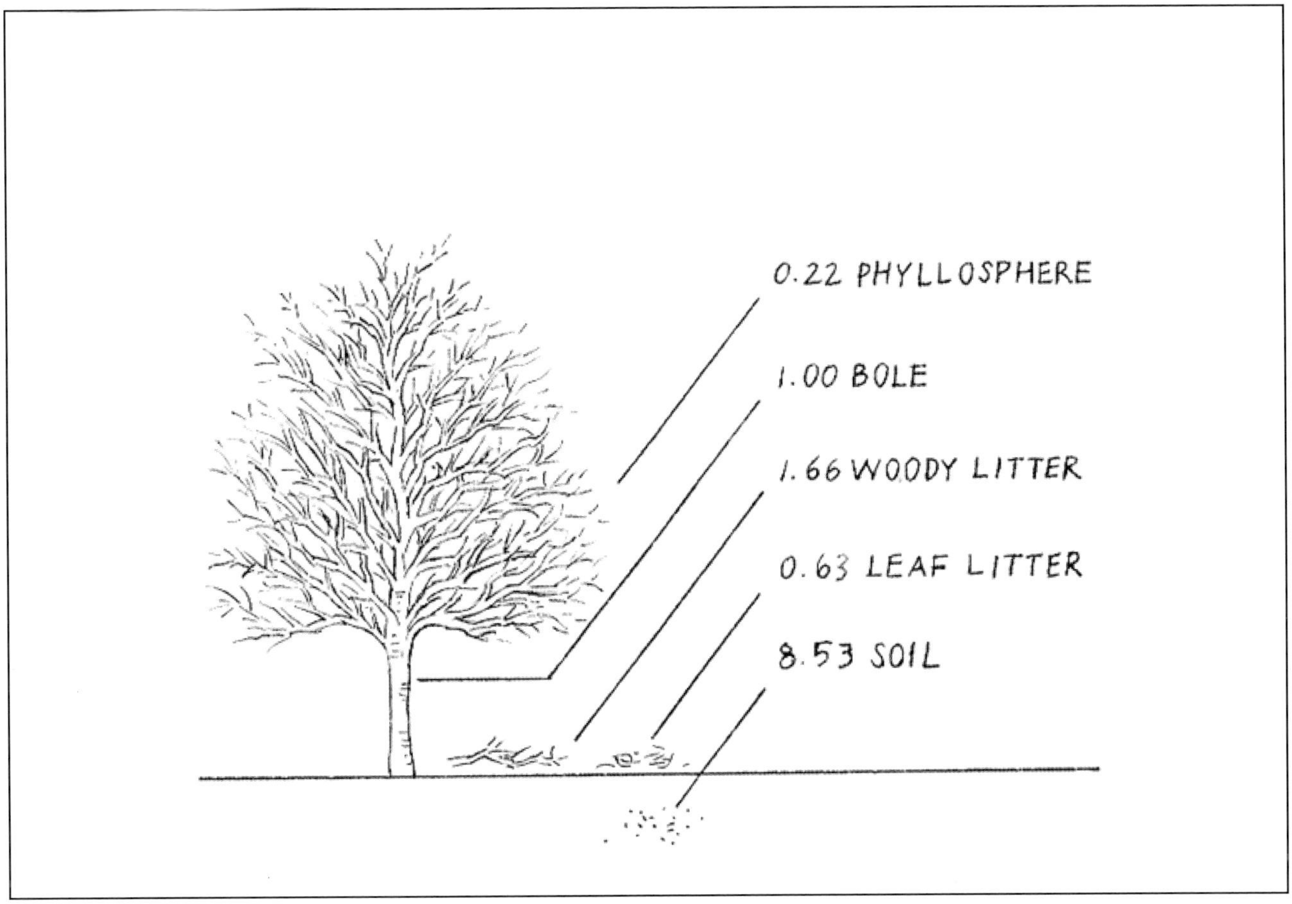

Figure 6. Here's another way to look at how much nitrogen is stored in temperate climates. This diagram shows in kilograms (1 kg = 2.2 pounds) the amounts of nitrogen generated per hectare (2.5 acres) per year, and where the element is located on various levels of a temperate forest. Notice that the wood/leaf litter (the "mulch" above the soil) contains five pounds of the total nitrogen, with about four times more (19 pounds) found in the soil itself. We also see how much of the nitrogen in a temperate zone is stored in the soil, rather than in the aboveground biomass. In a different view, the litter contains 2.29 kilograms of nitrogen, and the soil 8.53 kilograms, or about four times more than the litter. Also notice how little nitrogen is retained in the foliage. (*Bole* is defined as "the stems and main wooden axis (trunk) of a tree." *Phyllosphere* means the foliage and branches other than the bole.)

From: Nitrogen Fixation In A Deciduous Forest In The South Eastern United States, 1978. R.L. Todd, R.D. Meyer & J.B. Waide, *Ecological Bulletin*, (Stockholm) 26: p 172-177.

and see pink and white, you have a healthy plant with vital nodules. If the insides are brown, something's wrong; add some inoculant before planting again.

- "Here's a diagram that shows what amounts of nitrogen are found where in a temperate forest. Notice that the nitrogen in all the wood litter (the mulch above the soil) adds up to 88 pounds, as opposed to 2037 pounds in the soil."

Soil nitrogen is 23 times greater than the litter's nitrogen. See Figure 5.

- "Many [tropical] tree species grow extensive root systems that in some cases may extend much further than those of trees in temperate-zone climates."

As a general rule, trees in the temperate zone grow roots that extend only 1.5 to 3-5 times wider than the foliage canopy (dripline); this is not nearly as far as those of some tropical trees. Clay soils confine the roots to half-again wider than the dripline. Lighter soils and sandy soils will allow roots to grow 3-5 times, or more, wider than the foliage. See the *Root*

Dynamics chapter for the full story.

The practical application of this information is to distribute water, compost, mulch, and fertilizers at or beyond the dripline for the best effect on new feeding roots. Be aware that, in a neighborhood with small lots, what your neighbor does to his yard may be influencing the trees in your yard, and vice-versa. (Be sure to read the chapter on roots (Pages 283 - 319) for much more information.)

- "Biological Nitrogen fixation rates in tropical forests (15 to 36 kg N ha yr) are similar to or higher than estimates for their temperate counterparts (7 to 27 kg N ha yr)."

Designs that are actually based on tropical ecosystems may recommend less organic matter (mulch/duff) on the surface than what's needed in our temperate ecology.

- "The amount and rate of production of nitrogen available to soil microbes and plants are much greater in many tropical forest soils than in temperate counterparts, due to high rates of nitrogen fixation by free-living microbes and rhizobia associated with legumes, which are especially abundant in many tropical ecosystems.

When viewing a forest farming/gardening design, remember that tropical forests contain more leguminous trees and nitrogen-fixing lichens and bacteria, and thus will accumulate more nitrogen in plant parts above the soil than will temperate forests. This means that (some) tropical forests naturally manufacture more nitrogen than the soil in your backyard. In one paper (*Ecology of Nitrogen Fixing, Nitrifying, and Denitrifying Microorganisms in Tropical Forest Soils*, Brendan J. M. Bohannan.) it was said: "...tropical ecosystems had 88 kg N ha yr of fixed nitrogen compared to 12-14 in our deciduous hardwood forests." This means that some tropical systems can contain about seven times more nitrogen than our temperate gardens.

Nitrogen and Tree Pests and Diseases

You actually have to be careful that you don't have *too* much nitrogen in your soil, as it can promote pests and disease. In Village Homes in Davis, CA, for example, where the groundcover for most of the orchards is clover, the apple trees came down with mildew in a climate that is normally so hot and dry that mildew is seldom a problem. The theory about this problem was that the clover added more nitrogen that the trees actually needed—forcing unneeded growth. The addition of humid air from the use of sprinklers instead of drip irrigation added to the nitrogen surplus and generated mildew. Other examples include:

A study done with aphids in western Washington, found that "Honeydew [the clear frass/sh*t of an aphid] reaching the soil caused a reduction in... aboveground net primary production and nitrogen uptake by trees."

The hemlock woolly adelgid, *Adelges tsugae*, a tiny sap-sucking insect related to the aphid, is currently causing widespread death and decline of hemlock trees in the eastern United States.

"Hemlock woolly adelgid attacks trees of all ages and sizes and threatens to widely eliminate eastern hemlock." (From: *Nitrogen Fertilization of Hemlock* ***Increases Susceptibility to Hemlock Woolly Adelgid*****, M S McClure, 1991.)**

Elsewhere we read: "Aphid [adelgid] populations are enhanced by nitrogen fertilization. Eastern North American hemlocks have evolved chemical defenses to protect against chewing insects, but they are highly vulnerable to sap-sucking insect such as HWA [aphid] infestations." and: "Fertilized hemlocks had five times more adelgids, had inferior color, and produced 25% less new growth than unfertilized trees after a single adelgid generation."

It's similar with aphids. "Grain aphids on plants with high nitrogen produced 33 progeny per aphid, with low levels the progeny averaged 13 progeny per aphid. The pea aphid with high nitrogen had 69 progeny per aphid whereas on low nitrogen the average was 47." (*Origins of Pest, Parasite, Disease and Weed Problems*, 1977. J. M. Cherrett and G. R. Sagar, editors)

Attempting to match, in a temperate zone, the number of leguminous trees found in tropical forests can over-stimulate the formation of nitrogen in our garden or "food forest." As mentioned earlier, the amount/rate of production of nitrogen that's available to soil microbes and plants is much greater in many tropical forest soils than in temperate counterparts, due to high rates of nitrogen fixation by free-living microbes and rhizobia associated with legumes, which are especially abundant in many tropical ecosystems.

In my experience, I have seen very few fruit or nut trees that needed fertilizer or compost. Fruit trees are very good at scrounging enough nutrients to be healthy and productive. The chart in Figure 8, based on my 1986 book *Designing and Maintaining Your Edible Landscape - Naturally*, (page 229) shows that a reasonable amount of growth translates as no need for extra compost or fertilizer. Unfortunately, I have no hard research for this information, which was developed by my experience, that of our county's California Cooperative Extension Agent John Smith, and the 1953 book *Western Fruit Growing*, by Reed Brooks. Very few trees put out less growth than that reported in these measurements, thus many gardeners are wasting compost by applying it willy-nilly every year. At the minimum, get a soil test to see exactly how much (if any) fertilizer the tree needs. Any mulch needed may be a relatively unfertile material, like straw, pine needles, or rice hulls, that will conserve moisture while not over-fertilizing the tree. (See Figure 8.)

Pest Populations Increase with Excess Nitrogen Fertility

COMMON NAME	GENUS AND SPECIES	CROP
European red mite	*Panonychus ulmi*	Apples
Two-spotted spider mite	*Tetranychus telarius*	Apples, beans, peaches, tomatoes
Clover mite	*Bryobia praetiosa*	Beans, peaches
Greenhouse thrip	*Heliothrips haemorrhoidalis*	Beans
Green peach aphid	*Myzus persicae*	Brussels sprouts, tobacco
Greenbug	*Schizaphis graminum*	Oats, rye
Corn leaf aphid	*Rhopalosiphum maidis*	Sorghum
Spotted alfalfa aphid	*Therioaphis maculate*	Alfalfa

Figure 7. You can give plants too much nitrogen if you use too much compost or fertilizer. Always check your soil with a soil test to determine if you even need extra nitrogen.

From: *Manage Insects on Your Farm, A guide to ecological strategies*, Miguel Altieri, Clara Nicholls, with Marlene Fritz. Sustainable Agriculture Network, USDA, 2005, page 65.

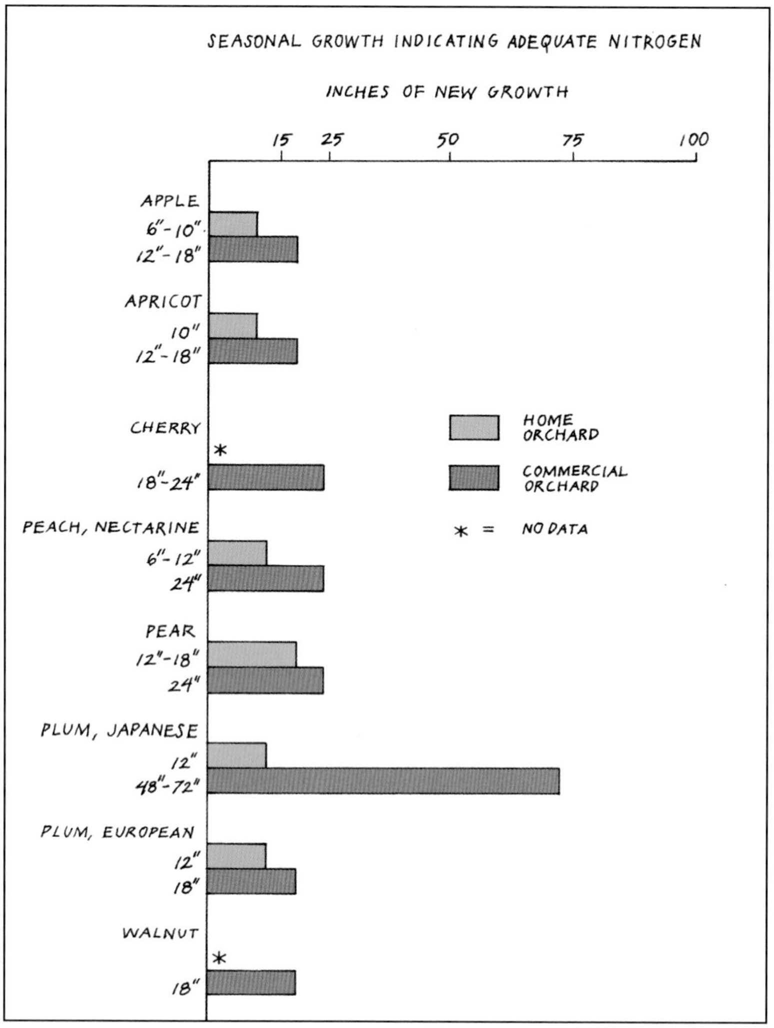

Figure 8. While consulting over the years, I've rarely seen an apple tree that needed fertilizer or even compost. Most apple trees grown with no amendments to the soil show at least six inches of growth annually—witness the growth of many 100-year-old abandoned apple trees. And I've never seen an unamended plum with less than 12 inches of growth. Save your compost for elsewhere in the garden, and don't waste your money on fertilizer.

NOTE: As we'll see in Chapter 9, the roots of trees grow 1.5 (in clay soil) to three times (in sandy soil) wider than the foliage. Any ferilizer used to grow vegetables in this wide area would feed the fruit tree at a rate greater than the tree needs. Separating the tree roots from the vegetable areas will make for healthier trees as they won't be overstimulated by the nutrients needed to keep the vegetables productive. The amount of fertilizer and compost needed by vegetables may cause pest and disease on the fruit tree. That's why I try to isolate fruit trees roots from vegetable beds. Some yards are too small to do this and must moderate the fertility of the vegetables as much as possible.

Bonus Chart:

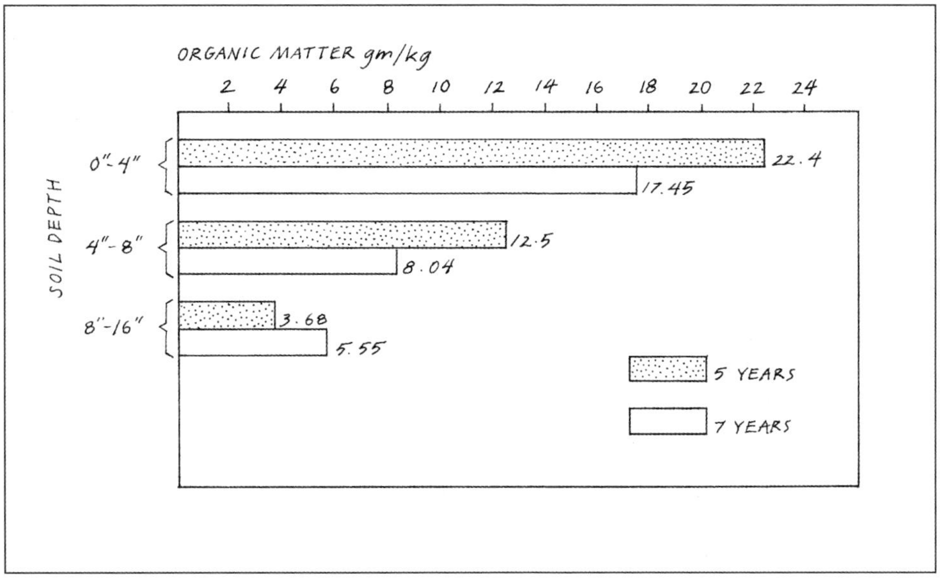

Figure 9. Organic matter slowly decomposes into humus, which helps to liberate unavailable minerals into a soluble form that can be absorbed by tiny root hairs—a process known as "mineralization." All this takes place in a matrix of soil life, the dynamics of the rhizosphere, and a web of moisture. In five years, the organic matter at eight inches deep or deeper is seven times less that in the top four inches of soil.

(From: "A Study on the Variation of Soil Microbial Activity Following Deforestation" by Zhang Chenge, Zheng Fenli and He Xiubin, 12th ISCO Conference, Beijing 2002.)

Dr. Weaver spent a lot of time in the 1930s researching the absorption of nutrients by plants from different layers in the soil. One of his studies showed that the absorption of nitrates by potatoes in the upper 1.5 feet of soil was about twice the amount absorbed at the deeper 1.5-to two-foot zone.

In another study by Weaver, done with corn; the roots in the deeper soils gathered progressively fewer nitrates (down to three, four, and five feet), and there were fewer roots in the lower soil.

Chapter 2
Food Gardens

A Curious Beginning – The Dump Heap

Here's a true story (from my own life) of how perceived "wisdom" can change in the face of scientific exploration.

Perhaps the greatest jump-start to my career as an environmentally sound horticulturist and organic landscape gardener was a chapter of Edgar Anderson's book *Plants, Man and Life*. It was called "The Dump Heap and the Origin of Agriculture."

Back in 1978, Jamie Jobb and friends (of which I was one, along with Paul Jackson, Diane Laufer, Lynn Ferar, Brian Weissbuch, Steve Parcells, Louis Ratzeberger, Cynthia Seiler, Tom Rall, Tom Bassett, Patrick Houck, Sue Stevens, John Forbes, and Nancy Beaubaire) were so impressed by this theory that for several years we put out a critically-acclaimed journal called the D.U.M.P. H.E.A.P. (Diverse Unsung Miracle Plants for the Healthy Evolution Among People.) To qualify for a bulk-mail permit, we mailed 350 copies all over the world—including to Bill Mollison.

This cutting-edge publication was, by the way, produced in the days before word-processing, spellcheck, layout programs, emails, and iPads! We were actually going to libraries and *reading books*, including:

- *Forest Farming: Towards a Solution to Problems of World Hunger and Conservation*, by J. Sholto Douglas and Robert Hart, Rodale Press, 1978. (This book is described in detail in Chapter __)
- *Tree Crops*, by J. Russell Smith, The Devin-Adair Company, 4th edition, 1977.
- *Sensitive Chaos: The Creation of Flowing Forms in Water and Air*, by Theodor Schwenk, Rudolf Steiner Press; Revised edition, January 1, 1996.
- *Underexploited Tropical Plants with Promising Economic Value*, Washington, D.C.: National Academy of Sciences, 1976.
- *Plants, People and Environmental Quality*, U. S. Department of the Interior, with the American Society of Landscape Architects Foundation, 1972.
- *How to Have a Green Thumb Without an Aching Back*, by Ruth Stout, New York: Simon & Schuster, 1955.
- *The Ruth Stout No-Work Garden Book*, Ruth

Above, berries make a wonderful addition to an edible landscape. Instead of mixing them inside a food forest amongst the trees, cluster the plants in rows to make management and pruning easier. With two rows of raspberries, you can cut down the canes of one whole row, eat berries from the uncut row, and eat berries from the first row the second season. Repeat.

Stout and Richard Clemence, Emmaus, PA: Rodale Press, 1971.

- *Intensive Gardening, Using Dutch Lights, Surface Cultivation and Composting for the Commercial Production of Crops, and Introducing a Motion-Study Routine,* Rosa Dalziel O'Brien. London, UK: Faber & Faber Limited, 1956.
- (And, of course, hours of watching Monty Python, who certainly had their way with shrubbery!)

These books (and many more) made for lots of late-night discussions and daily inspiration. This all eventually led to my conjuring up of the phrase "edible landscaping" for my kind of landscaping work, but it was Edgar Anderson and his dump heap who started it all.

The Origin of Agriculture? Well, Maybe.

In a chapter of Anderson's book called "The Dump Heap and the Origin of Agriculture," we read of a farmer in Santa Lucia, Guatemala, who grew many plants in what at first appeared to be chaos. Further study by Anderson over the years revealed that the garden was a collection of tropical- and temperate-zone plants (primarily squashes and pumpkins) grown in an apparently random fashion.

In fact, hundreds of years of trial and error had taught this farmer, through oral history and example, not only how to plan out his crops for the greatest ease of planting and care, but also how to best produce perennial food crops, how to use "random" plantings to help thwart pests, and how to cultivate *chayotes*, (a light-green, pear-shaped vining vegetable with a single soft, large, edible pit, and edible flesh and skin, described in detail on page 38, as both a horticultural crop and for shade.

The "Dump Heap" chapter presents the theory that the selection of new food crops occurred in the dump heaps of ancient societies, where all refuse was left to simply rot away. Here, in what scientists today call *middens* (an Old English word that's been around for centuries), natural selection led

Figure 10. The tiny white specks in the lower left-hand corner of this shot are bits of seashell found in a midden (dump for domestic waste) used by the Native American people in Big Sur, CA. The rich midden soil makes for great harvests.

to new plants being selected for cultivation, sort of like finding a hard-skinned zucchini—perhaps a zuke cross-pollinated with a pumpkin—in your compost pile.

In middens, it's been postulated, the crossing continued until a new useful vegetable was found. Eventually, new edible plants found their place in small home gardens, many grown in a vertical-layering configuration to make good use of limited space.

Several distinguished horticultural investigators seemed to agree with this theory:

From Carl Sauer, a contemporary of Anderson: (Agricultural origins and dispersals, *the domestication of animals and foodstuffs*; 2nd Edition, Carl O. Sauer. 1969 by M.I.T in Cambridge, MA)

"Man also was an agent of plant evolution by carrying plants to new places, along his paths, to his dooryard, on the refuse heap. New variations arose by such carriage, both as hybrids and mutants..."

Even Charles Darwin seemed to support the dump-heap theory long before it arose as a modern agricultural concept. In Jack R. Harlan's *Crops & Man* (2nd Edition, American Society of Agronomy-Crop Science Society, 1992, p 78), we read that Charles Darwin, among others, was convinced that nomadic people could not develop agriculture:

"...it struck me [wrote Darwin] that the possession of some property, a fixed abode, and the union of many families under a chief, were the indispensable requisites for civilization... the first steps in cultivation would probably result...from some such accident as the seeds of a fruit tree falling on a heap of refuse, and producing an unusually fine variety."

Further support comes from *Serendipitous Backyard Hybridization and the Origin of Crops*, published by The National Academy of Sciences USA, vol. 104 no. 36, Colin E. Hughes, 2007: "There is also evidence to suggest that hybridization has been important in many other Mesoamerican crops, such as *Cucurbita*, *Phaseolus*, *Capsicum*, *Lepidium*, *Hyptis*, and *Panicum*. It seems that, as suggested by Anderson, the simple step of bringing species together, consciously or casually, in dump heaps and informal backyard orchards has played a central role in Mesoamerican crop domestication." (Notice: *not* temperate America.)

Bump Heap

Alas, the Dump Heap theory has taken a beating since the mid-1970s; the thinking about Anderson's theory has changed dramatically, and has come under increasing criticism.

From: *On the Origin of Near Eastern Founder Crops and the 'Dump-Heap Hypothesis'* (*Genetic Resources and Crop Evolution*, 52: p 491–495, Springer Verlag, 2005):

"The accumulated evidence on the ecological preferences of the wild ancestors does not support a weedy origin of the domesticated forms. An interesting drawback of the dump-heap hypothesis, mentioned by J. G. Hawkes (*The Diversity of Crop Plants*. Harvard University Press, Cambridge, 1993, p 1–46), is that it does not explain the timing of the initiation of agriculture in the Near East circa 10,500 years ago, probably because the association of certain weeds with man's habitation sites is much older than that.

"This of course has implications on the cultural roots of Man/Plant association in general and Man/Crop Plants in particular. In this context, the "special mental leap" which is not part of the dump-heap model as rightly pointed out by Hawkes (1983), becomes fundamental in explaining the timing of the origin of Near East agriculture. The emergence of the specific crop package plants was a result of an intended selection rather than a mere 'environmental accident'—prehistoric man and the suitable plants involved just happened to be there at the same time. Such environmental deterministic views tend to underestimate the role of human initiative in the Neolithic transition."

Furthermore, from *Crops and Man* (Jack Harlan 2nd Edition, 2005, 3rd printing, *American Society of Agronomy*, Madison, WI): "Evidence accumulated since the Sauer-Anderson models were suggested has indicated that some of their presuppositions were incorrect. For example, sedentary life is not essential to the evolution of agriculture. In Mesoamerica there is good archaeological evidence that the people remained nomadic long after they were purposely growing plants for food. In the Near East, there is evidence that a nuclear center developed in an area not in the tropics and by people not necessarily dependent upon aquatic resources."

From *On the Origin of Near Eastern Founder Crops and The 'Dump-Heap Hypothesis'*, Shahal Abbo1, Avi Gopher, Baruch Rubin and Simcha Lev-Yadun (from: *Genetic Resources and Crop Evolution*, 2005, 52: p 491–495), we read:
"Many students of the subject of plant domestication have seriously entertained the hypothesis that man's first crop plants have originated from weeds associated with the disturbed habitats surrounding pre-agricultural ancient human dwellings and or with human refuse heaps – the so called 'dump heap hypothesis.' Contrary to the 'dump-heap hypothesis', we propose that Near Eastern farming originated as a result of a long-term interaction between humans and plants and was mainly driven by the nutritional features of the respective crops and cultural forces."

I'm still thankful that Anderson's book helped get me started, and it's pretty amazing how long it took science to question his theory. It's always disturbing, but helpful, to understand a dismantled and outdated theory. More dogma bites the dust (or perhaps the jury may still be out on this one).

Permaculture

My early excitement with sustainable and organic edible landscaping was accompanied by a real interest in permaculture. Permaculture is not a new brand of yogurt, but a combination of the words "permanent" and "agriculture," coined in Australia by Bill Mollison and David Holmgren. Early in my fascination with organic edible landscapes (circa 1976), I happened across the book *Permaculture One, A Perennial Agriculture for Human Settlements,* one of only 12 early copies that had been shipped to America.

At the time, I was preoccupied with the problems of integrating more perennial food crops into ornamental landscapes and of creating landscapes that were both environmentally sound and sustainable. Then, lo and behold, here was a book that stated on its back cover: "By carefully designing a system around functional relationships between plant and animal species, you can create a stable 'cultivated ecology'....Permaculture [is based upon] the underlying philosophy of working with, rather than against nature..."

In his more recent book, *Permaculture: A Designer's Manual,* (circa 1988, with Reny Slay), Bill Mollison gives another definition of permaculture: "the conscious design and maintenance of agriculturally productive ecosystems that have the diversity, stability and resilience of natural ecosystems....the harmonious integration of landscape and people providing their food, energy, shelter and other material and non-material needs in a sustainable way." (Notice, by the way, his use of the word "agriculturally.")

According to one permaculturist: "A mature ecosystem, such as ancient woodland, has a huge number of relationships between its component parts: trees, understory, ground cover, soil, fungi, insects and other animals. Plants grow at different heights. This allows a diverse community of life to grow in a relatively small space. Plants come into leaf and fruit at different times of year."

Here's what some permaculturists envision as the goal or design of a multi-layered garden.

The Eight Layers of a Food Forest or Garden:

1. The foliage or canopy (such as full-sized fruit and nut trees)
2. The lower tree-layer (such as dwarf fruit trees)
3. Edible shrubs
4. Herbaceous plants
5. Surface crops, i.e. cover crops, vegetables, herbs, soil-improvers
6. Root crops
7. Vertical plants: climbers, vines
8. Fungi and rhizosphere zone: the soil biota and mycorrhizae "layers"

In the 1970s, permaculture was developing in parallel with the work of numerous other organizations: the Farallones Institute Urban and Rural Centers in Northern California; the New Alchemy Institute in Cape Cod, Massachusetts; the Regeneration Projects of Rodale Press in Pennsylvania, Wes Jackson's work at the Land Institute in Kansas, the Aprovecho Institute in Oregon, Tilth in Seattle, Washington, and a myriad of other groups dedicated to working *with* nature in ways that conserved, recycled, and promoted natural resources.

I was a true believer in permaculture until the early 1980s, when I began double-checking its tenets against the scientific literature and my personal experience. I should mention that the goals of my work and those of permaculture are remarkably similar, although the paths to them are rather different. As I see it now, here are the overall limits to permaculture as a premise/ideology (there are many exceptions with regard to individual permaculture designers and designs):

- It's often not based on scientific facts or research. Almost no peer-reviewed

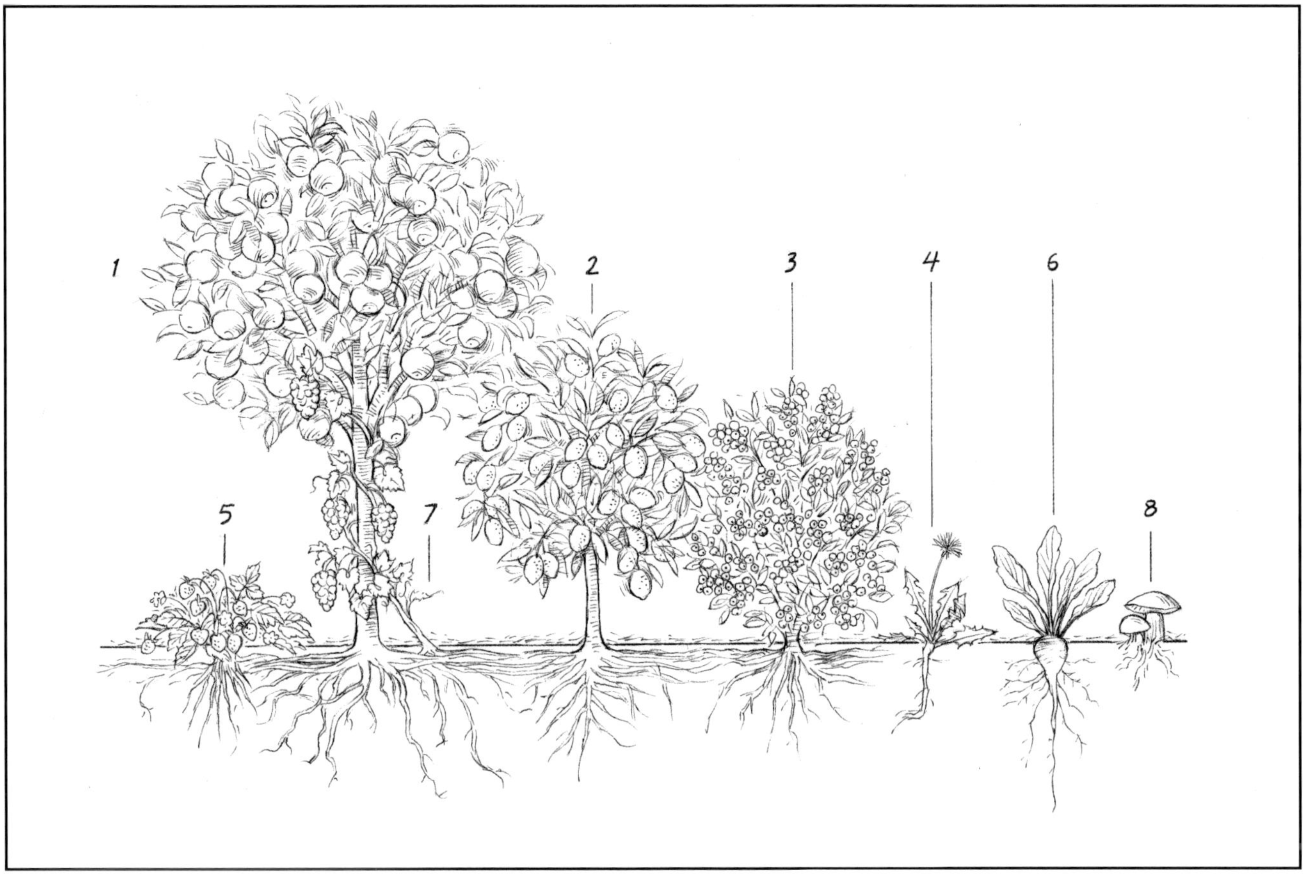

Figure 11. Here are the eight zones or layers of a classic food forest. 1. The foliage or canopy zone (examples: full-sized fruit or nut trees that, if left unpruned, can be over 30 feet tall); 2. The lower tree layer (semi-dwarf fruit trees, filberts, and pawpaws, mostly under 30 feet tall); 3. Edible shrubs and dwarf fruits under 12 feet tall; 4. Herbaceous plants; 5. Surface crops—vegetables, herbs, cover crops/soil-improvers such as clovers; 6. Root crops; 7. Vertical plants: climbers, vines; 8. Fungi and rhizosphere zone; soil biota and mycorrhizae layers.

papers.

- A vision of the future of plants and society, not the facts of now.
- Based on the sociology of indigenous (not urban or suburban) people.
- Often dependent on agrarian-like labor.
- May encompass well-intentioned wishful thoughts and desires not based on experience or scientific principles.

My opinion is that permaculture advocates, admirable as their belief system is, sometimes make grand leaps of disputable thinking in order to hold together the framework of an intellectual dogma. This not to say there are no practical *and* good-looking permaculture gardens. (For your consideration, in this chapter see the garden of Yeffi and Philip Vanatta in Saratoga, CA, near Silicon Valley.) There are many such, but in my opinion, those "permacultured" gardens that are attractive as well as functional are very similar to those that I and other edible-landscapers have designed since the mid-1970s.

I should explain that I first met Bill Mollison in 1980, when the Farallones Institute Urban Center sponsored him in his first lecture on the US mainland at the University of California at Berkeley. Afterwards, he spent a week at the 80-acre Farallones Institute Rural Center in Occidental California, where I was living at the time.

We went on many walks together, and I was amazed at how perceptive and intuitive Bill was about the landscape and environment around him. He spotted birds, plants, and snakes I'd never noticed. (He

Figure 12. A west-facing view of a food forest growing around a house near Silicon Valley in California. Cherry, apricot, peach, fig, plum, pear, and avocado trees fill the front and back yards, along with roses, herbs, milkweed, ground covers, and crocosmia (the orange/red flowers in the middle).

even located an old cannabis plot by its fencing in the Rural Center's "back forty.") He was very comfortable to be with, and had a delightful sense of wry humor—I loved it when he'd say to me, "Let's sit down and tell lies." (He died in 2016.)

Bill and I met a number of times after that, and I've always admired him and continue to be impressed by the brilliance of many of his teachings and concepts. Years later, however, after I'd encountered enough science and practical experience to disagree with him occasionally, I found that some of his guidelines didn't entirely hold up to scientific scrutiny, especially if applied to temperate zones, rather than the tropical climate he'd studied more extensively.

This book is in great part a survey of Bill's impressive work and some amendments to it that might actually be a better interpretation for more

Figure 13. This is an intensive vegetable- and flower-bed at the Esalen Institute in Big Sur, CA. The cabbage in the foreground will fill the space after the light-green lettuce is harvested.

Figure 14. This is the approach to the east side of the food-forest house shown in Figure 12. Fruit trees line both sides of the walkway, with a wildlife habitat under those on the right. At the left is an orange milkweed, and ornamental flowers line the walk. All is well mulched, and is watered with an extensive drip-irrigation system. The angled sticks and netting help deter squirrels.

temperate climates.

So here's my quick review of what I see as Mollison's major guidelines, followed by my commentary on them and by reference guidelines for the treatment of each in this book:

1. **Small Scale Land-Use Patterns Are Possible.**
   Yes, some of Mollison's models in the USA take the form of edible landscaping and suburban permaculture, but scaling down agricultural ecosystems to a small home garden can be difficult, fraught with potential problems, and may just plain fail. (So check out Chapter 6 of this book.)

2. **Use Intensive, Rather than Horizontal Patterns.** (As described on page 4 of the first Chapter.)
   Consider this photo (Figure 13) of a diversified planting crammed full of vegetables and edible orange-petaled calendulas (do NOT eat the vertical poisonous rose-colored foxglove!). This intensive pattern has been described by permaculturalists as "stacking in space and time," Too fancy for me, should just simply say: intercropping.

3. **Plan for Diversity of Plant Species, Microclimate, and Habitat.**
   I would agree, up to a point. Diversity, however, is limited by space (there are only so many niches to be filled) and by your gardening skills. (To get the lowdown see Chapters 1 and 3.)

4. **Plan for Long-Term Evolutionary Changes.**

Oak trees live a long time. They can't adjust to evolutionary times except by making seedlings. We can interfere in their evolution by killing the oaks in their youth before they make a lot of acorns—say the first 50-100 years. One oak tree I know has been irrigated in a moat around its base for over 40 years, contradicting all Mediterranean and California guidelines that caution not to water inside the dripline of a tree's canopy less than 10-20 feet from its trunk. This tree still looks fine, but such irrigation may, by crude estimation, take, say 50-100 years off a 300+ year-old oak tree's life. Who cares? Not most homeowners, but in the interests of the future, you might want to refrain from irrigating at the base of your oak trees.

5. **Wild, Undomesticated Plants Are Important Elements of Each Design.** (Numbers 1-4 are from p 6-7, *Permaculture: A Designer's Manual*)
This is a *very* touchy subject. See the discussion of invasive plants in Chapter__.

6. **Work with Nature, Rather than against It.** (p 15, *Permaculture: A Designer's Manual*)
A common trait with all environmentally sound landscapes. This is sometimes stated as "Pattern after natural systems."

7. **Make the Least Change for the Greatest Possible Benefit.** (p 15, *Permaculture: A Designer's Manual*)
This is a worthy goal, though often difficult for the novice gardener to interpret.

8. **The Yields of a System Are Theoretically Unlimited.** (p 15, *Permaculture: A Designer's Manual*)
In my opinion, the operative word here is "theoretically." In other words, based on hope, not reality, and most likely impossible. A more reasonable statement might be: "The longer-lived, and lower maintenance the system, the higher the total potential and long-term yields." (The important word is "potential," as this is a very elusive goal.)

9. **Everything Makes Its Own Garden, or Has an Effect on Its Environment.** (p 34, *Permaculture: A Designer's Manual*)
Goes without saying.

Consider the California coast oaks and redwood trees of Occidental, CA, where I

Figure 15. The owners of the house increase the yields of trees in their small front yard (see #8, to the left) by grafting many cultivars onto the same tree. Here a 'Nugget' apricot graft (marked by white paint) has been added to an existing apricot tree.

lived for many years. In the area's mixed ecotone of Douglas fir trees, oak woodland, redwood forest, and adjacent chaparral, the forest creates summer irrigation by capturing fog. In 10 years of record-keeping, one 100+-foot Douglas fir tree on my property captured, on average, one-and-a-half to three times more total rainfall each year than the adjacent open meadow, and the shrubbery below the dripline of the Douglas fir thrives. This is an outstanding example of how nature creates beneficial conditions; this degree of complexity, however, may be impossible to achieve in a garden.

Figure 17. Chickens are kept in Zone One, near the front of the green wall of the food-forest house (See below for a discussion of zones.)

10. Forests Are the Model for a Permanent Agriculture. (p 3, *Permaculture Two*)
This is a statement very much unsubstantiated in temperate climates, and requires a lot of discussion (see Chapters 1 and 6 for the use of the term "agriculture.") This concept is very difficult to implement by the average homeowner growing vegetables in a suburban yard. As mentioned, I live in an ecotone called the chaparral where trees are much less common and shrubs, herbaceous plants, and annuals proliferate. You won't find this situation described in permaculture books (nor much in mine for that matter), because they're often written to apply to an entire country or world, not just 15.3 million acres of California.

Figure 16. Illustration of a chayote (*Sechium edule*). This is #18 in the U.S. Department of Agriculture's Pomological Watercolor Collection in the Rare and Special Collections section of the National Agricultural Library in Beltsville, MD.

11. Consider the Site as a Set of Concentric Zones of Use, from the Most Intensive to the Most Diffuse. (p 53-60, *Permaculture: A Designer's Manual*)
This is just good old common sense. One exception to it is a realistic edible landscape. (See Chapter 12.) In *Permaculture: A Designers' Manual***,** Bill recommends analyzing a yard by zones and sectors as a way to determine a pattern of design that conserves energy.

Typically, the first zone is the area closest to the walls of the house. This is where "Components needing continual observation, frequent visits, work input, and complex techniques should be placed *very close to hand*, or we waste a great deal of time and energy visiting them." (p 49, *Permaculture: A Designer's Manual*) He defines this zone as that occurring within 20 feet of the home and claims that this is enough area for homeowners to "produce most of the food necessary for existence, with some modest trade requirements." (p 49, *Permaculture: A Designer's Manual*) This is the place

Figure 18. The east side of the food-forest house features fruit trees, squash vines, fennel for pest control, ornamentals, asparagus, and a birdhouse (also for pest control). This is part of the yard's wildlife habitat.

for rabbits, chickens, culinary herbs, the daily salad greens, dwarf and espaliered fruit trees, and annual bedding plants for seasonal color and fresh-cut flowers.

12. **Plant as Many Varieties of Plants as Possible.** (p 65, *Permaculture: A Designer's Manual*)
This interfaces with the subject of invasive plants "of worth." A certain amount of diversity (not just willy-nilly "endless" diversity for diversity's sake) is appropriate. (See the detailed discussion on the subject in Chapter 1.) In his first lecture on the mainland of the USA, Bill said: "Everything is going to be everywhere eventually; so why not speed it up?" This presupposes a way to maintain diversity with invasive plants without them becoming a problem.

13. **A Complete Permaculture Ecosystem Requires Animals.** (p 75, *Permaculture: A Designer's Manual*)
This comes from the farming scale of permanent agriculture, but keeping chickens and rabbits is now becoming legal in more and more of suburbia. (Do worms count?) For more on the subject, read the discussion on manure-free vegan gardens in Chapter 4 and

Figure 19. Across the street from the food forest is a clipped landscape with lawn, typical of the neighborhood. Here we also see part of the angled netting placed to foil squirrels.

Figure 20. As per #18, the food forest grows sea kale (*Crambe maritima*). Its leaves are blanched like asparagus.

Appendix 2.

14. **There Should Be No Placement of an Element without It Serving at Least Two or More Functions.** (p 16, *Permaculture Two*) and conversely—**Every Function Should Be Served in Two or More Ways.** (p 16, *Permaculture Two*)
    This can be done. Chayote is a great model (see Figure 16). There are many similar examples for many gardening zones. (See also pagesn 39 and 40.)

15. **It is Not the Number of Diverse Things in a Design that Leads to Stability; It is The Number of Beneficial Connections Between Components.** (p 32, *Permaculture: A Designer's Manual*)
    The uses of ornamental and edible plants that are beneficial-insect flower attractants are good examples of this tenet. The plants do *not*, however, need to be companion-interplanted. Beneficial insects can fly far enough to allow for colorful plantings that look like ornamental flower beds, while allowing the "good bugs" to terrorize the "bad bugs." (See "Companion Planting" in Chapter 10) Another aspect of this is Bill's "Diversity is Dynamic Stability."

16. **Stack as Many Plants Together as Possible; the Density Depends Upon the Rigors of the Environment.**
    This can be complicated, and not as practical in temperate climates as in the tropics. (See Chapter 1.) Another way is to say it is, "Stack in Space and Time." **(See #2 and Figure 11.)**

17. **Always Improve the Soil, Using No-Till Methods Whenever Possible.** (p 29-45, *Permaculture Two*)
    You bet. (See Chapter 4.)

18. **Replace Ordinary Annual Vegetables with Perennial Equivalents.** (*Permaculture: A Designer's Manual*)
    Following this advice can imply a radical shift in dietary patterns that are very difficult to change. Examples such as chives, Egyptian

top-set onions, garlic and society garlic aren't very impressive options. Asparagus, sun-chokes (Jerusalem artichokes) and globe artichokes (in mild climates) are good examples. Fruit is good but there's a limit to its nutritional value and how much you can eat. Nut trees are great additions, but often are too large for a small suburban lot, subject to disease and blight, and take five years or more to come into bearing.

19. A devote of permaculture, Michael Judd, succinctly defines what Bill talks about throughout his books: "Edge Effect: Edges are the Site of Dynamic Interplay. 1 + 1 = 3 (the sum of the parts is greater than the individual parts) Increase Beneficial Edges, Minimize Problem Edges." You can read in Chapter 3 here how difficult it is to take the original definition of "edge" (a macroscopic ecological concept) and apply it to a small garden.

Some Golden Rules for Edible Landscaping

I treat edible landscapes as a form of permaculture for the suburbs. Regardless of where you live, there are a few important guidelines to consider for a well-designed and productive edible landscape. As simple as the following recommendations may seem at first, these address the points where people most often make their initial gardening mistakes.

- **Plan in Advance; Make Your Mistakes on Paper, Not in Your Landscape.** Sketch out the broad strokes of the main elements of your "master plan." Set priorities.
- **Start Ever So Small.** You can always expand as you follow your master plan.
- **Plant Your Annual Vegetables and Herbs No Further From the Kitchen Than You Can Throw the Kitchen Sink.** Remember the maxim: "Out of Sight, Out of Mind."
- **Be Lazy; Let Nature Work for You.** An example: "Use beneficial insects as part of a strategy to control pests."
- **Time and Money Spent Early Mean Time and Money Saved Later**. Don't pinch pennies on a functional hardscape—the non-edible parts of your plan will prove to be very important for aesthetic and recreational reasons.
- **Try to Grow Plants that Serve More than One Use**. Example: "Cilantro and coriander are the same plant. Let a few cilantro plants go to seed and harvest the coriander. Allowing some of the cilantro plants to blossom also draws beneficial insects into the garden, as a way to hold back pests."

How to Begin Growing an Edible Landscape

Start with herbs. Dried herb seasonings are some of the most expensive, over-priced items in a supermarket—and the most tasteless when compared to their fresh counterparts. Plan to have some of your favorite herbs as close to the kitchen as possible. Many herbs will thrive in small containers on a deck, balcony, fire escape, or sunny south-facing bay window. Some examples of these are: rosemary, thyme, lavender, garlic chives, regular chives, true oregano, and various (well-contained, potted) mints. All are quite comfortable in terracotta pots near a sunny doorway.

Next comes the growing of fresh daily salads, an important part of a nutritionally balanced, tasty meal. Planting and harvesting your own salads can make a big dent in your food budget—especially in the off season, or in the dead of winter, when you're otherwise stuck paying ridiculous prices for Peruvian, Chilean, Australian, North African, or Mexican produce shipped halfway around the world.

Then move on to vegetables; in my experience, it's most efficient to designate a specific area for intensive vegetable cultivation. Too often, when attempting to create an edible landscape, people who scatter their salad greens randomly as accents throughout their flower bed often forget where all those greens are until they've poked up their flower stalks and are too bitter to eat.

For the intensive cultivation of vegetables, you can use boxed raised beds (usually rectangular, but they can be any geometrical shape that allows all areas to be reached for weeding and harvesting).The sides of these bed-boxes, which combine convenience with practicality, can be made of two pieces of 2"-X12" recycled redwood stacked 24 inches high, or constructed of the newer plastic "woods."

Make each box 3 to 3 ½ feet wide by 6 to 8 feet long. If you live where there are gophers, moles, and voles, you'll want to start with a layer of ½ -inch hardware cloth stretching under, and securely attached to, the bottom of the box. In general, the soil in a raised-box bed will drain faster in wet weather, and warm up sooner in the spring.

Figure 21. The 2"X12" boards of a raised bed contain and protect growing vegetables; the wire bottom of the box keeps out gophers. The yard is three feet above the flood level of a nearby creek.

For most edible landscapes, the surface area of raised wooden beds for all annual vegetables/salad greens, plus some tomatoes, squash, corn, etc., is usually less than 100 square feet per person. The best way to solve the design problem of the boxes' angular nature is to arrange them in some form of a sweeping curve to distract or "fool" the eye, which then doesn't focus on the angular portion of the plan. You may want to design a floral and herbal border along one edge of the sweeping curve to hide one end of the beds. This flower and herb border (which can also contain edible plants) will not only provide colorful highlights throughout the growing season, but can also include plants that help attract beneficial insects to pollinate nearby fruit trees and act as "warriors" against insect pests. (See Chapter 11 for much more on this.)

Fruit and Nut Trees

In the early days of gardening, I often mixed ground covers, herbs, ornamentals, edible perennials, and fruit trees all together, in a somewhat controlled form that I called (pre-permaculture) "jungle" gardening. I soon discovered that to harvest or prune a large tree, for example, I was either stepping all over the smaller shrubbery and crushing the groundcover plants or spending an inordinate amount of time gingerly stepping around each plant.

While this wild-and-woolly layering of many types of plants is personally intriguing to some, and once was to me, it's certainly difficult to maintain. Instead, one can arrange some of the elements of an edible garden to make its care and harvesting simpler, faster, and more convenient. With some space left beneath and around the fruit trees, you can harvest, prune, manage pests, and water with much greater ease.

Now I plant most of my fruit trees clustered together in groves. They needn't be laid out in straight, orchard-like lines, but in a somewhat random pattern with a mixture of tree sizes and varieties — more like a forest, but edible. I use a permanent mulch beneath the dripline of all the trees' eventual, mature canopy, at very least. Beneficial insects can easily fly from the nearby flower and herb border to the trees to gobble up pests. (Contrary to myth, the two plantings don't need to be right on top of each other for this benefit.

See "Companion Planting" in Chapter 10.)

Planning an Edible Landscape

Live in your home for at least one year before beginning to design your permanent edible landscape. You'll learn countless useful things about soil drainage, wind, sun and shade patterns, birds, pests, noise, rainfall patterns, and other subtle environmental clues (and it will give you time to read this book!). This act of willpower is often very difficult for enthusiastic new homeowners and avid gardeners, but you can start with annual vegetables and flowers and try not to be attached to where you grow them. The first year's bed of lettuce may ultimately be the best place for a dwarf fruit tree.

An initial "year of observation" also gives you time to grow cover crops like legumes and annual grasses that enrich the soil with nitrogen nodules found on their roots—which also loosen the soil by "composting" as they die. Potatoes are another good crop for rapidly improving the soil's friability. Any soil improvement you do will be greatly appreciated by any subsequent plant, so you can't lose.

Knowing the sun and shade patterns for your yard will automatically help you to determine where to plant. The best sunny spots should be reserved for vegetables and fruits that require lots of heat to fully ripen. Many salad greens are quite adaptable to various levels of shade (up to fifty percent of dappled sunlight, or less than four hours of direct sunlight). In lower-light levels, salad greens will grow more slowly, but their leaves may actually get bigger. (See Figure 22.) Contrary to popular belief, not all fruits require full sun to mature with a full flavor. While peaches are not forgiving of shade, apples and pears will ripen to full sugar in much less light than you'd think. Cane fruits, such as blackberry, raspberry, and ollalie berry, also are well adapted to less than full sun.

Shade Tolerant Veggies

2 Hours - Asian Greens, Mesculun Mix

3 Hours - Culinary Herbs, Mustard Greens

3-4 Hours - Arugula, Kale, Lettuce, Mizuna, Spinach, Swiss Chard

4-5 Hours - Beans, Beets, Carrots, Celery, Collards, Pak Choy, Peas, Radishes, Rutabagas

5 Hours - Broccoli, Brussels Sprouts, Cabbage, Cauliflower

Figure 22. Food you can grow in limited sunlight.

Take your time to watch the flow of water, people, and materials through the landscape. The most convenient and efficient edible landscapes I've seen are the ones where the watering system for the entire project was planned first, not as an afterthought. This doesn't mean all the fixed plumbing has to be put in at the beginning, but you should know where all the hose-bibs are going to go before doing any serious groundwork or structural development, such as pathways, arbors, and patios. You can always put major pipes and main lines in at the beginning and then add the water supply to each new area you cultivate.

Pathways are the next most immediate need. Make sure they follow a route that's comfortable for you. It's easy to figure out where to locate permanent paths. Look for the downtrodden grass, bare dirt, or scuffled mulch in your existing landscape or yard. Designing pathways at a drawing board is hazardous, because you may be inclined to draw circles, complex curves, and all kinds of trippy routes that look great as you peer down at the paper. Although fanciful patterns may be pretty, they're usually not the most practical patterns of flow in actuality.

It's best to develop a series of permanent pathways, whether covered with wood chips, pebbles, or brick. Main pathways for people should be at least 6 feet wide, which allows you to walk hand-in-hand down the proverbial garden path with your wife, husband, or (warning, California-ese alert) "intimate other." Main pathways, those that will also be used for bringing in amendments and hardscape materials and hauling out trimmings and produce, should be 12" to 18" wider than your wheelbarrow or garden cart. Side paths need only be 18" to accommodate a single person walking.

Stone gravel is my preferred path material, as it reflects sunlight for better plant growth, while its mass stores warmth to help the garden grow better at night in cool-summer areas. Use no more than three different materials or textures for the overall hardscape—the fixed part of your design. Choose, for example, wood, or plastic "wood," for the beds and arbors, wood chips under the fruit trees, and brick for pathways. If you use more than three predominant textures or materials, the design may be a bit too "busy."

In the end:

The Recipe for an Edible Landscape is Easy:

1- Start with a heap of hedonism.
2- Add a healthy dose of study, planning, and paper-sketching.
3- Combine with a modicum of patience, a mess of elbow grease, and a pinch of persistence.
4- To this, add a good measure of experimentation, play, and carefree flaunting of some of the rules you read in gardening books (even this one!).
5- Combine into the shape and form you desire.
6- Let the sun, soil, and elements slowly flavor the recipe for several years.

And presto. You've got a gorgeous, healthy edible landscape that will sustain your family's soul and palate for decades to come. Keep it personal; grow your favorite foods and build your garden around your favorite activities. (For more details see the last Chapter of the book, Chapter 13.)

Quandaries with Some Edible Landscapes

The difficulty with Ros Creasy's first book (*The Complete Book of Edible Landscaping: Home Landscaping with Food-Bearing Plants and Resource-Saving Techniques*) and my early inclinations was that we were both hung up on the idea of making virtually every perennial an edible crop. We subsequently discovered, from years of feedback, that this created too much work for the average homeowner. Later books by Creasy don't focus as much on this concept, and often show kitchen gardens and thematic gardens as part of edible landscapes.

And as I've mentioned, one of the drawbacks to planting vegetables creatively throughout the garden is the difficulty of finding the leaf or fruit before it's no longer edible. Too often, in the early days, I planted lettuces and other leafy crops that became noticeable only after their seed-heads rose above the nearby foliage, at that point too bitter to eat.

Dealing with Ugly Edible Landscapes

Another problem arises when the ugliness of late summer/fall needs to be hidden. Who, after all, wants to look at the browning stalks of corn and the mildew-covered leaves of winter squash and pumpkins? To hide this late-summer or early-fall unsightliness, these can be grown behind a visual barrier or hedge (although this might lead to such plants being located further from the house than other vegetables, ornamentals, and herbs).

After all these years, there are now (c. 2020) many books that recommend trading in your existing lawn and front yard for a food garden (in some cases, alas, without any attempt at ornamental beauty). Some just advocate installing rows or boxes of vegetables where there used to be turf—this is, frankly, far uglier than a well-designed edible landscape. Other books showcase very attractive front-yard gardens and edible landscapes.

Off-Season, Off-Kilter "Beauty"

Notice that no book on growing food plants at home—mine included—features photos of what a garden looks like after its various edible components are harvested. This is a mental image that needs to be projected during the designing and planting process. None of the photographs of the best-looking edible landscapes are taken in the winter, and the more annual food crops that are included in the design, the uglier the scene usually gets in the winter—anywhere.

This is why hardscape features are so important in an edible landscape. You can entertain the eye during the "slow" times of the year or around major harvest times with decorative elements such as: sculptures, brick walkways, colorfully painted trellises, stone walls, brick or flagstone patios, wattle fences, brightly painted raised beds, ornamental corner brackets for raised beds, gnomes (hey, some people love them), and much more. Evergreen hedges can also be used to define the structure of the garden, and an alleé of evergreen plants (edible or not) is a good way to lure the eye away from the tackier sections of the garden. Such elements also help to define the garden during the snowy or dreary times of the year. To insure a continuous year of harvesting in mild-winter areas, it's important to learn which cool-season vegetables to plant and when to plant them.

Ros Creasy has a good solution for this dilemma—hedges, but tiny ones. Here we see the diminutive hedge that lines her pathways. It serves the purpose of a hardscape element in that it unites the garden and attracts the eye in late summer/ early fall, when the garden isn't looking its best. (See Figure 1, Chapter 1.)

Another problem with the lawn-to-garden books is that almost none of the gardens are shown as they're being harvested. The following photo of a rainbow of lettuces, as seen from Ros Creasy's bedroom, shows that a small amount of the red lettuce has been harvested (she's already transplanted some seedlings to fill the gap, a good strategy for any gardener).

Other options for this situation include: have all the seedlings for the next planting on hand, to execute a quick removal of the old and planting of the new; or plant your lettuce or other salad greens behind a row of cabbage. Cabbages take so long to mature that you can grow and harvest two or more lettuce crops without seeing the "holes."

You might also consider, with a crop like lettuce or spinach, harvesting every other plant to maintain the pattern, or, as is common in many edible landscapes, surrounding the annual vegetables with perennial flowers—as seen here in the gardens of the Esalen Institute in Big Sur, CA. (Esalen is well-known in the "self-help" world, but unbeknownst to many, it has amazingly beautiful gardens that are very attractive

Figure 23. A view from Rosalind Creasy's bedroom of her rainbow garden. (I can't reveal how I obtained this picture.) The kitchen is equally close—15 feet away.

Figure 24. A close-up of lettuce in contrasting patterns in one of the Esalen Institute's many gardens. Young salad seedlings are planted among flowers for an attractive look (*not* for companion planting).

and provide a huge amount of the vegetables served up to its hundreds of guests.)

Elvin Bishop Stays Tuned to Edibles

Elvin Bishop *loves* to garden. And though perhaps he's best known for his gold-record R&B classic "I Fooled Around and Fell in Love," Bishop, 78, is serious when it comes to growing things. For over 30 years, along with his wife Cara and daughter Emily, he's been cultivating a medley of flowers, vegetables, herbs, berries, and fruits around his Marin County, California home.

Besides the art of the blues, Bishop has mastered something very few San Francisco Bay Area gardeners even attempt—the art of year-round food production. His floral kitchen garden even continues to produce large quantities of vegetables through Northern California winters (which, though moderate in temperature, can reach a low of 15° F and receive over 40 inches of rain). Working his flourishing half-acre has become a grounding note in Bishop's life, a steady rhythm anchoring life's more unpredictable riffs.

Figure 25. Elvin Bishop with his daughter Emily. Elvin grows everything from seed, and makes his raised beds attractive by planting ornamentals in the corners and letting them drape over the sides.

Figure 26. A vining winter squash in Elvin's garden frames a view of roses, vegetables, and sunflowers, with limbs of an Asian pear tree at upper right. The tastiest Asian pear is 'Hosui.'

Touring the Garden

Visitors and friends enter Bishop's edible landscape, through a trellised gate supporting a rambling kiwi vine that frames the garden's southern edge. The west-facing deck of his house overlooks a gently sloped and attractively terraced garden. Some might call this an edible landscape or *potager*, but Bishop is more pragmatic, "I'm just making it up," he says, "I read a lot of garden magazines for the growing tips and plant varieties, not for the trends." Vegetables dominate the 30-foot wooden-box terraces that run east to west with flowers capping each end. Below the deck and house, a cascade of rosemary, tall hybrid tea roses, four o'clocks, gladiolus, scarlet pimpernel (as an ornamental groundcover), spring tulips, sweet peas and climbing nasturtiums spills in colorful profusion.

To the west, a passion-flower vine slowly engulfs a Douglas fir tree, which towers above the vegetables as a backdrop to the terraces, while the heads of ornamental sunflowers sway over beds of robust dahlias, clambering sweet peas, and the loosely-trellised ornamental and edible squashes and gourds that cover young Emily's summer playhouse.

A few discreetly-placed fruit trees—apples, a loquat and Asian pears—separate the primary vegetable area from small garden "rooms" filled with flowerbeds. Mulch is conspicuously absent from the pathways and within the beds because, Bishop says, it provides too much habitat for earwigs, snails and slugs, and hampers his pest-control efforts.

Along the south-facing slope of the house, gooseberries, pineapple guavas, heirloom and modern apples, figs, raspberries, blueberries, strawberries, peaches and Asian pears abound—often with vegetables planted below, peasant-garden style. Beneath some of the trees, spring brings a riot of wildflower bloom, highlighted by blazing red poppies.

From the Midwest to the Far East

Bishop's interest in Asian vegetables was initiated by his marriage 25 years ago to Cara Wada, whose Japanese-American heritage and traditional cooking style brought more far-eastern vegetables, first into her husband's diet, then into his garden. "Cara's family made me aware of different tasty Asian vegetables. I figure if it tastes good, then it's worth growing. Besides, it's nice to have vegetables to give away to the family."

Many Asian vegetables are members of the cabbage or crucifer family (*Brassica* spp.) and thrive in moderate weather. In the Marin County "Mediterranean" climate, winter gardening makes good sense; plenty of winter rain keeps things moist and, after temperatures drop, the cold keeps most pests (including aphids and root maggots, which can thwart summer plantings, but not, unfortunately, slugs and snails) at bay.

The secret to a garden that can be harvested from mid-winter through the following spring is timing, and Bishop's sense of the rhythm of the seasons is finely tuned. His brief recipe for cultivating *Brassicas*: choose seed for early-, mid- and late-season crops to spread the harvest. Seed everything on June 1st in six-packs, using sterile potting mix. In two to three weeks, transplant the seedlings deeply into four-inch pots. Set out into the garden at the beginning of July. Shade for several days if the weather is hot. Water till the rains begin. Weed as needed and harvest when ready.

Reaping the Rewards

In addition to its other rewards, Bishop's garden offers an abundance of produce; each year he personally puts up about 300 jars of beans, corn, pickled beets, dill pickles, applesauce, apple juice, peaches, plums, pears, tomatoes, and various jams, including kiwi.

Another fond pursuit of Bishop's is fishing—which he's celebrated in several songs on his albums, but when asked to choose between fishing and gardening, he leans forward with a knowing smile, "Fishing is sometimes chancy. I'd stick with the gardening, because I'm pretty sure I'm gonna get something to eat from my efforts."

Besides, Elvin Bishop just *loves* to garden. He may sing the blues for a living, but in his garden, life is sweet.

Multiple Influence, Synergy

Permaculture proposes designing a landscape that "theoretically has unlimited yields" or "sustains itself and those who construct it." To this end, the design process emphasizes synergy—devising a garden in which the overall effect is greater than the sum of the individual parts.

The assumption is that in a permaculture garden "the sum of yields (of all the species of plants and animals) will be greater, simply because a single-species system (monocrop) can never use all available energy and nutrients." In *Permaculture One*, we read: "Symbiotic interaction in a well-designed and -controlled permaculture can further increase yields [beyond mere species diversity]. Plants and animals may not only co-exist without competition but the presence of one may improve the environment for others." This suggests that synergy promotes an ecological symbiosis. What's important to this healthy synergistic co-existence is not solely diversity itself, but "a diversity of *beneficial functional connections ...Complexity*, in the sense of some powerful interconnections between species, is what we (as permaculturalists) are really seeking in food systems."

This concept is summed up in two much-repeated designer guidelines: "There should be no placement of an element without it serving at least two or more functions" and "Every function should be

Figure 27. A view from Elvin's house shows a combination of ornamental sunflowers and a small deck for relaxation. The orange nasturtiums provide edible flowers, leaves, and seeds. A ground cover of thyme lines the garlic patch. Most of the vegetables are grown in beds in the center of the garden and surrounded by flowers.

served in two or more ways."

But how realistic is this concept of synergy created by one element serving multiple functions and backups for every function? The true answer is quite ambiguous.

One homely example of multi-use is **cilantro**: its leaves are used as a spice, and its blossoms in salads. Save some to bloom and you have an excellent beneficial insect attractor. Leave some flowers to ripen and you'll have coriander.

Another is the **chayote** (*Sechium edule*), an amazing plant that grows practicality as fast as kudzu (*Pueraria montana* var. *lobata)*, has edible young tendrils, makes a "fruit' that can be cooked like a zucchini, has very tasty soft seeds, produces large tubers that can be prepared like any potato dish; provides shade; has annual foliage (vines die to ground in winter) to let warming winter sun in, and perennial roots (providing you don't dig up all the immense tubers). Chayote is, in short, a fantastic multipurpose plant. Maybe. There's one hitch; it only grows in sections of USDA Zones 10b-11 that stay above 35^0F or (with lower yields) in very special microclimates in Zones 8 and 9. The vines are huge.

Yarrow is drought-resistant, a great cut flower, makes a great landing platform for beneficial insects, and comes in many ornamental colors. (It can also be somewhat invasive in some climates if left to its own devices.)

Attractive to bees, colorful, aromatic, and used as a seasoning or to make your laundry smell nice without chemicals, **lavender** is my favorite multipurpose plant. I grill with the foliage of Spanish lavender

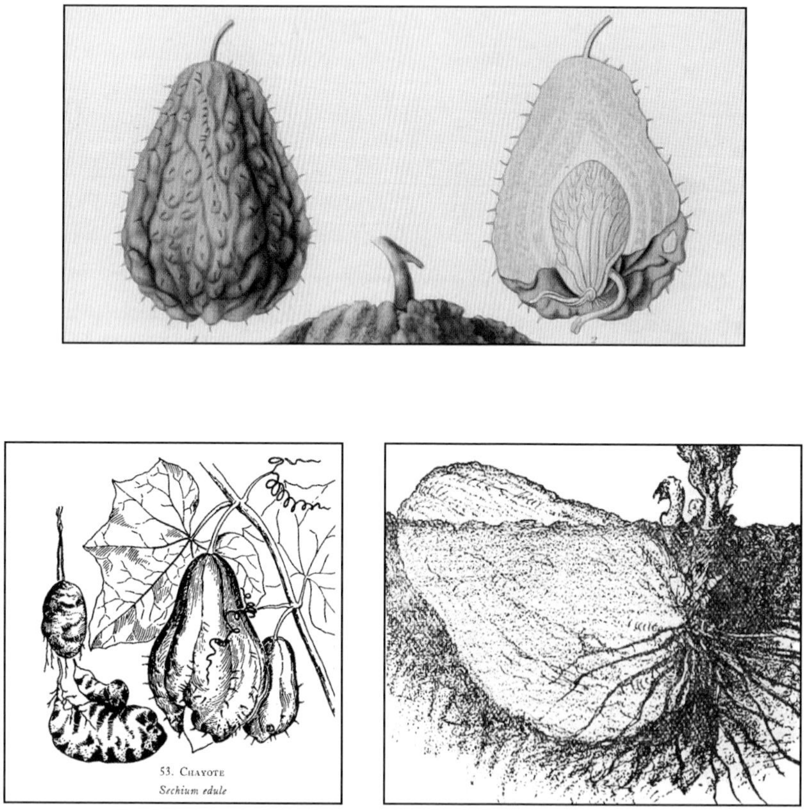

Figure 28. Chayote (*Sechium edule*) only grows in Zones 9-11, which is where I live. I love this multipurpose vine. The upper illustration includes the seed (at right), the tastiest part of the plant; it can be eaten like zucchini. Many chayotes have smooth skin. The tubers shown at lower left can be prepared any way you cook a potato. Leave some of the tubers in the ground and you'll have a perennial squash vine. The young tendrils can be steamed, and taste a bit like green beans. On the right: how to plant it. One vine can cover an entire garage.

(*Lavandula stoechas*) all summer long.

Let's not forget the edible leaves, flowers, pods, and pickled seeds of the trailing **nasturtium** (which also, however, provides a home for snails and slugs; if you live in a very-mild-winter climate, be careful to not let it escape).

But how realistic is the concept of creating synergy with each element serving multiple functions and backups for *every* function? The true answer, again is quite ambiguous. Take, for example the chart that appeared in my book *Designing and Maintaining Your Edible Landscape—Naturally* in 1986, while I was still somewhat under the influence of permaculture thought. (Pages 59-60.)

MULTI-PURPOSE EDIBLES **COMMON NAME (Botanical name)**	**Annual, Perennial, Biennial**	**Root**	**Bulb**	**Tuber**	**Young Shoots**	**Stem**	**Seed**	**Spice**	**Leaf**	**Tea**	**Flower**	**Fruit**	**Attracts Beneficial Insects**	**Attracts Bees**	**OTHER**
AMARANTH *(Amaranthus sp.)*	A						X		X		X		X		High in iron, calcium, potassium, and Vitamin C
ANGELICA *(Angelica archangelica)*	P					X							X		
ANISE *(Pimpinella anisum)*	A						X		X		X		X		Invasive
BEE BALM *(Monarda didyma)*	A								X	X	X			X	
BLACK LOCUST *(Robinia pseudoacacia)*	P										X			X	Fixes nitrogen, long-lasting fence posts, pods feed livestock
BORAGE *(Borago officinalis)*	A								X	X	X			X	Very good bee "food"
CARAWAY *(Carum carvi)*	A						X		X		X		X		
CELTUCE *(the Composite Family)*	A					X			X						
CHAYOTE *(Sechium edule)*	P			X	X		X					X			Fast vine for shade
CHICORY *(Cichorium sp.)*	P	X							X	X	X			X	Deep rooted
CHIVES *(Allium sp.)*	P				X			X	X		X				
COMFREY *(Symphytum officinale)*	P	X							X	X					Deep rooted; high in vitamins
CORIANDER *(Coriandrum sativum)*	A						X		X		X		X		
DANDELION *(Taraxacum officinale)*	P	X						X	X	X	X		X		Invasive
DAY LILY *(Hemerocallis fulva)*	P			X							X				
DILL *(Anethum graveolens)*	A					X	X	X	X		X		X		Deep rooted
EGYPTIAN TOP-SET ONION *(Allium cepa var. viviprium)*	P		X		X		X	X	X		X				

KEY: LIFE CYCLE (LC): A = annual P = perennial, B = biennal EDIBLE PARTS, USES: R = root, B = bulb, TB = tuber, Y = young shoots, S = stem SD = seed, SP = spice, L = leaf, T = tea, FL = flower, F = fruit INSECTS: PR = attracts predatory insects, P = attracts pollinating bees

continued

Figure 29. From my 1986 book *Designing and Maintaining Your Edible Landscape—Naturally*. I don't know why I forgot to list black locust and dill as invasive. I should also have added a footnote advising not to till or cultivate anywhere near comfrey. I've learned a lot since then.

COMMON NAME (Botanical name)	Annual, Perennial, Biennial	Root	Bulb	Tuber	Young Shoots	Stem	Seed	Spice	Leaf	Tea	Flower	Fruit	Attracts Beneficial Insects	Attracts Bees	OTHER
FAVA BEAN *(Vicia faba)*	A						X		X		X				Fixes 200 lb.of nitrogen per acre per year
FLORENCE FENNEL *(Foeniculum vulgare var. azoricum)*	A	X				X	X	X	X		X		X		Invasive
GARLIC *(Allium sativum)*	P		X		X		X	X	X		X				
HAMBURG PARSLEY *(Petroselinum crispum)*	B	X						X	X				X		
HYSSOP *(Hyssopus officinalis)*	P								X	X	X			X	
LEMON BALM *(Melissa officinalis)*	A								X	X	X			X	Invasive
LUFFA *(Luffa cylindrica)*	A				X	X			X		X	X			Good shade vine; sponges
LOVAGE *(Leviticum officinale)*	P					X	X	X	X				X		
MALLOW *(Malva neglecta)*	A								X		X				Invasive
MINER'S LETTUCE *(Claytonia perfoliata)*	A				X	X			X		X				
MUSTARDS *(Brassica sp.)*	A								X		X			X	
NASTURTIUM *(Tropaeolum sp.)*	A				X		X	X	X		X				Nurse crop for beneficial insects
ROSE HIPS *(Rosa rugosa)*	P						X	X		X	X	X			High vitamin C; barrier plant
ROSEMARY *(Rosmarinus officinalis)*	P							X	X		X			X	
RUSSIAN OLIVE *(Elaeagnus angustifolia)*	P											X			Fixes nitrogen; good wildlife habitat; good windbreak shrub
SIBERIAN PEA-SHRUB *(Caragana arborescens)*	P						X								Fixes nitrogen; good livestock food
SOCIETY GARLIC *(Tulbaghia violacea)*	P		X		X			X	X		X				
SWEET VIOLET *(Viola odorata)*	P								X		X				High vitamin A; shade-loving
VIOLA *(Viola cornuta)*	A								X		X				

Figure 4.8

KEY: LIFE CYCLE (LC): A = annual P = perennial, B = biennal EDIBLE PARTS, USES: R = root, B = bulb, TB = tuber, Y = young shoots, S = stem, SD = seed, SP = spice, L = leaf, T = tea, FL = flower, F = fruit INSECTS: PR = attracts predatory insects, P = attracts pollinating bees

Figure 30. Continued from the 1986 edition of my book *Designing and Maintaining Your Edible Landscape—Naturally*. I should also have listed Florence fennel as slightly invasive, and included Russian olive and rose hips as invasive on the list.

Ecological Edges

An oft-repeated concept of permaculture is: "Edge Effect. Edges are the site of dynamic interplay. 1 + 1 = 3. (The sum is greater than the two parts.) Increase beneficial edges, minimize problem edges." The whole "edge" concept originated with ecologists talking about large ecosystems, and comes from observations based on a macroscopic environmental setting, not from the conditions in the average suburban landscape or back yard.

One of the ecological definitions of "edge" is: "…a transition area between two different ecological communities that exhibits competition between organisms common to both." (Angus Forbes, University of California, Santa Barbara.) He goes on to say, "…One of the major issues in ecology is what is termed 'the scaling problem.' Ecological processes that seem important to a particular patch of forest turn out to be inconsequential to the forest as a whole."

So, can ecotones be scaled down to a small yard? Let's see.

Ecological edge is the scientific term now used for the boundary area between different plant and animal communities. One of the more common ecological-edge effects is created wherever human horticultural activity begins to interact with and impact a natural landscape. Such effects can be beneficial, in that they can promote plant growth and high habitat diversity. Increased light and wind penetration of a forest edge, for example, will change temperature and humidity to create a physical environment characteristic of both forest and meadow. These changes assist various species of plants and animals in their encroachment into the *ecotone* (the term applied to the actual zone of transition).

As a transition area, an ecotone may be narrow or wide, and it may be local (the zone between a meadow and some trees) or regional (the transition between a forest and a grassland). An ecotone may appear to the observer as a gradual blending of the two communities across a broad area, or it may manifest itself as a sharp boundary line. (The word was coined from a combination of *eco*(logy) plus *-tone*, from the Greek *tonos* or tension—in other words, a place where ecologies are in tension. And in case you've ever wondered where the word "ecology" came from, it derives from the Greek *oikos*; (house, household, dwelling place, family) and *logos* (word, language, language of reason). (The term "Oecologie" was coined by the German zoologist Ernst Haeckel in 1866 in his book *Generelle Morphologie der Organismen*.)

Ecological edges and ecotones can be created by human activity, or occur naturally. Reed beds, for example, are a common form of natural lake- or pond-side ecotone. The beds tend to accumulate organic

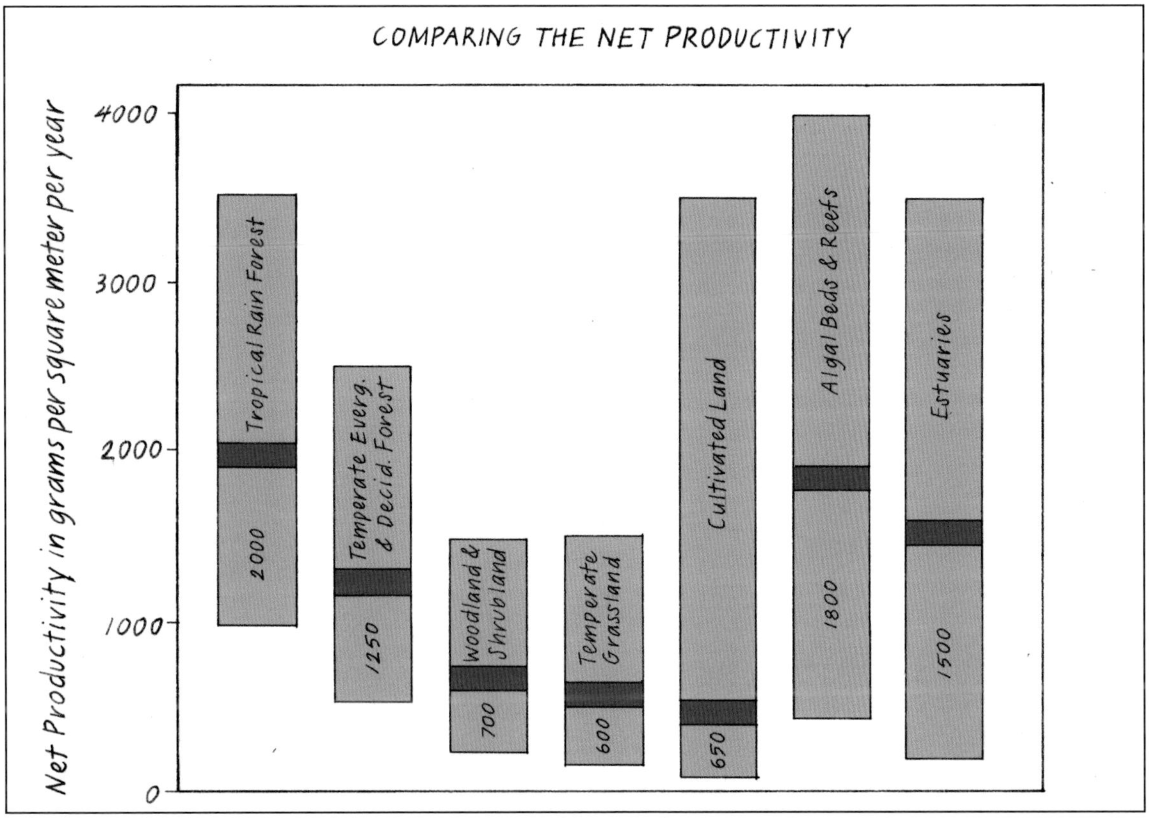

Figure 31. Robert Whittaker used a scale of grams per square meter per year to show that the most productive estuary rating (3500) equals that of tropical rain forests (which often have no effective edge), cultivated land, swamps, and marshes. The mean amounts (dark blue horizontal bars), show that estuaries (at 1500 grams per square meter per year) are less fertile than tropical forests, and only 250 g/sq. m/yr. more productive than temperate evergreen forests. The pale-blue bars show the range of best and worst productivity. The productivity of cultivated land—about 3500 grams—can equal that of the tropics. At its worst, farmland rates at only 100 g/sq. m/year.

matter, which is colonized, first by shrubs, and then by trees, moving the reed bed further into the lake and eventually, depending on the water source, creating either a marsh or solid ground.

An ecotone contains plants and animals from both adjacent ecosystems, along with species that only populate the ecotone. According to prominent ecologist Dr. Eugene Odum, "The tendency for increased variety and density is known as the edge effect." This concept has been eagerly adopted by many wildlife biologists, under the assumption that an ecotone is the best place for wildlife, but the thinking on ecotones, edges, and diversity has evolved over the years.

Here's a staple of permaculture thinking: where one type of large or macroscopic ecosystem meets another, there is a mixing of plants and animals indigenous to both. This produces a unique environment that contains some of the virtues and diversity of both ecosystems, as well as of plants and animals that thrive only in the mixed zone or "ecological edge." In *Permaculture: A Designer's Manual*, this mix is described as, "...normally richer in species than the contributing media themselves." It's also stated there that "...complex systems of land/ocean interface (edge), such as estuaries and coral reefs—show the highest production per unit area of any of the major ecosystems."

Well, it's not that simple. For instance, the assumption that estuaries show the highest net productivity of any major ecosystem is refuted by respected ecologists. In the book *Communities and Ecosystems*, by Cornell University ecologist Robert Whittaker, an informative table ranks the net primary productivity (a gauge of fertility) of various ecosystems. (See Figure 31) Using a scale of grams per square meter per year, the most productive estuary rating of 3500 *equals* that of tropical rain forests (which often have no effective edge), cultivated land, swamps and marshes. Looking at the mean amounts, estuaries (at 1500 grams per square meter per year) are *less* fertile than tropical rainforests and only 250 g/sq. m/yr. *more* productive than our temperate evergreen forests.

Since fertility is not a fixed number, but a fluctuating range of figures, it's misleading to state unequivocally that estuaries show the highest production of any of the major ecosystems. In addition, most of the accumulated fertility or productivity of an ecosystem must stay within that ecosystem in order for it to persist.

If the ecosystem is to remain self-sustaining, we can only extract a small portion of its productivity each year as food, timber, or other material. The reference to reefs as one of the most productive ecosystems in the world is accurate, but how many of us have the opportunity to "garden" a coral reef or algae bed? (Never mind the great uncertainties involved in trying to apply aquatic ecological dynamics to our terrestrial fields and gardens.)

Figure 31 shows the extremes and the mean productivity for seven ecosystems. The darker horizontal bar within each vertical bar is the mean productivity. A label below each dark bar lists the mean productivity for each ecosystem. The top of the vertical bar indicates the highest productivity. The bottom of each vertical bar indicates the lowest productivity. Notice that the mean for tropical rainforests exceeds that of estuaries, and the maximum productivity of cultivated land can equal that of estuaries. Also note, however, that the mean productivity for temperate grasslands are lower than all other ecosystems.

So, acting on the assumption that all "edges" embody the most fertile productive environments, permaculture focuses on creating more "edge" to increase the dynamic of the environment. In *Permaculture—A Designer's Manual* (1988), we read, "...a complex edge...can be considered the basis of the art of productive landscape design. And most certainly, increased edge makes for a more stimulating landscape. As designers we can also *create* a harmonic with plants, water, or buildings."

At this point, hard-core permaculture advocates attempt to construct an intellectually unifying pattern or theory based upon the many forms of a sine wave and the spiral form represented by the Fibonacci series (as in the spiral of a Nautilus shell and the various curves manifested in sunflower heads, which they offer as virtually universal forms that can/should be used to design a landscape.

Although this abstract discussion of edge takes on a very literal interpretation in many permaculture-book illustrations, hard science seems to refute it. Since, ecotones are represented by large environments, and not by microclimates, attempting to apply such dynamics to smaller areas is problematic. I believe such reductions from the macroscopic (e.g., a forest) to the microscopic—the home orchard—are not realistic. The interface of a single tree, or even an orchard, in one's backyard landscape doesn't necessarily mean that a vegetable plot next to it will be more productive. A reduction in scale does not necessarily apply.

In addition, the cross-sections of various planting schemes in permaculture books feature a polite progression of small plants to large trees, forming a convenient diagonal line, although a natural edge seldom takes this form. Arranging the tallest plants on the north side of a landscape and graduating each shorter plant toward the sunny south has little to do with the lofty and complicated theories of "edge harmonics"—it's just good old common sense.

Parabolic Curves versus Square Forest Ecology
Just how important is a sweeping curve or parabolic line in enhancing beneficial edge effects? This question was addressed in a revealing and well-written paper by Bart Johnson, Landscape Architecture Professor at the University of Oregon, in a paper entitled "*The Design Edge*," written at Cornell University in 1990.

While edges benefit some species, in Johnson's words, "A benefit to one species, however, usually involves a loss for another. For example, the creation of a rich habitat combining aspects of both the forest and field can endanger species that depend upon undisturbed habitat." (4) According to Johnson, there are four ways that certain species may be impacted:

1. Colonization: Both native and exotic plants and animals can colonize an edge, but sometimes an introduced exotic will overwhelm some native species. Edges tend to harbor generalist invaders—species adapted to colonizing a wide range of habitats.

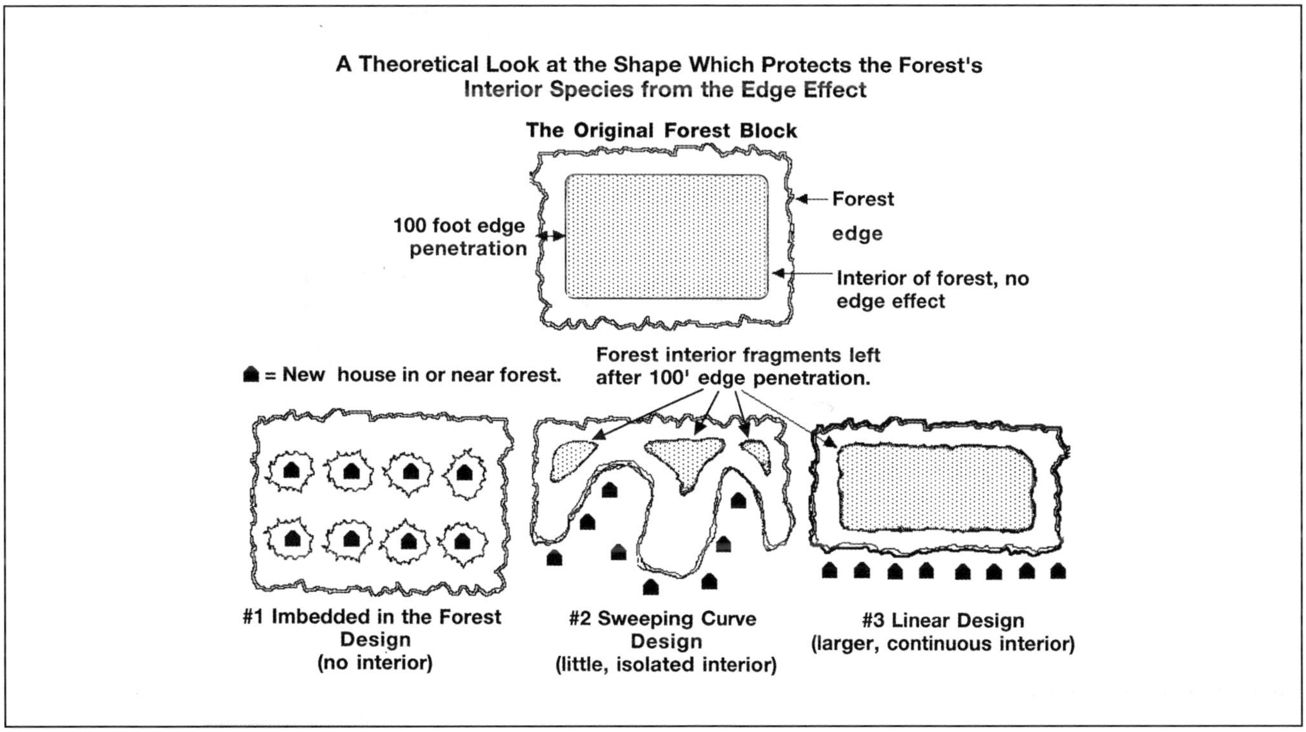

Figure 32. The theoretical illustrations (by Bart Johnson) pictured here show three ways that a continuous forest block could be developed for housing while leaving remnants of the forest for environmental quality.

2. Parasitism: When a forest is changed with the addition of edges and gaps, interior bird species (which prefer undisturbed forest only) may have their nests parasitized by invaders such as the brown-headed cowbird, which routinely lays its eggs in other bird's nests.

3. Predation: Predatory animals that thrive in the ecotone will also migrate into the forest and disrupt the reproduction and survival of interior species.

4. Herbivorous activity: As an example, browsing by deer can suppress young vegetation enough to eradicate tree growth in localized areas.

Johnson then asks the pertinent question: "At what scale are edge effects important?" His answer:

"There is no set distance to assign to edge effect. In one study, changes in vegetation structure were reported up to 40 feet from the forest edge, while disruption of bird nests occurred up to and beyond 200 feet. (1) Another study found vegetation change up to 170 feet (5) and one concluded that 2000 feet was required before a safe interior was found for forest birds. (1) In the Great Smoky Mountains, intensive [browsing] by deer was not seen beyond 0.6 mile from the forest/meadow edge." (9)

To further explore the topic, Johnson constructs a simple mathematical study, with a few provisos. "Let us assume a single edge effect width of 100 feet...in an isolated forest patch. [These] examples ignore real-life factors such as topographic variation and vegetative heterogeneity." The theoretical illustrations pictured here imagine three ways that a continuous forest block could be developed for housing with remnants of the forest left for environmental quality.

The books and papers used to explain this phenomenon are:

(1) *Permaculture One, A Perennial Agriculture for Human Settlements*, Bill Mollison and David Holmgren, Tagari Books, Stanley, Tasmania, 1978.

(2) *Permaculture—A Designer's Manual*, Bill Mollison, Reny Mia Slay and Andrew Jeeves, Tagari Publications, Tyalgum, Australia, 1988.

(3) *Concepts of Ecology*, E. J. Kormondy, Prentice Hall, NJ, 1959.

(4) *Communities and Ecosystems*, Robert Whittaker, Macmillan, NY, 1975, p 224.

(5) *Fundamentals of Ecology, 3rd Edition*, Eugene Odum, W. B. Sanders Co., Philadelphia, PA, 1971.

(6) "Edge effect and Conservation of Biotic Diversity," L. D. Harris, *Conservation Biology*, 1988, Vol. 3(10): p 82-84.

(7) *The Fragmented Forest*, L. D. Harris, University of Chicago Press, 1984.

(8) "Changes in Wildlife Communities Near Edges," R. H. Yahner, *Conservation Biology*, 1988, Vol. 2(4): p 333-339.

(9) "Impacts of White-Tailed Deer on the Vegetation of Cades Cove, Great Smokey Mountains National Park," S. P. Bratton, 1979. *Proceedings of the American Conference of the Southeast Association of Fish & Wildlife Agencies*, Vol. 33, p 305-312.

The edge effect even applies to insects, depending on species. In one study, wasps belonging to *Carabidae* [parasitic wasps] were very numerous on field edges, but the surrounding field was the most favorable site for linyphiid spiders and for parasitic ichneumonid wasps. (*The Effect of Companion Plants on Cabbage Root Fly (Delia Radicum L.) Occurrence and the Importance of Field Surrounding Flora for Beneficial Fauna*. D. Abanowskiej.)

For more on this topic, you can read:

• *Concepts of Ecology*, E. J. Kormondy, Prentice Hall, NJ, 1959.

• "Edge effect and Conservation of Biotic Diversity," L. D. Harris, *Conservation Biology*, 1988, Vol. 3(10): p 82-84.

• *Fundamentals of Ecology, 3rd Edition*, Eugene Odum, W. B. Sanders Co., Philadelphia, PA, 1971.

Zones and Sectors

Perhaps the most useful aspect of that first copy of *Permaculture One* was Bill's articulation of designing by zones and sectors. This is an example of a writer clearly describing something that I instinctively knew and applied in my own design work, but had failed to previously articulate. Most of what he wrote about was just practical sense, but articulated *so* well.

Viewing a yard as a series of zones, each with a different level of activity and use, is an important step in planning a thoughtful and conscious landscape. In *Permaculture: A Designers' Manual*, Bill defines analyzing a yard by zones and sectors as a way to determine a pattern of design that conserves energy. He defines zones as "...energy available *on site*: the people, machines, wastes, and fuels of the family...(these are) zones of use, of access, and of time available." The "energy entering or flowing *through the site*: wind, water, sunlight and fire may enter the site...[and further define] sectors." In my observation, after many years as a landscape consultant, a fundamental omission of many owner-built landscapes is the failure to think carefully about the flow of water, air, and materials throughout a landscape, and envision its future patterns of use and activity.

Figure 33. Rosalind Creasy's chicken coop is halfway from her house to the street, but only 30 feet from the kitchen, a good use of Zone Two. (The neighborhood kids love to gather the eggs for her.)

Sectors closely resemble what we commonly refer to as microclimates—areas of special conditions. Bill more carefully defines sectors as: "more site-specific than are the conceptualized zones. They outline the compass directions from which we can expect energy...the wild energies, the 'elements' of sun, light, wind, rain, wildfire, and water flow." An often ignored, yet special, microclimate, is the interaction between the home and the garden. The sunny southern walls of most homes offer growth-extending microclimates for coldframes, row tunnels, greenhouses, and cold-tender vines, trees, shrubs, and vegetables. Conversely, the deep shade of the north side of the house may provide an excellent microclimate for elements such as compost bins, worm bins, mushroom logs, and a potting shed.

Typically, the first zone is the area closest to the walls of the house. This is where, as outlined in *Permaculture: A Designers' Manual*: "Components needing continual observation, frequent visits, work input, or complex techniques should be placed *very close to hand*, or we waste a great deal of time and energy visiting them." He defines this zone as within 20 feet of the home and claims this is enough area for homeowners to "produce most of the food necessary to existence, with some modest trade requirements." This is the place (as seen at Village Homes in Chapter 6), for, rabbits, chickens, culinary herbs, daily salad greens, intensive vegetables, dwarf and espaliered fruit trees, and annual bedding plants for seasonal color and fresh-cut flowers.

Practical Zones

As with any recommendations, there are always exceptions. The difficulty with perfect circles of zones drawn on paper is that they don't often translate well to reality; the drawings in permaculture books often conveniently show only south-facing houses with an east-west aligned rectangle.

Take, for an example, keeping chickens. You can make an argument for placing the coop in one of two zones: 1) in Zone Two so manure smells don't drift into the house, and potential fly problems are reduced, or 2), in Zone One, so it reminds owners (and makes it more convenient) to close up the coop every night to keep out furry predators. I would prefer to put my coop where I could see it from the television. As I settled in for the evening news (I guess this dates me) I would see that the door of the coop was either open or closed.

Another example is growing corn in a beautiful edible landscape. Let's face it, as corn matures it can get brown and ratty looking. A view of it planted in Zone One would make for an ugly landscape. Perhaps placing it in Zone Two, behind an evergreen hedge, would solve this problem, but you'd have to remember to pick the ears at the right time. (I once planted corn, at a client's request, behind the house near a dog pen, far from the kitchen, and ended up eating some of the abandoned corn on my lunch breaks—it was out of sight, out of mind to my clients, and they just never bothered.) In my first tiny edible-landscape design in 1977, the client had a direct view of the herbs and vegetables from the window above his kitchen sink, and they never went to waste.

Figure 34. This front-yard walkway from the street through the patio to Rosalind Creasy's house is lined with herbs and edible and ornamental flowers. On her way home, she can pick herbs for cooking, edible flowers for the salad, and ornamental flowers for the table.

Sunlight will also often impact the use of circular zones. If your kitchen is on the north side of a south-facing house and the porch or entry deck is on the south side, you won't be able to see the herbs and vegetables from the vantage point of your kitchen. You could put vegetables that tolerate some shade near the north-facing kitchen to keep them in Zone One, or further out in Zone Two or Three where the winter sun still shines. (See page 32 for examples.)

The herbs can be planted in containers on the south-facing porch or deck and picked as you come home, or in zone two or three where the sun shines all year— but you may tend to forget them. Maybe try some shade-tolerant herbs in zone one near the kitchen. Examples to try in half shade are: bee balm (*DO NOT* let it go to seed, as it becomes a very nasty, invasive weed; better yet, skip it altogether), sweet violet (edible flowers), sorrel, stinging nettle (yes edible, if you pick it ever so delicately and steam it like a potherb), huckleberries, day-lilies (edible flowers), persimmons, pawpaws, and spearmint *(DO NOT* plant it the ground, as it can also become a very nasty weed—containers only). A greenhouse attached to the home for convenience and to heat the house, might push the vegetables into Zone Two.

Another problematic example is/are ornamental trees. Growing them next to the house means you can't grow sun-loving plants there. Trees planted on the west side of the house to shade it from hot afternoon sun may eliminate that zone for vegetables like corn and tomatoes, which would then have to be moved to Zone Two or Three.

Theoretically, cisterns to catch roof-runoff water far from the house could be placed at a distance from it so as to not take up valuable "real estate" in zone one (near the exterior walls of the house), but piping to the tank may not be practical. You could bury the tank near the house, but that's very expensive, and the required access to it would still eat up some vegetable-growing space.

So we all need to face the fact that dealing with life's curveballs (and life *always* throws curveballs) means that your perfect circle drawings of zones will, in all probability, need to be adjusted to a playing field of irregular shapes.

The Victory Gardens of World War II

Sometimes the Covid pandemic feels like a war, but gardening has given many people a pleasurable break from the boredom, tedium, isolation, and lack of human contact while providing fun, exercise, tasty food, and the satisfaction of raising one's own vegetables, fruits, and nuts.

During World War II, "Victory Gardens" were part of a movement to grow food as a way to conserve resources; save metal (using glass canning jars and fruit-drying techniques); feed people; save money; save fuel (the produce didn't have to be shipped miles away); promote a healthy diet; support the war by helping gardeners extend food rations; and help counter the loss of food shipped to the troops overseas. As one handbook read: "You may not carry a gun or drive a tank, but you can grow food for Victory!"

Victory gardens had nothing to do with beautiful edible landscapes—they were row-crops all the way, straightforward and very productive.

The government issued a wide range of handbooks to help the novice gardener, canner, or food-dryer achieve self-sufficiency. One handbook, the 32-page *Victory Garden Guide*, can be found on the Internet. It offers a wide range of information, and even promotes compost piles. Google™ *Victory Garden Guide* for a free PDF read. (Original copies go for $20.) It's interesting how much of the information in the handbook is still applicable today.

The interest in food gardening has really blossomed due to the pandemic. In fact, some resources like seed, seedlings, and compost have become somewhat scarce in some areas. Here are 40 tips about how to be more Victorious in your garden:

1. Read gardening books in the winter so you'll have unrestricted time in the garden all summer. Buy and read all 18 of my books.
2. To make the most of your space, plan your garden on paper first.
3. If your outdoor space is tiny, consider joining a community garden or allotment. Find out the average dates of first and last frosts and the average rainfall in your area.

4. Start a compost heap. If rats show up, you'll have to build a secure box.
5. If you buy large amounts of compost, order in late summer when the demand is low. Tarp it for the winter and it'll be even better for the garden come spring. (This avoids the long spring waiting lists.)
6. Use a worm bin for kitchen scraps. You can lock it so raccoons don't get inside.
7. Share seed. Most seed packets have more seed than you'll use the first season. Coordinate with friends to divvy up the packets of seed.
8. Start a "seed library" at your local library to make a wider range of vegetables available to all.
9. Form a seed exchange with local gardeners. It works like a seed-based version of a swap meet.
10. Contact heirloom-seed organizations and seed companions in very early winter to get seed before the spring rush. For example, the Baker Creek Seed Company offers literally thousands of choices. The Seed Savers Exchange has been saving rare and endangered seed for decades. You can arrange to swap seeds with its members.
11. A number of non-profits and schools in my area hold veggie-transplant sales to raise money. Ours even take orders on the Internet for curbside delivery when you drive up.
12. Better yet, save your own seed once you get growing.
13. Grow your own seedlings for transplant. This takes some practice, but it's well worth it.
14. You can recycle used shower doors from the dump or a contractor to make the top of a cold frame, or even a small greenhouse.
15. Gather sheets of cardboard through the winter to be ready to mulch your garden paths, and between large plants like squash and tomatoes.
16. Look for arborists with truckloads of arbor mulch (a mixture of wood chips and shredded leaves) and ask them to dump it at your place for free. (They'll save plenty of money by not having to pay to dump the chips elsewhere, and you get free mulch.)
17. During the winter, sharpen all the tools that require it. You can also use indoor time to

paint/mark tool handles with bright colors so they're harder to lose.

18. Straw bales for mulch or straw-bale gardening can also be bought in late summer thru the winter to beat the spring rush. (See page 101.)
19. To avoid the cost of wire cages, use tree branches to hold up plants like beans and tomatoes. (I think they look better—more rustic, and they blend into the landscape.)
20. Put in a drip-irrigation system with a timer to save water and time, *and* to increase yields. (See pages 144 - 145.)
21. Learn the timing for having seedlings on hand ready to transplant right after you harvest a crop. (This is called succession planting.)
22. In mild-winter areas, learn which Asian vegetables will grow through the winter and you'll have salad greens for the whole year. (Elvin Bishop—see pages 35 - 37 — is able to harvest greens every day of the year from his garden.) Elsewhere, use cold frames to start early and harvest late.
23. In some areas you can "store" root crops in the ground until you need to dig some up for a meal. In places where the soil can freeze, you can pile plastic bags full of leaves over your carrot patch, and just lift the bags to dig out carrots during the winter.
24. In areas where there are periodic summer rains, a rainwater catchment barrel or tank can get you through some tough times.
25. If your rain-barrel or catchment tank is above ground, cover it with climbing crops like green beans to help keep them a bit cooler.
26. Keep a journal; writing down (or typing into a computer) the history of the garden will not only raise your awareness, but will make it easier to plan for the following year.
27. Buy canning jars and lids in the winter when they're less in demand.
28. Order a dehydrator in the winter for the same reason. You can now buy effective solar versions.
29. Order bare root fruit trees in September so you'll have a choice for delivery in the spring. In the spring, so many people are buying trees that some varieties run out. Often, a fall order is complete.
30. Neem is a natural herb that comes from the *Azadirachta indica* or Indian lilac tree and has many different traditional uses. It's known for its pesticidal and insecticidal properties, and an extract from its seeds can help get many common pest and disease problems under control.

31. Keep some granular or liquid fertilizer on hand for supplemental feeding in the summer.
32. Buy a roll of a non-woven polypropylene or polyester cloth-like material called Remay^{tm}, and cover your plants to protect them from early fall frosts and exclude flying insects. Remay transmits 70% of available light, allows rain and overhead irrigation to reach plants and soil, keeps heat in and flying insects out, and is an excellent windbreak for young transplants.
33. To be Victorious over critters like gophers and moles, line the bottoms of your raised beds with ½-inch aviary wire or wire hardware cloth. For plants placed far apart; like squashes and tomatoes, use four-foot-wide rolls to make as big a basket as possible.
34. For fruit trees, you may want to buy translucent cloth bags to cover individual fruits for protection from some pests.
35. To get a harvest in two years or less, plant some very dwarf fruit trees.
36. Plant fruit trees that bear on new growth so you don't have to worry about pruning. Examples are: persimmons, walnuts, quince, figs, pineapple guavas, citrus, kiwis,

and Chilean guavas.

37. Flush your drip-irrigation system at least once in the spring before using it.
38. In areas where the winters freeze the ground, put your main drip-irrigation assembly and timer away in the garage or house. Open the ends of the tubing and drain all water out of it in the garden. Or, blow the water out with a small air-compressor.
39. Keep a supply of natural Bt (*Bacillus thuringiensis*) spray on hand to deal with caterpillars like tomato hornworms—it doesn't kill anything else. As mentioned, another useful natural insecticide is Neem, which can kill any insect it lands on, so you'll need to carefully target your spraying to avoid killing beneficial insects. (You should wear a mask.)

Bonus illustration:

Figure 35. An inner view of a landscape I designed for Lou and Susan Preston of the Preston Farm and Winery in Healdsburg, CA. When viewed from the lawn at right, the lavender hedge blocks much of the rigid geometry of the raised beds, and helps lure the eye away from "holes" in the raised-bed crops during the year. The boxes are in the first Zone of the landscape, with the kitchen just inside the railed entry. (The cat isn't dead, only napping.)

Chapter 3
Diversity, Stability, and Invasives

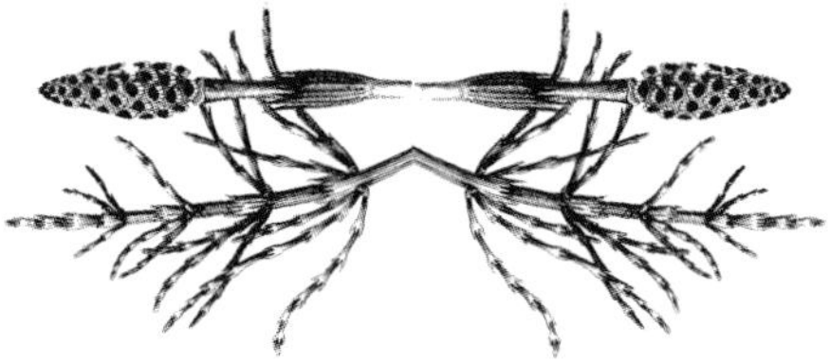

Horsetail (*Equisetum arvense*), is a *very* invasive plant. Do not grow it in the garden's soil. Grow it only in a container.

An oft-repeated theme in the sustainable horticulture world is the premise "Diversity equals Stability." Personally, I wish the continual references to this meme in the popular press, and in gardening books and websites would just disappear. While diversity within an ecosystem allows for a complex interaction of elements, plants, animals, and people it's not a panacea. Randomly chosen diversity or complexity doesn't necessarily provide any special benefits to a garden, and may even cause problems.

Stability...Sometimes Not

Sustainability and stability are often thought of as synonymous. For stability (the supposed outcome of diversity), forests are the commonly mentioned model. Forests don't really represent the stability of a "mature ecosystem." They're in constant flux, sometimes even invading adjacent grasslands, meadows, and gardens (so well described in pages 51 and 56 of Michael Pollan's book *Second Nature*). Trees, in the course of their natural lives, die or fall down, opening up "holes" for a new round of natural succession.

This also happens when tornadoes, fires, and storms knock down or incinerate forestland, replicating (on a larger scale and/or a slower basis, as in the chaparral of Mediterranean ecologies) what often happens to a single tree in its lifetime. Holes may be opened up in the canopy by exotic diseases (such as sudden oak death [S.O.D] syndrome in California and elsewhere), and humans create such openings whenever they cut down one or more trees (as seen to such a disturbing degree with clear-cutting). All of these events may also eliminate standing dead trees that alter the insect ecology.

As Dan Botkin put it in *Discordant Harmonies*: "Change now appears to be intrinsic and natural at many scales of time and space in the biosphere. Nature changes over essentially all time scales, and in at least some cases these changes are necessary for the persistence of life..."

Philip Van Soelen, former co-owner (for nearly 40 years) of California Flora Nursery, has good insights into the differences between stability and sustainability: "I have sometimes used the simplified animal example of an island with rare nesting seabirds. Add feral cats and rats to that mix and you will momentarily increase species diversity in simple terms of numbers of species, and a decrease in stability. But the cats and rats will drive the rare birds to extinction on the island. Stability will be re-established, but at the cost of real-world biodiversity and long-term sustainability."

Many perceived "mature" forests are actually part of a succession. For example, where oak woodlands meet Douglas fir and redwood forests near Occidental, California, the fir trees and redwoods will eventually kill off all the oaks with shade. The oak trees are not a "mature climax" forest—the end result of an

Figure 36. Two coast live oak trees (*Quercus agrifolia*) frame a view of the same Pacific-coast redwoods (*Sequoia sempervirens*) that will eventually (in many hundreds of years) crowd them out.

ecological progression—but are just part of the transition to redwood or Douglas-fir forests.

(Native Americans in California and elsewhere, unbeknownst to Europeans, managed the oak woodlands in order to harvest the acorns, using fire and other means to prevent other trees from shading out the oaks, thus holding up the natural progression to a mature forest.)

In nature, annual plants attempt to escape predators and pests by sprouting in different places every year. This is a strategy that maintains an overall stability, but not the stability of the individual plant. The meadow may be maintained. If vegetables in a small garden are planted in virtually the same place every year, the plants might as well saying to the pests: "Hey, here's breakfast, lunch, and dinner."

Trees, however, are stuck living in the same place year after year. Their protection, from pests and disease (one form of strategy for stability) is internal. They manufacture various alkaloids, tannins, glycosides, and other chemicals that are toxic to pests. (This happens except when the tree is under a lot of stress, such as in a severe multi-year drought). Fruit trees have been selected over centuries for fruit production, not for their alkaloids and tannins, so the modern fruit tree is more dependent upon our care to thwart pests and diseases. In a small orchard, fruit trees are not diversified enough to be classified as a stable population.

This not to say that our gardens can't be diverse. Most gardens have plenty of room for a more well chosen diversity that may enrich the biological atmosphere and enhance the environmental dynamics to create a more self-modulating garden. Such gardens are perhaps more natural, but are still a far cry from actual native ecosystems.

In his books, permaculture advocate Bill Mollison also often referred to: "...the conscious design and maintenance of agriculturally productive ecosystems which have diversity, stability and resilience of natural ecosystems...." Complexity, [another word for diversity] in the sense of some powerful interconnections between species, is what we [as permaculturalists] are really seeking in food systems."

Or, as Robert MacArthur Award-winning ecologist Simon Levin writes: "Environmental heterogeneity [diverse nature] provides a diversity of resources that can lead to coexistence among competitors that could not coexist in homogeneous environments." (http://www.ecologyandsociety.org/issues/view.php?sf=1)

...Sometimes Diversity *Does* Further Stability

Here are some proven ways to evolve toward some measure of stability:

- **Crop Rotation** – In gardens, if simple monoculture is to be changed to a higher level of diversity, you need to start with crop rotations that interrupt weed and pest cycles, while providing fertilization to subsequent crops.

- **Farmscaping** (using hedgerows or hedges, as outlined in Chapter 11) – Diversity can be further increased by providing habitat for natural enemies of pests (beneficials) with borders, windbreaks, and special plantings.

(See *Farmscaping to Enhance Biological Control* for information on special plantings for beneficial insects. This is available from Appropriate Technology Transfer for Rural Areas [ATTRA] P.O. Box 3657, Fayetteville, AR, 72702.)

- **Intercropping** – Intercropping is the practice of growing two or more crops in proximity to each other to promote their biological interaction. This is also known as "companion planting," and will be discussed in Chapter 10, which provides more information on intercropping of vegetables.

In Space and Time

As we read in ATTRA's *Intercropping Principles and Production Practices*:

"There are four basic spatial arrangements used in intercropping. Most practical systems are variations of these.

- **Row intercropping** – growing two or more crops at the same time with at least one crop planted in rows.
- **Strip intercropping** – growing two or more crops together at a wide enough distance to permit separate production but close enough for the crops to interact.
- **Mixed intercropping** – growing two or more crops together with no distinct rows. The closest idea to companion planting.
- **Relay intercropping** – planting a second crop into a standing vegetable at a time when the standing vegetable is at its reproductive stage, but before harvesting."

Moving and Sustainability

It's hard to make a true sustainable garden if you move a lot. The U.S. Census Bureau reported that the average American moves 12 times during his/her lifetime (every 6.6 years if you live to age 80). Tough economic times will probably slow down this activity, but many people will still move too often to cultivate a truly mature, sustainable garden.

This instability also greatly influences which edible tree crops the average homeowner will grow. Although the temperate-climate almond comes into fruition in about three to four years, trees such as chestnuts take five to seven years to even *begin* bearing. Commercial walnut production begins in 10 to 15 years, but doesn't reach peak levels until 25 to 30 years.

Standard-sized (non-dwarf) fruit trees, depending on the variety planted, take about two to seven years to come into bearing. Non-dwarf apple trees take two to five years to begin to fruit, and sweet cherries, for which there's no very effective dwarfing rootstock (except one genetic-dwarf cherry, a poor bearer), take four to seven years to fruit. Persimmon trees (with occasional climate-influenced exceptions) usually take five to seven years or more.

Figure 37. Persimmons, with their ornamental and tasty fruit and fantastic fall color, make attractive edible-landscape trees. They are nearly pest-and-disease-free. The variety shown here is the 'Fuyu', which is eaten when still firm.

While many dwarf fruit trees start bearing in their second year, some, such as apples grown on rootstocks called M27 and M9 (originally from East Malling Research Station, Malling, England) will fruit the first year. This fruit, however, is often stripped off to allow the tree to grow a year or two, as fruiting virtually stops this type of tree from growing new shoots. It would seem then, that for the average person, dwarf fruit trees are the ideal choice, especially because the care and harvesting of the shorter tree is so much easier than with a standard-sized tree. When you're picking fruit, shorter is better.

Perennial vegetables, however, take fewer years to come into harvest:

Globe Artichokes – 2nd year
Asparagus – two to three years, four for a *really* decent harvest
Horseradish – year number one or two
Rhubarb – 2nd year

Invasive Diversity

In 1980, in the course of his first lecture on the mainland of the USA, and before a crowd of 200-plus people at UC Berkeley, Bill Mollison proclaimed: "Everything is going to be everywhere, why not speed it up?" A bit vague, but I think he was referring to plants—perhaps weeds, or at least edible weeds. But, at the risk of repeating myself, blatant diversity does *not* guarantee stability. A sustainable and efficient garden diversity must be, to the best of our current knowledge, composed of the proper plants.

If you want to start a heated conversation with sustainable gardeners, just mention the word "invasives." There are many definitions of "invasive," the first being: a plant that's not native to the area in question. To that you can add non-native plants that spread and choke out native plants. Another definition includes plants that are especially successful at colonizing disturbed, but potentially productive, sites and at maintaining their abundance under conditions of repeated disturbances.

As gardeners push the boundaries of this term, the mavens have come up with some new pseudonyms—"plants that create functional diversity," "wildish" and "opportunistic plants" are three. These PC (Permaculturally Correct) terms actually seem to me to be a fruitless attempt to avoid the harsher—and I would say more realistic—term, "invasive."

Invasive plants can outcompete other plants. One of the great concerns of horticulturalists is that there are non-native plants that choke out native/indigenous plants. Invasive plants in the vegetable garden are usually called "weeds." (One amusing definition of a weed is: "A highly proliferating plant that hasn't figured out how to grow in rows.") Edible or otherwise, weeds spread easily in a garden and create extra work for the gardener, so PC or not, I'll stick with the more traditional "invasive."

Figure 38. This garden of squash, red orache (*Atriplex hortensis*), nasturiums, and many ornamental flowers makes a good example of a diverse ornamental vegetable garden. The flower border maintains a variety of blossom that captures the eye when the vegetables aren't looking good. Good intercropping.

In "The Population Biology of Invasive Species" (*Annual Review of Ecology and Systematics*, Vol. 32 (2001), pp. 305-332), we read: "...Invasive species have severe negative consequences for biodiversity."

Not all naturalized non-natives become invasive, however. For example, Naked Lady bulbs (*Amaryllis belladonna*) thrive when planted in my area but seldom "escape" from captivity. They just make more bulbs, not fertile seed. In fact, lines of Naked Ladies can be used to detect the location of outside walls of vanished former houses and barns.

(Fred Pearce in his book; *The New Wild* (Beacon Press; 2015), makes a distinction between "alien" and "invasive." He proposes that the label "alien" refers to the fact that a plant actually comes from or belongs elsewhere, and "invasive" describes its behavior. He prefers to use "alien" instead of "invasive" throughout his book.)

Pioneer Plants

Many plants we characterize as invasives tend to colonize bare or disturbed ground. Pampas grass, discussed in more detail later on, tends to show up first on earth disturbed by landslides, as can be occasionally witnessed on Highway 1 along the west coast of California. The term for this phenomenon is "pioneer plants." In nature, pioneer plants are found on disturbed soils such as those that remain after a forest or grassland fire. In my area, some native plants that colonize burnt soil actually *need* the fire's heat, moisture, and the chemical signature of smoke

Figure 39. Manzanita (blooming) (*Arctostaphylos* spp.) and California lilac ground cover (*Ceanothus* spp.). Two California pioneer plants.

to help germinate the seed—the native California Matilija Poppy (*Romneya coulteri*) is an example.

- Examples of pioneer plants in California include the manzanita (*Arctostaphylos* spp.), California lilac (*Ceanothus* spp.), coast redwood (*Sequoia sempervirens*), and scrub oak (*Quercus berberidifolia*).

- In New England, examples include redstem ceanothus (*Ceanothus sanguineus*); shagbark hickory (*Carya ovata*), and lodgepole pine (*Pinus contorta*).
- In Colorado, it's juniper (*Juniperus* spp.), ponderosa pine (*Pinus ponderosa*), spruce (*Picea* spp.), and lodgepole pine (*Pinus contorta*) trees.

Some pioneer trees, like redbuds (*Cercis canadensis*) are nitrogen-fixing trees that improve the soil until it's fertile enough for other plants to move into it (takes many decades). So as far as nature is concerned, pioneer plants are considered natural, rather than invasive.

Invasive Plants

Too much diversity, or the wrong kind, only promotes invasive plants. Here are many of the invasive plants, as listed in Mollison's *Permaculture Designers' Manual* (1988), that can choke out native plants:

Australian Pine (*Casuarina equisetifolia*)

Autumn Olive (*Elaeagnus umbellata*)

Blue gum (*Eucalyptus globulus*)—unbelievably invasive in Sonoma County, CA.

Cape Weed (*Arctotheca calendula*)—unbelievably invasive in Sonoma County. (See Figure 14.)

Cardoon, Wild Artichoke (*Cynara cardunculus*)

Chinaberry Tree (*Melia azedarach*)

Chinese Tallow (*Triadica sebifera*) has spread in the Gulf States by about 370 percent over a 16-year period.

Comfrey (*Symphytum officinale*), See page 40.

Gorse (*Ulex europaeus*)

Figure 40. A view of cape weed (*Arctotheca calendula*). Less than five years earlier, this was a grass meadow. Only the road stops the spread of the cape weed. Notice how intensively the grasses have been suppressed.

Honeysuckle (*Lonicera* spp.)

Kudzu (*Pueraria montana* var. *lobata*)

Multiflora Rose (*Rosa multiflora*)

Pampas Grass (*Cortaderia selloana*)—unbelievably invasive in Sonoma County.

Russian Olive (*Elaeagnus angustifolia*)

Saltcedar (*Tamarix* spp.)

Scotch Broom (*Cytisus scoparius*) *Very* invasive in Sonoma County.*

Tree-of-Heaven (*Ailanthus altissima*)

*On his first tour of the US, Masanobu Fukuoka (of no-till fame) saw Scotch broom, recognized it as a legume and suggested it be planted everywhere. Little did he know that it's an invasive plant that's as flammable in the summer as a gasoline-soaked torch.

Here, as listed in other permaculture and gardening books (including my 1986 edible-landscaping book!; indicated by an *) are some of the invasive plants that choke out native plants:

Acacia (*Acacia* spp.)

Amur Honeysuckle (*Lonicera maackii*)

Autumn Olive (*Elaeagnus umbellata*) *

Bamboo (*Phyllostachys* spp.)

Eucalyptus (*Eucalyptus* spp.)

Cardoon (Wild Artichoke) (*Cynara cardunculus*) *

Casuarina (*Casuarina* spp.)

Chicory (*Cichorium intybus*) *

Comfrey (if the roots are tilled) (*Symphytum officinale*) * (see page 74, Figure 54.)

Figure 41. Cardoon (*Cynara cardunculus*). This specimen is 10-12 feet high. Their seed is wind-blown.

Forget-Me-Not (*Myosotis sylvatica*) (See page 75.)

Foxglove (*Digitalis purpurea*)

Honeysuckle (*Lonicera* spp.)

Horsetail (*Equisetum* spp.) *

Kudzu (*Pueraria montana* var. *lobata*) (was NOT in my book!)

Lemon Balm (*Melissa officinalis*)

Nasturtium (*Tropaeolum majus*) * (See page 41.)s

Red Clover (*Trifolium pratense*) *

Red Valerian (*Centranthus ruber*)

Russian Olive (*Elaeagnus angustifolia*) *

Scotch Broom (*Cytisus scoparius*) (NOT in my book! See page 61.)

Spanish Broom (*Spartium junceum*)

Sudan Grass (*Sorghum bicolor*) *
Tansy (*Tanacetum vulgare*) *

Tree-of-Heaven (*Ailanthus altissima*)

Vetches (*Vicia* spp.) *

Green Ash (*Fraxinus pennsylvanica*) is sometimes used as a windbreak for homes and gardens, but ash trees can represent a problem in much of New England and in adjacent Middle-Atlantic states. The Emerald ash borer (EAB; *Agrilus planipennis*), is an exotic beetle first discovered in southeastern Michigan near Detroit in the summer of 2002. It is also established in Windsor, Ontario; was found in Ohio in 2003, northern Indiana in 2004; northern Illinois and Maryland in 2006, western Pennsylvania and West Virginia in 2007; Wisconsin, Missouri and Virginia in summer 2008; Minnesota, New York, Kentucky in the spring of 2009; Iowa in spring of 2010; and Tennessee in the summer of 2010. Cutting down Agrilus-infested trees for firewood can spread the borers if the firewood is moved off the property. (It can be burned on the property without spreading the bugs, but this isn't always possible in suburbia.)

Costly Invasives

The United States is thought to suffer $1.1-120 billion in economic losses *per year* due to exotic invasive species.

California alone is home to 4,200 native plant species, and is recognized internationally as a "biodiversity hotspot." Approximately 1,800 species of non-native plants also grow in California.

Here's a list of some invasive plants found in California. (Data is from the USDA NAL National Invasive Species Information Center.)

Air Potato (*Dioscorea bulbifera*)

Autumn Olive (*Elaeagnus umbellata*)

Beach Vitex (*Vitex rotundifolia*)

Brazilian Peppertree (*Schinus terebinthifolius*)

Canada Thistle (*Cirsium arvense*) (See page 40.)

Chinese Tallow (*Triadica sebifera*)

Cogongrass (*Imperata cylindrica*)

Common Teasel (*Dipsacus fullonum*)

Dalmatian Toadflax (*Linaria dalmatica*)

Diffuse Knapweed (*Centaurea diffusa*)

Downy Brome (*Bromus tectorum*)

Fig Buttercup (*Ranunculus ficaria*) Also known as (*Ficaria verna)*

Garlic Mustard (*Alliaria petiolata*)

Giant Hogweed (*Heracleum mantegazzianum*)

Hairy Whitetop (*Lepidium appelianum*)

Houndstongue (*Cynoglossum officinale*)

Japanese Stilt Grass (*Microstegium vimineum*)

Japanese World Climbing Fern (*Lygodium japonicum*)

Japanese Honeysuckle (*Lonicera japonica*)

Japanese Knotweed (*Fallopia japonica*)

Japanese Spiraea (*Spiraea japonica*)

Johnsongrass (*Sorghum halepense*)

Kudzu (*Pueraria montana* var. *lobata*)

Leafy Spurge (*Euphorbia esula*)

Medusahead (*Taeniatherum caput-medusae*)

Mile-A-Minute Weed (*Persicaria perfoliata*)

Multiflora Rose (*Rosa multiflora*)

Musk Thistle (*Carduus nutans*)

Old World Climbing Fern (*Lygodium microphyllum*)

Oriental Bittersweet (*Celastrus orbiculatus*)

Princess Tree (*Paulownia tomentosa*)

Purple Star Thistle (*Centaurea calcitrapa*)

Quackgrass (*Elymus repens*)

Russian Knapweed (*Rhaponticum repens*)

Russian Olive (*Elaeagnus angustifolia*)

Saltcedar (*Tamarix* spp.)

St. John's Wort (*Hypericum perforatum*)

Scotch Broom (*Cytisus scoparius*) (See page 61.)

Scotch Thistle (*Onopordum acanthium*)

Figure 42. Yellow star thistle, used by some people to make honey. (USDA)

Spotted Knapweed (*Centaurea stoebe*)

Tree-of-Heaven (*Ailanthus altissima*)

Tropical Soda Apple (*Solanum viarum*)

Whitetop (*Lepidium draba*)

Witchweed (*Striga asiatica*)

Yellow Star Thistle (*Centaurea solstitialis*), to the left.

Yellow Toadflax (*Linaria vulgaris*)

Grasses, Flowers, Vines, & Groundcovers

Bittersweet (*Celastrus orbiculatus*)

Chinese Silver Grass (*Miscanthus sinensis*)

Common Daylily (*Hemerocallis fulva*)

English Ivy (*Hedera helix*)

Five-leaf Akebia (*Akebia quinata*)

Fountain Grass (*Pennisetum alopecuroides*)

Japanese Honeysuckle (*Lonicera japonica*)

Periwinkle (*Vinca* spp.)

Purple Loosestrife (*Lythrum salicaria*)

Yellow Flag Iris (*Iris pseudacorus*) (below)

Figure 43. A double-whammy of invasive plants. The yellow flag iris (Iris pseudacorus) in the foreground and the Scotch broom (*Cytisus scoparius*) across the water are both very invasive. The iris will spread to fill the entire bank of the pond, and the bases of the plants will silt up and begin to fill in toward its center. This iris can even choke out cattails.

It's strange that this list from the National Invasive Species Information Center does not include two major invasive plants in my Northern California area: **iceplant** (pictured here, *Mesembryanthemum* spp.) and **pampas grass**, (*Cortaderia selloana*). Iceplant (the bright red plant in the photograph below) was introduced to stabilize sand dunes. It has escaped from captivity and has spread with a vengeance.

Figure 44. Here, iceplant (*Mesembryanthemum* spp.) has choked out all native plants. While it's very attractive to the eye, many gardeners spend hours trying to get rid of this "alien," which was introduced by the government to stabilize sand dunes.

NOTE: The 3:1 rule: For every year you delay controlling an invasive plant infestation, it will take three years for you to regain control.

The kudzu vine of the southern US is theoretically a multipurpose plant, a nitrogen fixer with many edible parts and useful fibers. It was introduced from Japan at the 1876 Philadelphia Centennial Exposition as an ornamental and forage crop, and has been around so long that some people think it's native to the southeast.

Adding kudzu would certainly supply another unit of diversity to your garden, but you better leave town before your neighbors come after you, as kudzu vines grow rapidly, may extend to 100' in length at a rate of about one foot per day, and entangle everything that's moving slower than 25 MPH! Kudzu is an extreme example of an exotic plant gone wild. There is even a permaculturist who started a group to promote the use of kudzu and research ways to use it (including recipes). I just hope he doesn't want us all to plant it.

Hybrid Veggies and Disease Resistance

Hybrid vegetable seed is often selected for providing resistance to specific diseases and pests. When buying tomato plants, for instance, the letters V, F and N tell the buyer that the tomato is resistant to, respectively, *verticillium* wilt, *fusarium* wilt and nematodes. (Heirlooms are technically, in the horticultural sense, hybrids, in that they're the result of crosses of two varieties at some time in the past.) Some heirloom tomatoes have resistance to some of these same factors, but seldom to all three in one plant.

The difficulty with conventional, "modern" hybrids is their genetic similarity. If your plants are attacked by a disease that the hybrid isn't "programmed" or selected for, then the plants may be sitting ducks, and the whole crop, or a great portion of it, may succumb to the disease. The dogmatic opponents of hybrids often mention this lack of genetic diversity as the Achilles' heel of hybrid seed. They can certainly point to a few notable examples of the vulnerability of a form of agriculture dependent upon hybrid seed. As a notable example, 15 percent or more of the US's 1970 feed- and shell-corn crop was wiped out by a strain of Southern corn-leaf blight that attacked hybrid plants that were not bred to resist this disease. It's thought that the farmers lost a billion dollars of revenue from diseased corn.

In all their testaments to the evils of hybrid seed, however, heirloom proponents can only dredge up a handful of scary agricultural examples. More notable for the home gardener, the Achilles'-heel argument only works well on an agricultural scale. Five hundred acres of one type of wheat is a far cry from the dozen or two types/varieties of vegetables found in the average home garden. If all your tomato plants die, you'll be depressed—but you probably won't starve.

Heirloom Diversity

OK, the setting is 1982, at a nonprofit organization (we'll call it "The Institute") in Northern California.

In the early 1980s, its young gardening interns got swept away by the growing tide of interest in preserving genetic vegetable diversity, and started growing rare and unusual heirloom varieties in their well-tended organic gardens. To sample the various flavors and save seed, they grew around 25 varieties of heirloom tomatoes per season. One year, they decided to hold a contest between hybrid tomatoes and open-pollinated heirloom varieties, monitoring both to see which manifested the highest incidence of disease-hardiness. They may have been shamed into it by the pestering of older pro-hybrid (or at least not anti-hybrid) gardeners. Or, the Church of Heirlooms may have decided to prove its superiority to the rest of the heretics. Whatever.

Summer passed. Evening conversation turned to other events or topics, and the first frosts blackened what remained of the tomato plants. Finally, I pulled aside one of the Heirloom advocates. "Where's the report?" I asked. "What were the results?" In a hushed tone, he said "the VFN-resistant hybrids were more disease-resistant than the heirlooms, so we didn't do a report." So much for the integrity of this particular flock of the Church of Heirlooms. The results failed to confirm their expectations, so they suppressed them.

It should be noted, however, that the world's food security could actually be threatened by a failure to conserve wild plant species that are genetically related to the crops currently grown for food. Genetic information held in certain crop varieties is crucial to the development of the heat-, drought-, salinity-, pest- and disease-resistant, fast-growing, high-yielding new varieties, that will most likely be necessary to reduce food insecurity in the face of climate change.

According to Jacques Diouf, the FAO (The Food and Agriculture Organization of the United Nations) Director-General, "There are thousands of wild relatives of current domestic crops that still need to be collected, studied, and documented because they hold genetic secrets that enable them to resist heat, droughts, salinity, floods, and pests." And the reverse can happen. Tomato plants are also a common host plant for nasty nematodes. (There are "good" nematodes.) You *must* rotate nematode-infested tomatoes with another crop to prevent infecting more tomato plants and other susceptible plants. Try doing it with plantings of Brassicas (the cabbage family can handle nematodes) and onions, which are also resistant to the pest.

As mentioned earlier, most of our common agricultural crops are composed of alien species. Luther Burbank's introduced blackberry (*Rubus armeniacus**, the European blackberry) is now growing all up and down the west coast, and is one of the best wild-forage food plants that anyone has developed (the birds agree). So the European blackberry is loose on the coast in the Pacific Northwest, an outlaw plant that feeds people (even as they spray it, curse at it, and dig it up). It's mainly distributed by birds that keep eating the berries and pooping the seeds everywhere. Resourceful people just make the best of it—great blackberry jam and pies.

* (**Himalaya Blackberry**; *Rubus armeniacus* Focke = *Rubus discolor* Weihe & Nees = *Rubus procerus* P.J. Muell. In many places the scientific name is still given as *Rubus procerus,* or even *Rubus discolor.* But the correct name is now *Rubus armeniacus,* or more particularly, *Rubus armeniacus* Focke. *Rubus Armeniacus*.)

In the environmentally-sound development Village Homes in Davis, CA (discussed in Chapter 6), the exotic plant Dutch white clover was used as a ground cover under fruit trees, and was watered with overhead oscillating sprinklers. With the extensive use of this clover (a leguminous plant) as a ground cover, it was supposed that the fruit trees were receiving free nitrogen and phosphorous fertilizer in mild amounts at various times during the season. (After each mowing, the "shock" of the mowing causes some of the nitrogen nodules to be "shed" from the root, and to be available to adjacent plants' roots as a nitrogen source.)

This seemed like a good idea until the apple trees came down with mildew, in a climate that's normally so hot and dry that this was seldom a problem. The theory that the gardeners eventually developed was that the clover added more nitrogen that the trees actually needed, forcing unneeded growth, and the addition of the humid air from the sprinklers to the nitrogen surplus generated mildew.

Figure 45. A clover groundcover grows beneath an almond tree in front of a Village Homes (Davis, CA) solar house. The clover provides more nitrogen than the tree requires, and this may result in mildew and/or aphids.

Invasive "Weeds"

In the disturbed-by-cultivation soil of a garden, it's another story. Many plants that we call weeds invade and colonize disturbed soil. Tilled soil is a bit like a landslide in that it's exposed and available to seed, and too often to the seed of invasive plants. Some permaculture advocates find themselves in a dilemma on the subject of invasives, grounded in the fact that some of these colonizers/pioneer species are both edible and easy to "cultivate" with no attention. While worthwhile as food, certain plants can spread enough to be a nuisance and, as mentioned above, some like pigweed (*Chenopodium* spp.) and mallow (*Malva* spp.), can be carriers of fusarium wilt that can infect tomatoes, eggplants, and peppers.

According to Linda Chalker-Scott of the Washington State University Extension, writing about a certain best-selling permaculture book: "The author seems not to understand that there may be unfilled niches in certain ecosystems that can be exploited by invasives, endangering native species whose niches may overlap.... The author's naive tolerance of invasive species is a poor example to follow and certainly not based on current mainstream science."

French Intensive/Biodynamic Invaders

According to Alan York of UC Santa Cruz: "... [Alan Chadwick] taught organized methods in his impassioned lectures, and they worked—not just now and then but every time. Chadwick's methodology involved a marriage of the universal principles of classical French-intensive horticulture with organic gardening. What we all learned, working for him, was that when the discipline of order was followed, the results were spectacular."

Chadwick's gardens, like his personal character, were constantly on the move, and well-tended in every detail. He subsequently transferred his work from Santa Cruz to a new garden in Covelo, CA. I visited the garden Chadwick had abandoned in the Round Valley (Covelo), California several years after he left; it had become an *enormous* weed patch consisting mostly of lemon balm (recommended by Chadwick as a perennial for its herbal medicinal value) and comfrey that had grown from bits of roots left in repeatedly dug soil and partially decomposed compost. (Grown for enhanced composting fertility, Comfrey is discussed in more detail in Appendix 1.)

Chadwick' s formerly well-defined paths and raised beds had been overwhelmed with beneficial but invasive plants. With no gardeners, chaos had overtaken the garden within a year or two.

Here are some more useful but invasive plants that might show up in your garden:

Figure 46. Each green lump around this budding persimmon tree is a mound of lemon balm, spreading widely in less than three years after planting this forest garden. The permaculturists who planted this say they will harvest all the lemon balm, or cut it for mulch. They should of not *ever* let it go to seed.

Tansy

Tansy (*Tanacetum vulgare*) has been recommended as a beneficial insect attractant. In 1986, I recommended tansy in my edible-landscaping book as an accumulator of potassium, but I sure wouldn't now. I don't know what I was thinking when I planted it beneath an apple tree to "keep ants away," after one of the books I'd read at the time had recommended the herb as a way to repel ants, flies and mosquitoes; I believed this tip until I bought a potted plant that turned out to be harboring an ant colony in its soil!

By the time I noticed that the tansy had spread through my orchard, it was too late. (See the discussion on companion planting for a lot more about the limits of influence of diversity of so-called companion plants.) Tansy continues to show up in the literature and on the Internet as a companion plant. Luckily, more and more references also mention its invasive nature—especially in areas with summer rains.

Pampas Grass

This exotic ornamental grass (*Cortaderia selloana*), originally from Argentina, Brazil, and Chile, is found up and down California Highway 1 and throughout the west coast. It tends to heavily colonize disturbed soils, sprouts long before many native plants, and can quickly shade an entire landslide, so that few natives can re-grow in its shadow. It may "escape" from the landslide area but, fortunately, not with too much vigor. Although it can be used for floral displays, thatching, and papermaking, this plant is so aggressive that it's banned from sales at many plant nurseries.

Lambsquarters

Lambsquarters (*Chenopodium album*) appeared in Great Britain in the late-glacial and post-glacial periods. The Romans and Europeans used it extensively, and lambsquarters was once regarded as one of the most delicious of all wild vegetables. It's also very nutritious—more protein, vitamin C, vitamin A, and calcium than spinach.

Lambsquarters loves disturbed, fertile soil and, as an invader of open habitat, it colonizes many gardens. It's so prolific that most of us couldn't possibly eat all of it that might invade our growing space.

You may think twice about nurturing lambsquarters if you're growing lots of heirloom tomato plants, because this "wild edible" also harbors *verticillium* wilt—the Achilles' heel of many non-hybrid tomato plants.

The Spread of Pest and Disease

Verticillium wilt is caused by the soil-borne fungi *Verticillium dahliae* and *V. albo-atrum*. Both infect a very wide range of garden plants, entering through the roots and growing up through the water-conducting tissues, causing wilting of the upper leaves (due to stress from lack of water) at any time from spring until fall. Edible plants most often affected include zucchini, potato, tomato, cucumber, and strawberries.

Tomatoes without the hybrid label FVN label (meaning, as mentioned above, *fusarium-*, *verticillium-*, and nematode-resistant) may be infected by invasive plants if they're carriers of the diseases, and many heirloom tomatoes aren't resistant to them. Careful observation is critical to detect any sign of *verticillium*. If you do spot it, *don't* compost the infected plant. Send it to the dump, not to the community compost (green waste) program. Rotate the tomatoes with broccoli, as there's increasing data that broccoli helps to reduce the quantity of *verticillium* in garden soil.

Weeds that Spread Pests and Diseases
Redroot pigweed (*Amaranthus retroflexus*)
Wild carrot (*Daucus carota*)
Chicory (coffee-weed) (*Cichorium intybus*)
Dandelion (*Taraxacum officinale*)
Western salsify (*Tragopogon dubius*)
Wild mustard (*Brassica kaber, B. arvensis, Sinapis arvensis*)
Common lambsquarters (*Chenopodium album*)
Yellow woodsorrel (*Oxalis stricta*)
Broadleaf plantain (*Plantago major*)
Stinging nettle (*Urtica urens*)

Figure 47. Stinging nettle (*Urtica urens*) is an edible "weed," but can transmit pests and diseases.

These are all edible weeds that can transmit a wide variety of viruses to vegetables.

Other examples (mostly from: "Interactions Between Weeds, Arthropod Pests, and Their Natural Enemies in Managed Ecosystems," Jan. Feb 2000. Robert F. Norris and Marcos Kogan, *Weed Science*; Vol. 48, No. 1, p 94-158.):

- **Crimson clover** attracts *Lygus* bugs that attack strawberries as an example.
- **Shepherd's purse, wild mustard**, and **winter cress** (yellow rocket): harbor black rot.
- Cucumber Mosaic Virus can come from **chickweed** and **groundsel**.
- All members of the *Chenopodium* family (**amaranth, fat hen, pigweed, and saltbush, etc.**) carry the potential of spreading pests to Swiss chard, and attract armyworms, flea beetles, leafhoppers, Japanese beetle (can be 10-fold increase in the grubs), and thrips.
- **Groundsel** is a carrier of rust and attracts greenhouse whitefly and aphids.
- **Common lambsquarters** attract cabbage aphid, Mexican bean beetle, *Lygus* bugs, corn borer, potato leafhoppers, northern and southern corn rootworms, and cabbage whitefly.
- **Wild carrot** can, of course, bring in carrot fly.
- Green-peach aphids and spider mites are attracted to **mallow**.
- **Amaranthus** also can bring in flea beetles.
- **Dock** harbors black cutworm.
- **Stinging nettle** is a very good pot-herb if steamed, but will attract carrot flies.
- African and fall armyworms and the Hessian fly are attracted to **Bermuda grass**.
- The pesty weed, **California burclover** harbors leafminers.

Living with Invasives

One Saturday years ago, I joined a group that had assembled to pull up invasive Scotch broom (*Cytisus scoparius*) plants that had infested parts of the Golden Gate National Recreation Area north of the Golden Gate Bridge. After an entire day of work, the seven of us had pulled perhaps several hundred plants, but several thousand acres in the 80,000-acre preserve remained infested with Scotch broom.

Native to northern Africa and parts of Europe, this plant was first introduced to North America on the East Coast and was later brought to California as an ornamental. From the 1850s through the early 1900s, it was frequently planted in gardens, and later it was used for erosion control along highway cuts and fills. So many of these invasive plants have taken over the native California landscape that it's hard to imagine controlling, much less eradicating, them.

Sometimes it's the introduced ornamentals that really impact native plants. One example is the Sudden Oak Death (SOD) problem, especially in California, caused by the import of rhododendrons from Europe. SOD is slowly killing massive numbers of the native tan-oak trees (Not a true *Quercus*, but *Notholithocarpus densiflorus*.) and some native California Coast live oaks (*Quercus agrifolia*). Sometimes the situation is flipped. Native California oak trees can harbor an equally native fungus (*Armillaria* spp.), that when artificially irrigated can become aggressive and kill wine grapevines. Warm-weather irrigation is what causes *Armillaria* to become out of-balance. (Incidentally, some people really hate vineyards and think of them as invasives or aliens.)

Restoration Ecology

In *Foundations of Restoration Ecology* (edited by Donald A. Falk, Margaret, A. Palmer, and Joy B.

Figure 48. Today, the hills in much of California are golden brown each summer (upper photo); this is in part due to invasive Eurasian grasses that have taken over from the native species, especially in the case of overgrazing by large herds of other Eurasian imports—sheep, cows, goats, and horses.

Early Spanish explorers in the late eighteenth century found springtime California covered from San Francisco to San Diego in spectacular carpets of wildflowers growing among widely spaced native bunchgrasses. Today, invading Eurasian plant species have pretty much devastated this nearly forgotten botanical heritage. *Stipa pulchra*, once one of the most common native bunchgrasses, if left uninvaded, would be deep green in spring and a more subdued green in the summer. On semi-shaded hillsides, *Festuca californica* would show up as a smokier blue-green. Sadly, this spectacular mix of evergreen grasses and perennial and annual wildflowers is long gone, except for a few rare places like Figueroa Mountain. The lower photo (© Michael Marzolla) shows, on a spring day, the remarkable palette that much of the whole state used to reveal.

Zedler; with foreword by Richard J. Hobbs, Island Press, 2006) we read: "Ecological restoration is an attempt to return a system to some historical state, although the difficulty or impossibility of achieving this aim is widely recognized. A more realistic goal may be to move a damaged system to an ecological state that is within some acceptable limits relative to a less disturbed system. In this sense of the term, ecological restoration can be viewed as an attempt to recover a natural range of ecosystem composition, structure, and dynamics. Ecological restoration has been practiced in some form for centuries. For instance, many indigenous peoples tended lands to sustain natural ecosystem services, such as production of basket-weaving materials, food crops (acorns), or forage for game animals, and they continue to do so."

Professor Arthur Shapiro, a Distinguished Professor of Evolution and Ecology at UC Davis, takes a more radical view toward the phrase "restoration ecology." "Restoration ecology" [that which seeks a balance between the native and "constructed" environments] is a euphemism for a kind of gardening informed by an almost cultish veneration of the "native" and abhorrence of the naturalized [non-native plants that have become well-established over time, another euphemism for "invasive"], which is commonly characterized as "invasive." Let me make this clear: neither "restoration" nor conservation can be mandated by science—only informed by it. The decision of what actions to take may be motivated by many things, including politics, esthetics, economics and even religion, but it cannot be science-driven. In the case of "restoration ecology," the goal is the creation of a simulacrum [**similarity**] of what is

•BIOLOGICAL ZONES•	AGRICULTURE	NATURAL ECOSYSTEM (REMNANTS)	WEEDY GRASSLAND (Abandoned Croplands)	CONSTRUCTION ZONE (WEEDY)	NEW SUBURB (0-15 YRS.)	LANDSCAPED PARKLAND	OLDER SUBURB (15-50 YRS.)	OLDEST SUBURB (>50 YRS.)	CITY CORE
•NET PRODUCTIVITY•	HIGH P>R**	STABLE P=R*	LOW P>R	MINIMAL	LOW P>R	VARIABLE	HIGH P>R	LOW P=R	MINIMAL
•BIOMASS•	LOW	HIGH	LOW	VERY LOW	LOW	LOW-HIGH	MEDIUM	MEDIUM-HIGH	LOW
•PLANT DIVERSITY•	LOW	HIGH	HIGH	LOW	LOW	LOW-MEDIUM	HIGH	MEDIUM	LOW
•TYPES OF PLANTS•	CULTIVARS	NATIVE	NATIVE/EXOTIC	WEEDS	EXOTICS	NATIVE/EXOTIC	NATIVE/EXOTIC	NATIVE/EXOTIC	WEEDS/EXOTIC
•No. OF BIRD SPECIES•	N.A. ☆☆	25-31	5	N.A.	5	24	19	13	13
•No. OF BIRDS per 247 ACRES•	N.A.	381-438	247	N.A.	158	626	742	162	796

REDRAWN FROM: "Ecology and Mgmt. of Disturbed Urban Land," R.S. DORNEY, LANDSCAPE ARCHITECTURE, MAY 1979.
** P>R: BIOMASS accumulates/growth > decay. * P=R: decay = growth. ☆☆ NOT AVAILABLE.

Figure 49. This is a chart resulting from a study of the diversity and productivity of urban, suburban, and rural areas respectively. When photosynthesis (P) is greater than respiration (R), more plant tissue is accumulating than decaying.

believed to have been present at some (essentially arbitrary) point in the past.

" I say a simulacrum, because almost always there are no studies of what was actually there from a functional standpoint; usually there are no studies at all beyond the merely (and superficially) descriptive. Whatever the reason for desiring to create such a simulacrum, it must be recognized that it is just as much a garden as any home rock-garden and will almost never be capable of being self-sustaining without constant maintenance; it is not going to be a "natural," self-regulating ecosystem. 'Restoration ecology' is like plowing the sea."

For Example, Bird Diversity

In this graphic, the study summarized found that there were more birds (pigeons?!) in the urban center than any other region. However, the number of *species* of birds in the urban core was about 50 percent that of the remnants of the natural ecosystem and about even with that of suburban areas that were 0-15 years old. (See Figure 49.)

Mark Davis (the former Dewitt Wallace professor and chair of biology at Macalester College in Saint Paul, MN) writes: "People, including conservationists and scientists, apply these terms [invasive, exotic] indiscriminately," he said. "It is best to call them 'recent arrivals' or even 'non-native and native.' But words like 'invasive,' 'alien', and 'exotic' suggest these are bad things and shouldn't be there." Here's the way Davis looks at it: "You should worry about introduced pathogens, viruses, bacteria and fungi. Besides harming people, they can decimate food production crops, livestock and wild populations. Then next on the list are the introduced insects that can devastate and threaten society. There is not enough emphasis on that and way too much preoccupation worrying about non-native plants." Statements such as these are not well tolerated by native-plant enthusiasts who point to examples such as iceplant (*Carpobrotus edulis*), which has thoroughly smothered many California coastal hillsides in a sort of weedy, invasive monoculture.

Toby Query, who has worked as a natural-resources ecologist for the city of Portland Oregon, makes the point that some non-native plants can serve a useful function. "During the process of removing the plants (reed canary grass and introduced blackberries) that they no longer desired, they realized that a rare bird, the willow flycatcher, liked to nest in the blackberry bushes. Before realizing what was occurring, they cut out the blackberry bushes in the middle of the birds' nesting season, not realizing they were habitat for a species of concern."

Flying in the Face of Invasives

In California, 32% of native butterfly species are now feeding or breeding on non-native plants. According to UC Davis's Professor Shapiro: "Over 40% of the butterflies have no known native hosts in the urban–suburban environment. Twenty-nine of the 32 butterflies known to breed in Davis, CA do so at least in part on alien plants; 13 have no known native host plants in Davis. Were certain alien weeds to be eradicated or their abundance greatly reduced, the urban-suburban butterfly fauna would disappear. This might be regarded as an unfortunate, and perhaps intolerable, side-effect of such programs."

As an example, now that native Monterey cypress trees have been decimated along the Monarch-butterfly migration routes, eucalyptus (a tree imported to America from Australia in the 1800s (for lumber—it didn't work out) is a major roosting tree for Monarchs in coastal California.

According to Toby Hemenway (author of *Gaia's Garden*): "Eucalyptus (*Eucalyptus globulus*) may have prevented great reduction of Monarch populations." I've seen groves of eucalyptus supporting many thousands of Monarchs at the Esalen Institute in Big Sur California, while, just .25 miles away, there were just dozens of Monarch on the native Monterey cypress. The question for me is, why did the Monarchs switch to the invasive, fire-prone eucalyptus? And, does it change their body chemistry for good or bad? (Believe it or not, I, in the interest of science, tasted Monarch honeydew, and it tasted of eucalyptus. I didn't die!)

Now we're entangled in the paradox of people who call themselves environmentalists fighting to save blue gum eucalyptus from removal because of, in part, the Monarchs. In *Losing Paradise: The "New*

Figure 50. Monarch butterflies in Big Sur, California are shown resting on an invasive import, eucalyptus, instead of the native Monterey cypress trees.

Normal" of California Wildfires, by Glen Martin, we read: "UC Berkeley has proposed removing more eucalyptus, but the effort has met considerable resistance from local 'tall trees' advocates who, so far, have been successful in halting further eucalyptus removal through legal action."

Eucalyptus trees, however, are a supreme fire risk each summer and fall. The "fuel load" (in tons per acre) of eucalyptus woodland is three times that of a native California oak woodland. This invasive import has created a fire hazard in many areas of the state. Others defend the blue gum because it's "naturalized"—another word used to mask the concept of invasives. This is the double-edged sword of dealing with an exotic plant.

When asked about the invasives debate, Toby Hemenway said: "We [permaculturalists] restrict our activities to planting valuable wildlife species [that can include invasives, ed.] in yards, farms, and other highly developed areas. I think we can safely conclude that native-plant enthusiasts have done far more damage to native insect populations than any permaculturists—real damage at scale, versus hypothetical damage." Not a safe conclusion as far as I'm concerned.

Nature's Salvation?

Fred Pearce, mentioned earlier living with invasives to the extreme. The subtitle of his book, *The New Wild*, is *Why Invasive Species Will Be Nature's Salvation* (Beacon Press, 2015). First, here's a bit about his assumptions. (Remember that he prefers to use "alien" instead of "invasive" throughout his book.) His view is that alien plants inherently produce greater diversity, the ultimate goal, according to him, of an ecosystem. He supports the view that the goal of a stable ecosystem is increased diversity, and goes so far as to say that a greater number of aliens means more natives, not fewer. Such a system means there is "biotic resistance"—augmented ecosystems where greater diversity protects against "threats from future incomers."

Also, like the author Daniel Botkin (author of *Discordant Harmonies*), he supports the concept that nature is always changing, and "constantly out of balance." These two concepts seem, in my reading, unreconciled in his book. Another remarkable assumption is that he believes aliens are not at all "bad," and are indistinguishable from natives—as both groups can do harm or good. Both, says Botkin, need to be treated only on their merits. In summary, he states "Alien species, the vagabonds, are the pioneers and colonists in constant renewal. Their invasions will not always be convenient for us, but nature will re-wild in its own way. That is the new wild."

The Rambunctious Garden

Living with invasives is also the thrust of a book titled *The Rambunctious Garden, Saving Nature in a Post-Wild World* by Emma Marris (Bloomsbury Publishers, 2011). Marris argues that trying to eradicate all non-native plants in the larger landscape is probably both a lost cause *and* unnecessary. In a garden setting, it makes sense to control those that thwart the gardener's goals, but not to remove naturalized plants that aren't causing any harm. Some can even be welcomed as food for wildlife—or humans. This is a radical approach that is gaining *some* limited traction in the native plant world—but *not* among the purists.

Marris believes that: "restoring the complex ecosystems we have destroyed may be, at the moment, just too hard." She goes on to say "I hope the guild of restoration ecologists grows and flourishes. But not everywhere.... We [should] keep parks around for those who just love wildness in the raw.." She concludes with: "The rambunctious garden is writ across the whole Earth. In different places, in different chunks...we can manage nature for self-willed wildness, for ecosystem services, for food and fiber and fish, (exotic) trees and frogs. We've forever altered Earth, and so now we cannot abandon it to random fate. It is our duty to manage it." She's basically saying we live in a hybrid of wild nature and human management.

This offers, in my opinion, the most realistic view of living with non-native plants.

Biodiversity, Natives, and Exotics

In 2015, Linda Chalker-Scott wrote a very important paper reviewing the literature on whether non-native and noninvasive woody shrubs and trees were a problem for, or a benefit to, biodiversity. Her conclusion is summed up in the title of the paper: "Nonnative, Noninvasive Woody Species Can Enhance Urban Landscape Biodiversity." (*Arboriculture & Urban Forestry*, 41(4), July 2015. P 173-186.)

She recommends:

- Reducing lawns and replacing with vertical vegetation.
- Increasing shrubs and trees where houses are distant from natural forests.
- Using both native and noninvasive exotic species for increased biodiversity, both local and regional.
- Planting fruit-bearing plants for humans and animals.
- Wherever possible, adding a permanent water source.

Her summary is thought-provoking, "'Natives-only'...reduces the planting palette and ultimately the community biodiversity [for humans, mammals, insects, birds, lizards, pollinators, and more] of urban areas."

Beneficial Exotics?

Native California Plants Are *Not* the Most Drought-Resistant Plants

(With great pride I published this article in one of my *Bob's Honest-to-Goodness* newsletters in 1996. Little did I know I would find out something I hadn't thought of. Read on.)

One of the myths to crumble as I landscaped through the California drought of 1977 was the idea that California native plants are superior in their drought-resistance. (And while certain California desert plants are among the world's most drought-hardy, many won't survive the 58-inch winter rains at my house.)

There are a handful of Mediterranean climates throughout the world besides California—the Mediterranean basin, of course, the middle elevations of the Chilean mountains, portions of South Africa and parts of Australia. Most of these climates are cousins to our Californian-Mediterranean ecosystem, and droughts are more severe and the rainfall considerably more sparse in some of these places. Plants from these areas will adapt to many other aspects of the climate and soil in parts of California and still retain their superior drought-hardiness. A virtually unrecognized article entitled, "A Trial of Drought Tolerance," in *Pacific Horticulture* magazine, discussed Barrie Coate's (nurseryman and staff of Saratoga Horticultural Research Foundation, Green Valley, CA) observations of various plants installed during the drought of 1975 on a south-facing site in Santa Cruz County. The plants were irrigated during the first summer only. The study evaluated both the size and the condition of the plant relative to how it would *look* as a landscape plant after a summer without supplemental irrigation—in other words, sheer survival was not the objective. The unirrigated plant's ornamental appearance was an important criterion. According to Coate, "The California native plants were not generally the best performers in this hot, dry site. Some exotics from climates similar to that of the test site performed better than many of the natives. Of nine plants that performed best, six are native to the Mediterranean region and two are from Australia."

The "winners" of this small trial were, in the order of best ornamental value:

Number one? - **Pride-of-Madeira** (*Echium candicans*)
2 - **Spanish lavender** (*Lavandula stoechas*)
3 - **Jerusalem sage** (*Phlomis fruticosa*)
4 - **Rockrose** (*Cistus parviflorus*)
5 - **California fuchsia** (*Zauschneria californica latifolia* now *Epilobium* spp.)
6 - **Peppermint tree** (*Agonis flexuosa*)
7 - **Sage-leaf rockrose** (*Cistus salviifolius*)
8 - **Hypericum** (*Hypericum coris*)
9 – **Water gum** (*Tristania laurina)*
10 – **Manzanita** (*Arctostaphylos manzanita* 'Dr. Hurd')

(Of this list, only the California fuchsia and manzanita are native to California.)

The "winner" of this small trial, Pride-of-Madeira (*Echium candicans* [*syn. fatuosum*]), is an exotic from the Canary Islands, now localized and slightly invasive in my part of northern California.

But, and it's a big but, *Echium candicans* is only invasive in our *very* local moderate, coastal environment. Very close to the Pacific Ocean, this shrub spreads some, but overwhelms very few native plants. But according to the Strybing Arboretum of San Francisco: "*Echium* spp. has become widely naturalized along our West Coast, especially at the edge of the bay or along sunny coastal bluffs above the ocean. Called "Tower of Jewels" or "Pride-of-Madeira," depending on the species, *Echium* spp. is an energetic member of the Borage family with invasive tendencies that have elicited warnings of "do not plant" from as far away as Australia [I did see *massive* stands of *Echiums* (not the *E. candicans* species) in Australia, not yet duplicated an California] and New Zealand where it has become a nuisance." It took me years to put these two facts together. I have seen the *E. candicans* spread in microclimates similar to the coast, but not with much vigor or threat to very many native plants. Pure stands are scattered and rare.

Figure 51. Pride-of-Madeira (*Echium candicans*) in mid spring; the Monarchs *love* echiums.

Deer and Exotic Beauties

Although I now live in the suburbs with a fenced garden, I lived in a rural area with deer for over 30 years. I didn't build a fence, just learned how to live with the foraging ungulates. My garden did well because it had almost *no* native plants. In my area at least, those people that insist on promoting natives must not be living on the edge of forests where deer thrive.

One reason that my garden did so well was that I used Mediterranean plants (non-invasive selections) from other summer-dry climates. I suspect what made the native plants into fodder for the deer was their history of growing along with foraging deer for hundreds of years. The munchers' systems have simply become adjusted to the alkaloids, tannins, and other chemicals that the native perennial plants use to reduce browsing.

I like to think that many plants we see in the forest are just a tiny fraction of what tried to grow there. Between deer, drought, rabbits, and other "plagues," many seedlings die before reaching maturity. I often refer to this as "the silent screams of the forest." We just have no idea as to how many plants died to reveal the one we see. If we buy a plant, we're very attached to its survival. We just can't afford to buy five or six or more plants at $10 each to have only one survive—perhaps. Almost every native I planted, except several local species of salvias (*Salvia clevelandii and S. leucophylla*) and wild ginger (until August when they were grazed to the ground, only to recover in the winter rains), was eaten by deer, if not the night after planting, soon enough to make me angry.

Since local deer didn't evolve with many types of plants from other Mediterranean zones, these plants have, I figure, an alkaloid and chemical mix that deer can't tolerate. Lavenders, for example, are from the islands and nearby mountains of the Mediterranean Sea, and I've never had deer (or gophers for that matter) eat lavenders. (Pride-of-Madeira (*Echium* spp.), being apparently completely deer- resistant, has naturalized by the coast along California's Highway 1.

Some of the exotics growing happily in my garden included: 10 varieties of **rosemary** (*Rosmarinus* species with white, pink, and many shades of blue blossoms); 20 types of **lavenders** (traditional blue-lavender *Lavandula* spp. flowers, plus pure white, pastel-pink, and lime-green flowering varieties); various culinary and ornamental species of **thyme** and **garlic chives** (which usually became deer salad by late summer); **foxglove** (*Digitalis purpurea*); various euphorbias (*Euphorbia* spp.); **ornamental grasses** (my favorite is *Stipa gigantea* with its towering 8-foot-tall seed-heads); four types of **santolina** (*Santolina* spp.); and **rockroses** (*Cistus* spp.).

There are exceptions. **Red-hot poker** (*Knifophia* spp.) plants were deer-resistant at first in my garden; after ten years they became Bambi snacks. The deer adapted to some other Mediterraneans, but the exotic Mediterraneans were usually more reliable than native plants. They're also very adapted to our rainless summers. I believe, in fact, that some are more drought-resistant than Northern California natives because they originally came from harsher Mediterranean climates. *Echiums* are my favorite example, native to the Canary Islands where it may rain only 5.7 inches per year. They thrive in my 58-inch winter rains if the drainage is ideal, and tolerate all dry summers—even severe droughts.

In my area, I've observed that most native-plant gardens can only be found in urban and suburban yards or gardens with deer fences.

Figure 52. Here's a triple whammy of invasives at The Sea Ranch in Northern California: *Echium candicans*, iceplant, and Monterey cypress—over 200 miles north of its native California habitat.

Bonus Charts and Photos:

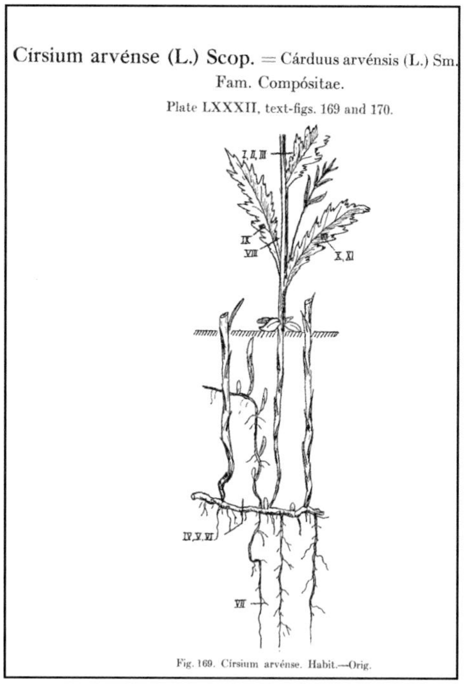

Figure 53. A drawing of Canada thistle (*Cirsium arvense*). The growth of the roots shows why it's so difficult to get rid of this invasive plant. (Drawing by Emil Korsmo, *The Anatomy of Weeds*. 1954.)

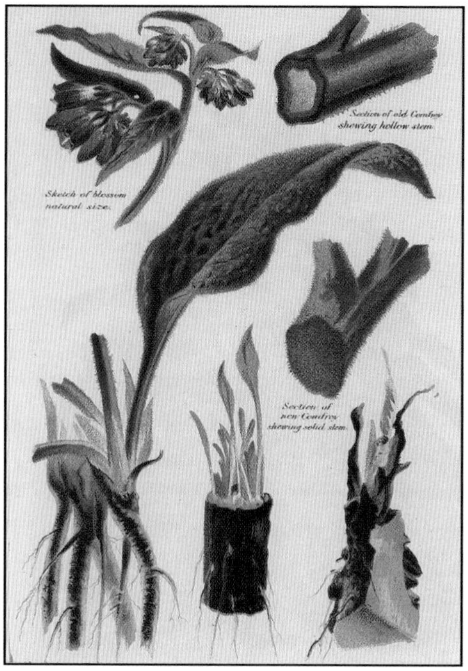

Figure 54. Comfrey (*Symphytum officinale*) is a beneficial plant, but the smallest pieces of its roots can sprout to infest a garden. Do not till or dig anywhere near comfrey.

Figure 55. Even though it fixes nitrogen, Scotch broom (*Cytisus scoparius*) is a very invasive legume and should always be removed. It tends to grow at the edge of forests, and is a very serious fire hazard in the dry summers of California. Its seeds can stay viable for up to 80 years.

Figure 56. While deer may nibble some on pale-blue Forget-Me-Nots' (*Myosotis sylvatica*) flowers, they certianly can't keep it from spreading. It often crowds out grasses over time. Here it is encroaching on the Northern California native huckleberry (*Ribes vaccinium*). But it will not spread under the dense shade of the huckleberry.

Figure 57. Nasturtium (*Tropaeolum majus*) easily grows in very mild-climate areas. It does not like freezing weather. Where it is happy it'll spread—but not with much vigor. In a home garden, it's pretty easy to control. All parts (except the roots) are edible, so you can eat your problem.

Chapter 4
Nurturing Soil

Before you plan or plant anything in your garden, you need to do two important things:

1. Dig a trench.

2. Take a soil sample.

This is the best way to learn about your soil's horizons. (A *soil horizon* is a layer parallel to the soil surface, with physical, chemical and biological characteristics that differ from the layers above and beneath.) Sampling your soil will give you a sense of 1) how deep the topsoil is and 2) the location of a dense and distinct layer of clay soil (otherwise known as a hardpan, or a claypan), that's largely impervious to water and is usually found somewhere below the uppermost topsoil layer.

Knowing the depth of your topsoil will help you understand where your roots will thrive. It also lets you know how deeply you can dig and whether it might be advantageous to start off with plants with roots that can help improve the soil (more on this in Chapter 8).

A drip-irrigation system is a boon to a sustainable garden, and hopefully you've begun to set one up. Here's an easy way to determine, early on, how long

Figure 58. You'll need to dig a trench (but not one this big) to discover how deep the topsoil in your garden is. Here a cross-section at The Sea Ranch, CA shows the depth of the dark-brown topsoil there. This is the "A" horizon where a majority of plant roots are found. The gray layer is the "B" horizon where only some roots grow. At the lower right is the orange parent rock where roots usually don't grow.

Clover (*Trifolium* spp.)—pictured above—can fix 100 to 150 pounds of nitrogen per acre. Or, .22 to .34 pounds of nitrogen per 100 square feet. The foliage is 34 percent nitrogen and makes a good green manure.

Figure 59. A summary of soil horizons. Raw litter consists of un-decomposed leaves and other detritus. In Figure 59, the litter is very shallow and doesn't show up in the photograph. Most roots prefer the A horizon, and 50 percent or more of the roots will grow in the top 12 inches of topsoil. If the B horizon is a layer of clay, it forms a barrier beyond which roots won't easily grow. The shallower the A horizon, the wider the roots will grow beyond the edge of a tree's foliage canopy.

your periods of irrigation will need to be. While your soil-test trench is open, place some drip emitters near its edge. Run your drip system for one hour and note how widely the moist spot spreads below the ground. Except in very sandy soil, the belowground width will be wider than the wet spot on the soil's surface. Run the system for another hour, and again note the width of the moist spot. This tells you how far apart the emitters can be placed and still irrigate your plants efficiently.

If, for example, you can place emitters 24 inches apart in a heavy clay loam instead of one foot apart in a regular loamy soil, you'll save 50 percent on the cost of emitters.

While you're digging that trench, be sure to gather soil for testing. Without a soil test, you'll just be guessing about how much fertilizer you need. In my area, for example, it's uncommon for soil to need extra phosphorous, so I can save money by not dumping a bunch of colloidal phosphate fertilizer on my garden. Take enough reasonably dry soil from the top zone—six to 12 inches deep (avoid the top six inches of soil as it has too much organic matter)—to fill a one-gallon Ziplock™ bag.

Send it to a lab (Google™ "soil testing near me.") to test for the macronutrients NPK (Nitrogen, phosphorus and potassium), calcium, salts, organic matter, and pH. If your house was built earlier than 1978, the year lead was banned as an ingredient in paints, be sure to add a test for lead. When sending a sample to a lab, include clear information about what kind of crops you'll be growing, e.g. vegetables, fruit trees, etc. Ask the lab to be sure it will interpret the results in terms of *organic* fertilizer and compost application rates. Free tests available through county or state agriculture cooperative extensions may not interpret results as they apply to organic gardening, or may only recommend chemical fertilizers and provide application rates of pounds of chemicals per 1000 square feet or per acre.

Take a look at this soil test I did for a client:

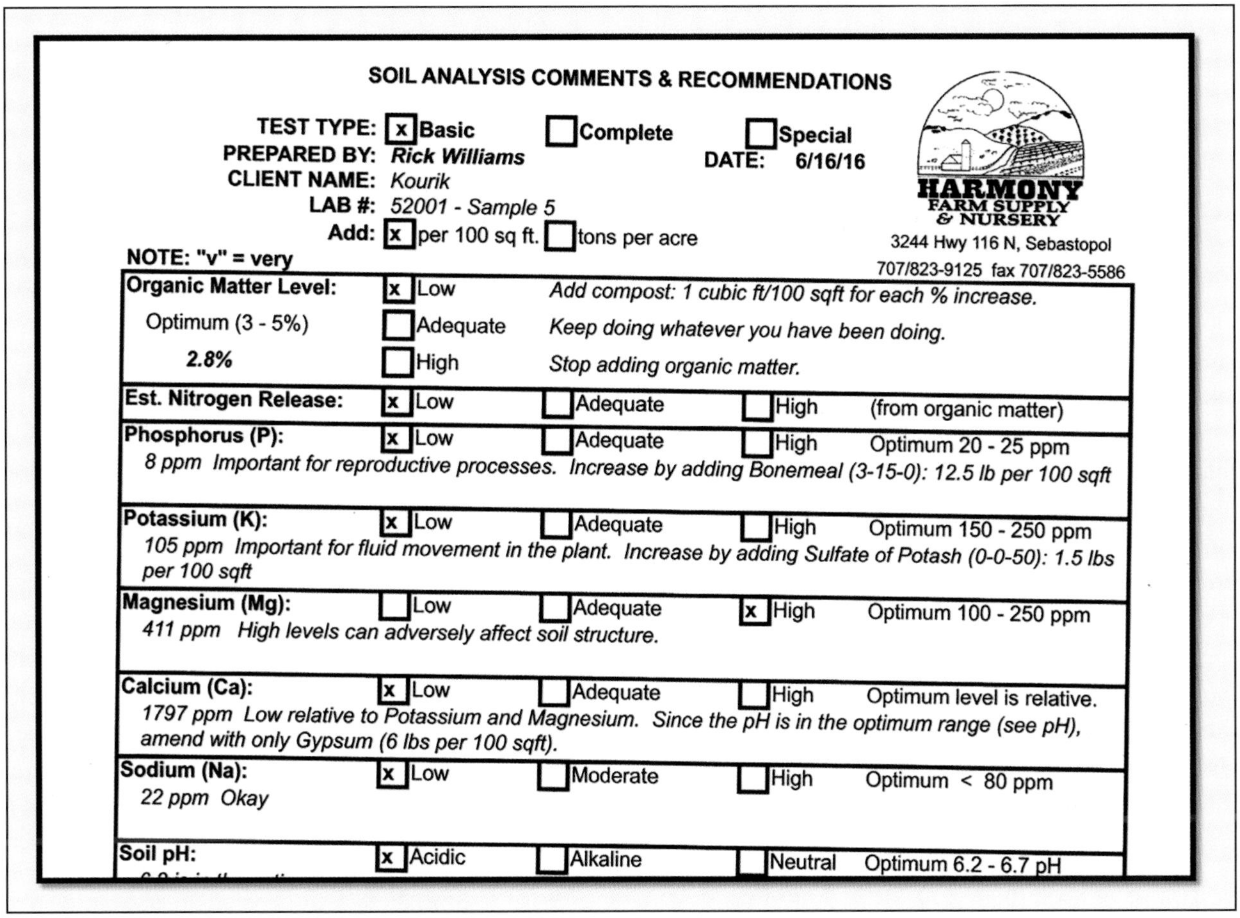

SOIL ANALYSIS COMMENTS & RECOMMENDATIONS

TEST TYPE: [x] Basic [ ] Complete [ ] Special
PREPARED BY: Rick Williams DATE: 6/16/16
CLIENT NAME: Kourik
LAB #: 52001 - Sample 5
Add: [x] per 100 sq ft. [ ] tons per acre

HARMONY FARM SUPPLY & NURSERY
3244 Hwy 116 N, Sebastopol
707/823-9125 fax 707/823-5586

NOTE: "v" = very

Organic Matter Level:	[x] Low	Add compost: 1 cubic ft/100 sqft for each % increase.		
Optimum (3 - 5%)	[ ] Adequate	Keep doing whatever you have been doing.		
2.8%	[ ] High	Stop adding organic matter.		
Est. Nitrogen Release:	[x] Low	[ ] Adequate	[ ] High	(from organic matter)
Phosphorus (P):	[x] Low	[ ] Adequate	[ ] High	Optimum 20 - 25 ppm
8 ppm Important for reproductive processes. Increase by adding Bonemeal (3-15-0): 12.5 lb per 100 sqft				
Potassium (K):	[x] Low	[ ] Adequate	[ ] High	Optimum 150 - 250 ppm
105 ppm Important for fluid movement in the plant. Increase by adding Sulfate of Potash (0-0-50): 1.5 lbs per 100 sqft				
Magnesium (Mg):	[ ] Low	[ ] Adequate	[x] High	Optimum 100 - 250 ppm
411 ppm High levels can adversely affect soil structure.				
Calcium (Ca):	[x] Low	[ ] Adequate	[ ] High	Optimum level is relative.
1797 ppm Low relative to Potassium and Magnesium. Since the pH is in the optimum range (see pH), amend with only Gypsum (6 lbs per 100 sqft).				
Sodium (Na):	[x] Low	[ ] Moderate	[ ] High	Optimum < 80 ppm
22 ppm Okay				
Soil pH:	[x] Acidic	[ ] Alkaline	[ ] Neutral	Optimum 6.2 - 6.7 pH

Figure 60. Always test your soil. If it seems noticeably different in various areas, take more than one sample. This sample result shows that the soil needs to be improved with at least two cubic feet of compost per 100 square feet in order to bring it up to around five percent organic matter. Nothing can be done to lower the high amount of magnesium except to buffer it with added compost and calcium. Notice that bonemeal is suggested to increase the phosphorous level. The nitrogen release is low due to lower levels of organic matter, but can be improved by adding compost.

Notice that it tells you how much compost to apply per 100 square feet in order to boost the organic matter by one percent. (Six to eight percent organic matter in the soil is plenty for home gardeners; farmers get excited if their soil's organic-matter content is over four percent.)

The Helpful Fungus Among Us

Wouldn't it be great if there were something that acted as a conduit for pulling water and nutrients directly into roots? Well, actually, there is—*mycorrhizae,* or beneficial fungi. *Mycorrhizae* are the vegetative part of special fungi. They consist of a network of fine white filaments, known as *mycelia* or *hyphae*, that permeate the soil. *(Myco* means fungus and *rhiza* means root.*)*

Mycorrhizae gather nutrients (especially phosphorous) and moisture, and feed them either directly into root cells or into spaces between the cells where they can also be absorbed. This is coevolution at its best: the mycorrhizae acquire sugars, starches, proteins and lipids from the roots of a plant, so that they can proliferate and feed the plant itself. It's not a parasitic relationship, but rather a mutually beneficial association.

The nature of this coevolution is here detailed by Dr. Linda Chalker-Scott, of Washington State University:

"Roots of plants under *mild* nutritional stress release chemical cues, such as organic acids, that stimulate mycorrhizal growth. If the hyphae encounter these receptive roots, they penetrate the cell walls, pressing up into the cell membrane and creating chemical passageways between the two partners. Multiple infection points and hyphal branching create a cottony sheath around the roots that extends far into the surrounding soil.

The mycelia of these cooperative fungi grow in association with most plants, especially conifers. It's a delicate balance; as much as 20 percent of the total carbon (in the form of carbohydrates) integrated into trees and plants by photosynthesis may be transferred to their fungal association. This transfer of carbon to the fungi, which might be considered a drain on plants, is, however, often balanced by an increase in photosynthesis following the mycorrhizal inoculation, thus compensating for any carbon "lost" to the fungi.

Occasionally, plant-growth suppression has been attributed to mycorrhizal colonization, but this usually occurs only under low-light or high-phosphorus conditions."

We read from *Soil Microbiology, Ecology, and Biochemistry,* edited by Eldor Paul, 4th Edition p. 82: "In reality, most forest trees can't live without mycorrhizae. Many trees need phosphorous, but are not able to absorb soluble phosphorus on their own, and so are dependent upon the mycorrhizal relationship. The extensive mass of fungal mycelia and its subsequent huge surface area allow the mycelia to "mine" a much greater amount of soil and duff than the root hairs of the tree would be capable of on their own. As an example, the mycorrhizae can increase the absorbing surface area of pine seedlings by 80%. (David Sylvia of the International Mycorrhiza Society.)

Ninety-five percent of the world's plants depend on mycorrhizal association. Some examples of the five percent of non-mycorrhizal plants include: the *Brassica* family of vegetables—broccoli, cauliflower, kale, cabbage and Brussels sprouts; blueberries and other ericaceous plants (belonging to the heather family) such as cranberries, huckleberries, and lingonberries; spinach and beets. The non-mycorrhizal "weeds" include: *Echium vulgare*, lupines, and *Rumex* spp. or docks. Only five percent of plants are what is known as *Ectomycorrhizal* (see below). These include most nut trees, madrones, pines, and willows.

Plant Root Cells: Outside and Inside

There are two primary types of mycorrhizal associations: *Ectomycorrhizal* association (abbreviated as EM) is an interaction in which tiny mycelia of certain kinds of

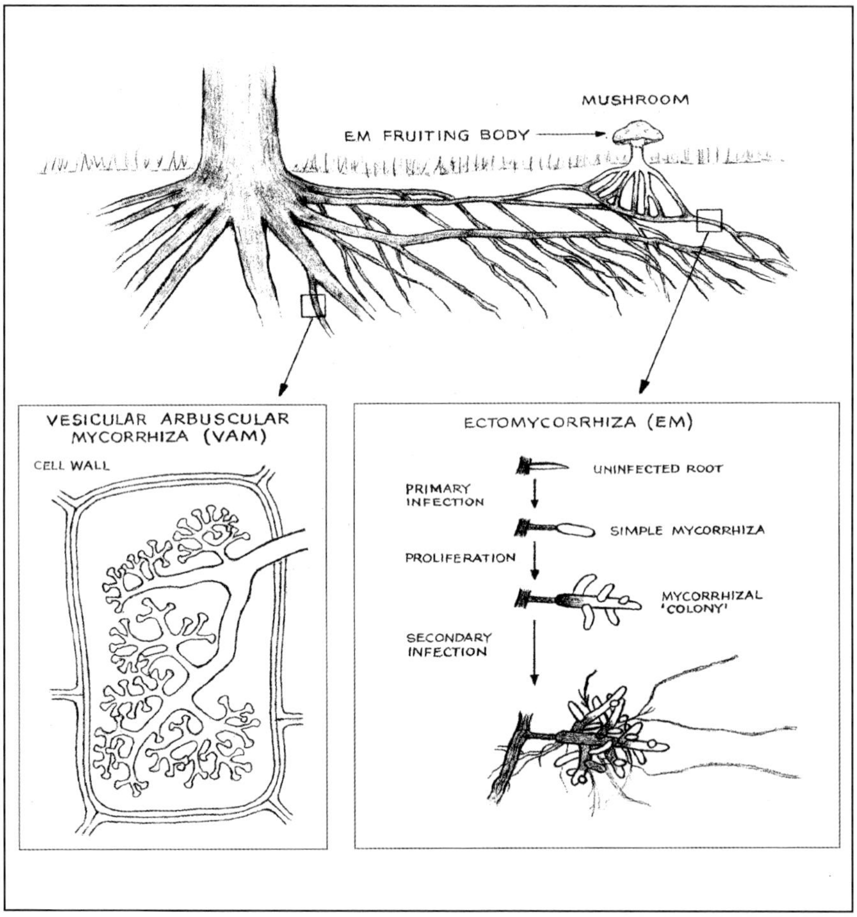

Figure 61. There are two types of mycorrhizae. The *ectomycorrhizae* coat the outside of the roots, forming what look like the nodules of a nitrogen-fixing legume. The *VAM mycelia* actually pierce into a cell to transfer nutrients—mostly phosphorous—directly into it. Only the ectomycorrhizae produce a fruiting body (mushrooms). The mycelia prefer the upper soils, so adding mulch and not tilling are important measures to protect them.

Effect of mycorhizas on the growth of pine seedlings		
Indexes of growth	With	Without
Length of shoots of seedlings, cm	35.5	17.5
Increase in length after 2 years, cm	18.0	3.0
Length of sprouts of 2nd order, cm	10.0	0.3
Weight of shoots part, gm	17.0	3.1
Weight of roots, gm	11.0	4.5
Number of leaves	42	12
Total area of leaves, cm^2	591.0	96.0

Figure 62. This chart looks complicated, but it's really not too complex. Do the ratios. For example, the total area of leaf growth is about six times greater when mycorrhizae are added to a sterile soil mixture. (With mycorrhizae = 591 and without = 96.) Another example is the weight of the shoots (in grams); when mycorrhizae are added it's about five times greater than with sterile potting-soil mix.

(From *Soil Organisms*, Part IV. *Interaction Between Soil Microorganisms and Plants* by N. A. Krasil'nikov, Academy of Sciences of the USSR Institute of Microbiology, 1958.)

fungi coat the cells of the root-hairs of a plant. An association in which mycelia actually penetrate *into* the cells of the feeding roots is known as *arbuscular mycorrhizae* (AM or *endomycorrhizae*). (This form of mycorrhizae is also known as VAM, which stands for Vesicular-Arbuscular *Mycorrhizae*.) Some trees, such as alders *(Alnus* spp.), willows (*Salix* spp.), poplars *(Populus* spp.*)*, and *Eucalyptus* spp. can maintain both AM and EM associations on the same tree. (*An Introduction to Agroforestry*, by P.K. Ramachandran Nair, 1993.)

"Mycorrhizal fungi added to a sterile planting mix can increase the absorbing surface area of pine seedlings by 80%. AM association has been known to increase the stem weight of tree seedlings by two- to 80-fold when compared to the roots of seedlings without the fungi. At the same time, the root weight of the seedlings with the beneficial fungi was increased by four- to 70-fold by [V]AM." (*An Introduction to Agroforestry*, by P.K. Ramachandran Nair, 1993.) Some sources estimate that the absorption of nutrients by trees can be increased 100- to 1000-fold. (Dr. Mercola, Mycorrhizal Applications, Inc., PO Box 1029, Grants Pass, OR 97528.)

The chart at the left (note the 1958 Soviet Union spelling of mycorrhizae), based on the growth of pine seedlings started in a sterile potting mix, provides a number of ways to look at the influence of sterile soil inoculated with beneficial fungi. (From *Soil Organisms,* Part IV. *Interaction Between Soil Microorganisms and Plants* by N. A. Krasil'nikov, Academy of Sciences of the USSR Institute of Microbiology, 1958.)

Nitrogen is also transferred from the soil to trees via mycorrhizal association. EM helps in the mineralization of nitrogen in locations where plant litter is rich in lignin and tannins. Although only a few mycorrhizal fungi can mobilize nutrients from these primary sources, EM fungi can also obtain nitrogen and other nutrients from other sources of organic matter, such as the biomass left in the soil from dead microbes. Thus mycorrhizae may play an important role in nitrogen cycling in acidic and highly organic soils. The fertilizer needed for nitrogen and phosphorous can be reduced by 30-40 percent with VAM fungi.

There are even more benefits to having mycorrhizae in your garden soil:

- The association promotes growth and yields in fruit and nut trees.
- In a California native ecosystem, there are ten times more fungi than beneficial bacteria. This means the fungi are more important in furthering growth than much of the other soil life.
- They moderate drought and high soil temperature.

- They also provide micronutrients, such as zinc, copper, molybdenum, and potassium.

- Mycorrhizae can also be a source of food. A large number of mushrooms growing in an area indicates the healthy presence of mycorrhizae. *Boletus edulis* (which includes porcini), morels and the chanterelles are the primary edible mushrooms that indicate the association, and lobster, hedgehog and matsutake mushrooms also are indicators. (NOTE: **Don't** eat a mushroom unless you have a very experienced mushroom-hunter confirm that it's edible.)

- Mycorrhizae may have an important role in nitrogen cycling in acidic and highly organic soils, as nitrogen is transferred from the soil to trees via mycorrhizal association. As mentioned above, EM also helps in the mineralization of nitrogen where plant litter is rich in lignin (the cells that make trees woody and sturdy) and tannins (astringent plant products). In *An Introduction to Agroforestry* by P.K. Ramachandran Nair, 1993, we read: "Mycorrhizal fungi are also often associated with the roots of nitrogen-fixing trees. Endomycorrhiza, which penetrate the host roots, ...the vesicular arbuscular mycorrhizae (VAM) are the most common and are the most important for plant nutrition. Nodulation and nitrogen fixation require a high P level in the host plant, which can be facilitated by the mycorrhizal symbiont."

- The hyphae promote the growth of hormones and vitamins that boost plant resistance to some pathogens.

- The mycelia produce chemicals that hinder pathogenic bacteria, hostile fungi, and hungry herbivores. As an example, VAM can help prevent root rot (*Phytophthora cinnamomic*).

- The fungi can help lower the pH of alkaline soil

Some Mycorrhizal "Don't"

- *Don't* use mycorrhizal inoculums. They're basically a waste of time and money, as most garden soils already contain enough fungal spores. Simply do things to encourage their growth, like maintaining a healthy amount of mulch, and improving the soil from the top down. (More on this later.)

- *Don't* water within the perimeter of the dripline of trees, or the fungi will disconnect from the root systems.

- *Don't* till or disturb the soil. Another important use of no-till management.

- *Don't* use chemical or organic fertilizers on a tree, or the mycorrhiza will disconnect, which encourages the invasion of competitive, non-mutualistic alien fungal species. The tree will also be more susceptible to diseases.

- *Don't* use insecticides or fungicides—organic or chemical.

- With California oaks, *don't* allow grass or weeds under the dripline of the tree, as they'll crowd out and replace the litter layer from which the fungi absorbs nutrients.

And my favorite example: David Sylvia of the International Mycorrhiza Society refers to the work of M.H. Miller and coworkers from the University of Guelph in Canada, "[They] documented an interesting case in which disturbance of an arable, no-till soil resulted in reduced VAM development and subsequently less absorption of phosphorus by seedlings of maize in the field. Disturbance reduced both mycorrhizal colonization and phosphorus absorption by maize and wheat roots, but did not reduce phosphorus absorption by two non-mycorrhizal crops, spinach and canola." The authors concluded that under nutrient-limited conditions, the ability of seedlings [that participate in mycorrhizal association] to release phosphorous can be very helpful for crop establishment.

Organic versus Conventional Care of the Soil

Studies of the influence of fertilizers on arbuscular mycorrhizae in farms and forestry have shown that conventional husbandry, with the use of pesticides, fungicides, and chemical fertilizers, harms AM. On the other hand, it's clear that the practice of organic farming enhances the effect of VAM.

In these studies, the AM "infection" rate was found to be much higher at biodynamic farms than at farms using conventional chemical applications. (*Arboriculture: Integrated Management of Landscape Trees, Shrubs, and Vines*, [4th Edition] by Richard W. Harris, James R. Clark and Nelda P. Matheny, January, 2003.)

In typical organic farming, the number of varieties of AM was also much higher than in conventional farming. (From "Differences in Mycorrhizal Colonization of Rye (*Secale cereale*) Grown in Conventional or Organic (Biological-dynamic) Farming Systems," October 2008, B. Sattelmacher, S. Reinhard and A. Pomikalko, *Journal of Agronomy and Crop Science*.) As one study summed it up: "... Organic management in agro-ecosystems [another word for organic farming] contributes to the restoration and maintenance of these important below-ground mutualists." ("The Effectivity of Arbuscular Mycorrhizal Fungi from High Input Conventional and Organic Grassland and Grass-arable Rotations," July 1998. J. Scullion, W.R. Eason and E. P. Scott, *Plant and Soil*; Volume 204, Issue 2, pp 243-254.)

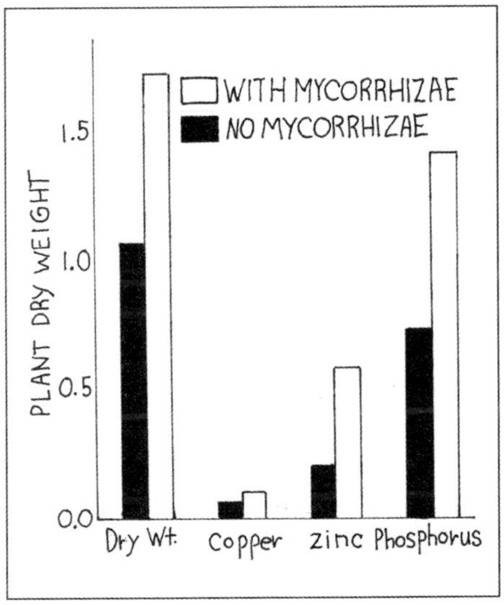

Figure 63. Mycorrhizal fungi primarily gather phosphorus from the soil and deposit it directly into the roots. Here we see that roots without mycorrhizal fungi absorb about 50 percent less phosphorus, but acquire close to three times more zinc with mycorrhizae. (Soil Nutrient Bioavailibility, 1984. Lambert et al., *Soil Science Society of America Journal*, 43: 976-980.)

Elsewhere, a study in Europe found that "...Organic management in agro-ecosystems (organic farms) contributes to the restoration and maintenance of these important below-ground mutualists. The VAM richness (numbers of species of the fungi) increased significantly with the time since conversion to organic agriculture." ("Positive Effects of Organic Farming on Below-Ground Mutualists," 2010. E. Verbruggen et. al., *New Phytologist*, 186, pp 968-979.)

"[Vesicular] arbuscular mycorrhizal fungi was investigated at two sites, cultivated using conventional or biological-dynamic farming methods. The [V]AM infection rate and infected root length were significantly higher at the biologically-dynamic [better known as biodynamic] cultivated site [as compared to conventional farming]." (*Arboriculture: Integrated Management of Landscape Trees, Shrubs, and Vines*, (4th Edition); by Richard W. Harris, James R. Clark and Nelda P. Matheny, January, 2003.

In another study we read: "In the least fertile soil, inocula from organic farms were more effective than those from conventional farms. Intensive [e.g., conventional—Ed.] farming practices may reduce the effectiveness of indigenous arbuscular mycorrhizal populations, particularly where fertilizer inputs are high and inherent fertility is low." ("Differences in Mycorrhizal Colonization of Rye (*Secale cereale*) Grown in Conventional or Organic (Biological-dynamic) Farming Systems," October 2008, B. Sattelmacher, S. Reinhard and A. Pomikalko, *Journal of Agronomy and Crop Science*.)

One more: "The average number of AMF [Another former term for AM—Ed.] taxa was highest in grasslands, intermediate in organically managed fields and significantly lower in conventionally managed fields. Moreover, AMF richness increased significantly with the time since conversion to organic agriculture. AMF communities of organically managed fields were also more similar to those of natural grasslands when compared with those under conventional management...We suggest that organic management in agro-ecosystems contributes to the restoration and maintenance of these important belowground mutualists." ("The Effectivity of Arbuscular Mycorrhizal Fungi from High Input Conventional and Organic Grassland and Grass-Arable Rotations," July 1998. J. Scullion, W.R. Eason and E. P. Scott, *Plant and Soil*, Volume 204, Issue 2, pp 243-254.)

A Canadian study found that: "...the organic system performed better than the conventional system ...mycorrhizal colonization [was] (30 to 50% higher)." (Influence of Organic Management with Different Crop Rotations on Selected Productivity Parameters in a Long-Term Canadian Field Study, M. Entz, J.W. Hoeppner, L. Wilson, M. Tenuta, K.C. Bamford and N. Holliday, University of Manitoba, Winnipeg, Manitoba, Canada, R3T 2N2.)

Some researchers used *Trifolium repens* clover to grow the fungi and then inoculated elephant garlic with its spores. They found that "In the least fertile soil, inocula (spores) from organic farms were more effective than those from conventional farms." ("The Effectivity of VAM Fungi from High Input Conventional and Organic Grassland and Grass-Arable Rotations," 1998. J. Scullion et al, *Plant and Soil*, Vol. 204, Issue 2, pp 243-254.)

And, in spite of many similar references, I'll add just one more. Since 1906 (a study that ended in 1946, a study/comparison has been done in Austria on fields containing control plots (with no fertilizer), plots with farmyard manure, and plots with chemical soluble fertilizer, with winter rye used as a cover crop each winter. The results really substantiate the importance of organic farming and its impact on VAM.

"We found the soil organic matter to be 96 percent higher in the plots fertilized with manure." VAM colonization of the rye's roots was increased by 46 percent. They concluded that farmyard manure increased the levels of organic matter, increased the frequency of VAM, and enhanced the crops' uptake of nutrients due to VAM in a dry form. "...enhanced nutrient uptake by VAM promoted crop yield stability and sustainable plant growth." ("Yield and VAM of Winter Rye in a 40-Year Fertilization Trail," 2010. M. J. Gollner *et al*, *Agronomic Sustainable Development*; 31:373-378.)

Where mycorrhizae are concerned, many benefits of organic farming carry over to the management of landscape trees. For example, fertilizing a lawn around a shade tree with organic methods will further the health of the tree's mycorrhizae.

Indicators

There are many types of Mycorrhizal fungi. One researcher came up with 101 species of fungi forming mycorrhizal associations with Norway spruce (*Picea abies*) trees and 117 species interacting with a Scots Pine (*Pinus sylvestris*). As mentioned, this interdependence is easy to spot by the presence of certain mushrooms growing in a given area. Thirteen examples of the types of mushroom to look for are (especially under conifers and pines):

1. *Amanita muscaria* – fly agaric. A relative of the infamous "death cap".
2. *Boletus piperatus, B. luteus, B. badis, B. varigatus, B. granulatus and B. bovins* - The bolet or boletus mushrooms. Some species are edible. (**Note**: Never eat a mushroom without the guidance of a skilled mushroom hunter!)
3. *Lycoperdon pulcherrimum* - A species of puffball. Some species are edible. **See Note above.**
4. *Agaricus (Inocybe) eutheloides* – A species of mushroom related to the common supermarket mushroom. **See Note above**.
5. *Clitocybe rivulosa* – A very poisonous mushroom.
6. *Calvatia saccata* – Another type of puffball. Some species are edible. **See Note above.**
7. *Cortinarius argentatus*
8. *Tricholoma personatum* (*Clitocybe saeva*) – blewits. Some species are edible. **See Note above**.
9. *Lepiota naucina* – Poisonous.
10. *Lactarius deliciosus*
11. *Russula* spp.
12. *Scleroderma citrinum* - common eartball
13. *Rhizopogon luteolus*

Consult a local mushroom society for help with identification of these mushrooms.

Purchased Inoculants

Some garden and agricultural suppliers now offer beneficial microbial and fungal inoculants to "boost"

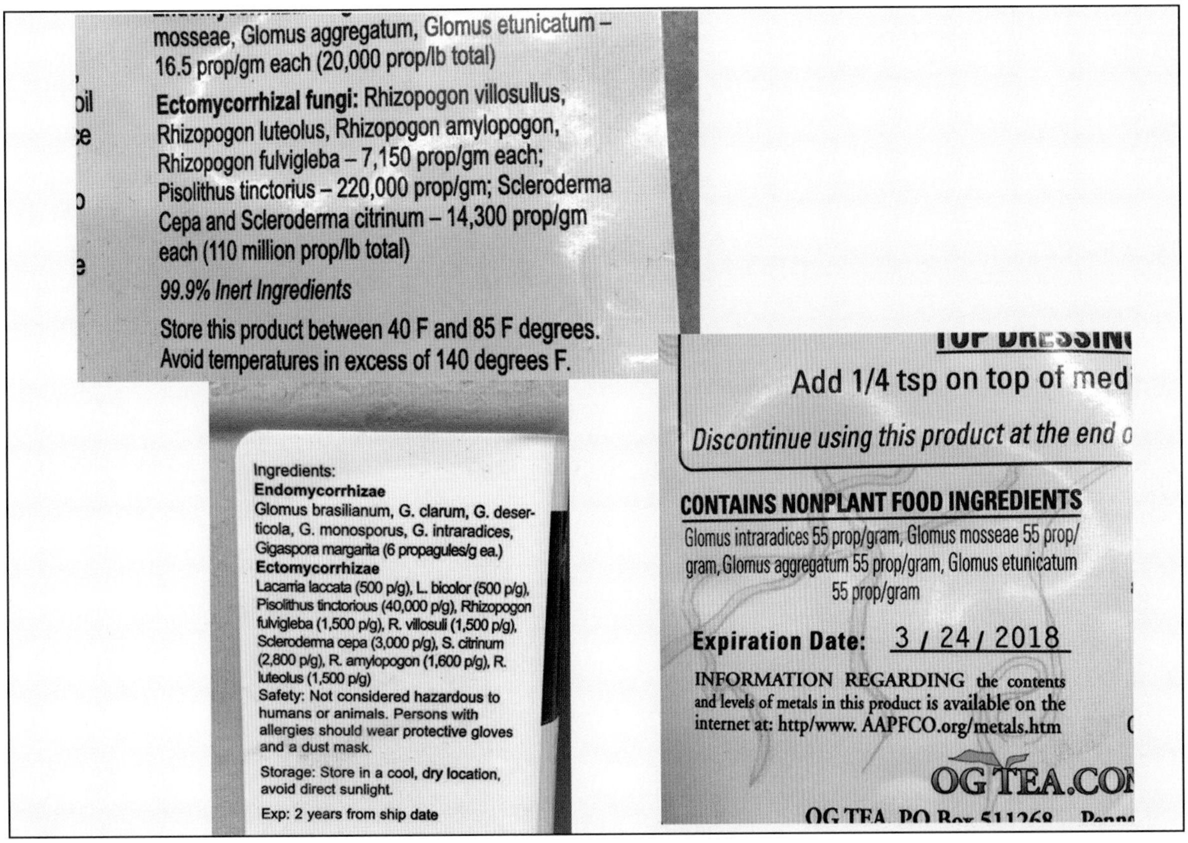

Figure 64. These three different labels show how inoculant products vary. The label should list the expiration date (as at lower right). The lower-left label says "2 years from shipping," but gives no shipping date. Unlike the others, the lower- right product doesn't give storage temperatures. The Genus *Glomus* spp. is an important endomycorrhiza. On the labels "prop" stands for the number of spores—propagules (283 propagules per ounce), but sometimes it's listed in grams and at other times in pounds. (One pound = 454 grams.) So 20,000 prop/lb equals 44 prop/gram. Thus the lower right product has more *G. aggreatum* than the upper left. Good luck with the math.

the soil's fertility, touting the benefits of mycorrhizal inoculants for plant roots The most common species of beneficial mycorrhizae sold for this purpose include *Glomus fasciculatus*, *G. mosseae* and *G. etunicatus*. In one study, the *G. etunicatus* produced the most growth of roots and shoots, so you might want to look for this spore when choosing an inoculant product. ("Corn [*Zea mays*] Growth as Affected by Soil Compaction and Arbuscular Mycorrhizal Fungi," 2013; M. Miransari, *Journal of Plant Nutrition*; Vol. 36, P 1853-1867.)

The hot trend at a local organic-supply nursery in my area is the use of mycorrhizal inoculants. The prices range from an innocuous $16 to an extravagant $52 per bag. Since these fungi are already prevalent in most gardens, I consider it a waste of money. On one occasion, this led to a protracted discussion of the topic with a staff person at the abovementioned nursery. I maintained that the inoculants were only needed in harsh or sterile places like beaches, urban lots far removed from a natural meadow or forest, sterile soil mixes in container plants, "dirt" *highly* degraded by the overuse of pesticides, fungicides, herbicides, and/or soil fumigation, toxic mine tailings (heaven forbid), or when house construction bulldozes topsoil down to sterile clay, and I gave him the following reference information:

I contacted soil scientist Dr. Phillip J. Craul, Senior Lecturer at the Graduate School of Design, Harvard University, and author of *Urban Soil in Landscape Design* and *Urban Soils: Applications and Practices*, with the question "What do the new packaged microbial and fungal products offer gardeners?" His

Figure 65. The left side of this label shows a wide range of VAM and ectomycorrhizae. The right side includes a list of beneficial bacteria. As an example, *Bacillus thuringiensis* (Bt) is a species of bacteria that lives in soil; it kills some insects, mostly leaf- and needle-feeding caterpillars, but does not harm mammals. *Azotobacter chroococcum* is a nitrogen-fixing bacterium that lives in the soil and is not dependent on legumes.

response was: "I'm going to be frank, most of the time the home gardener doesn't need to fuss with (inoculants). Such products are only helpful in special cases where you're working with sterile material; (however), most of the biological analyses show enough organisms are usually present under artificial conditions to repopulate the soil—even in urban areas."

Afterwards the nursery staff person with whom I was speaking admitted that a lot of the inoculant products were sold to cannabis growers who were trying to go organic.

Jim Downer, **Ph.D. Cooperative Extension, Ventura County, California**, remarks: "Other standard planting practices such as mulching may provide greater benefits toward the establishment and growth of newly planted woody ornamentals than application of mycorrhizal inoculants." (***Snake Oil, Horticultural Myths, Horticultural Urban Legends, and Persuaders in our Industry***, by Jim Downer Ph.D. Cooperative Extension, Ventura County, California.)

Dr. Linda Chalker-Scott, of the Washington State University Extension, who has dismantled a great number of myths, sums it up: "From a practical standpoint, two realities emerge from all the research done thus far:

- "Healthy soils naturally contain indigenous mycorrhizae. Adding packaged mycorrhizae to such soils is a waste of money and resources."
- "If soils are impaired to the point where indigenous mycorrhizal species can't survive, mycorrhizal amendments alone won't help." ("Mycorrhizae: **So, What the Heck Are They, Anyway?" 2009. Dr. Linda Chalker-Scott, WSU Science Editor, *Master Gardener*).**

In 2005, it was reported in several publications that: "[Research] indicate(s) that mycorrhizae not native to a site may not function at that site because they are not adapted to grow there." ("Mycorrhizal Fungal Inoculation of Established Street Trees," March 2003. Bonnie Appleton, Joel Koci, Susan French, Miklos Lestyan, and Roger Harris, *Journal of Arboriculture* 29(2)) And "Existing Mycelia can inoculate the roots to produce better growth and greater yields." (***Snake Oil, Horticultural Myths, Horticultural Urban Legends, and Persuaders in our Industry*, Jim Downer Ph.D. Cooperative Extension, Ventura County, California.)**

So, in summation, unless one is planting in the kinds of tortured soils mentioned previously, many of the microbial and fungal elixirs offered for sale are probably a waste of money. Healthy soil's vast complex of biota defends its own territory from "invading" microorganisms, and existing microbes and mycorrhizae can be very good at fending off inoculants. According to Martin Alexander in *Introduction to Soil Microbiology*, "Microorganisms inoculated into non-sterile soil lead to poor growth, and often the seeded species is eliminated in a period of days or weeks...(due to) a rivalry for limiting nutrients, the release by one species of products toxic to its neighbor and direct feeding of one organism upon a second." (*Introduction to Soil Microbiology*, Martin Alexander, John Wiley & Sons, Inc., 1961, p 429.)

And the good news is that there's lots of this good fungus to go around. As quoted above, one researcher came up with 101 species of fungi forming mycorrhizae with Norway spruce (*Picea abies*) trees and 117 species around Scots pines (*Pinus sylvestris*).

Mycorrhizal plants are better able to endure environmental stress than are plants without mycorrhizae. Nutrient-poor or moisture-deficient soils show improved plant growth and reproduction when in association with mycorrhizae. [mycorrhiza.ag.utk.edu/] As a side note, trees such as alders *(Alnus* spp.), willows (*Salix* spp.), poplars *(Populus* spp.*)* and *Eucalyptus* spp. can have both AM and EM associations on the same tree.

In *An Introduction to Agroforestry* (1993), by P.K. Ramachandran Nair, we read: "Mycorrhizal infections are known to increase the absorption of phosphate ... Mycorrhizal fungi are most often associated with the roots of nitrogen-fixing trees. Endomycorrhiza, which penetrate the host roots, are more common than ectomycorrhiza, which remain external to the roots. Among the endomycorrhiza, the VAM are the most common and are the most important for plant nutrition. Nodulation and nitrogen fixation require a high Phosphorous level in the host plant, which can be facilitated by the mycorrhizal symbiont."

And another favorite example: David Sylvia of the International Mycorrhiza Society refers again to M.H. Miller and coworkers from the University of Guelph, Canada, "[They] documented an interesting case where disturbance of an arable, no-till soil resulted in reduced VAM development and subsequently less absorption of phosphorus by seedlings of maize in the field. Disturbance reduced both mycorrhizal colonization and phosphorus absorption by maize and wheat roots, but did not reduce phosphorus absorption by two non-mycorrhizal crops, spinach and canola." The authors concluded that under nutrient-limited conditions, the ability of seedlings with mycorrhizal association to release phosphorous may very helpful for crop establishment.

(Beneficial fungi are not alone in releasing nutrients, as discussed in *Modeling Green Manure Additions in Alley-Cropping Systems: Linking Soil Community Dynamics and Nitrogen Mineralization*; Y. Carrillo, and C. Jordan: "Soil biota play an important role in regulating nutrient mineralization...[It's] estimated that protozoa feeding on bacteria were responsible for up to 95% of the total N released into the soil in two arable farming systems." (From *Advances in Agroforestry*, Volume 4, Series Editor: P.K.R. Nair, 2008.)

Soil Depth

As is the case with roots, mycorrhizae mycelia and spores are found with descending frequency in soil. The deeper down, the fewer fungi, as there are fewer roots, and less oxygen in deeper soils. See the following graph ("Soil Depth: An Overriding Factor for Distribution of Arbuscular Mycorrhizal Fungi" 2013; A. Shukla1, D. Vyas and Anuradha Jha, *Journal of Soil Science and Plant Nutrition*, 13(1), 23-33. Illustration from page 129: *Understanding Roots*, Robert Kourik, Metamorphic Press, 2015.):

The top four inches of soil contains about one-third more spores (and therefore mycelia) than the soil just eight inches deeper. Even more remarkably, over 50 percent of VAM is found in the top *two inches* of the soil. This explains why mulch and no-till techniques are so important to protect the fragile mycelia.

The simple act of walking on bare soil compacts and destroys beneficial mycelia, because they grow so close to the surface. This is yet another case for using adequate amounts of mulch, which helps to prevent such damage by acting as a protective layer.

Drought

Beneficial mycorrhizal relationships also help plants survive or thrive during times of drought. Dry soil reduces the absorption of nutrients—especially

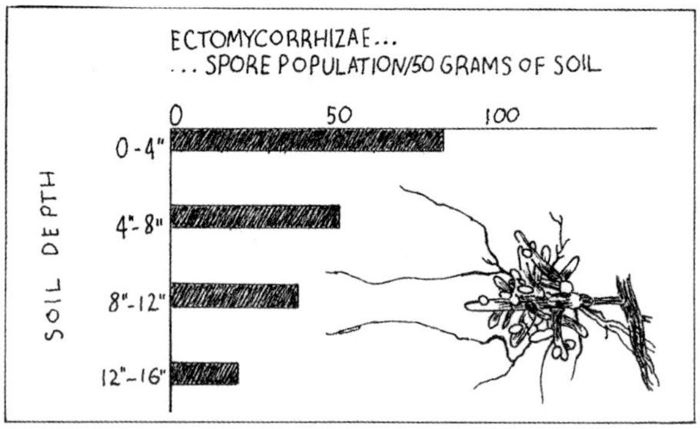

Figure 66. The top four inches of soil contains about 25 percent more mycorrhizae spores than the next four inches, indicating how close to the surface the mycelia like to grow. This is another reason why no-till gardening is so important.

phosphates. Arbuscular mycorrhizae have been shown to reduce the reaction of plants to drought, free up more nutrients, and increase absorption:

"Water stress usually causes a decrease of nutrient uptake, while a decline in soil moisture produces a reduction in the diffusion rate of nutrients, particularly phosphorous, from soil to root." (Pinkerton and Simpson, 1986) "The ability of mycorrhizal hyphae to transport nutrients into host plants (Koide 1991) can improve plant water relations under drought" (Fitter 1988)... The majority of studies of the influence of [AM] on drought avoidance and tolerance have shown that [AM] increase plant resistance to water stress.

As mentioned, the soil's vast complex of biota defends its own territory from "invading" microorganisms, and existing microbes and mycorrhizae can be very good at fending off inoculants.

If you want to inoculate seedlings before planting them in containers with a sterile soil mix, you might try one of the inoculants on the market such as "Power Organics Mycorrhizal Root Booster," from <www.fungi.com/mycogrow/> or search "Mycorrhizal Root Inoculant" via a Google™ search and it will reveal many other suppliers. Try the key words: mycorrhizae inoculants, mycorrhizal fungi inoculants, and commercial mycorrhizal fungi.

Here are five things to look for when buying an inoculant:

1. There should be a *lot* of genera and species of mycorrhizae listed on the label. Look for *Glomus etunicatus,* other *Glomus* spp., and both types of fungi. I saw one product that had only *one* ectomycorrhiza.
2. Check the expiration date, as the spores have a limited shelf life.
3. Look for the temperature range at which they should be stored, and check against possible temperatures and storage conditions at the plant nursery. (Don't order inoculants by mail during the hot summer or cold midwinter months.)
4. Check the price against similar products; some inoculants are very expensive.
5. Check the spore counts for each species listed. While the math is almost unfathomable, some inoculants have much higher spore counts than others.

Nurturing What You Have

If most soils already have these beneficial fungi, how do we protect and enhance their growth? Here's how:

- **Mulch, mulch, mulch**. The fungi thrive on an endless supply of decomposing organic matter. (Four inches of mulch is usually adequate to provide a layer of decomposition and contain the moisture necessary for mycorrhizae.)

- Be sure to **spread the mulch over the entire root system** of a plant, remembering that tree roots in a clay soil may grow at least 1.5 times wider than the canopy dripline of the tree, and up to three to five times wider than the canopy in, respectively, a loam or sandy soil.

- As much as possible**, prevent soil compaction**. Deep mulch (three to four inches) is important, as are permanent pathways in areas with significant foot traffic. Use permeable concrete if parking or road hardscapes are required.

- **Keep the soil as aerobic as possible** by avoiding compaction; promote an aerobic environment by keeping the pore space of the soil undamaged, thus allowing for good drainage. (Pore space consists of the small voids betwen soil particles. In healthy soil, the pores are large and plentiful enough to retain the necessary water, oxygen and nutrients that plants absorb through their roots.) The top four inches of soil may contain one-third more fungi than the next four inches. Under most normal growing conditions, up to 50 percent or more of a tree's roots may be found in the top 12 to 24 inches of soil.

- **Avoid using high levels of phosphate fertilize**r. While mycorrhizal association increases the available phosphorous for tree growth, too much added phosphate, especially in the form of soluble fertilizers, is detrimental and can even kill mycorrhizae. Do not add any form of phosphates unless soil testing shows a deficiency.

- **Protect the beneficial bacteria of the soil** as well, with the use of mulch, by avoiding compression, and maintaining an aerobic environment. Do not use *any* bactericides.

- **Maintain a diversity of plants** and trees to provide an environment that encourages many different species of fungal mycorrhizae, each with its own function in the underground life of plants.

When to Use Inoculants

- If planting in a container with a sterile potting mix.
- When a planting site has been scraped to subsoil.
- In very sandy or beach properties.
- To amend soil that contains too much salt.
- On toxic tailings from mining. (I hope this doesn't apply to you!)
- When sterile topsoil has been added to the site.
- If certain fungicides have been used in the planting area. (See: http://mycorrhizae.com/ faqs: What effect do fungicides have on mycorrhizal fungi?)
- After soil fumigants have been used.

Again, unless you're planting in the above circumstances, most of the fungal elixirs are probably a waste of money.

Wood-Wide Web

It turns out that these beneficial fungi serve another great purpose—they help plants "talk" to one another. The mycelia literally act as conduits for plants to communicate, sort of like the Internet of the forest. It's remarkable. Mycelia from one plant link up with and support others—for example, nutrients from alder and birch trees are passed on to hemlocks, cedars, and Douglas firs.

This interaction was being studied at least as far back as 1997, when Suzanne Simard found that the *hyphal link* (the mycelia of ectomycorrhizae) between a birch tree (*Betula papyrifera*) and a Douglas fir (*Pseudotsuga menziesii*) transferred three to 10 percent (a six percent average) of photosynthesis carbon uptake from the birch tree to the Douglas fir. ("Net Transfer Of Carbon Between Ectomycorrhizal Tree Species In The Field;" 1997. Suzanne W. Simard, David A. Perry, Melanie D. Jones, David D. Myrold, Daniel M. Durall & Randy Molinak. *Nature*, vol. 388 pp 579-582.)

Intelligence?

The network of mycelia demonstrates an even more remarkable and slightly spooky dynamic: the network can connect one plant to another as an early-warning system for pathogens. Through this network, it's actually possible for one plant to "listen" to another. When one plant is invaded by a harmful pathogen, its neighbor is "warned," and can react by activating its defense systems.

And it's not just trees that communicate. Fava beans (*Vicia faba*), as an example, can receive a signal (via the mycelia) that aphids have attacked other plants in the area. The aphid-free plant reacts by exuding the obscure compound methyl salicylate, which attracts a type of beneficial wasp that kills aphids.

With mycorrhizae mycelia linked together and most of the roots linked to the mycelia and all of them talking to each other, the forest really *is* like one living entity. Cool.

Now here's the dirt on nurturing soil, from the most intensive cultivation to the least.

Double Digging

Perhaps the most widely accepted "new" idea in gardening in the past 30-plus years has been double digging. John Jeavons, via a group called Ecology Action of the Mid-Peninsula, and in his many books, has been quite successful in popularizing raised-bed gardening. (His version of the original raised bed is in the form of a low mound of composted soil, *not* soil contained in planting boxes.) Jeavons actually meant to promote a very astute program of gardening called the Biodynamic/French-Intensive method, based on Alan Chadwick's work at the UC, Santa Cruz campus in the 1970s.

In 1999 Jeavons registered his own growing system as the GROW BIOINTENSIVE® method, an integrated set of techniques, tools, plants, and environmentally sensitive management. Alas, the popularization of the method has led to some very diluted and confused concepts, all of them more or less re-interpretations of Chadwick's work

The GROW BIOINTENSIVE method uses a variety of techniques to deeply texturize the soil so that more plants can be grown more closely together, but it doesn't "cultivate," in the sense of turning soil *over*. The true GROW BIOINTENSIVE method of cultivation loosens soil without inverting any of its layers, moving it laterally in such a way that the upper horizons of the soil stay relatively in the same zone as before the digging. (See Figure 67.) The goal of the method is the development of good soil structure through deep soil preparation. NOTE: The soil is ***not*** dug and churned deeply, as implied by so many misinformed books, articles, newspapers, and television shows.

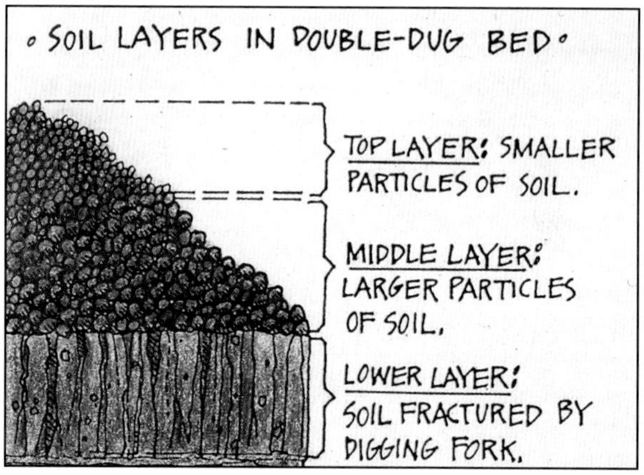

Figure 67. The GROW BIOINTENSIVE method, while it uses a variety of techniques to deeply texturize the soil so that more plants can be grown more closely together, doesn't "cultivate," in the sense of turning soil *over*.

True GROW BIOINTENSIVE allows the beneficial bacteria to thrive in the zone of soil for which they spent thousands of years evolving and adapting. The simplistic media-misinformation of the method has led many to assume that the more you churn and the deeper you dig, the better the garden will be. This is absolutely *not* the case; in fact, excessive cultivation is often detrimental. As mentioned, any digging, in the sense of mixing and churning the soil, kills microbes and slows the absorption of nutrients, subtle but important effect.

As with any gardening method, this approach has its pluses and minuses when it comes to the concept of import/export economies.

Benefits:

- As already mentioned, the conservation of resources.
- The "in-house" production of compost without significant external inputs promotes a true sustainable garden.
- The dense planting creates mature coverage by foliage; this suppresses weeds, conserves moisture, and promotes higher yields.
- Promotes the cultivation of fertile foliage for the compost pile, as a way to grow all needed nutrients within the garden. Jeavons' method uses crop rotation, nitrogen-fixing plants, green manuring, and cover-cropping, and uses no imported manure.
- Lofts the soil into "raised beds" to allow for good drainage, aerobic soil conditions, the storage of moisture, good root growth, and warmer soil earlier in the season.

- Reduces the carbon footprint, especially when compared to conventional gardening.
- Creates a slightly greater surface area to help increase yields.
- Works well for small yards where space is too small for conventional planting.

Limitations:

- Requires *lots* of effort. Jeavons estimates that the initial preparation of a 100-square-foot bed takes six to 14 hours. Two to four of these hours are spent double digging. (Subsequent re-digging in following years might take three hours.)
- Not needed in sandy soil, as all the benefits of drainage are already present. (Compost can, as in no-till gardening, be added to the surface.)
- Not as easily adapted to big crops like corn and vining squashes or pumpkins (although Jeavons does pull it off). Even tomatoes require, according to Jeavons, 18- to 24-inch spacing. This makes for a *lot* of digging to create a bed big enough for a reasonable tomato crop.

Figure 68. "D"-handled spade on the left; long-handled shovel on the right.

Working With "D" Problem

For many gardeners, especially "Baby Boomers," the short-handled tools used in classic double digging are very hard on the back. In England, where the arts of spading and shoveling have long been refined into a science, a spade is used for slicing through and lifting sod; edging lawns and beds; skimming weeds; and opening straight-sided holes or trenches. Their spades are built for such tasks, bulldog sturdy, with sharp cutting edges, an almost-straight-line angle from handle-tip to blade-edge and, often, short D-handles.

Their shovels, on the other hand, are built to load, unload and spread sand, gravel, soil or mulch, with "round-pointed" or flat-edged blades. In the U.S., we tend to call spades and shovels alike by the name of shovel and to use them both for much the same kind of digging tasks.

In 1975, when I started landscaping, I bought cheap, long-handled shovels at the flea market. In 1979, I moved to the Farallones Institute Rural Center, an appropriate-technology community where Bio-Dynamic Intensive gardening was practiced with a fervor akin to religious ecstasy. I was swept away by the enthusiasm of the disciples of Alan Chadwick, the high priest of intensive raised-bed gardening, for short, D-handled spades, and finally shelled out the money for what many considered to be the Cartier of digging implements.

But in spite of the praise by my gardener friends and a certain mail-order catalog—that had gained fame in great part from its eloquent, even romantic, praise of these heavy, hand-made, forged tools—I kept returning to my trusty long-handled shovels for most tasks. Finally, I ran across research about the ergonomics of tools that confirmed what my body had already whispered—light, long-handled shovels are best for most garden tasks. At best, D-handled spades are only suited for work literally "in the trenches."

Ergonomics is the science of adapting tools and physical working conditions to suit the worker, which often involves redesigning tools to better fit the human body. Much of ergonomics means rethinking what sometimes turn out to be arbitrary designs of common everyday tools. The study that confirmed my body's psychic hints was called: "The Ergonomics of Shoveling and Shovel Design—an Experimental Study." (1986; *Ergonomics*, Vol. 29, No. 1, p 19-30) by Andris Freivalds of the Department of Industrial and Management Systems Engineering at Pennsylvania State University.)

In a study of digging, using sand to simulate cultivating a garden plot, the short, D-handled shovel, with a pointed blade, "was universally ranked lowest for every dependent measure used [perceived exertion of shoulder and arms, low back and the overall shovel], due to its short handle, requiring a stooped posture." Perhaps most important to the real health of the gardener, Freivalds found "low-back compression forces [with the short-handled shovel] were higher than for most other shovels."

Figure 69. Shown here are a D-handled spade (left) and a digging/spading fork. I use the spading fork to crack open the ground before building mounds (See page 177.) Taken while enjoying the great view at the Esalen Institute's central garden.

Of the long-handled shovels, the single most important criteria for ease of use was weight— the lighter the shovel, the easier the task of digging. One study cited by Freivalds found "shovels weighing less than 29 percent of shovel load to be 20 percent more efficient" and the researcher recommended a shovel weight of between 3.3 and 3.96 pounds. Unfortunately, most of the English-forged spades weigh from 4.5 to as much as six pounds. Another researcher found long-handled shovels increased productivity 25 percent over short-handled shovels.

While short, D-handled spades forged in ancient English foundries may be "dense with durability and intelligence," they're sure tougher on the body. So, I'll continue to use my lightweight, long-handled, round-pointed, hollow-backed shovel for most garden tasks.

Oh yes, there *is* one task where the short-handled spade was rated better, by 10 percent, than a long-handled shovel—in the confined workspace of a coalmine. So, if you've got a coalmine to excavate during these times of uncertain energy supply, the short- D-handled shovel or spade may be just the tool for you!

Misusing the Shovel

In the process of texturizing the soil, the soil's surface will be raised naturally because of the incorporation of some air and organic matter into it. The raised-bed look of the cultivated area is a byproduct of the soil-improving techniques—not a single-minded goal. Conversely, heaping a bunch of dirt to make the shape of a raised bed may not produce the same results as following intelligent guidelines for soil improvement.

For example, I tried to warn some interns at the Rural Center about tilling soil —especially clayey soil—when it's too wet. Some students, however, insisted on the early-spring cultivation of some raised beds, and managed to churn enough air and compost into the wet clay to form a bed some eight to twelve inches higher than the paths.

By the end of the crop cycle the bed's surface was actually inches *lower* than the surrounding paths. Their heap, artificially pumped up with a lot of cold air, had destroyed the clay's plate-like structure by tilling it when it was too wet. Yes, sometimes the mere heaping or boxing of a good topsoil will provide enough extra drainage for a healthier and more productive crop, but take note: an improperly heaped-up raised bed may look like a GROW BIOINTENSIVE garden, but, speaking ecologically, it's only a distant imitation.

Raised beds in boxes will often show a net gain over old-fashioned rototilled dirt gardens, so I'm happy to see even the bastardized versions of raised and double-dug beds proliferate around the country. I often prefer to use raised-bed boxes when planting vegetables in edible-landscape designs, appreciating them for their tidiness, the gopher-excluding wire mesh across their bottoms, the enhanced drainage, their convenience, and their permanence. I don't pretend, however, that I've duplicated the delicate ecological dynamics of a GROW BIOINTENSIVE® garden.

Pluses and Minuses of Various Raised Beds

The simplest and most inexpensive way to make raised vegetable beds is to rake soil up into a mounded shape. At the risk of repeating myself, this type of "bed" should in no way be considered a French Intensive-Biodynamic raised bed. Many gardeners prefer to construct formal raised beds with masonry, cinder blocks, brick, horse troughs, corrugated metal, plastic "wood", or actual wooden sides. All of these containers retain the soil while imparting order and efficiency to the kitchen garden and helping to define it. Formal raised beds one to two feet tall also provide a number of other advantages:

1. Enhanced drainage for better growth.
2. Soil which warms slightly sooner in the spring.
3. Containment that helps to keep soil or mulch from spilling into paths.
4. Pathways between beds can be cleaned and tended without disturbing plantings.
5. A structure that allows wire mesh to be added to the bottom of a bed to exclude tunneling pests such as moles and gophers.

Figure 70. A "D"-handled digging fork goes to work. A poster from the Victory Garden days of World War II. I have a digging fork but do not use it to turn over the soil, just to crack it open by rocking it back-and-forth.

There's always an exception. In this case, it's carrots, which prefer a deeper soil with good tilth and drainage, and free of rocks and other obstructions that can produce deformities. In less than ideal soils, you need to cultivate as deeply as possible before seeding them. Carrots are best grown in double-dug beds (with vigilant trapping of gophers), or in wire-bottomed boxes with sides that rise at least 24 inches above the soil line. A box 24 inches deep can hold at least 80 to 90 percent of most vegetables' root systems.

If your area is home to gophers or voles, you'll need to build garden boxes with half-inch aviary wire or half-inch hardware cloth (it comes in four-foot-wide rolls) stretched across the bottom. One-inch chicken wire, while less expensive, may allow baby gophers to sneak into the box. The aviary wire also contains more galvanized metal than chicken wire and thus lasts longer in the ground. Even with this protective barrier in place, any roots that grow through the wire at the bottom of the bed will be eaten. The Cadillac (or Tesla!) bed-bottom version is the use of half-inch hardware cloth, because it doesn't rust through as quickly. When preparing beds for planting, be sure to work the soil with a flat-bottomed spade so you don't damage the wire.

Raised beds also have certain drawbacks:

1. A formal raised bed with solid sides will cost more and require more time and effort to construct than a simple mounded bed.
2. Some gardeners find the angular geometry of the raised-bed structures unaesthetic. (Beds can, however, be masked by perennial plantings or low evergreen hedges—see the last chapter in this book.)
3. When watering a raised bed with drip irrigation, extra effort must be made to bring the water supply into the bed unobtrusively.

There are many materials that can be used to construct a raised bed. Each has its own unique mixture of attributes and imperfections:

Used Bricks (Recycled)
Pluses:
1. Very good-looking; construction has character when first built.
2. Since bricks are fairly narrow, the finished wall doesn't take up excess garden space.
3. Wire mesh is easily added to the bottom.
4. Uses recycled materials.

Minuses:
1. Can be expensive if the bricks are purchased new, rather than scavenged.
2. Requires some masonry skill or practice to construct.
3. In areas where the ground freezes, requires a poured-concrete base.

Cinder Blocks
Pluses:
1. Wide enough to sit on while gardening.
2. If not mortared, easier to work with than brick.
3. Can be built on a base of packed gravel.
4. It's easy to add a wire-mesh bottom.

Minuses:
1. Not easily found as recycled material.
2. Best used only for square and rectangular beds.
3. Thick blocks take up garden space.
4. Looks gray and "industrial," like concrete.
5. The large holes in the blocks, usually placed facing up, tend to fill with mulch and soil from the bed.
6. Unless mortar and a concrete base are used, the blocks will often settle in a slightly skewed position.

Plastic "Wood"
Pluses:
1. Easy to work with, like lumber. No splinters.
2. Made from recycled consumer plastic waste.
3. Has a long life and won't rot.
4. Easy to add a wire-mesh bottom.
5. Easier construction of odd-angled shapes such as octagons or pentagons.

Minuses:
1. Up close, looks very fake.
2. *Expensive* compared to lumber.
3. For sturdy attachment, requires screws placed close together instead of nails.
4. Needs more frequent support along the lengths of the "boards".

Recycled, Untreated Rot-Resistant Lumber
The use of the two main rot-resistant lumbers available, redwood and cedar, is controversial due to environmental issues. They are, however, still the easiest material for building raised beds. (Use of recycled wood lessens the environmental impact.)
Pluses:
1. Attractive natural surface.
2. Available as recycled material.
3. Easy to work with; no mask, gloves, or goggles required when using hand tools.
4. Easy to add wire-mesh bottom.
5. Can be used to construct odd-angled shapes.

Minuses:
1. If in contact with the soil, may have to be replaced within five to15 (plus or minus) years. (Note: redwood heartwood can last this long, but white redwood sapwood should be

avoided, as it will usually rot within four to five years.)

Cinder Blocks, the Gopher Solution
Here in Northern California, I live with "ground rats"—gophers. Raised beds with wire bottoms are the only effective way for gardeners to coexist with pocket gophers.

An 18-inch-tall by 10-foot-long raised bed made out of cinder blocks will initially cost about three times as much as a redwood box with the same dimensions, but will last virtually forever. (See the illustration of cinder-block bed construction.)

Sonoma County Cinder-Block Recipe:
• Lay down hardware cloth to cover an area slightly larger than the planned outside dimensions of your box, There should be several inches protruding on the ground outside the cinder blocks when you construct the box.
• Set the blocks in place with the holes facing up. For the second row, stagger the blocks by one-half. Higher is better to protect your plants from relentless gophers, and two to three stacked layers of cinder block are ideal.

• Fill the bed and the top holes of the cinder blocks with a composted soil mix to kick-start your garden.

• Install three or four lengths of half-inch in-line drip-irrigation tubing along the length of the bed.

• Install quarter-inch tubing along the top of the blocks to irrigate the soil pockets, which are a good place for useful herbs, or leave the top unirrigated to use as a sitting bench.

Salvaged Broken Concrete (Urbancrete)
Urbancrete construction is the reuse of broken pieces of concrete, such as jackhammered sidewalks.
Pluses:

- It's usually free.
- Once it gets covered with moss, it can be very attractive.
- It's easy to stack for retaining walls or raised beds.
- It makes attractive patios; an herbal ground

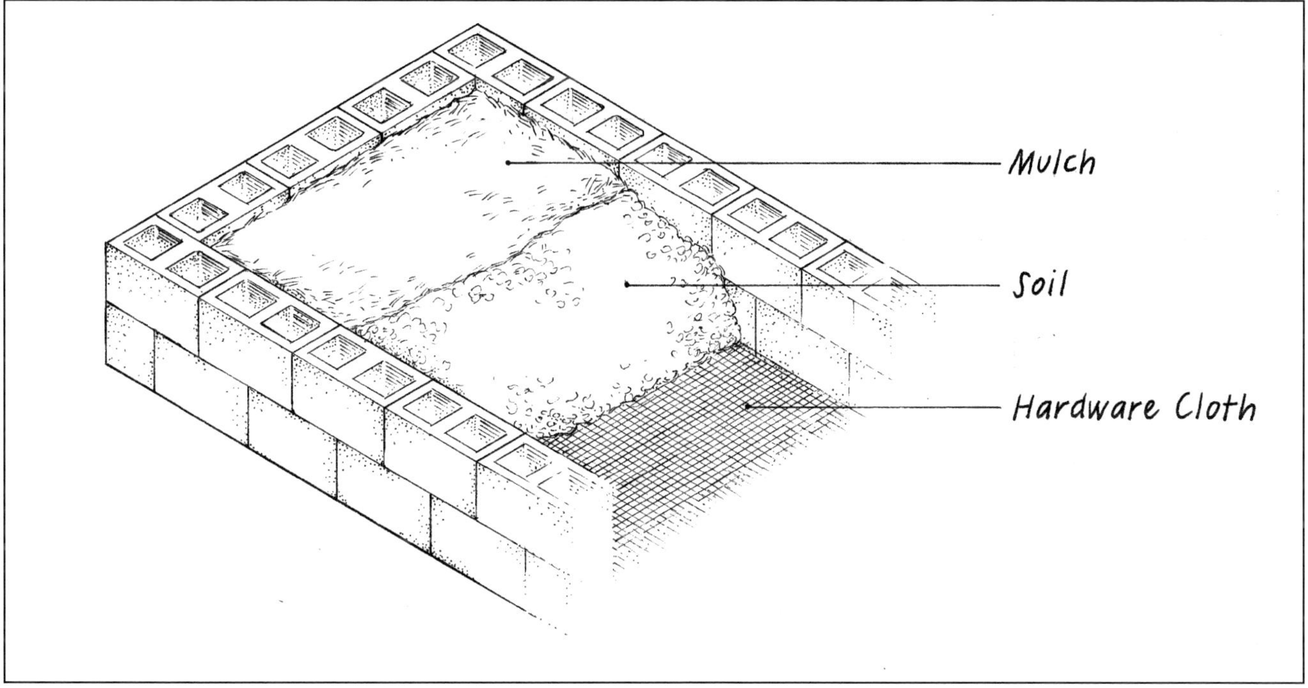

Figure 71. This is a good way to create a permanent raised bed without using wood that rots. The 1/2-inch-wire hardware cloth will keep out gophers. You can also layer lots of compostable goodies with sheet mulching or "lasagna gardening," just be sure to heap the layers well above the top of the cinderblocks, as they will settle considerably. (The soil shown here has already settled.)

cover like thyme can be planted between the pieces.

- It's basically permanent.

Minuses:

- Heavy and hard on the back if the pieces are too big. May require costly massage therapy or physical therapy following construction.
- Takes a long time to get covered in moss. (Try mixing dried moss in buttermilk and apply the
- solution after every rain.)

Bastard Trenching

In my 1986 book, *Designing and Maintaining Your Edible Landscape—Naturally*, bastard trenching seemed like such an elegant and easy way to dispose of household kitchen waste.
From <mytinyplot.com>: "Bastard trenching is basically like a conveyor belt of soil. You mark out an oblong of land (or bed) into small squares (you can do this mentally if you like). You dig out the first square to a depth of one spit (spade depth). You dump this in your wheelbarrow and take it to just beyond the end of your oblong and empty it. Then you proceed to fork up the bottom of the pit that you've just made, add some manure and dig it in well. Then you move to the next square and do the same again. But instead of putting the removed soil in the wheelbarrow you dump it in the first square.

"This is genius – why? Because all the soil stays in the right place. The topsoil stays on top, the sub-soil is cultivated and manure is added in the right place. Then everything is put back in the right order like a giant jigsaw."

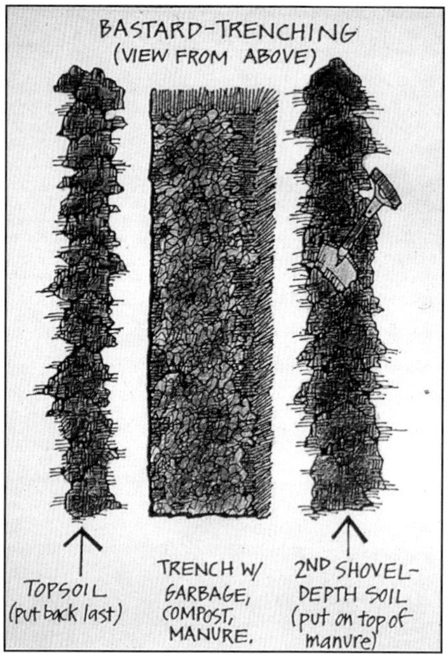

Figure 72. When bastard trenching, make sure your trench is at least 18 inches deep (or deeper) to keep out raccoons and rats.

Now I'm not so sure about bastard trenching. If I had it to do all over again (and if I didn't have my worm bin), I'd make sure that the upper soil layer stayed at the same "elevation" as the original soil line, with the lower zone thrown forward to cover the kitchen scraps. All this is irrelevant if raccoons, dogs, or skunks keep digging up the scraps, in which case you need to bury them deeply enough to outwit the critters, or use an enclosed compost bin or worm bin with a locked lid.

Deep Mulch, No-Till

Tillage, according to Professor of Biology Tyler Volk of New York University, is a large contributor to the surplus of atmospheric carbon dioxide and thus to global warming. Unfortunately, even if every gardener in America were to switch to no-till gardening, it probably wouldn't offset the gases produced by commercial cultivation, but the concept is still important.

Our planet's soil contributes ten times more carbon dioxide to the atmosphere than all of humankind's activity. That's 60 *billion* tons of carbon dioxide per year, folks. This odorless gas is produced by the earthy lifestyle of the myriad of creatures that inhabit the soil—microbes, bugs, worms, fungi, and algae—as they breathe, "pass gas," and expire. Paul Kaiser is a no-till agriculture proselyte who farms at Singing Frogs Farm in Sebastopol, CA. "The Department of Agriculture says that 31 percent of greenhouse gases come from agriculture," he said. "Tilling releases nitrogen and other gases into the air, and the organic compounds of the soil are destroyed by it.

When soil is stripped of its living cover in order to grow crops, up to one-fourth of its carbon is released, since the carbon dioxide and most of the carbon in the surface layer is seldom replaced. The warming of the bare soil in the sunlight also speeds up the activity and death of soil microorganisms, thus increasing the process of outgassing.

The depletion of the supply of organic matter by cultivation is well illustrated by the report of a 60-year-long study made by Hans Jenny of the Missouri Agriculture Experimental Station in central Missouri, in which an undisturbed virgin prairie soil was compared to an adjoining field cropped to corn, wheat, and oats without the addition of manure or fertilizer. "No erosion took place, yet 38 percent of

Figure 73. Here Chester Aaron adds more rice straw to his garlic beds. Piled 12 or more inches high at first, the straw will settle down to four to six inches deep within a few months. Chester uses rice straw because there's a grower of organic rice nearby, and the straw contains no weed seeds.

the organic matter represented by the virgin soil had been lost during that period because of cultivation." (From the 1938 USDA Handbook *Soils & Men*.) The depletion of organic matter by cultivation was known all the way back in 1938! Yet "proper" tillage was the focus of the *Soils & Men* handbook. (I suppose that either there weren't too many women farmers back then, or sexism ruled. You guess.)

In your own very tiny way, you can minimize the loss of carbon dioxide and depletion of organic matter from your garden by experimenting with no-till methods.

Ruth Stout's No-Till Method

Ruth Stout (1884-1980) was known in America as the guru, or grandmother, of no-till gardening, using deep straw mulch. She published nine books, filled with many anecdotal musings, on the subject.

Ruth, however, had a little-known secret. She had actually stoutly tilled and manured her garden plot for 14 years before switching to her till-free technique, which involved applying a deep mulch of spoiled hay or straw. (See *The Ruth Stout No-Work Garden Book*; Rodale Press Publishers, 1971.) On page 24 she writes: "..For about 12 years, (I did) about everything experienced growers told me. ....(my plot) was plowed each year." On page 117, "For the first 14 of my gardening years I covered my asparagus with manure, cultivated it each spring..."

So go ahead and till your soil—even double-dig—if the ground is clayey, has a hardpan barrier near the surface, or in any situation where a few more years or seasons of dig-and-till will contribute to conditioning your soil to require less frequent tillage while maintaining good yields. There are some advantages to tilling for a few or more years; it speeds up the process of getting to a no-till regime.
Nowadays, there are plenty of advocates for no-till gardening. Many say it mimics nature's way of improving soil from the top down, while others maintain that they're relying on earthworms to do the "tillage" at the speed of nature.

An interesting fact is that artificial tillage can actually impede the storage of moisture in the soil. Research by Dennis Linden, soil scientist at the USDA Agricultural Research Station in St. Paul, MN, found that the common earthworm (*Aporrectodea tuberculata*), working under tilled soil, forms a "less desirable pattern (of) meandering tunnels" which slows the absorption of moisture by the soil. Worms under plots with no tillage, however (in this study, managed with herbicides), formed "a network of vertical burrows, due to night crawlers, and depressions, due to common earthworms, (that) tends to funnel water rapidly downward through (and into) soil layers." (*ARS Quarterly Report*; Oct. 1 to Dec. 31, 1997.)

Linden cares for his own garden with layers of organic mulch and compost made from his household food waste. According to him, using no-till methods "means that the soil absorbs water faster and encourages keeping water in the soil as a reservoir. This stored moisture in the soil provides another measure of drought resistance...."

In a thesis on mulch ("The Effect Of Mulch Type And Thickness On The Soil Surface Evaporation Rate," by Michael McMillen, June 2013, Horticulture and Crop Science Department, California Polytechnic State University, San Luis Obispo, California), it was determined that if the mulch you're using is loose (like wood chips and arbor chips), layering it more than four inches deep isn't worth the expense in regards to water savings—each additional inch accounts for proportionally less water conservation.

(For weed control, however, a depth of six inches or more of arbor chips is appropriate.)

In another study, it was found that "For both [pine and oak], stem growth was greater with the three-inch depth of mulch [chipped pine and shredded hardwood chips] than with other mulch depths [six and 10 inches] or the control."
("The Effect Of Wood Mulch Type And Depth On Weed And Tree Growth And Certain Soil Parameters," September 1995; Katrina M. Greenly and Donald A. Rakow, *Journal of Arboriculture*; 21(5))

As an aside, it was found that "... mulching with high C:N [carbon to nitrogen] ratio (125:1) shredded wood pallet chips stimulated microbial growth while adding little nitrogen to the soil. These results are consistent with the hypothesis that soil microbes out-compete plants for nitrogen, and that addition of organic matter with high C:N ratios can induce nutrient deficiencies in plants by stimulating microbial growth." ("Effects of Organic Mulches and Fertilization on Soil Microbial Activity, Nutrient Availability, and Growth of River Birch," 2002. *International Society of Arboriculture Research*; Daniel A. Herms, John E. Lloyd, and Benjamin R. Stinner.) Daniel Herms, The Davey Tree Expert Company, Kent, OH (330) 673-9511

Wisdom from Barbara Fick

Barb Fick, former community horticulturist with the Benton County office of the Oregon State University Extension Service, says she's maintained her large vegetable garden by the no-dig method for decades. Not only does it use less water, she says, it discourages weeds, improves the soil and reduces labor for tilling and weeding.

Fick's explanation (as described by David Richards): "The crux of no-till gardening is to pile on enough mulch to keep weeds from germinating and growing up through it. If you use leaves, grass clippings or straw, you might need as much as eight to 10 inches of mulch. If you start out on the bare ground with something more solid, like cardboard or 10 sheets of newspaper, you'll need fewer inches of leaves or grass clippings on top."

NOTE: Everything in gardening is a double-sided coin. According to Linda Chalker-Scott, Ph.D., Extension Horticulturist and Associate Professor, Puyallup Research and Extension Center, Washington State University, in her blog "*The Myth of Paper-Based Sheet Mulch,*" the flip side of newspaper mulching is:

- When wet, the layers of paper can create an impermeable barrier to water and air exchange.
- Garden pests, moles and gophers can hide under newspaper and cardboard sheet mulches. In Fick's experience, termites preferred cardboard over wood chips. (A friend of mine finds gopher snakes hiding under big sheets of cardboard mulch each spring. Must be the free lunch.)
- If the paper is allowed to dry out, irrigation water may sheet away rather than seep through the layers, especially in arid climates or well-drained soils. (I put my drip-irrigation system down first and then cover it with newspaper or cardboard.)
- Deep organic mulches work better than cardboard and newspaper to prevent weed growth.

Fick suggests that newspaper and cardboard sheet mulches can be effective for annual beds if they're properly maintained. She says that she piles on aged mint straw (compost) in the fall and leaves it on as mulch the following spring and summer.

"Whatever you use, don't skimp on mulch," is her advice.

A heavy layer of mulch not only keeps weeds from growing, it also keeps the underlying soil moist, greatly reducing the amount of watering you need in the summer.

Another advantage to "no-till" is that you don't turn over most of the soil in your garden, which prevents you from bringing old weed seeds up to the surface where the conditions are good for them to germinate; this reduces the weed population even more.

To establish a new no-till garden, start in the fall or winter by outlining your beds with landscaper's spray. *Don't* use Roundup™ , Ortho™ , or other toxic sprays; go with sprays made with vinegar, fatty acids or essential oils; these are non-toxic and wash away with a little rain. You can also use a garden hose or rope to delineate edges and curved borders. Give this step plenty of thought and planning—if you keep this configuration permanently, or at least for a long time, the paths between your beds will be compacted by foot traffic, making them inhospitable to weeds.

Here's some good news: since you won't be tilling your garden, you don't need to confine your garden design to straight lines and rectangles; curving lines are often more pleasing than straight ones. You might even want to incorporate some vegetable beds into your existing non-vegetable garden.

In any case, be sure to lay the vegetable beds out so that you can easily reach any part of the bed from a nearby path while kneeling, as it's important to avoid stepping into the bed and compacting the soil.

If you put your new no-till garden into an existing lawn area and want the paths to remain as lawn, don't forget to make them wide enough for your lawnmower.

When you're ready to plant in the spring, just push the mulch layer aside in the spots where you want to put your seeds or starts. For a little boost in the first year or so, you may need to dig out old roots and add a little topsoil or compost in the planting holes.

Over time, the mulch layers you keep adding will form enough soil to support your garden, with the added benefit that the soil formed by the addition of so much organic matter will likely be loose, full of earthworms and teeming with the healthy microbes that make nutrients available to your plants.

When one of your crops is finished, just incorporate the dead vegetation into the existing mulch. Try to avoid letting the plants go to seed, unless you want "wandering" vegetables. As an example, I let my arugula go to seed and it sprouts randomly around the garden. Less work.

The best way to water a no-till garden is with drip irrigation, further cutting water waste.

If the prospect of a vegetable garden blanketed under huge mounds of mulch doesn't fit your vision, Fick recommends that you remember what the usual reality is out in your garden – "soil baking in the sun and plants pathetically turning gray as they head toward dehydration."

In a 2005 paper, *The Importance of Soil Organic Matter; The Key to Drought-Resistant Soil and Sustained Food and Production*, Alexandra Bot

Figure 74. Damp newspaper mulch shown before holes were poked through to plant garlic cloves. Slick colored paper should be avoided, but other newsprint colors and black inks are now soy-based and safe.

(FAO Consultant) and José Benites (FAO Land and Plant Nutrition Management Service) of the Food and Agriculture Organization of the United Nations in Rome found "a 200–300-percent increase in population size of root nodule bacteria on legumes in a zero-tillage system, as compared with conventional tillage." They also found a five- to ten-fold increase in population sizes of the same bacteria, when compared with cropping systems without soybean.

Tilling greatly affects nitrogen in the soil. Too much tillage causes the nitrogen to convert to a gas and escape into the air. (Most of the soil's nitrogen, as the following chart indicates, is in the upper layers of the soil where tillage reaches.) The top six inches of topsoil, well within the reach of a rototiller, contain 52% of its nitrogen. (Variations in microbial community composition through two soil depth profiles, 2003; Noah Fierera, Joshua P. Schimela, Patricia A. Holden; *Soil Biology & Biochemistry* 35 pp 167–176.)

Sheet Composting

Sheet composting (now often whimsically referred to as "lasagna gardening") offers an easy way to improve the garden's soil without strenuous digging. The technique is akin to Ruth Stout's deep-mulching practices, and consists of the use of thin layers of compostable material laid out over the soil like a thick mulch. By layering high-carbon wastes with nitrogenous plant refuse, you essentially construct a thin, wide compost pile.

Use a blend of dry brown leaves or woody stems (chipped, chopped, or not); fresh grass clippings; green-manure crops such as buckwheat, vetch, bell beans and clover; wet kitchen garbage or scraps (no meat scraps, as they'll tend to attract hungry animals and/or smell bad), and green weeds from the garden (before their seeds ripen).

Experiment with a ratio of one part green matter to three or four parts dry dead matter. As you layer, water the dry materials well. At each planting location, make a large planting pocket in the native soil to receive the roots. NOTE: sheet composting *doesn't* generate enough heat to kill weed seeds, diseases, or pathogens, so after planting, water everything again, then cover the area around the plants with five to ten sheets of newspaper followed by a weed-free mulch; this should take care of the nasties.

Sheet composting slowly helps improve soils from the top down, while allowing the gardener to quickly dispose of large quantities of compostable materials and avoid unnatural tillage. The sheet-composted area will require more irrigation the first year or two as the roots grow into the uncultivated soil.

Cardboard is not needed on top of grass when using sheet mulching to kill a lawn. When the layers above the grass are two- to four-inches thick, it's thick enough to suppress and rot the grass.

Again, please note there are limits to using newspaper and cardboard. See the previous comments by Linda Chalker-Scott.

To continue with no-till methods in a new planter box, layer compost and compostable materials as needed. Be sure to include a layer of B&W newspaper and/or cardboard (with all plastic packing tape removed) on top to thwart the sprouting of stray seeds in the compost or soil mix. Add a weed-free mulch to hold down and hide the newspaper.

Straw-Bale Gardening

Growing annual vegetables or ornamental annuals on straw bales is a way to rapidly add lots of compostable organic matter to the soil's surface in a short period of time. I first tried it in 1978 and it worked fine—very friable soil after two seasons. This is the *single* best technique to quickly develop good soil tilth and fertility. Here's the recipe:

1- Place six or more bales of straw in a cluster. Place the bales with the end grain facing up and leave the strings or wires on. The more bales the better, as this will create less surface area on the perimeter for evaporation.

2- Thoroughly soak the cluster of bales. (Takes a *lot* of water if you're starting with dry bales. Or, wait until the bales are rained on.)

3- Layer several inches of fresh manure, such as chicken manure, on the top (the exposed end grain) of the bales.

4- Add four or more inches of soil to cap off all the bales. (All right, I know this requires some digging. But it's only a one-time effort to start off a remarkable process. You can also use a fertile potting soil.)

5- Water all the layers again.

6- Plant and mulch with plenty of loose straw mulch on top—six inches or more.

7- Keep the plants and bales moist. If it doesn't rain, use drip irrigation to keep the bales moist.

Et voilá! The bales will grow plants above the ground where no gophers can get to them—at least the first year! Then you can go back to sheet composting, mulching, tillage, and/or layering with finished compost.

Surface Cultivation Yields

Advocates of surface cultivation methods are enthusiastic, but few produce actual numbers about yields to guide aspiring no-till gardeners, so here are some statistics for the results of surface cultivation: These measured yields were taken from articles written by a man with the delightful name of Ben Easy, in the Jan. 1952/Oct.1954 issues of the English publication *Mother Earth*.

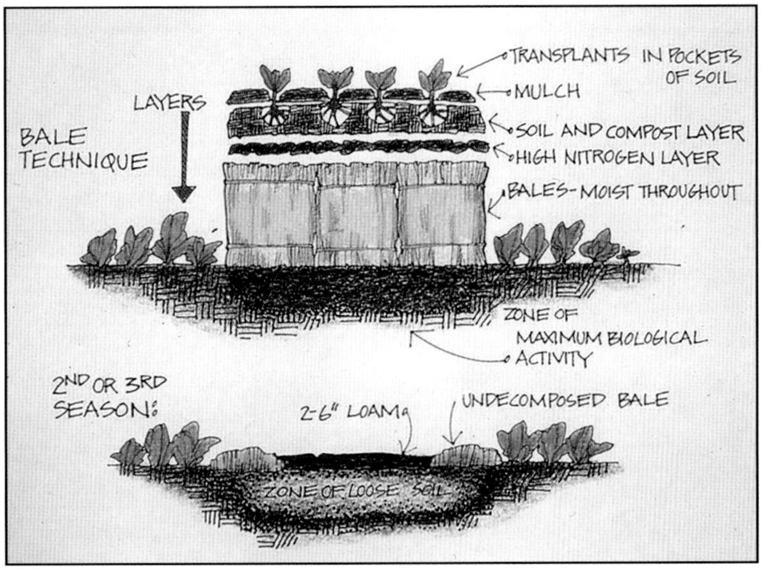

Figure 75. Straw-bale gardens can even be grown on concrete—I first saw them done that way. In England, they're often used in greenhouses, and are easily removed to scrub down a concrete floor.

Easy quotes an informal seven-year study done by J. H. L. Chase in England, a comparison of yields between a single-dug plot and a surface-cultivated plot. As Figure # 76 (full citation from *Designing and Maintaining Your Edible Landscape—Naturally* page 109) reveals, some surface-cultivated crops such as cabbages and cucumbers actually out-performed cultivated crops in terms of both the highest yields and the average yield. In the first year of the study, the surface-cultivated plot actually produced a two percent greater yield than the cultivated plot.

You'll notice that over time, the total yields from the surface-cultivated plot in the study drop to only 70% of those of the "control" plot that's tilled once each year. This simply means you have the choice between vertical or horizontal gardening ecology. You can always add more nutrients via single digging to restart the surface-cultivation process, or spread out your planting area to compensate for the reduced yields.

Composting

Seen many cookin' compost piles lately? While plain old heaps of garden refuse are hidden away in the dark corners of many gardens, carefully constructed hot and active compost piles (the kind that were made as if with a recipe, achieved temperatures of 140^{0}F, and decomposed even the starchiest corn stalk within four to six weeks) are a rare garden fixture nowadays. Active composting has slipped rapidly into obscurity, as our lives swept into the 21st century and mortgages and student debt approach amounts we thought only the national debt could achieve. Some ardent proponents of composting have survived in the gardening backwaters of the times, but many former "Captains of Compost" have simply let their active piles dry up.

I propose keeping a "cold" (mesophilic, or room-temperature) compost pile. That is (if space allows), make a heap hidden somewhere in the background of the garden—out of view and not too far out of mind. If you want to save your garden waste instead of putting it in the garbage company's green-waste bin, simple pile it up and let it rot over years. If this is too slow for you, or you haven't the space, buy finished, high-quality compost. While it may cost up to $54 per cubic yard (as in my area), you get a jump on the garden, have the satisfaction of using recycled green waste, and don't risk stressing your back.

Commercial bulk compost is often turned or churned, but because of the surface area required in the composting yard to do so (and sometimes because the demand is so high) it's usually not cost-effective to properly turn *and* age the compost. As a result, commercial compost can be delivered to you in an unfinished state, and you may discover that the load you've purchased is still hot and steamy. Such compost should be allowed to "mellow" until it has a dark, loamy feel. Cover it with a tarp to hide its distinctive smell and keep rain off. If you have the space, let it digest and mature for as much as a year or more.

If you don't have the space, don't feel guilty—just buy a few sacks of aged compost. *Don't* get the

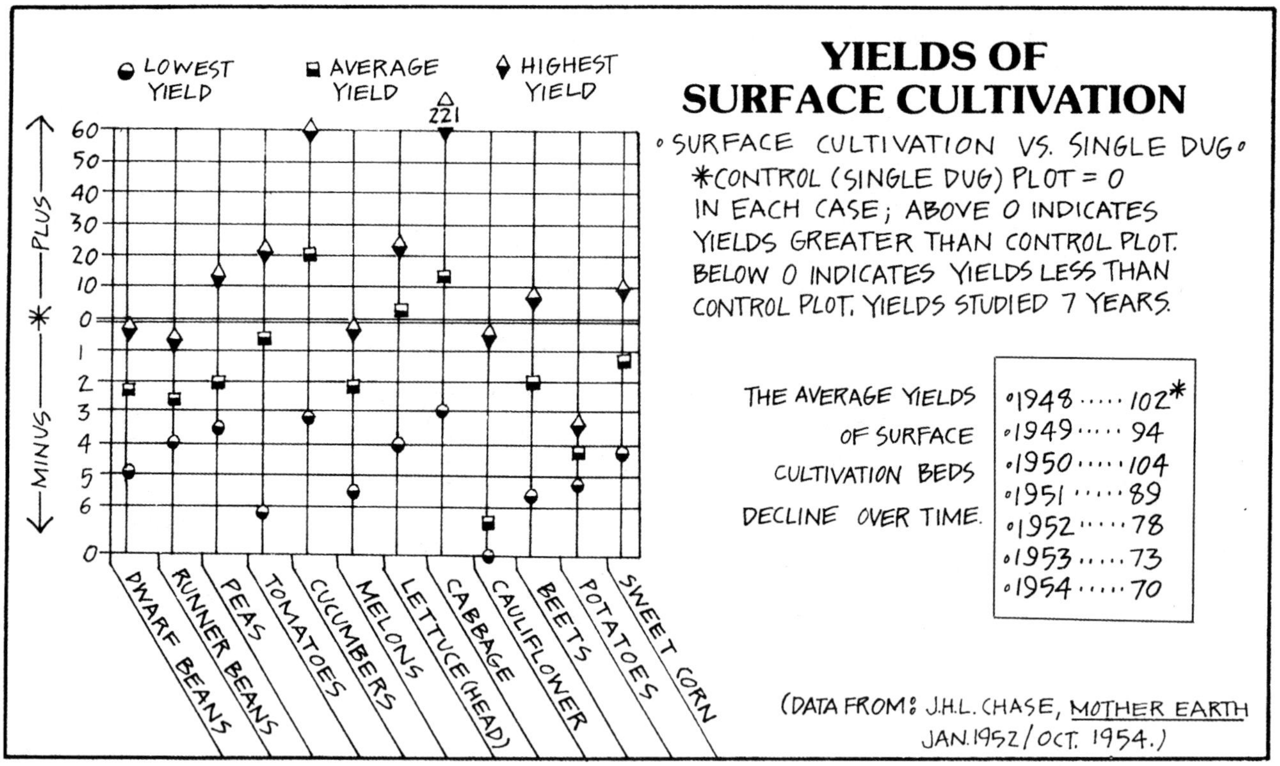

Figure 76. The yields produced by surface cultivation can be slightly better than those of cultivated gardens, but go down over time. To maintain optimum fertility, I recommend double digging or cultivating every six or seven years.

imposter "composts" made with mostly redwood or fir chips (often called "nitrolized wood shavings"), as there's usually not enough nitrogen in the mix to fully digest the wood. This stuff will rob nitrogen from your garden soil, and steal even more from a container planting.

The Scourge of Compost

It turns out that the methods of intensive composters have, ironically, encouraged an evil follower called *symphylans* (*Scutigerella immaculata*), which are about 1/4-inch long and look like very fast-moving, tiny white centipedes. They thrive in tilled soils with large amounts of organic matter and regular moisture levels, the same conditions often created in organic vegetable gardens. The symphylans eat the root hairs of many vegetable plants, stunting or even killing the plant, and they're nearly impossible to banish by any organic method(s). They are, however, easily controlled by adopting "no-till" methods.

Some organic gardeners and farmers think that symphylans become a problem when unfinished compost is added to the soil in great quantities. In some places, avid composters are seeing more and more of this horrible pest, although in other places with well-managed compost, the symphylans still invade. When they're identified as the cause of wholesale death of plants, panic ensues. One way to test for these critters is to put pieces of cut potato on the soil at night. In the morning, lift the potato pieces up and look for the centipede-like symphylans. Once you're able to see the damage of stunted, wilted and dead plants, it's too late to do anything but replant or put the area into a fallow where no food crops are grown.

Research has pointed to some possible ways to deal with symphylans, but there is no "silver bullet." No organic pesticide has been shown to be worthwhile in controlling these little beasts that can burrow seven to ten feet down into the soil. There are a number of organic-control options promoted in addition to or in place of a fallow period. 'Wheeler' rye; 'Monida' spring oats and 'Micah' barley, all used as a winter cover crop, have been effective at reducing symphylans in some studies. ("Aggressive" soil tillage to disrupt the pathways they use, and packing the soil after transplanting to collapse their tunnels, though procedures advocated by some, are not particularly effective, especially with symphylans levels that are high enough to prevent seedlings from forming roots.)

What works best to get rid of this pest is adopting nature's method of creating a healthy soil with good structure and a balance of micro and macro-organism predators; in other words, "no-till." Symphylans are found in forest systems and prairies throughout the continental US, Hawaii, and the United Kingdom. You have to ask yourself, why aren't they killing the forests or grasslands? The simple answer is that nature for the most part doesn't till, (s)he mulches.

Tilling disturbs the permanent underground burrows that predatory insects and organisms use to find and capture prey. Many pests reproduce much faster than predators, so when you till the soil you favor the pests—in this case, symphylans. Simply stop tilling and mulch with some good compost. What was barren soil due to severe symphylans feeding on plant roots will soon be a verdant paradise—and think of all the hard work you're avoiding as well.

Figure 77. A photo from the 1970s when hot compost piles were popular. In this case, two three-foot bins, each containing a cubic yard of organic materials, were turned periodically to make hot thermophilic compost.

Figure 78. A compost pile made by the gardeners at the Occidental Arts & Ecology Center in Sonoma County, showing various layers of green foliage, brown waste, and manures. It takes practice and skill to make the vertical sides.

To further discourage symphylans, keep the organic matter of the soil at around six percent or lower. (You can check this with a soil test and lab report). Inspect all potted plants, compost, and soil you bring into the garden. All this may help keep things under control, but you most probably will always have some symphylans.

When Compost Goes Bad

Compost can go awry for another obtuse reason. Tim Dundon, aka "Zeke the Sheik" (the "compost king"), of Los Angeles County, made and accumulated compost over a 17-year period, accumulating a pile 40 feet tall and 200 feet long. Twice in one month the heap spontaneously combusted, and the fire department arrived to douse the flames and smoke. While Zeke defends composting as "the key of energy that will eventually set mankind free from misery and gravity," I'd say his pile shows what happens when a zealous intellect elevates a natural process to unnatural proportions, turning a beneficial process into a hazard.

The Carbon Dirt Print

Carbon dioxide is one of the greenhouse gases. While not as harmful as methane, is does linger in the atmosphere and help cause global warming. One source of this gas is tillage. In part because the organic matter "combusts." Like an invisible flame, the organic matter oxidizes—dehydrates and desiccates. And the soil's bacterial life releases carbon dioxide when they die due to tillage. Two more reasons to avoid tillage.

Adding compost and the roots of cover crops helps tie up carbon in the soil. And a deep mulch keeps the soil slightly cooler so "combusting" of the organic matter near the surface of the soil is slightly lessened.

Figure 80 shows how much carbon dioxide escapes from the soil with four weed barriers compared to bare soil. This chart shows carboard slows the release of carbon dioxide compared to woodchips.

Since temperate soils store so much biomass, they also store more carbon than a tropical ecosystem—21 percent more. "Average carbon storage by agroforestry practices has been estimated as 9, 21, 50, and 63 Mg [one Megagram = 1000 kilograms] of Carbon per hectare in semiarid, sub humid, humid, and temperate regions." (*Carbon sequestration: An underexploited environmental benefit of agroforestry systems* F. Montagninil and P. K. R. Nair, Yale University, School of Forestry and Environmental Studies.)

The Traveling Carbon Footprint

In my area (Sonoma County, CA), the carbon footprint of a landscaper installing a "sustainable" landscape is considerable. Here's a total of just the travel time to get supplies for a (theoretical) job 15 miles from the landscape contractor's office (they live on the outskirts of town); it includes visits to all of the sustainable/organic suppliers in the county.

Mileage

- Round trip to the job site 12 times or more 180
- Pick up one truck load of organic compost 20
- Pick up one load of recycled or Trex™ timbers for a retaining wall and hardware 20
- Pick up organic plants from two suppliers and organic fertilizer 40
- Pick up stone for landscape feature 25
- Pick up drip irrigation supplies and timer 20
- Misc. hardware trips 25
- To CAD print-out shop 30
- All over creation for fish and fish-pond supplies 30

TOTAL 390 miles

And then there's the travel for the maintenance gardener, forever. Where's the sustainable in that? It's certainly a large sooty footprint.

Centralized companies that carry all sustainable-landscaping products are gradually becoming available. I can now go to just one store to get drip-irrigation and sprinkler hardware, organic fertilizers, vegetable and cover-crop seeds, sacks of mulch, potting soil and bagged compost, organic vegetable starts, organically-grown perennial landscape plants, bare-root ornamental and fruit trees in season, and supplies for rainwater catchment. My footprint is shrinking.

Bonus Chart and Photos:

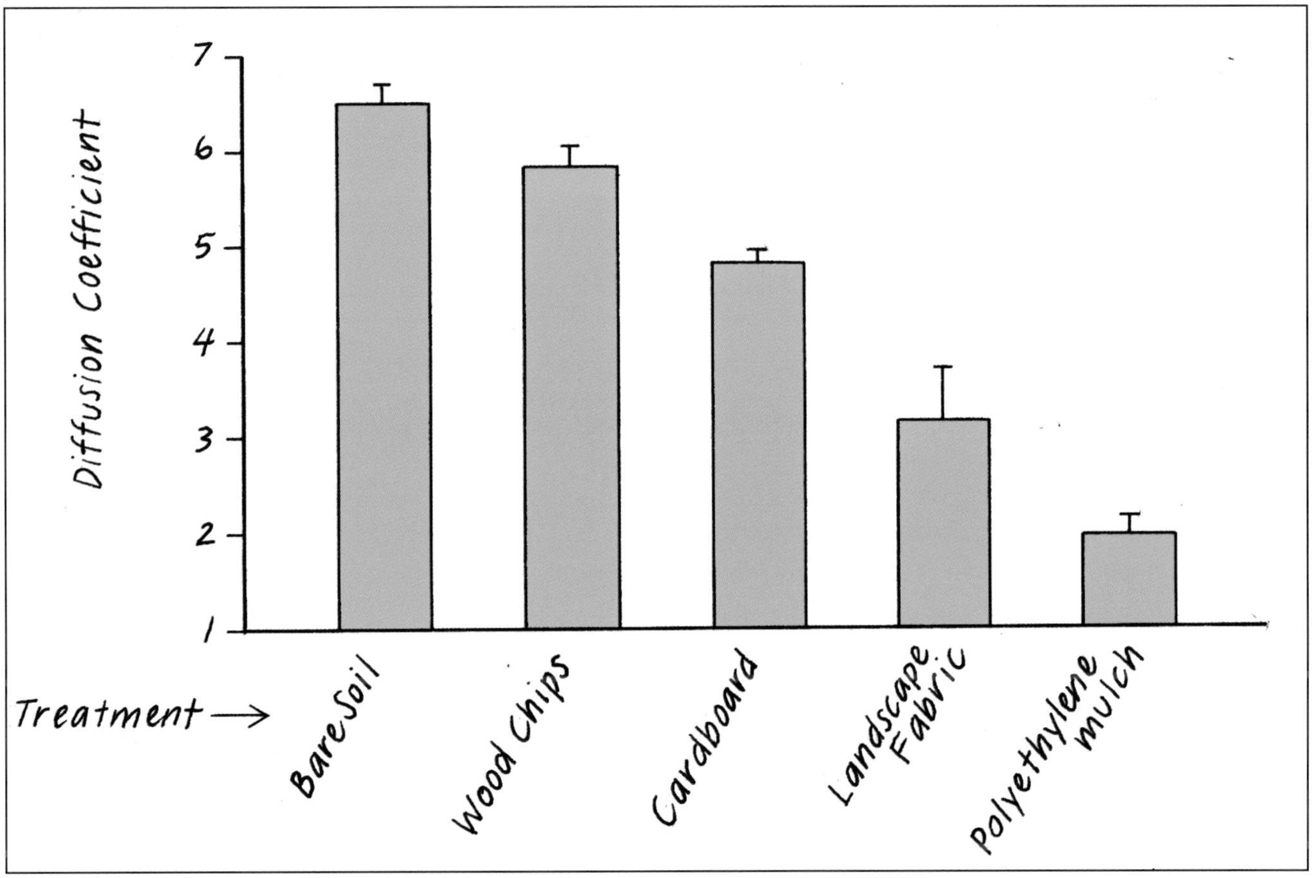

Figure 79. This comparison of how much carbon dioxide escapes (diffuses from) the soil with various mulches illustrates how cardboard blocks more carbon dioxide than wood chips, but less than plastic mulch. (The greater the number, the greater the outgassing.) It also shows the general migration of all gases to and from the soil, and reveals how greatly cardboard reduces the exchange of gases when compared to a mulch of wood chips. To keep the soil life "happy," use wood chips instead of cardboard. To block less gas, choose cardboard over landscape fabric. Wood chips are better than bare soil.

(From: Carbon dioxide and oxygen exchange at the soil-atmosphere boundary as affected by various mulch materials, 2019. Shahzad, Khurram, Bary, Andy I., Collins, Douglas P., Chalker-Scott, Linda, Abid, Muhammad, Sintim, Henry Y., Flury, Markus. *Soil & Tillage Research,* v.194 pp. 104-335.)

Figure 80. I saw these hay bales growing tomatoes in a greenhouse during my 1998 tour of gardens in the UK. This was at the Henry Doubleday Research Association. After the crop is over, the bales are pushed out of the greehouse so the concrete floor can be scrubbed with bleach—thus avoiding nasty chemicals.

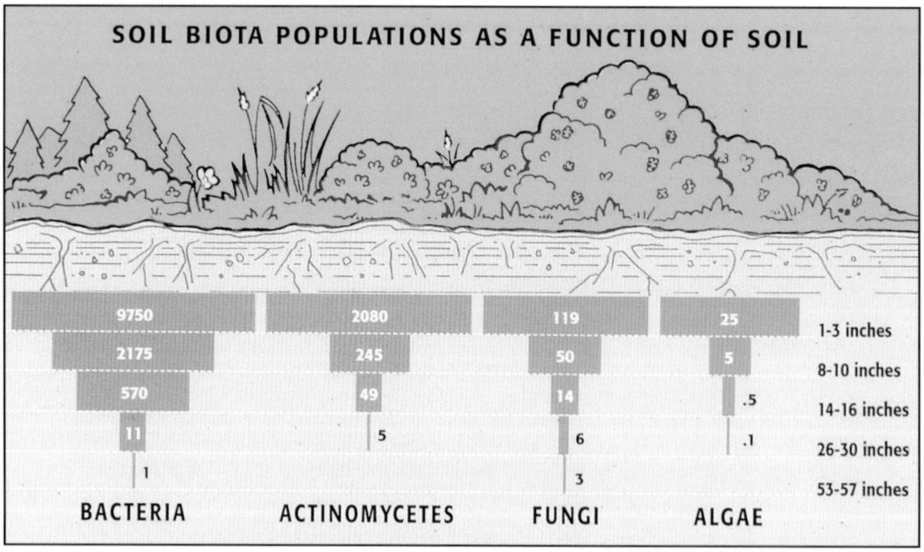

Figure 81. The top three inches of soil is so much more important than lower soils.With beneficial fungi, going from three inches to 10 inches drops the number by more 50 percent. With actinomyces it is a drop of 90 percent. This is another arguement for no-till gardening. Disturb the top three inches of soil as little as possible.

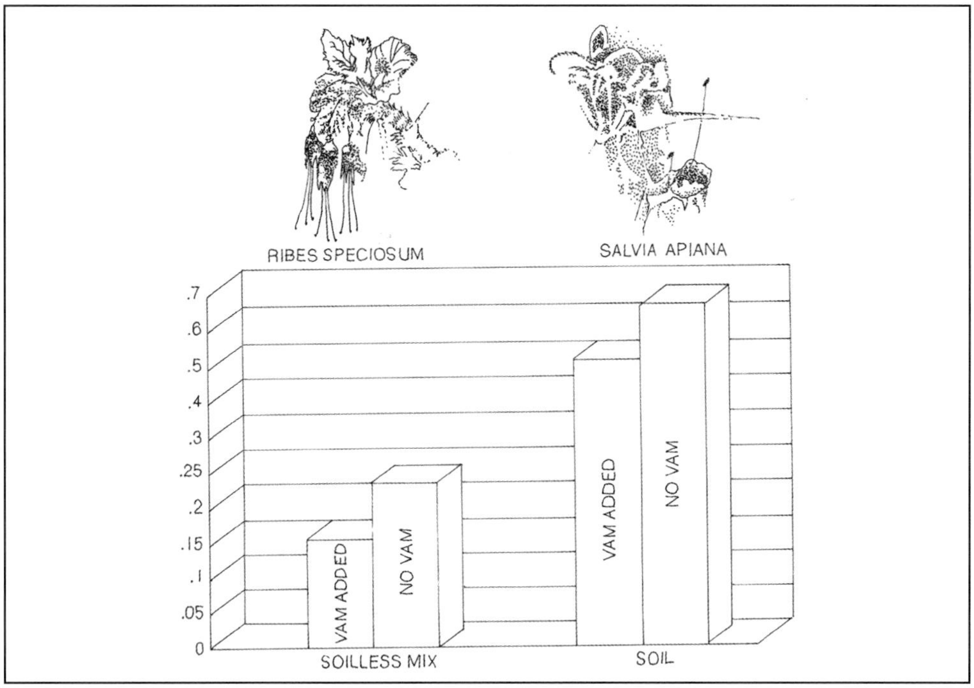

Figure 82. A study of two California native plants. It's an interesting example of where the inoculation with mycorrhizae had less of an influence than no inoculant at all. *Ribes speciosum* is a native currant, *Salvia apiana* is the native white sage. Why? Who knows? I ceratinly don't. (This ranks the shoot's dry mass per gram.)

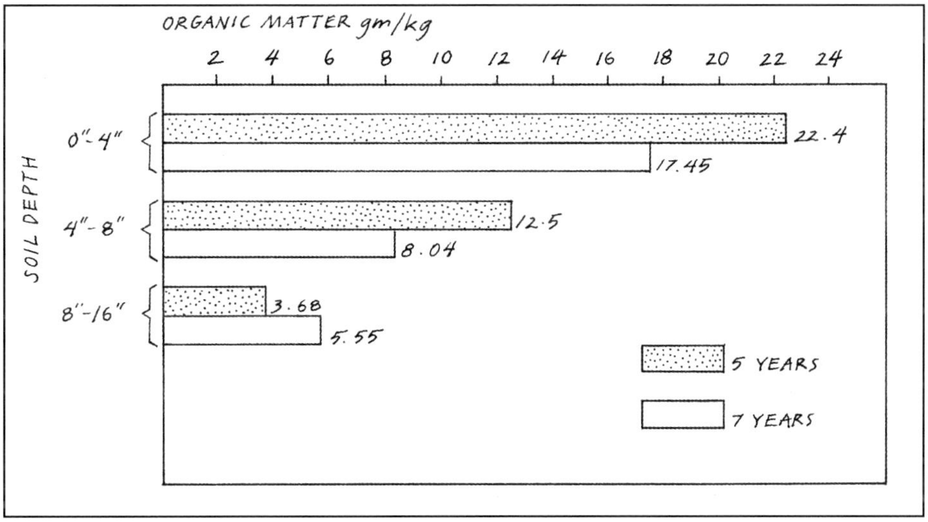

Figure 83. The top four inches of soil in this study has about twice as much organic matter in gm/kg than the next four inches. Tillage helps oxidize (burn up) organic matter. Another reason for no-till gardening.

Chapter 5
Conserving Resources

The Edge of Parabolic Windbreaks: Practical or Just Pretty?

Curvy, sexy aerodynamic shapes are practical in the domain of racing cars, but *not* for horticultural windbreaks. Windbreak designs in early permaculture books show tree groupings, walls, fences, and hedges carefully set out in aesthetically pleasing parabolic curves. In Bill Mollison's *Permaculture - A Designer's Manual*, the *only* configuration recommended for a windbreak is a crescent—a portion of a parabolic curve, elegantly extracted from earlier permaculture discourses on "Golden" (Fibonacci) spirals; smoke- and water-flow patterns, eddy/currents phenomena, and vortexes, the latter taken from the book *Sensitive Chaos*.

This aesthetic fixation on curved windbreaks, while it may create pleasing design elements, ignores most of the researched windbreak data, not to mention the extensive information gained in the process of establishing windbreaks after the environmental disaster of the American "Dust Bowl" in the 1930s.

In *Plants, People, and Environmental Quality*, published in 1972 by the US Department of the Interior and the ASLA (American Society of Landscape Architects), we read: "Studies have shown that shelterbelts and windbreaks are most effective when placed *perpendicularly* (ed. emphasis) to the [flow of the] prevailing winds." (Shelterbelts are wide swaths of trees, usually planted on farms. Windbreaks are composed of one or two lines of trees, and are more appropriate for home landscapes.)

Early permaculture enthusiasts, it seems, had chosen the pleasing mathematical form of a parabolic edge and then inferred the biological implications. While any curved line can lend artistic beauty to a design, it's a stretch to say that each curve represents an optimal biological design, or that the ecological dynamic has been enhanced, especially in the context of home landscaping.

This is what they ignored:

- Every windbreak, it turns out, generates turbulence at its ends. This turbulence back-eddies around the windbreak, often emerging from its ends at speeds up to 20 percent *faster* than the original wind speed.
- The traditional recommendation has been that, in order to leave a large area relatively unaffected by end turbulence on the broad leeward side of the windbreak, a natural windbreak should be 11.5 times wider than its mature height.

Consider the two plantings illustrated in Figure 84, based on real models and actual numbers. The straight windbreak, sited perpendicular to prevailing winds, is drawn the same length as that of the parabolic wind-

---

Mulches, as we know, save water. But there are mulch facts and then there are mulch myths. See pages 154-159.

STRAIGHT VERSUS CURVED WINDBREAKS

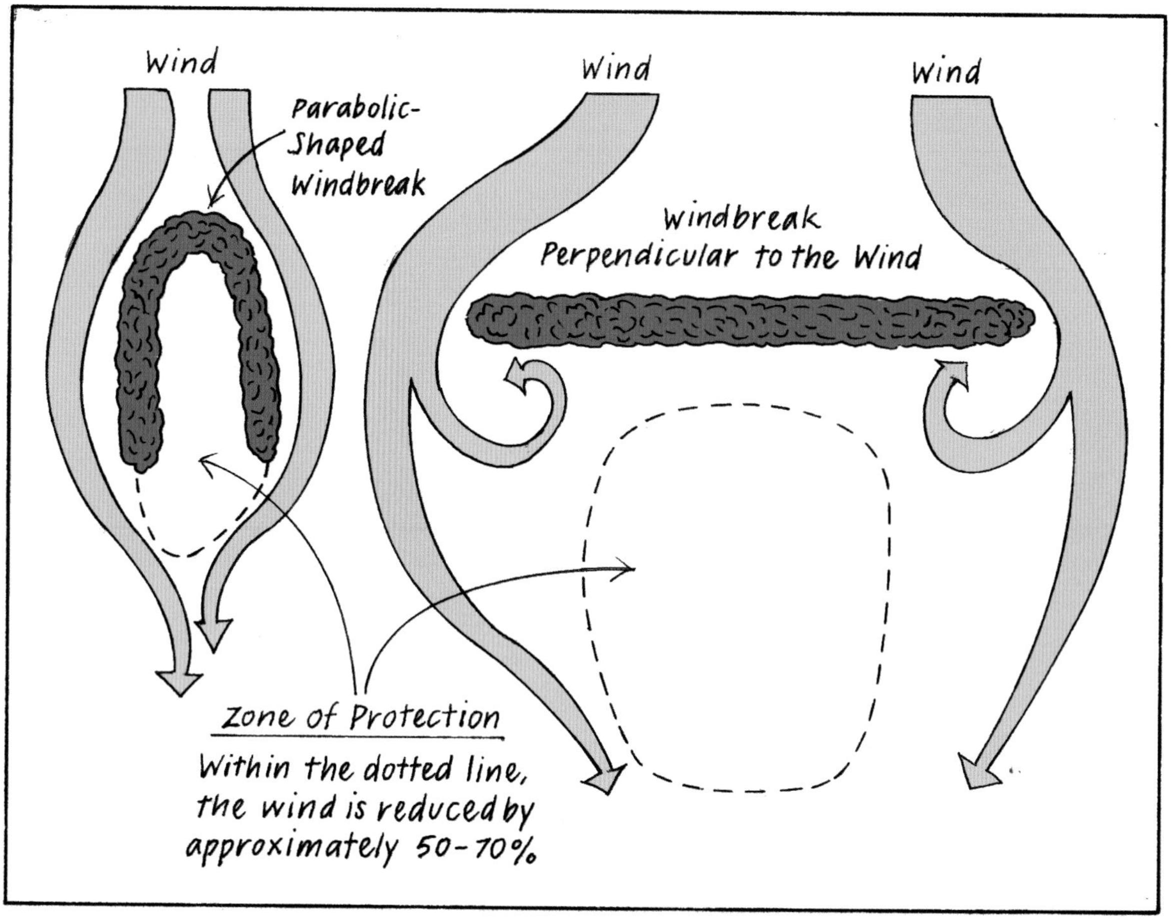

Figure 84. Parabolic-curved hedges, while attractive, are not the most effective windbreaks, as they will divert wind to whip around their edges at an even greater speed than in the open. A straight windbreak that's perpendicular to prevailing winds will protect a much larger area. In this illustration, the zone of protection on the right is nearly four times larger than that of the parabolic planting on the left. A parabolic curved hedge is, however, a good way to "trap" the warmth of the sun.

break. Using graph paper to chart representations of wind flow in the perpendicular windbreak, the area with a 50-70 percent reduction in wind speed represents about 1350 units/squares of the paper.

The parabolic shape, on the other hand, doesn't sustain nearly as much wind reduction on the leeward side as the perpendicular windbreak. The protected area (actual wind reduction unknown) covers only about 350 units/squares of graph paper.

This shows the area of protected space behind a straight windbreak to be *four* times greater than that behind a parabolic curve, demonstrating clearly that the best use of a given number of plants for a windbreak is the creation of a straight line perpendicular to the highest-velocity winds.

From *An Introduction to Agroforestry* by P.K. Ramachandran Nair we read: "Often, for best results, both conifers and broad-leaf species are grown together in windbreaks. The most commonly used windbreak species in North America include silver maple (*Acer saccharinum*), saltbush (*Atriplex canescens*), hackberries (*Celtis* spp.), Russian olives (*Elaeagnus* spp. [note: *very* invasive, at least in New England]), ash (*Fraxinus* spp.), honey locust (*Gledistia* spp. [can be invasive]), black walnut *Juglans nigra*, juniper (*Juniperus* spp.), spruce (*Picea* spp.), pines, sycamore (*Platanus occidentalis*), poplar, Douglas-fir, and bur oak (*Quercus macrocarpa*)." (1993, Kluwer Academic Publishers, 101 Philip Drive, Norwell, MA 02061, U.S.A.)

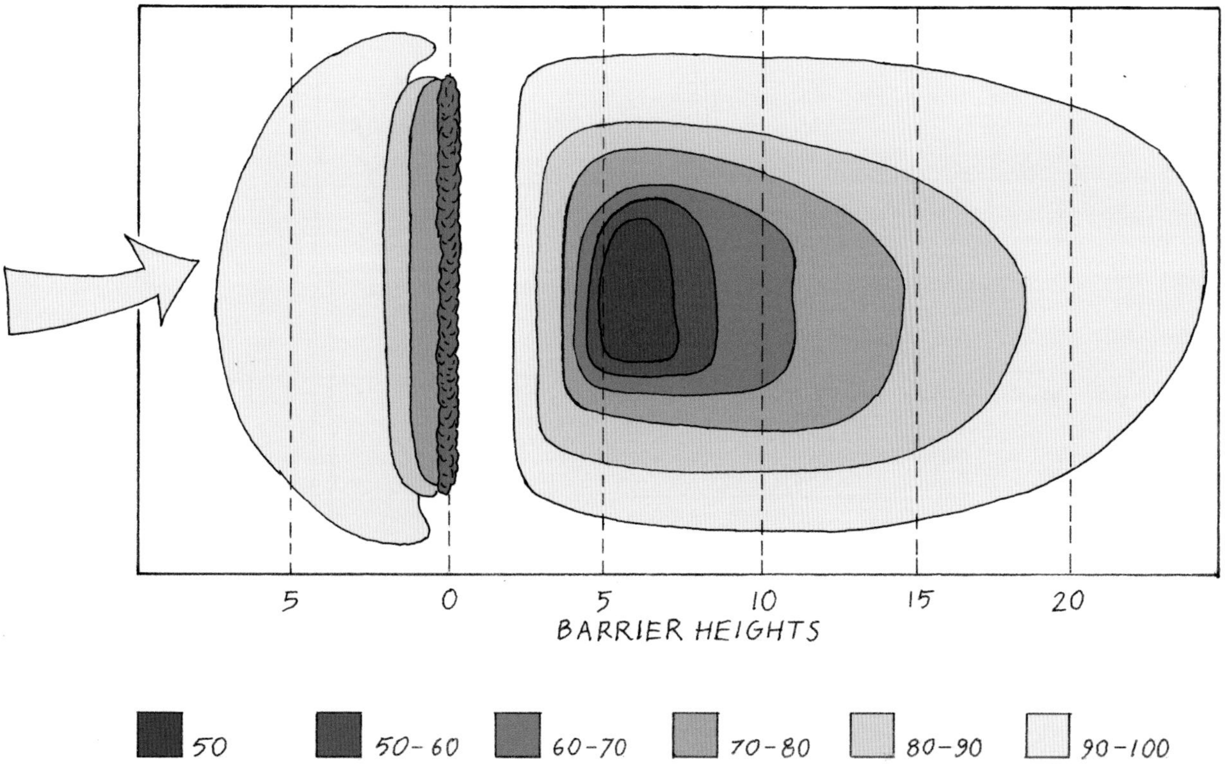

Figure 85. This is a view looking down on a five-foot-tall windbreak. The area protected by a windbreak is dependent on its distance from the hedge/trees. The "Barrier Heights" from zero to 20 feet are indicated on the right. At five feet from the windbreak, the speed of the wind is reduced by 40 to 50 percent. At a 20-foot distance, the wind has been reduced by only 10 percent. Notice that the wind is also slightly reduced in the front (windward) of the plants. This is because the air layers up (laminates) to lift the wind over the barrier, and a "pocket" of less speed remains. Here the wind speed has been reduced 10 to 30 percent on the windward side.

How and Where to Shape Winter Sunlight

Placing trees for summer shade and winter sunshine is usually more complicated than it would first appear. Deciduous trees placed on the southern side of a home can often actually *increase* energy bills. A study in Sacramento, California, determined that all homes, both solar and conventional, would pay more for energy year-round if both deciduous and evergreen shade trees were planted along the southern side of the house. Some varieties of trees in the study actually *doubled* the energy costs when compared with a house with no trees. The closer the shade trees are to the house, the higher the yearly utility bills. Solar houses with solar hot water systems and photovoltaic panels on the roof were more severely impacted by shade trees near the southern wall.

Does this seem baffling? Here's an explanation. As Figure 86 shows, sunlight coming through the south-facing windows in the winter can be deceiving. Even after their leaves have fallen, the branches of deciduous trees block the sun. It's just that we often don't see this wintertime shadowing, as the shade "fades" before it hits the ground. To demonstrate this, you can take a light-exposure-meter reading under a winter-bare tree and another out in the full sun; the difference in the results is often dramatic. Many kinds of trees, when leafless, actually block 50 percent or more of the sunlight—on this chart, alder is an extreme example. One of my favorites, the gingko, blocks only 30 percent of the winter light.

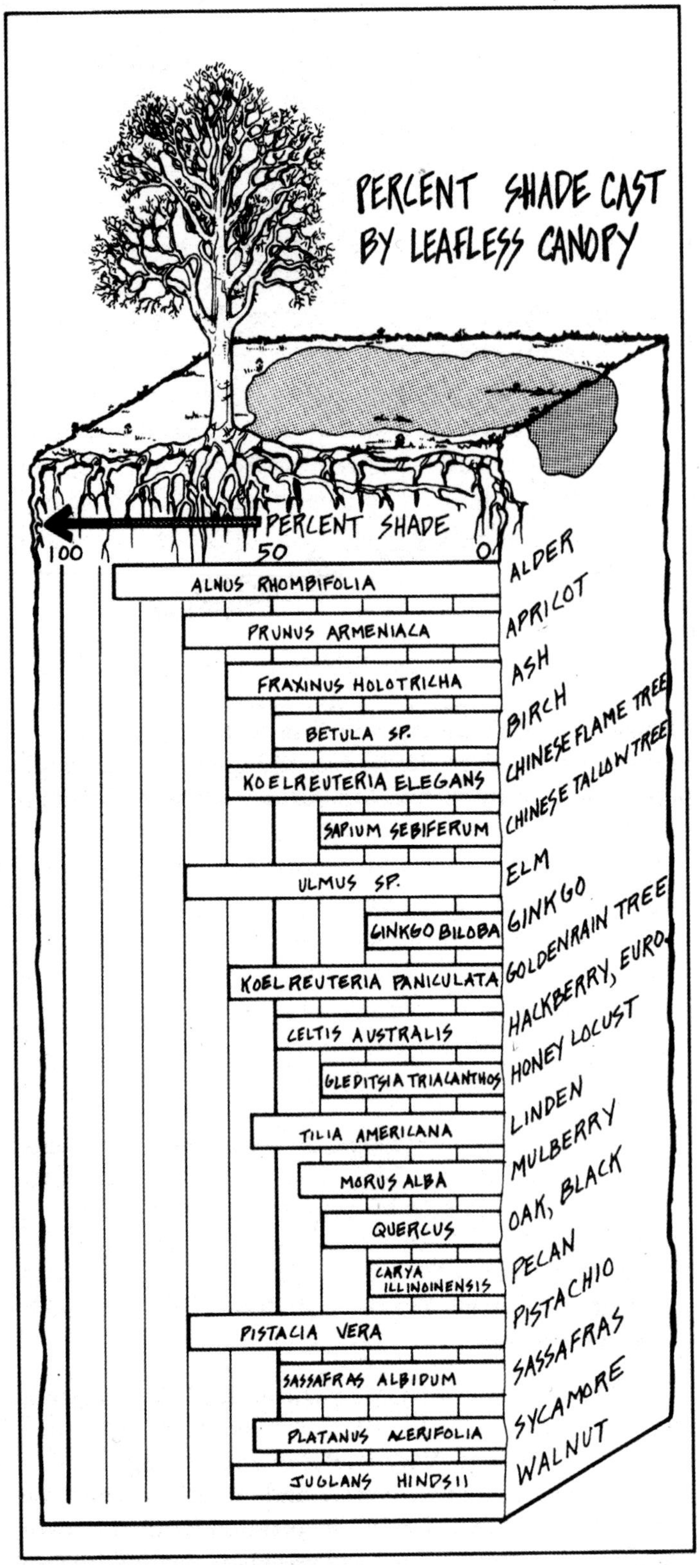

Figure 86. Most people see sunlight under a deciduous tree in the winter, and assume they're seeing all of it. The bare winter branches actually block some of the daylight, but the shadow they cast is so tenuous that it doesn't reach the ground, and a light meter is needed to see the difference. (There's a light-meter app for my iPhone that suits this purpose.) Here someone has measured the winter "shadows" of 19 trees. The results range from 30 percent less sunlight for pecan and ginkgo trees to 85 percent less for alder trees.

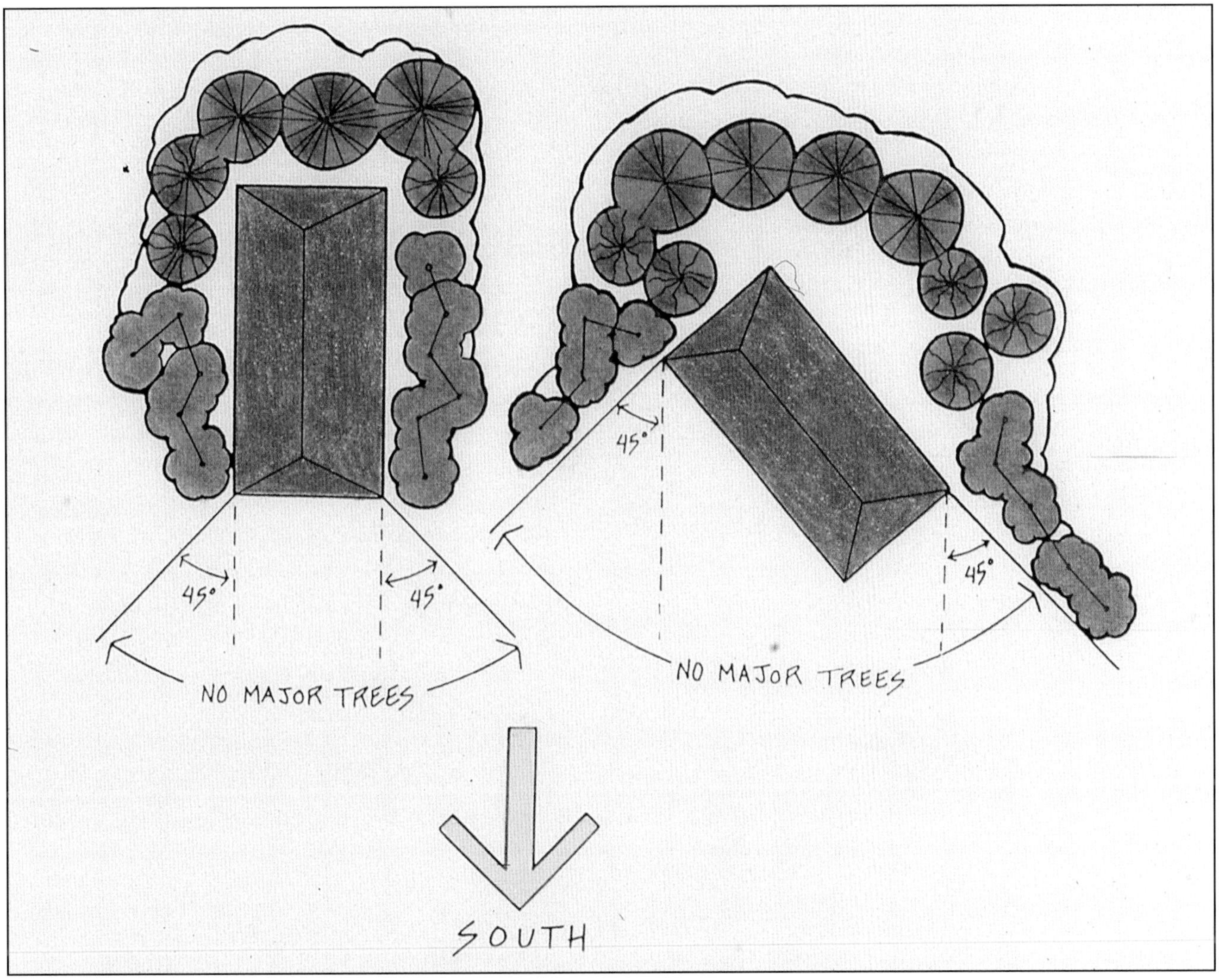

Figure 87. Because so many deciduous trees cast a lot of shade (see the preceding page), they can block a lot of light on the south side of a solar house and to south-facing rooms. Contrary to what many think, a lack of tall trees south of the house will allow for more warmth in the winter. Plant only shrubs that get as tall as the windowsill close to the house; taller trees should be placed further away.

Unexpectedly, the solution to this blocking is to leave a solar "window of opportunity," that is, a space with no shade trees, on the south side of your home. Here's how to determine the area to leave free of trees: on a scale drawing of your property, draw dotted lines due south from both the southeastern and southwestern corners of the house. Next, starting from the same corners of the house, draw solid lines at 45-degree angles to the dotted lines. One solid line will be east of the southeastern dotted line, and the other will be west of the southwestern dotted line. (With a solar house facing due south, these solid lines will match the dotted lines.) The area south of the house between these two solid lines should be free of trees tall enough to shade the wall and the roof.

To beat the heat of summer, don't forget the eastern and western sides of the house, especially if they have lots of windows. One study found that, when it comes to reducing heat gain, a western wall is six times more important than a southern wall, and an eastern wall is *fourteen and a half times* more significant than the southern wall and almost two and a half times more critical than the western wall. While these numbers may vary considerably, windows may contribute to overheating a house in the summer more than most people realize.

During the heat of summer and cold winds of winter, landscaping can provide another comforting layer of protection from the oscillations of unpredictable weather. You were born with a protective covering of skin; clothing adds another layer of comfort; the insulated "skin" of the house provides yet another measure of defense; and finally, you can enjoy an energy-conserving shelter of attractive foliage wrapped around your comfy home. Here are some guidelines:

- Place trees away from the house to prevent them from blocking, scratching, and otherwise damaging solar panels.
- Plant far enough from the exterior walls that you can easily repaint them when it's needed.
- Avoid planting trees near the west- or southwest-facing sides of the roof, as these directions receive the strongest sunlight and are the best options for solar-panel placement.
- Properly maintain trees that are already planted. Regular tree trimming will minimize neighbor disputes and protect solar photovoltaic panels.
- Remove dying, unstable, and unhealthy trees, which pose a threat to nearby structures and solar-paneled roofs.

Quick Fixes for Climate Control

The major drawback to using shade trees for comfort is the period of time required for them to grow large enough to have a real effect. Many shade trees won't have much of an effect until their second decade. At the same time that you plant a young shade tree, you can also plant perennial vines, which usually take only two or three years to cover a trellis as tall as the walls of the house. This trellis should *not* be located under the eaves of the house, which can block any hot air that might build up between the trellis and the

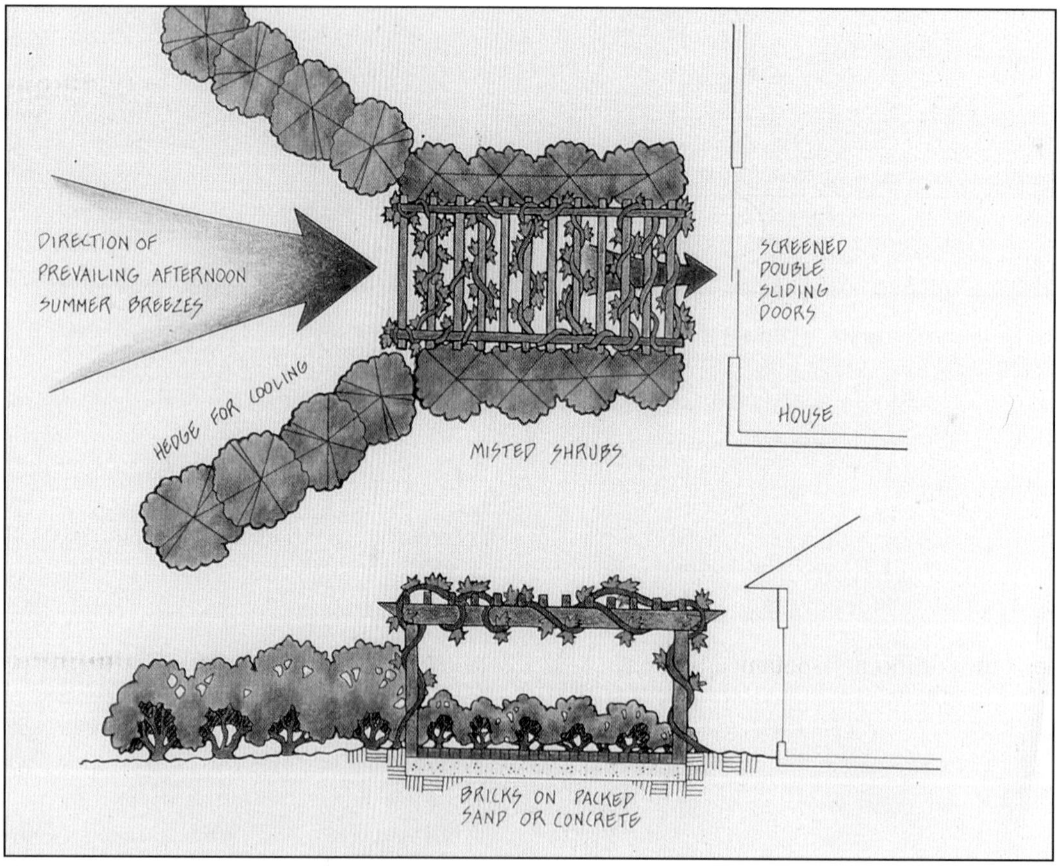

Figure 88. You can block wind to keep a house warmer in the winter and cooler in the summer. If your garden and house get too hot, try this arrangement: the hedges form an angle toward the patio and funnel the wind into the patio and the arbor to cool things down. If you have unglazed brick, rock, or flagstone flooring, you have yet another way to chill out: spray down the floor and the evaporative cooling will help cool the house. The arbor with vines shades the patio to keep it a bit cooler.

siding, and can prevent it from venting out of the top of the trellis. Edible perennial vines for summer protection include grapes, chayote, kiwi, and passion fruit. Evergreen ornamental options include honeysuckle, jasmine, potato vine (not the plant of Idaho tuber fame, but a flowering perennial that should be only planted in containers, as it becomes invasive), and trumpet and *hardenbergia* vines.

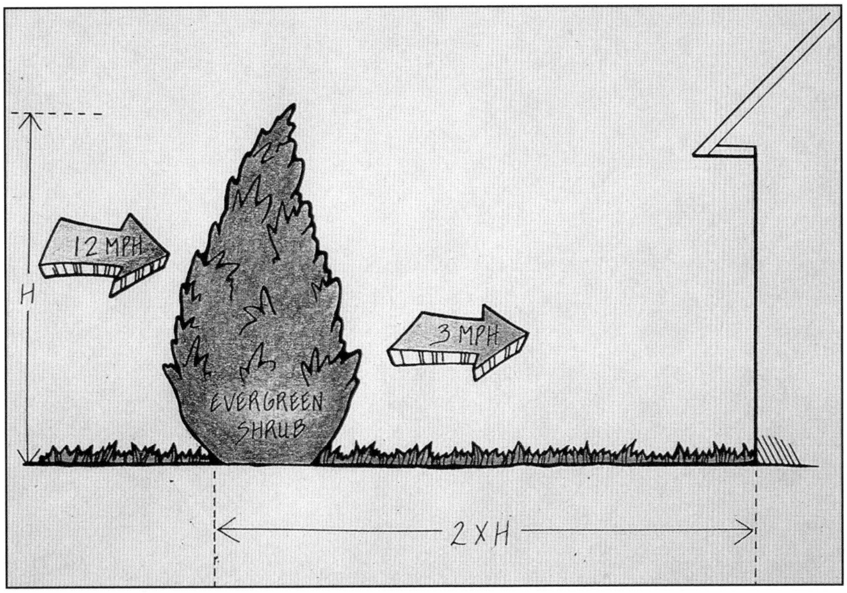

Figure 89. A windbreak should be 11 times as wide as it is tall—not a practical fit in many back yards. You don't need a windbreak to help keep the house warmer in the winter; instead use evergreen shrubbery placed at twice the height of the plants away from the walls, and you'll only get a quarter of the wind.

Annual vines grow even more quickly than perennials, and can cover a large area by mid- to late summer. Annuals are also best suited for use very close to the walls of the house, and those grown on a wooden-lattice trellis will lend a measure of shade from the very start. A trellis should be kept at least two feet away from the house; this allows enough air circulation to keep the siding free of mildew and rot and allows space to repaint the house when needed. Make the trellis easily removable or able to be bent back during re-painting activities.

You can plant vigorous, fast-growing edible vines such as scarlet runner beans, winter squashes, and luffa squashes (luffa or loofah—botanically *Luffa aegyptiaca*—is a vine-grown member of the pumpkin, squash and gourd family, *Cucurbitaceae*), or ornamental vines such as hyacinth bean and moonflower.

Annual and perennial vines and large shrubs can do a lot to shade and cool the walls of your home while your shade trees grow to a more mature form. You can also plant small trees or large shrubs near an existing air conditioner to shade it during the hottest months of the year, making sure, of course, that foliage or falling leaves and twigs won't interfere with the conditioner's fan or airflow.

Along with more money in your pocket from lower utility bills, a mature landscape with energy-conserving shade trees can, by some estimates, increase your property's value by as much as 15 percent. The

Figure 90. This south-facing home features an arbor covered with grapes, which makes for a cooler patio in the summer. The vines can be cut way back each autumn to let the fall and winter sun through. The mulberry tree helps to shade the arbor, but it may also be close enough to the house to shade it in the winter by 45 percent. (See Figure 86.)

American Forestry Association places the increased value anywhere from 3 to 20 percent, with a single fifty-year-old shade tree adding $57,000 to a home's worth.

Figure 91. Grapevines form a pleasant green arbor above a second-floor deck. The barrel is planted with squash; behind it is the south-facing living room of this solar house. The almond tree on the right is far enough away to prevent its branches from shading the house in the winter. Ornamentals and sweet peas on the left provide highlights of color. This is one of the 225 solar-heated houses in Village Homes (Davis, CA), a development established in 1975. Many of the homes include gardens, which may collectively provide as much as 25 percent of the produce used by the community.

Figure 92. Windbreaks need sturdy roots to stand up to heavy gusts, and root-bound transplants can easily blow over. The solution is to buy small plants like these tube-grown trees. The hole at the bottom of the eight- to 12-inch tube kills off the taproot by allowing it to dry out. This causes many lateral roots to form, proportionally more than are usually found in container plants. The trees grow quickly, and form sturdy root systems. Buy them from revegetation nurseries, or from the Internet. My tube-started trees grew to be 10 feet tall in less than four years.

An Agricultural Water Model versus A Garden

Agricultural models often don't "downsize" well to the home landscape. For example, it's been popular among some landscape designers to recommend swales (low or hollow places between ridges) and berms (artificial ridges or embankments) to slow water moving through a residential landscape and increase well-water supplies. I saw one such design that had been constructed at a 100-foot elevation above a

150-foot-deep wellhead. The designer had installed six 100-foot-long berms and swales in an attempt to capture more rain runoff in an aquifer (water-bearing rock that readily transmits water to wells and springs) He actually told the client that this would increase the production of the well. I think not. The rainwater would have to drain 250 feet from the water garden before it even got to the pump servicing the well, and the landscaping was done on top of a rock formation called Franciscan that does not form aquifers, just fractured rock and isolated pockets of water.

There's no scientific data on the subject of berm-and-swale construction on a home landscape. I asked one of the major proponents and teachers of such designs about this, and he said that he knew of no scientific studies, and had perhaps seen *one* example of the design's effectiveness during the Northern California drought of the 1980s. In his training program he continues to discuss the swale-and-berm concept for water storage in aquifers, and has his students digging 50-foot-long swales by hand.

There are plenty of situations where an aquifer close to the surface can receive some recharge water from swales, which are certainly useful on large parcels to slow down water, reduce erosion, and help keep silt out of nearby waterways. The unknown factor in all this, however, is how many swales of what size are needed for any *significant* recharge, and there's certainly not enough information to create a real context that fits every situation. Where rains are common and plentiful, swales may not be needed. As with so much in permaculture classes, a concept has been accepted as universally appropriate without scientific backup.

A guideline for improving the storage of water in the soil and increasing the capacity of ponds and lakes on a farm was outlined in great deal in the 1938 issue of the USDA Yearbook *Soils & Men* published after the Dust Bowl. (USDA Yearbook of Agriculture, Washington DC, USDA, 1938) While it focused on ponds and lakes, many of the dynamics can be seen and used in a smaller home setting.

Systems of catchments for surface storage and groundwater recharging *are* possible on a community-wide basis. At Village Homes in Davis, CA, with 225 homes on 70 acres, a series of swales were built on previously flat land prior to the construction of the homes. These swales meander throughout each

Figure 93. Here are two views of the drainage swales of Village Homes before any houses were built. All houses back up to this winter-rain drainage feature, and each takes advantage of it it in an individual way. The photo on the left shows a food garden growing alongside the swale, while the other shows an ornamental border.

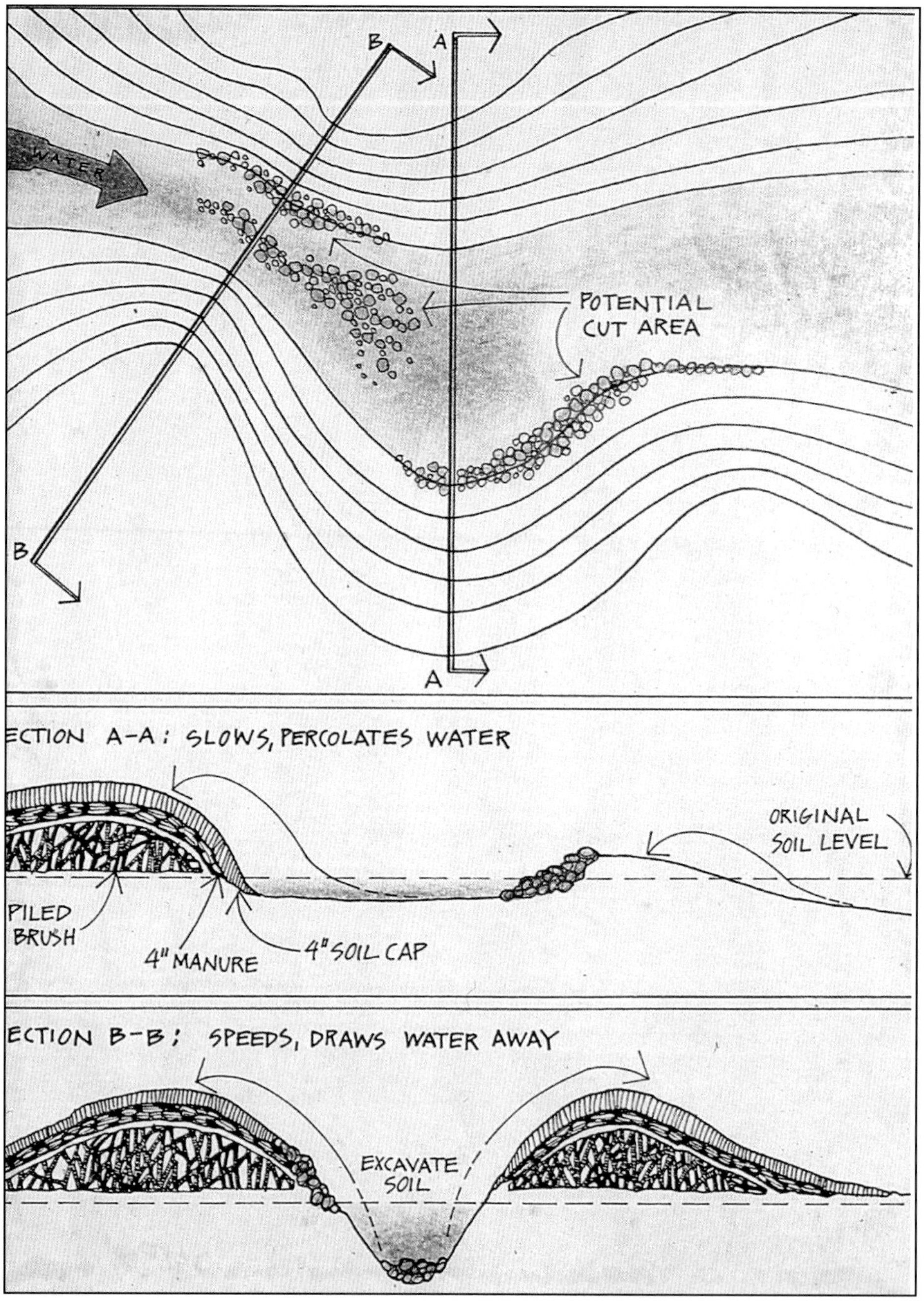

Figure 94. I built this berm-and-swale feature to drain a wet spot in my garden. I started with mixed wood chips and leaves from an arborist's truck, and piled them up to form the high parts of the mounds (berms). I then excavated some of the adjacent soil to make a swale for drainage, using it to make a soil cap on the wood-chip berm, To create a more natural look, I formed both narrow and wide sections, placing rocks to shore up the curves where the soil bank might erode. Transplants went into the soil cap on the berm.

neighborhood and add a dynamic complexity to the landscape. They also fill up with heavy winter rains, trapping any soil that might otherwise be washed away, and allowing time for the rain to percolate deep into the subsoil.

While most suburban areas just dump the rain into a storm-drain or sewer, the swales at Village Homes keep the winter rains within the suburb and store 90 percent of the water in the soil for the landscape to use into the dry summer. (They have *deep* topsoil there—10 feet or more!) While this doesn't eliminate summer irrigation, it does delay the time at which watering needs to begin. Since the community-owned lands and most of the private landscapes of Village Homes are tended with organic methods, the groundwater isn't contaminated by manure, herbicides, or pesticides. It's not entirely recycled, but is plumbed to Davis city potable water.

Way before the current interest in creating swales, I spent a lot of time building them to moderate the passage of water over soil surfaces in the gardens and landscapes I designed. I use berms next to drainage swales for a very simple reason: the roots of native Californian and Mediterranean plants are often sensitive to various forms of root rot, like *Phytophthora* spp., *Pythium* spp., and *Rhizoctonia solani*. The drainage provided by a berm keeps the moisture away from the top of root zones. On page 78 of my 1986 book, *Designing and Maintaining Your Edible Landscape – Naturally*, I illustrate the concept of berms used in conjunction with plants with Figure 94. Nothing fancy, just simple practicality.

Rainwater Gardens

Rainwater gardens, as they're called when installed in home landscapes, can be important additions, as they help keep rain out of overloaded sewers and allow the water to soak into the ground. Fortunately, there's a growing interest in using rain gardens in home landscapes, a practice often described by the slogan, "Slow it, spread it, sink it."

A well-designed rain garden is an organic, low-tech method for controlling the runoff from rooftops, driveways, patios, and other nonporous surfaces; it accomplishes this by channeling and directing water away from buildings and towards a lower area of the yard or landscape. In the process, impurities and pollutants are washed away by the garden's natural filtration and cleansing process.

In order to retain runoff and help control soil erosion, a rain garden should be dug into a shallow bowl-like shape. The planting beds should be sunken, in order to capture the most runoff and get the water to the landscape plants. Excess water will filter down through the soil, which helps to get rid of salts and chemical residues before they get into underground aquifers.

However (as I caution all clients), rainwater gardens, being water-based, are a *very* different ecosystem to take care of, requiring special attention to detail when installing the ponding areas, especially on slopes. They're more costly to install than, say, lawns with perennials. A rainwater garden, even if you do all the labor, can cost five or more dollars per square foot. They also require more weeding than many lawns.

Figure 95. The blue circle in the middle of the photo is the downspout exit to this rain garden, and the surrounding rocks protect the slopes from erosion. The chartreuse-blooming euphorbia likes dry soil and therefore is planted on top of the berm.

Because these gardens are wet in one season (or after a rain) and then drier, only limited types of plants can tolerate such extremes. Many people want to line the bottom of the swale with rocks to make it look like a little creek, but weed seed will inevitably blow or flow in, and the weeds are hard to get rid of without herbicides. I don't recommend rocks and thus it's harder for me to make a swale look natural. I don't end up designing a rain garden for most clients, but some cities are actually mandating them now. Check with the city planner's office.

So, here are the basics of building a rainwater garden:

Before doing anything, you *must* call 811 or 1-888-DIG-SAFE to mark the locations of all utility pipes and wires (don't just assume that they're 18 inches belowground) at least 10 days before you start digging. The information is free.

Placement

Rain gardens require special placement; they must be:

- 10 feet away from the foundation or basement of a dwelling.
- Three feet from any sidewalk or patio.
- Two feet (and downhill) from any building, concrete slab, or crawl space.
- 10 feet away from any retaining wall.
- Five feet from all property lines. (NOTE: double-check this with your planning department.)
- 100 feet from a well.
- At least 50 feet away (at the minimum) from any septic field
- Not installed where the soil is already soggy.
- Not placed under the drip line of any trees.
- Located downhill from any gutter downspouts.

(With all these constraints, I hope you can find a spot.)

Designing a Rainwater Garden

To determine the size of your prospective garden in square feet, multiply the length and width of the portion of the roof that feeds the downspouts of your house, or of the shed, driveway, sidewalk, etc. from which you plan to collect runoff. Each 100 square feet will generate 62 gallons of water from one inch of rain. To get the size in square feet of your prospective garden (which should be smaller than the square-footage of the source) multiply the square-footage by 0.1 (10 percent).

Before installing a rainwater garden, you *must* check your soil's drainage. When the soil is moist, dig a hole that's one foot wide and one foot deep. Fill the hole with water and allow it to drain completely. Fill the hole with water again, note the starting time, and note how long it takes for the water to completely disappear. If the water is draining very slowly, measure (in inches) how much water drains in each 30-minute interval. Repeat this two more times. The third time, check the water level again after 15 minutes and note the new water level. Multiply the change in water level in inches by four to get the number of inches of infiltration in an hour.

If it takes only 0-4 minutes for the water to soak into the bottom of the hole, you have fast-draining soil, which is good for most trees and plants. Five-15 minutes = good-draining soil, also ideal for most trees/plants. If the water takes 16 – 60 minutes to

Figure 96. New Zealand flax (*Phormium* spp.) - in the backgorund - takes a wide range of irrigation. But is a very good vertical background accent for the top of a dry berm.

disappear, but is draining at least 1 inch per hour, it will work for some trees/plants, but isn't a good area for those that need well-drained soil. And if the hole takes more than 6 hours to drain, it's *not* a good area for most trees and many kinds of plants, except for those that normally grow along streams and rivers.

A rate of 1/2" or more per hour indicates that your soil will drain within 24 hours. Drainage of less than half an inch per hour means that you'll have to find a different spot. (Some people use the simpler—though less accurate—method of digging a hole six inches deep, filling it, and coming back 24 hours later. If the water is gone, you're set to go.)

The depth of the garden's "pond" or sunken area should be between eight and 26 inches. The faster your soil drains, the shallower the pond needs to be. (If the rain drains at two inches per hour or faster, make the pond about eight inches deep— Counter-intuitive.) The bottom must be level no matter how much the yard slopes.

Use a hose to mark the shape you have in mind for your pond; this allows you to change the prospective layout as you find necessary.
You'll need to bring the water from the downspout to the pond by means of a connecting pipe. Some municipal ordinances require this pipe to be at least 12 inches below the ground. In colder climates, it may have to be even deeper, and you may have to use the white "plastic" pipe called schedule-40 PVC, or the black pipe called ABS. The pipe should drop at least

Figure 97. The foliage of rushes (*Juncus* spp.) in the left foreground are round, while the leaves of sedges (*Carex* spp.) have edges; all pictured here and are used in the rainwater garden designed by Josh Williams of Sebastopol, California. (Photograph © 2020 Philip Soelen)

Botanical Name	**Common Name**	**Grass**	**Shrub**	**Per**	**State**	**Sun/Shade**
Achillea millifolium	Yarrow			X	OR	F
Agastache foeniculum	Lavender Hyssop		X	X	VT/OR/MI	PS
Andropogon gerardii	Big Bluestem	X			VT/OR/MI	F
Aquilegia canadensis	Columbine			X	VT	F/PS
Asarum canadensis/caudatum	Wild Ginger			X	VT/OR	PS/S
Ascepias spp.	Milkweed			X	VT/OR/MI	F/PS
Aster spp.	Aster			X	VT/OR/MI	F
Baptisia spp.	Indigo			X	VT/OR/MI	F
Carex densa	Dense Sedge	X			OR	F/PS
Carex rupestris	Curly Sedge	X			OR	F/PS
Carex spp.	Sedge	X			VT/OR/MI	F/PS
Carex testacea	Orange New Zealand Sedge	X			OR	F
Ceanothus americanus	New Jersey Tea		X		MI	F
Convallaria majais	Lilly of the Valley			X	VT	F/PS
Cornus amomum	Silky Dogwood		X		MI	F
Deschampsia coepitosa	Tufted Hair Grass	X			OR	F/PS
Dicentra spp.	Bleeding Heart			X	VT/OR	F/PS
Echinacea purpurea	Purple Coneflower			X	VT/OR/MI	F/PS
Geranium maculatum	Wild Geranium			X	VT	F/PS
Helianthus angustifolius	Swamp Sunflower			X	VT	F/PS
Iris versiclor	Blue Flag Iris			X	VT/OR/MI	F/PS
Juncus effusus	Common Rush	X			VT/OR/MI	F/PS
Koeleria macrantha	June Grass	X			MI	F
Lobelia cardenalis	Cardinal Flower			X	VT/OR/MI	F/PS
Lupinus spp.	Lupine		X	X	VT/OR/MI	F
Monarda spp.	Bee Balm		X	X	VT/OR/MI	F/PS
Pancium virgatum	Switchgrass	X			MI	F
Polystichum munitum	Sword Fern			X	OR	S
Rubeckia hirta	Black-Eyed Susan			X	VT/OR/MI	F
Soldago spp.	Goldenrod			X	VT/OR	F/PS
Sorghastrum nutans	Indian Grass	X			VT/OR/MI	F/PS
Verbena hastata	Blue Verain		X		VT/OR/MI	F

Figure 98. These plants are used in rainwater gardens in Vermont (VT), Oregon (OR), and Michigan (MI). They can be grown in full sun (F), partial shade (PS), or shade (S). Grasses are important for such a garden, as they can usually survive both flooding and dry spells. (Per stands for perennials.)

Figure 99. Common grasses used in Sonoma County, CA in swales are rushes (*Juncus* spp.) and sedges (*Carex* spp.), both shown here. The vertical foliage shown is a variety of rush. Be aware that seeds will eventually blow into the crevices between the stones, making them very hard to keep weed-free without herbicides. Notice that the boulders are buried almost halfway for a natural look.

one inch from level per 10 feet of length.

The slope of the berm above the "pond" should be less than 12 degrees. To calculate the proper slope, pound two stakes into the ground, one uphill and one downhill. Stretch and level a string between the stakes, parallel to the ground. Measure the run (the horizontal distance between the two ends of the slope), and the rise (the difference in elevation between the high and low ends). (See Figure 100.) The slope, in degrees, is the rise divided by the length (run) times 100. Once you know the degree of the slope, you can find the corresponding number in the chart in Figure 100 to determine how deep the pond should be. The soil type and the speed of drainage will give you other factors for figuring the depth. Build a berm at the spillway at the edge of where the water exits that's at least two inches higher than the upper level of the deepest water.

Be sure to install an overflow pipe for heavy rains. Its entrance should be located in a one- or two-foot wide trough in the soil, two inches below the top of the berm.

Planting a Rainwater Garden

Carefully chose plants by separating them into three categories (use as many native plants as possible that fit into these categories):

1. For the bottom of the pond, choose plants that like wet soil and can survive standing water (includes a lot of sedges and *Carex* spp.).
2. For the slope to the berm, select those that need some moisture, but prefer a somewhat drier soil.
3. Plants for the top of the berm should be in the slightly drought-resistant category.

The chart in Figure 98 gives a list of sample plants for Oregon, Michigan, and Vermont.

Some other things to keep in mind:

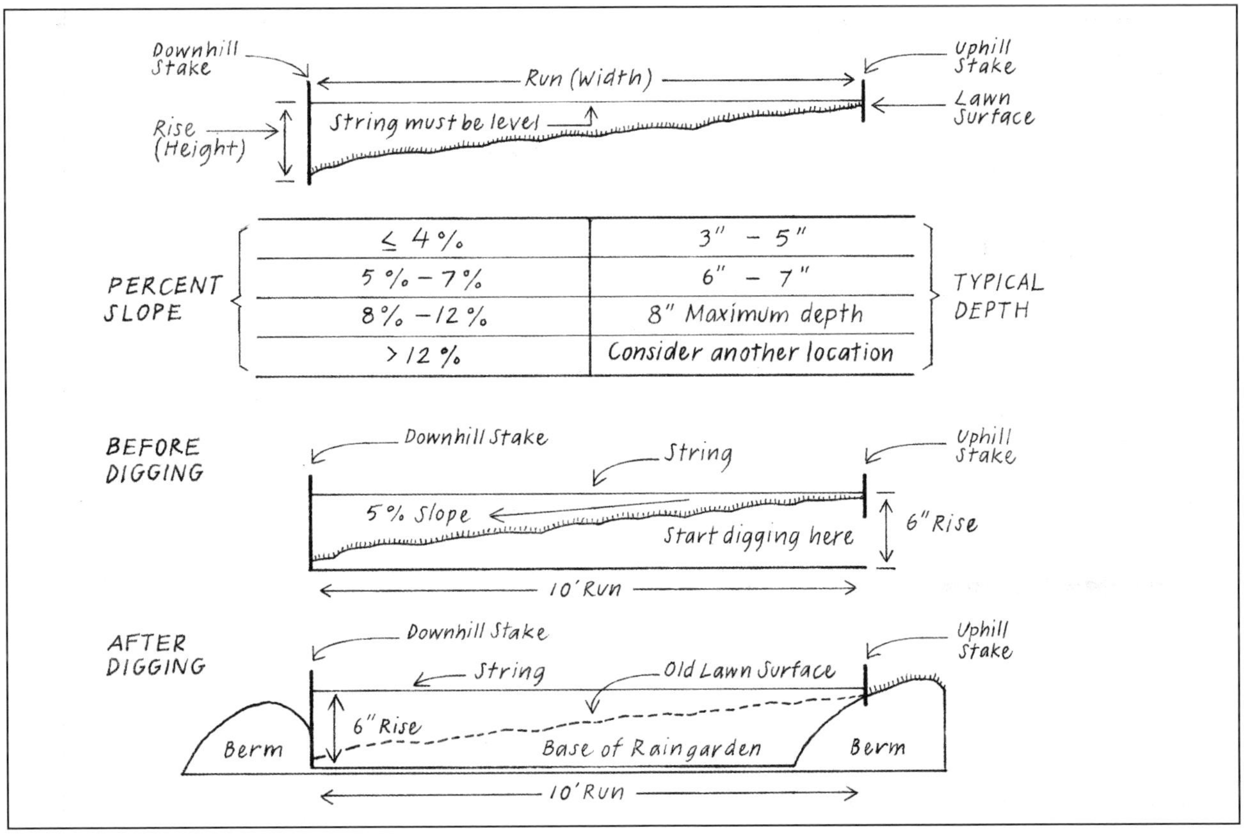

Figure 100. Here's how to gauge a slope and construct a level "pond" bottom in a rainwater garden. The rise (six inches) divided by the length (120 inches) X 100 equals the angle of the slope in degrees, which determines the depth of the basin. Here the slope is five percent, so the depth of the basin should be around six inches.

- Find out if you can legally build a rainwater garden; they may actually even be required by a new ordinance in your area.
- Check with your local water district or department to see if they can provide a pamphlet that covers building *and* maintaining a rainwater garden.
- Don't let water stand in the garden for more than 24 hours. Mosquitos can breed in water left standing for several days.
- Consider doing a soil test to see if you need to add nutrients or adjust the pH of the planting areas. (7.0 pH is about the best.)
- Don't put a rain garden under trees, as you'll disturb the roots. (Stay at least outside the drip line (edge) of the foliage canopy.)
- Avoid areas that are already wet or easily accumulate standing water.
- Remove any lawn areas before digging, or use black plastic sheets for one or more weeks to kill the grass.
- *Avoid invasive plants such as cattails*! Check with the Master Gardeners for a list of invasive plants for your area. Cattails are a cliché, and they spread near and far and become aquatic weeds.
- Use mulch to a minimum of two inches deep. (Four inches maximum.)
- As much as possible, choose native plants for the wet bottom (standing water during a rain) of a swale, as well as for the moist sides of the berm and the drier soil on the top. Water them until they're established. Note: in the Midwest, deep-rooted native Prairie grasses can absorb 30 percent more water than a lawn.
- Make sure salty water from snow removal doesn't drain into the garden.
- Don't let compost, manure, fertilizers or pesticides flow into the garden.
- Be sure to weed and mulch the garden as needed. Add plants to cover bare soil or mulch.
- Consider reading *Water Storage: Tanks, Cisterns, Aquifers and Ponds*, Art Ludwig (2005). Publisher: Oasis Design

Cisterns

One of the criteria for a sustainable landscape is reliable access to water, which often means a rain barrel or cistern. Cisterns are a hot topic in the realm of sustainable landscaping and gardening. (The modern phrase for cisterns is "rainwater catchment," or "RWC." One person in the trade told me that the difference is that cisterns are belowground and rainwater catchment uses an aboveground storage tank or barrel. I guess I'm old-fashioned, and will use "cistern" throughout. (Cisterns have a great history in much of the country, but have a questionable application in the arid west with no summer rains.)

Jean de Florette, a film by French director Claude Berri, is set in the scrubby olive tree-covered hills of southern France. As the film begins, a severe drought has dried up a family's cistern—an underground stone storage tank to capture rain from a roof covered by terracotta tiles. A farmer, played by Gerard Depardieu (of *Green Card* and *Cyrano* fame), watches helplessly as his family's crops begin to wither and their cash crop of meat rabbits begins to die of thirst. Then two neighboring peasant farmers plot to force the family into bankruptcy, cause the early death of Depardieu's character, destroy his family, and acquire the farm for much less than the pre-drought price. All this tragedy occurs because their cistern was too small to store enough roof and spring water for the driest years.

For better or worse, cisterns are used throughout the world to catch rainwater from the roof for drinking, washing, and irrigation. They come in all shapes and sizes—rain barrels, round wooden tanks, stone wells, ugly black plastic drums, and nondescript gray concrete tanks. In the 1800s and early 1900s, cisterns were common in America, even in cities, but, interestingly, the uses of cisterns and the reasons for building your own have changed since then.

Cisterns are experiencing a welcome and well-deserved revival, now taking center stage in a renewed interest in landscape water independence. This revival is also due to dwindling sources of water during droughts, not to mention the realities of dealing with runoff, trying to slow the velocity of water pouring from a roof, and/or reducing waterway silting.

Cistern systems [now loosely defined as any wa-

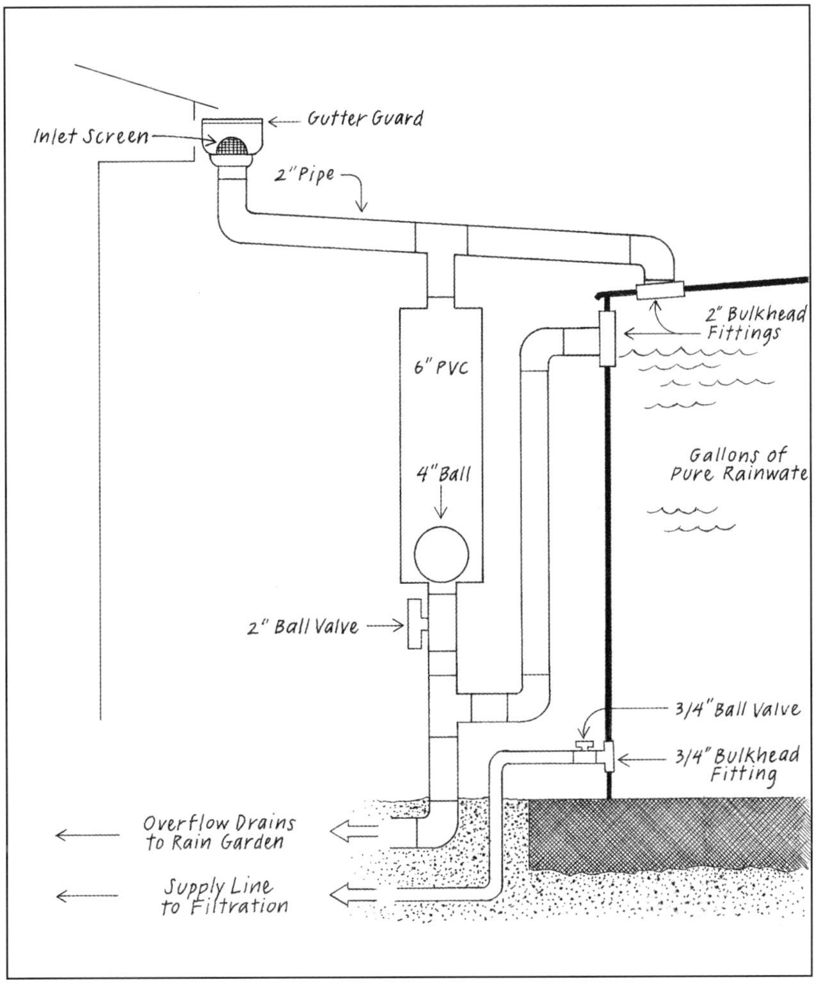

Figure 101. The downspout in the photograph above must be manually switched to drain the dirty water from the roof during a first rain (or until the water is reasonably clean) before switching the valve to direct water back to the cistern.

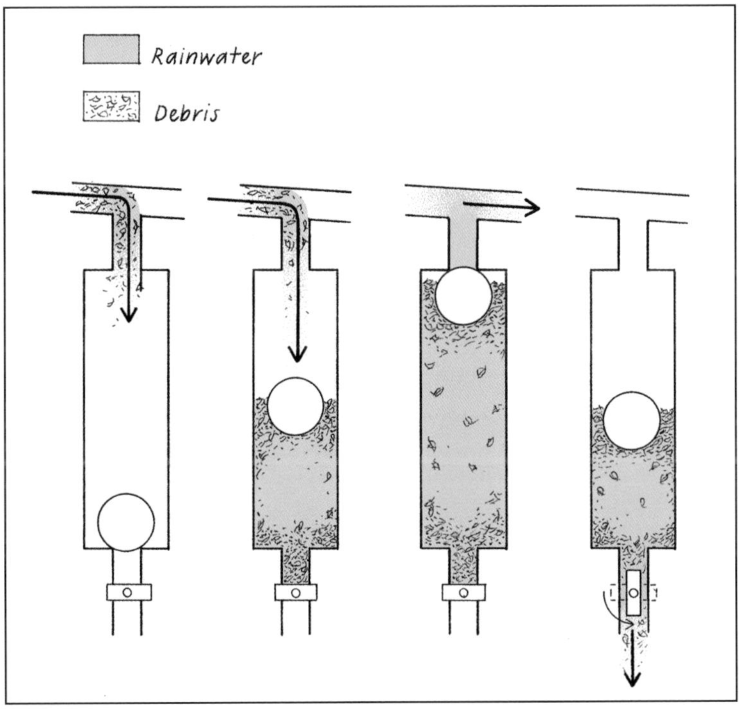

The drawing to the left depicts the Brazilian Ball Pre-Filter, an automatic way to collect debris from the first rain(s). The WATER Institute learned of it during a tour of Brazil as part of the 2007 Permaculture Convergence. It serves as a backup to the manual system when the runoff contains lots of tiny debris such as pine needles.

From left to right: A four-inch ball sits on top of the two-inch exit of a six-inch cylinder. As the first rain begins, debris (blue here) is flushed automatically into the six-inch cylinder. When the cylinder is filled with debris, the ball seals the hole at the top and fresh water spills into the cistern. The valve at the bottom of the six-inch pre-filter cylinder can be opened at the homeowner's convenience to flush out the accumulated debris.

You can find a four-inch ball where juggling equipment is sold.

If you have a large and/or dirty roof, the six-inch cylinder with the ball can be as high as six feet. The top should be level with the full tank of the cistern. All fittings are PVC.

ter-holding tank in or above the ground] are remarkably simple when compared to many modern technologies and to complicated water infrastructures such as dams, wells, and huge man-made lakes. With a cistern, the rain collected from surfaces like the roof of a home, barn, shed, or carport is simply gathered, stored, and drained or pumped to the landscape.

Typically, a gutter is installed at the edge of the roof to funnel the rain into a downspout, hose, or pipe that leads to the cistern itself. A diverter valve is built in the downspout or pipe so that the first rains after a dry spell can be directed away from the cistern to flush any accumulated filth off of the roof until it's clean. A simple device or grate is usually installed to screen out twig and leaf debris that may have gathered on the roof. Modern cisterns are frequently simple tanks made of metal, fiberglass, or polyethylene.

Modern Reasons for a Cistern

Most gardeners will readily tell you that their plants

Figure 102. A full 5000-gallon storage tank/cistern weighs about 40,000 pounds (or 20 tons!) when full, and needs a concrete pad to support the weight. This is far more than a gravel base could support without the gravel piercing the bottom of the tank.

look and grow better with rainwater. Repeated use of municipal water, especially in arid climates with alkaline soils, does seem to have a mild adverse effect over time—browning of the leaf edges and, perhaps, yellowing of the plants themselves.

[Indoor plants can be affected even more so. The salts, minerals, and chlorine found in many water supplies build up in the limited soil volume of the pot, and can be seen as a white crust on top of the soil. Over time, chlorinated household water can even kill sensitive houseplants. For some gardeners, a rain barrel under their home's downspout is a *must* for watering indoor plants.

Alas, in spite of the romantic scenarios of bygone homesteading days, I *don't* think cisterns should be used to collect drinking water. In some areas, you can get permits for using rainwater for drinking, but the cleaning process is very complicated and costly.

For example, in a Pennsylvania study of water quality in cisterns, it was discovered that 70% of the 40 systems studied had, on at least one occasion, accumulated lead and cadmium sediment that exceeded drinking water standards "as direct input (from) precipitation." Particulate filters for roof runoff reduced the lead input from 2610 ug/L (micrograms per Liter) to 102 ug/L. The sediment from unfiltered water contained 348 ug/L lead and the sediment in the filtered system measured only 23 ug/L. Interestingly, the chemistry of the rainfall, the corrosive action of acid rainfall on plumbing, and sediment in the water supply [either one or a combination of all three] caused 11% of the tap-water samples to exceed the 50 ug/L drinking-water limit for potable water. The conclusion of this study was that "The presence of lead and cadmium in the tap water of cistern water supplies in western Pennsylvania was sufficient to constitute a hazard to users of such systems (and) users...were advised to discontinue use of the cistern [for drinking water]..."

Happily, for landscape plantings, it's a different story.

Droughts are often the source of motivation for the installation of cisterns. Recent changes in the climate seem to be making many areas more parched than in years past. The 1990s drought in my area gave rise to a renewed interest and perception of need for water storage. In the mid-1990s I researched the idea of

Figure 103. This cistern holds less than 5000 gallons, and thus can be placed on a gravel pad. It's advisable to put a thin layer of sand on top of the pad so the gravel doesn't pierce the bottom of the tank.

doing a book on cisterns, but stopped, as I couldn't see any reasonable way for homeowners to afford to build cisterns in the arid west, which I thought was my only market. Boy, was I wrong! I sure wish I had published it then!

Cisterns in regions with no summer rain whatsoever *are* expensive propositions. In such an area, you must store up enough water during the winter for as much as six months of landscape irrigation. Consider, for example, a two-acre estate in a wealthy neighborhood in Santa Barbara, California, planted with water-loving ornamentals, some drought-resistant plants, a large half-acre lawn, and a vegetable garden. This landscape, based upon the conservative figure of outdoor use equaling 53 percent of all water use, had a staggering historical year-round usage rate of 500,000 gallons! A cistern big enough to hold that much water would need to be 27 feet tall and measure a whopping 50 feet on each side! (Try and hide that in your average backyard.) Such a cistern would also cause a severe drought in your cash flow.

Cisterns in Santa Monica, CA

As a more practical example, let's take a 1000 square-foot roof in Santa Monica, California. For some unknown reason, the homeowners used Hawaii as a guideline and example. This is especially puzzling since much of Hawaii gets some or a lot of summer rain, while in Santa Monica the average rainfall is 14 inches per year. Since 1000 square feet can accumulate and disperse 623 gallons during each one-inch rainfall, a 1000-square-foot roof can accumulate almost 8,722 gallons in Santa Monica during the winter (14 X 623).

The cost of a cistern to hold *all* the runoff water from 1000 square feet would range from $ 0.35 to $1.50 per gallon for a polypropylene tank to $0.38 to $1.50 for fiberglass, and $.60 to $1 per gallon for galvanized steel with a liner. In other words, $3,038 to $13,083 (without tax) to store 8,722 gallons. Not cheap for the average homeowner.

A rebate program at the time from the city for tanks over 500 gallons offered a one-time payment of $500, a figure that was meant to include design, labor, and materials.

In addition, a pumping system to irrigate the plants is required for large cisterns, especially in relatively flat Santa Monica. This can add up to a thousand dollars to the cost, depending on whether you use a sump pump (less expensive) or a pressurized-tank system.

There's also a small lag between winter rains and the beginning of landscape watering, when some water is conserved, and may help with flushing toilets in the very few areas where runoff water is approved for this use.

Cistern Water for Landscapes

However, the water captured—still using the example of Santa Monica—can only go so far in the garden. The scientific term for the water lost through the foliage of a plant and by evaporation from unmulched soil is *evapotranspiration* (abbreviated as ET), the water loss occurring from the processes of evaporation and transpiration. 99.9% of the water that's actually used by an irrigated crop or lawn is pulled up through the roots and transpired out through the leaves, with only a small amount, maybe 0.1%, actually used to produce plant tissue

The *ET rate* is a term used to describe the water consumed by plants over a period of time, and is based arbitrarily on the growth of un-mulched turf grass measured in inches per month or year. The yearly ET rate for Santa Monica is 44 inches, but the average rainfall there is only 14 inches. The water used by plants in a 1000-square foot Santa Monica garden equals about 30,500 gallons per summer when the full ET rate is taken into consideration.

This would require a roof measuring about 3,500 square feet to capture enough water. You would need six 5,000-gallon tanks (a footprint of 11 feet in diameter on the ground for *each* seven-foot-tall tank) To hold all those tanks you would need 207+ square feet of ground. In some areas, a tank bigger than 5,000 gallons requires a permit and a code-approved concrete pad (which sometimes, depending on local ordinances, must even reviewed by an engineer)

In situations without extravagant water usage, Santa Monica city water costs about $42.76 per month for the average family. At this rate, even with the $500 rebate, the payoff time for the cost of the six 5,000 tanks (not including the plumbing or pumping or concrete pads) would be 58 years! Is this a worthy investment? Makes groovy ecological sense, but may lack money cents.

Where rains are far and few between, and sporadic, the math can easily work out in your favor. Check with the local newspaper or fire department for a sense of when and how much it rains in your area.

Using drought-tolerant plants will also greatly reduce the payback time (you would probably use nearly the full evapotranspiration amount of water if you planted an edible landscape). Although the payback period is still considerable—14.5 years—some plants such as lavender (*Lavandula* spp.), yarrow (*Achillea millefolium*), 'Powis Castle' wormwood (*Artemisia* 'Powis Castle'), bearded iris, and New Zealand flax (*Phormium* spp.) use as much as 85 percent less water than lawn grass. Some other plants with low water use, according to California's WUCOLS report (WUCOLS is the acronym for Water Use Classifications of Landscape Species), include:

Figure 104. BlueBarrel Rainwater Catchment Systems™ makes kits to fit a series of 55-gallon recycled barrels in small spaces where a large tank won't fit. The barrels can be strung together in rows as long as needed. Since they don't include a sump pump, they only irrigate at (flat) ground level or downhill. The water can be turned on and off with a timer (pictured above) that doesn't need water pressure to activate the valve. Always use a water filter. (Photos courtesy of bluebarrelsystems.com.)

Aloysia triphylla	Lemon beebrush
Aquilegia spp.	Columbine
Artemisia spp.	Wormwood
Erigeron karvinskianus	American fleabane
Eriogonum spp.	Buckwheat
Eriogonum grande rubescens	Red-flower buckwheat
Euphorbia spp.	Euphorbias
Gaillardia x *grandiflora*	Blanket flower
Gypsophila paniculata	Baby's breath
Helianthemum nummularium	Sun rose
Iris douglasiana	Douglas iris
Lantana spp.	Lantanas
Limonium perezii	Perez's sea lavender
Linaria spp.	Toadflax
Lobelia laxiflora	Sierra Madre lobelia
Mentha x piperita	Peppermint
Oenothera spp.	Evening primroses
Salvia spp.	Sages and salvias (many, but not all; check with your nursery)
Santolina chamaecyparissus	Lavender cotton
Sedum spp.	Stonecrop
Sisyrinchium californicum	Golden blue-eyed grass

Figuring Out the Size of Your Cistern

In moderate rain areas, cisterns are very practical. Instead of building the cistern to accommodate the maximum rainfall your roof can capture, there are two more practical approaches to determining size. In many areas, a "drought" lasts only two to four weeks, and you can simply store all the water you'll need to get through each dry spell.

Typically, you'll need to know the average length of the dry spells in your area. Talk to the weather reporter for the local newspaper, radio station or television station, consult the weather records at the closest fire or forestry station, or contact the Master Gardener program through your Cooperative Extension.

Next, estimate your landscape's irrigation needs. Look at past water bills and calculate the total water usage for the irrigation season. (Many bills present water use in HCFs, or hundreds of cubic feet; each HCF equals 748 gallons.) Now calculate how much of your total water use goes to the garden. In temperate climates with summer humidity, exterior irrigation use may account for only 30 percent of total water use. Much of that may be wasted on a lawn instead of perennials and food gardens. In the dry, hot summers of the West, the landscape may use 50 to 70 percent of the total. Check with your local water department's water-conservation director for the average percentage for your climate.

Even with a cistern, water conservation should be practiced in the garden. Even though a cistern gives you a measure of water independence, there's no sense in squandering this important resource.

Greywater

For years, respectable and otherwise law-abiding people all over the country have been turning themselves into criminals. How? Simply by recycling their used laundry, bathtub, or shower water (known as " graywater" or GW) into their landscapes. Since greywater has long been defined by many State Boards of Health as sewage, this type of water thriftiness has been, up until recently in some areas, officially illegal.

Payback Time

Alas, many greywater system(s) (GWS) are not cost-effective. They require so much hardware, except for the very simplest gravity systems, that the payback time is horrible. In 1990, I participated in a study of eight GWS for a pilot/research program directed by Bahman Sheikh for the city of Los Angeles. There were many reasons for the pilot-systems study; documenting pathogens was one important aspect, but none were typically found in the top nine inches of the soil. However, the cost of systems, not including labor, ranged from about $500 (for a drum, sump-pump, and drip system) to over $8,000 for a large tank with an automated sand filter.

Water is still cheap in most places. In 1990 in Santa Barbara, CA, the amount of GW retrieved from all sources, using the simplest system ($1,131, not including labor) meant a payback time of about 25 years! A $5,400 expensive system (not including labor!) might take 120 years to break even!!

A Basic Schematic of an "Ideal" House Replumbed for a Grey Water System, with a Surge Tank and a Sump Pump.

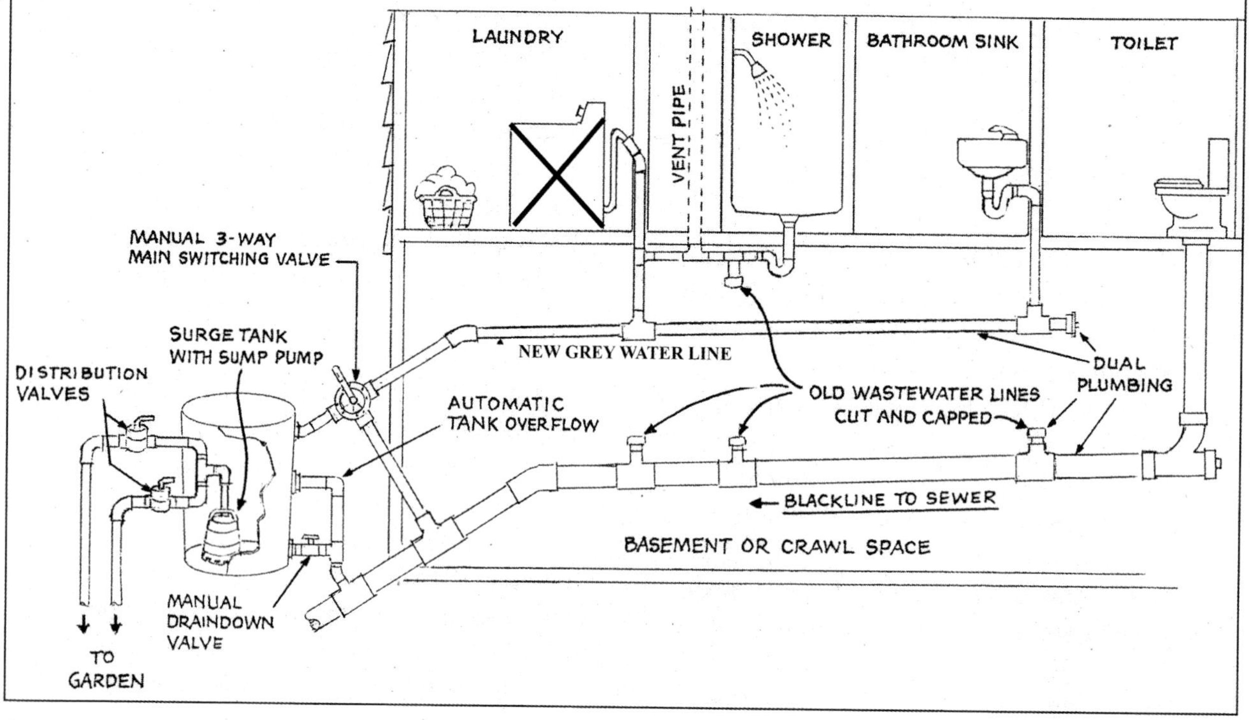

Figure 105. Here's a theoretical home, showing the plumbing for a greywater system. (It's never this easy in real life!) The manual 3-way valve allows you to switch the greywater to the sewer or septic tank when it rains or during the winter. The X on the washing machine indicates that using its greywater is very difficult and only for the most eco-exacting person. The two distribution valves to the garden allow for switching back and forth between two zones of irrigation. (See details for surge tank on the next page.) All plumbing should drain at least one inch every 10 feet. The toilet water should continue to drain to the black-water pipe and the sewer. (From my book *Drip Irrigation for Every Landscape and All Climates; 2nd Edition*; (Metamorphic Press, 2009)

All of the above figures come from Sheikh's 2010 *White Paper on Graywater*. He also reported that: "...the amount of water saved in six (of the eight) residential sites over a 12-month period ranged from a mere 2.2 percent to 11 percent of the total water use at the sites. ...The costs of the systems installed far exceeded the value of the 20-year water savings."

In contrast, Susan Carpenter, a writer for the *Los Angeles Times*, studied sustainable systems for her house in 2010-2011, and reported. "Financially, this (greywater) system is paying for itself, just slowly. LA DWP charges me less than half a penny per gallon, so technically graywater has saved me only $95 in water costs so far. But it's also reduced my sewer charge by about one third, saving me an extra $3.30 per month. In drought-prone Southern California, greywater feels like the right thing to do. Overall, it's been the easiest, most sensible, hassle-free, sustainable system I've put in place at my house." She did all the work herself, and the cost, including the systems for laundry-, shower- and bathroom sink-to-landscape was $1,988. The payback for this system was a very reasonable one year and four months.

The payback time for the laundry-to-landscape system described below is very rapid, especially if you install it yourself.

So the use of GW, in some cases, is for the environmentalists, not the bean counters. Here's more infor-

Figure 106. This is a 20-gallon surge tank located in the crawl-space of a home. The inlet is in the upper left of the tank, and the upper black pipe on the right is the overflow port, which is lower than the intake port. The white valve allows the water to be shut off from the tank and switched to the sewer or septic tank. The smaller white pipe coming off the top of the tank runs from the sump pump to the garden. The electrical cord is plugged into a GFI outlet above the tank.

mation for the green/grey gardener.

The Basics

Greywater must be collected via a separate set of plastic pipes so as to prevent any mixing with "black" water (from the toilet; it looks brown to me!). To reduce pipe clogging and limit bacteria-stimulating fats and food particles, the recycling of wastewater from kitchen sinks, garbage-disposal units, or dishwashing machines is contraindicated in all settings. In addition, a manual or electrically activated 3-way swing-gate valve or two brass ball-valves are required wherever GW pipes join; such valves allow the homeowner to switch GW back to

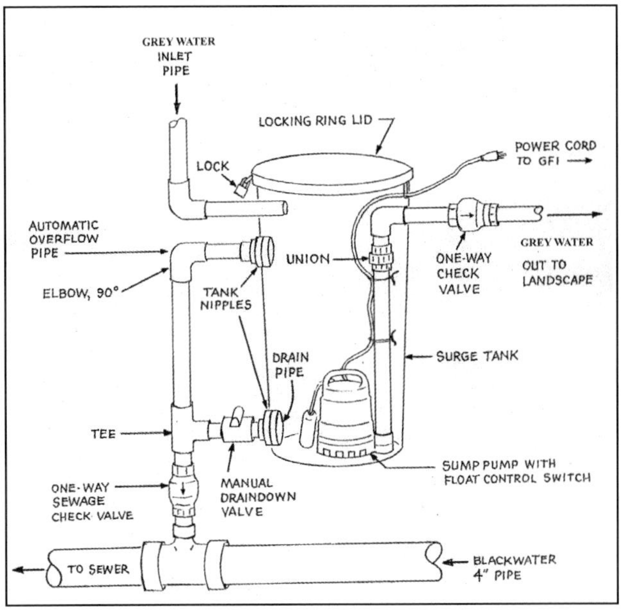

Figure 107. Details of the sump pump in a surge tank. You must install one-way check-valves to prevent the greywater from draining back into the surge tank and the sewer water from backing up into it as well. Be sure that the float-control switch doesn't hang up on the side of the tank. Always install a valve to drain the tank when needed, and lock the lid to keep out animals and people. (From my book *Drip Irrigation for Every Landscape and All Climates*; 2nd Edition; 2009.)

the sewer or septic tank in certain situations: when heavy rains saturate the soil; if someone is ill with an infectious disease; during the winter; after a heavy rain "event;" or if a new owner prefers not to recycle GW.

Most GW must pass through a drum or tank called a surge tank, which buffers extremes between the amount of GW generated and that dispersed by means of a protective overflow port to a sewer or septic tank. This drum is *not* a storage tank, as leaving GW in it for more than a few hours or a day at the most invites the multiplication of bacteria and dreadful septic fumes. A manual drain-down port allows for cleaning. (See Figure 107.)

Back in 1975 some gardeners were naive enough, and adventurous enough, to use GW with crops like strawberries that touch the soil. This was not a potential health issue, but a cautious person would not get the water on the foliage or use it with root crops.

The Limits of Greywater

A GWS doesn't reduce the size of your septic tank or city infrastructure, because the GW water has to go back to the septic tank or sewer when:

- The rains are heavy enough to make the GW "daylight"—show up on the surface of the ground.
- It rains (in the west) or freezes (elsewhere) in the winter.
- Harsh cleansers are used to clean the tub, shower, or bathroom sink.
- Bleach is used in the laundry.
- Someone in the house (or a visitor) appears to be or feels sick.
- Dirty diapers are included in the laundry. (But who does that now? Even most environmentalists use disposable diapers.)

All pipes, septic-tank size, and sewage system must remain the same so the water can be disposed of when the GW isn't being recycled.

Years of Illegal Experience (since 1975)

I had an illegal GW system at my rural house for over 25 years with no problems, no need to replace the pump, and no clogging. I used it to start young fruit trees, which then became mostly unirrigated, except by the limited amount of GW I produced. I didn't use a washing machine (didn't have one, hello laundromat) and gathered GW only from the shower, bathroom sink, and one side of a two-chamber kitchen sink. The kitchen sink is illegal in all current GW regulations, and it required persistent observation on my part to use the GW side to wash veggies, rinse already-washed dishes, and drain non-toxic liquids like tea and coffee dregs. Since I had a low-flow showerhead, took short showers, and produced few dirty dishes, that GW system was turned on only every four to six days.

The drought of 1975 hit Northern California just as I was starting my garden maintenance company, so I had to quickly adjust and adapt. There were no laws about GWS back then, but desire for the systems was immense. Five of my clients wanted GWS so they could save at least a part of their landscape plantings. Another ran four-inch flexible pipes from roof gutters to their swimming pool to turn it into a "cistern."

Time to go to the hardware store and plumbing supply company. Somehow each system I created had a feature that became standard in the drought of the 1990s, a fifty-gallon drum; I got mine from a supplier of plastic-lined food-grade drums in San Francisco. I plumbed several together to capture from several sources and hold GW until it got used in the garden, and quickly discovered that storing GW for any length of time produces a god-awful sewage smell. Thus I learned one of the first rules of GW—use it as soon as it's generated, and *don't* store it.

At the same time, the Farallones Institute Urban Center, in Berkeley, CA (an urban display house for alternative energy, intensive food production, and alternative technologies for the protection of the environment) was also experimenting with GW. In an attempt to strip some of the phosphates and other nutrients from the GW to prevent the rank smell, they proposed a fifty-five gallon drum sitting above a barrel filled with gravel. The theory was that the gravel would accumulate a coating of algae and bacteria that would prevent the stench. Alas, it was a total failure; this system failed to act like a scrubber *and* it was also enormously difficult to get rid of the gravel.

Is Greywater a Health Risk?

The potential for GW to transmit pathogens has long been the concern of most public health departments. Greywater can contain the same types of pollutants and pathogens as black water—but, usually in very tiny amounts, if at all. However, to date, and in spite of various theoretical scenarios, there has not been a single illness that has been documented to be caused by GW.

Most pathogens can survive outside of the body of a host for some period of time, but each disease has a different life-span in the soil. Some of the survival times for various pathogens are:

Organism	**Survival Time (in days)**
Ascaris spp. ova [roundworm]	up to 7 years
Entamoeba histolytica	6–8
Enteroviruses	8–175
Hookworm larvae	42

Salmonella spp. [typhoid fever]	1–120
Streptococci spp.	26–77

Most of these illnesses are transmitted exclusively via the "fecal-to-oral" path. The important caveat to the above list is that, in order for you to get sick from a GWS, someone else in the household must have a disease; fecal matter contaminated with the disease organism or parasite ova must get into the GW; you must come in contact with the GW (or soil that has been watered with GW); and your hand must come in contact with your mouth. If nobody's ill (including visitors), then there's nothing to transmit via a home GWS. Also, all systems must have a manual or automatic valve that can divert the GW to the septic tank or sewer whenever someone is ill. All these scenarios and facts are irrelevant when the GW is distributed both automatically and out of reach below the surface of the mulch or soil.

Remember, with a GWS, you and your family are now acting partially as your own wastewater treatment "plant," and its judicious maintenance and management are your responsibilities.

In 1992, the City of Los Angeles Office of Water Reclamation did a one-year study about the concerns of public exposure to pathogens. The report, entitled *Gray Water Pilot Project,* states: "if [the] data can be generalized, the implication is that gray water irrigation—below the surface of the soil—does not by itself elevate the health risk from handling the garden soil, as long as sanitary practices are followed." Furthermore: "Disease organisms, normally capable of surviving in the soil for a few days, were not present in gray-water-irrigated areas." As to the risks in cleaning the filter bags: "Individuals assigned the task of cleaning gray water filters—some doing so without protective gloves, in spite of instruction to the contrary—did not report any adverse effects."

In a thesis study involving carrots, lettuce, and peppers, Sara Finley found that: "Heavy metals were not detected in the greywater, but both fecal coliforms and fecal streptococci were present in high levels, averaging 4 x 105/100mL and 2 x 103/100mL of greywater, respectively. Despite these high counts, no significant difference in contamination levels was observed between crops irrigated with tap water, with untreated water, and with treated greywaters. Fecal coliform levels were highest in carrots and fecal streptococcus levels highest on lettuce leaves. However, contamination levels for all crops were low and do not represent a significant health risk." (*Reuse of Domestic Greywater for the Irrigation of Food Crops,* Sara Finley, Department of Bioresource Engineering McGill University, Montreal, Canada. 2008)

Even though you won't have any of these organisms in the garden if you're healthy, the prudent person will avoid using GW with plants that require working the soil several times per season, annual ornamental bedding plants, root crops or vegetables that touch the soil. The safest use of GW is in a well-designed landscape planted with perennial shrubs and trees that won't be moved or replanted (or fruit trees, if you refrain from eating the fallen fruit). Also safe are edible shrubs with crops well above the soil (may require sufficient pruning), corn, trellised tomatoes (properly managed), trellised beans and peas, sunflowers, etc. It also means you *can't* rotate crops by planting root crops in soil previously used by the GW crops.

Caution dictates washing your hands—just like Mom always said—after gardening where GW is distributed.

The California 2010 Code for a GWS

The current (2010) California GW law stipulates that a homeowner (single- or two-family households only) can plumb the discharge from a washing machine without a permit to a well-mulched sump or "moat," to any other well-mulched area, or to a branched drain system. Guidelines include the following:

- A city or county or any local government can, after a public hearing, restrict or prohibit the use of GWS.
- The guidelines allow the homeowner to direct the flow from the home only to a specific irrigation area.
- The GW must be contained within the irrigation area.
- The GW must be used on the yard where it is generated.

Table 4.1. State Analysis of Graywater/Wastewater Regulations

States without Formal Graywater Regulations			States Allowing Graywater Reuse		
States allowing wastewater reclamation that define graywater as wastewater (4.1.1)	**States not defining graywater (4.1.2.1)**	**States treating graywater as septic (4.1.2.2)**	**States permitting graywater using a tiered approach (4.2.1)**	**States regulating graywater reuse without a tiered approach (4.2.2)**	**States allowing residential irrigation only (4.2.3)**
Alabama	Illinois	Connecticut	Arizona	Florida	Hawaii
Alaska	Kansas	Kentucky	California	Georgia	Idaho
Arkansas	North Dakota	Maryland	New Mexico	Montana	Maine
Colorado	Ohio	Michigan	Oregon	Massachusetts	Nevada
Delaware	South Carolina	Minnesota	Washington	North Carolina	
Indiana	Tennessee	Nebraska		South Dakota	
Iowa		New Hampshire		Texas	
Louisiana		New Jersey		Utah	
Mississippi		New York		Virginia	
Missouri		West Virginia		Wisconsin	
Oklahoma				Wyoming	
Pennsylvania					
Rhode Island					
Vermont					

Table from "Treatment, Public Health, and Regulatory Issues Associated with Greywater Reuse. *Guidance Document*. By Sybil Sharvelle et. al. for WERF

Figure 108. US states that use a "tiered" approach apply different regulations to differently sized greywater systems. (From the Water Reuse Foundation, 6666 West Quincy Avenue, Denver, CO 80235.)

- Daylighting (exposure to sunlight, meaning exposure to people and animals) or runoff is prohibited.
- Greywater requires at least two inches of mulch, rock, or soil covering the area being irrigated.
- All GWS must be designed to prevent contact with all people and pets. (No mention of wildlife.)
- The GW can't be used on root crops or directly applied to edible portions of food crops. (Notice that they *don't*, as they should,

consider the tendency for gardeners to rotate crops.)

- All laundry water used to wash diapers must be diverted to the sewer or septic tank.

- The GW can't contain hazardous chemicals such as cleaning fluids, residue from car parts, water from washing greasy, or oily rags, fluids from photography darkrooms, or any liquids from similar sources.

- You can't use a drinking-water connection to supplement the GW.

- You must produce an operation and maintenance manual so others can use the GWS safely. The manual is to be kept in the house throughout the life of the GWS and indicate that, with a change of ownership or occupancy, the new people must be notified that there is a GWS.

- Any GW storage tank must be marked as GW and not potable. Gloves should be handy for use when working with the pipes, tank, and/or sump pump.

It's also important to know that GW should not be distributed during heavy rains as runoff can easily occur. During the winter, the system must be shut off and the water rerouted to the sewer or septic tank (because of seasonal rains in the northwest and southwest and freezing elsewhere). So, every system needs a valve plumbed in just before the pipe reaches the surge tank, in order to control the GW in the off-season. Unfortunately, this means your septic tank cannot be sized smaller due to the use of GW, as it must hold the amount of GW produced during the off-season.

Where Laundry-Water GWS Don't Work

Most washing machine discharge exits the machine through a curved pipe, and up and over into an open discharge pipe. Since most laundry facilities are located against an external wall, this source of GW is usually available.

However, other sources of GW are precluded in all slab-constructed houses. The GW sources are "entombed" —plumbed into the black-water pipes before the sewage water leaves the slab. Second-floor laundry machines are also out, as the GW is usually connected to the black-water line before reaching a downstairs basement or crawl space—if there is one.

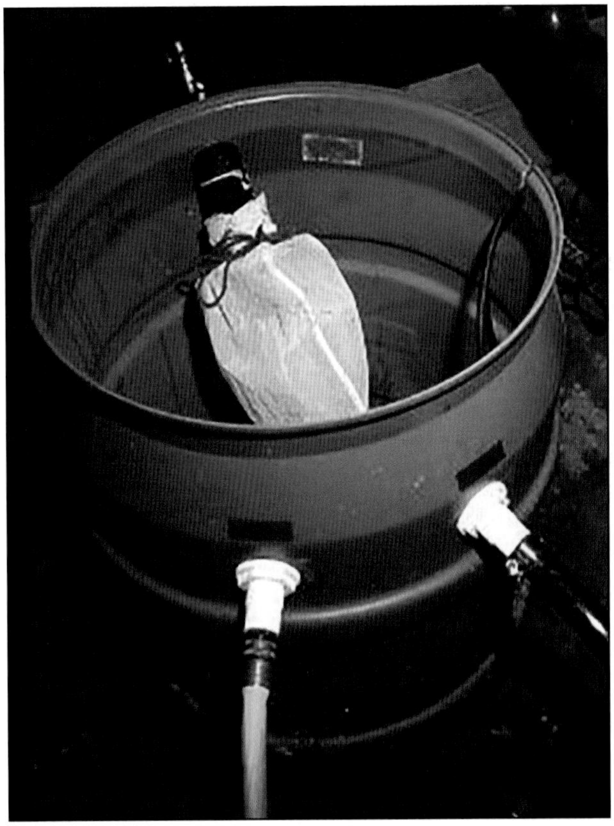

Figure 109. This is a system (shown with the lid off) that was connected to a washing machine. A 100-mesh filter bag is a must, and you can usually find them at beer supply stores. The blue hose on the right is the overflow tube, and the green hose goes to the landscape. This system only works if you are religious about cleaning the filter bag.

There are significant limitations to washing-machine pumps, which are only designed to get the water up, over, and into the inlet of the waste-pipe that goes to the sewer, and are not geared toward providing pressure without breaking. The most you can count on with a washing-machine motor is lifting the water no more than three feet (some modern machines can lift the water eight feet) and pushing it (on level ground in a two-inch pipe) up to 50 feet. Call your machine's manufacturer to confirm these numbers. Basically, washing-machine pumps really should only be used for level or downhill applications. Be sure to install a check-valve and vacuum-breaker to keep the water from draining back into your machine.

Most renters can kiss a GWS goodbye, unless they have a very environmentally minded landlord who allows the laundry-to-landscape system.

A Sump Pump for Hilly Sites

Even when using a dispersal system with a washing machine, extra pumping is often required. Where the total up-and-down elevation exceeds five feet, it's best to protect your washing machine's pump with a sump pump installed in a surge tank (often a 55-gallon plastic drum). A sump pump is equipped with valves that sense escalating water levels or pressure. When the water gets too high, they automatically pump excess water out through a discharge line. This kind of pump is required to re-pressurize the water when the landscape you want to irrigate is level with the surge tank but more than 50 feet away, or if all the plants are uphill from the tank. Sump pumps are designed to be submerged in water, and to automatically turn on when the water reaches a predetermined level. Each pump has the capacity for moving the GW a maximum height and distance—measured as the head (the vertical lift) and the run (the horizontal length). The flow rate of each sump pump is rated in gallons-per-minute (gpm) or gallons-per-hour (gph).

- Sump pumps are the most convenient pumps for homeowners.

- Sump pumps are basically designed to evacuate a large amount of water as quickly as possible. They can move lots of water in a short period of time, but often can't lift the water very high (they have what's called a low head).

- The float control on a sump pump should be set to drain the tank to as close to empty as possible. The less GW left to turn septic in the surge tank, the less pungent odor there will be.

- The pump must be wired with grounded direct-burial wire to a GFCI outlet that detects imbalances in the electrical current and shuts off power to that outlet to prevent damage and diminish the risk of shock. This outlet must be located *inside* the house. All electrical work must be done by an electrical contractor, and meet UBC guidelines.

- Choose a sump pump constructed of metal parts inside of a metal housing. (Plastic-bodied pumps often need four or more inches of standing water in them at all times, to keep the pump from overheating.) The slotted or perforated grate for the intake of the GW and the impeller that ejects the water out through the pump may be plastic.

- Sump pumps have one of two types of automatic switches for level control: a round or cylindrical mercury-float switch tethered to an electrical cord, or a vertical-path float switch with a float that slides up and down a pin passing through a vertical channel in the float. With a mercury-float switch, make sure that the tether doesn't get caught on any of the pipes or the pump itself.

- The threaded outlet port of the sump pump should be as large as the pipe or tubing from the washing machine.

- The sump pump must be rated for water heated up to at least 130° F.

- The horsepower rating of the pump is the least important specification. The important pump ratings are the head (the lift in feet) and the flow rate, in gallons-per-minute (gpm) or gallons-per-hour (gph).

- To choose the right sump pump for uphill pumping, you'll need to know the desired head.

- Typically, the head equals the gross difference in elevation. As a safety factor, the head for a GWS is determined by adding the total changes in the elevation, both up and down, from the surge tank to the point of dispersal.

To the head add at least 15% more feet of head. (In other words, if the total change in the elevation from the tank to the uphill shallow leach fields is 13 feet, than the sump pump must be rated for at least a 15-foot head. More is better.) Don't expect your pump to last long if it's always pumping at the uppermost limit of its rated capacity.

- You may want to check the pump manufacturer's chart of lift and flow rates to determine how much water can be pumped at the required head. Many pumps can discharge a huge volume of water when the point of release is just a foot or two higher than the pump, but then the gpm drops off rapidly with increased elevation. Washing machines discharge a lot of water very quickly. To prevent ejecting GW via the surge tank's overflow port, make sure the sump pump can discharge enough GW at your required elevation to keep up with the maximum discharge (in gpm) of the washing machine.

- There must be a flapper check-valve installed on the threaded discharge port of the sump pump, and it must be of the same diameter as the discharge pipe. This will prevent any water from siphoning back into the surge tank to turn on the pump again or leave GW to turn septic and smelly.

- The check-valve leads to the surge tank's discharge pipe. Where it passes through the walls of the tank, the discharge pipe must be PVC schedule 80 pipe or iron pipe. (PVC pipe inside of the tank can be schedule 40.)

- Repair or cleaning of the pump will be facilitated if the discharge pipe has a union (see Figure 107) located somewhere along its length after the check valve and before the tank intake. Only use threaded fittings within the surge tank (and always use gloves dedicated only to GW use.)

Gravel Mini-Leach Pits with Flower Pot "Emitters"

During the 1990 drought in Santa Barbara, a clever fellow named Larry Farwell developed a simple homemade disposal system. He made "emitters" from a number of upside-down flowerpots or other hollow containers placed over a pocket of gravel. (Versions of these are also called "branched drainfields.") By pairing the upside-down flowerpots with simple drip-hose valves, it's possible to adjust the flow of each pot to a reasonable rate. This is best used on relatively flat land. (See Figure 110.)

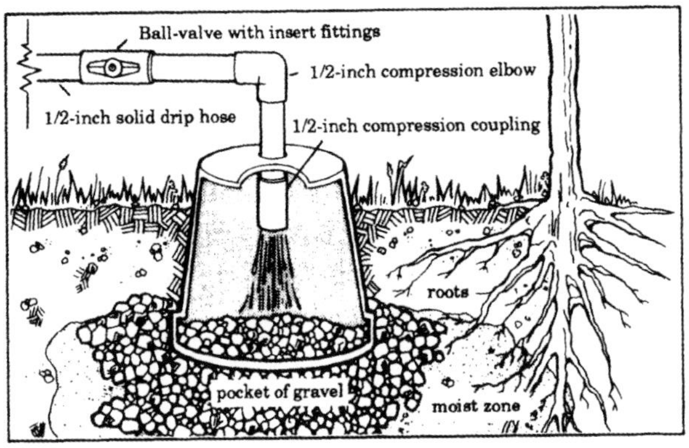

Figure 110. An "emitter" developed by Larry Farwell during the 1990 drought in Santa Barbara, CA. Use a plastic one-gallon nursery pot for the "emitter." The ball-valve is to adjust the flow to all "emitters."

Benefits – Fairly simple to construct, the upside-down flowerpot emitter has many benefits: it creates an air gap between the GW hose and the gravel so that roots won't grow back into the irrigation hose; provides more even distribution of GW than a mulched moat; spreads GW to more plants; can be used with shrubs and trees; doesn't require much gravel to be hauled in; is fairly simple to add to an existing planting; works well downhill from a gravity system; and works on gentle slopes.

Limitations – It takes some gravel, which may have to be carted in, and it takes time to adjust the ball valves to equalize the distribution of the GW. Any exposed tubing is ugly. Aboveground tubing may be chewed by dogs, and gophers may chomp on belowground tubing. If there's no wire bottom on the pots, gophers may also fill them from below with a "throw"—the dirt from tunneling through the soil. The pot-based emitter is also more capable of clogging than a water moat or a shallow leachfield and is not appropriate for lawns and most ground covers.

Two Zones

Each GW disposal system must have at least two separate areas (zones)—two tree moats, two lines of flowerpot "emitters"— between which to alternate on a regular basis. (See Figure __.) The rotation between the two zones allows aerobic bacteria in the soil to destroy pathogens and consume nutrients, permits the soil to drain, and helps to maintain a good aerobic soil condition. If one zone is watered with GW on a daily basis, the soil may become saturated and anaerobic, and its structure may degrade.

The soil type, plant type, length of disposal pipe or tubing and amount of GW will determine how frequently you need to switch between zones. For example, a single person dispersing GW to a sandy soil might only need to switch between two drought-resistant plantings every few weeks, while a big family of four, irrigating a heavy, clayey soil might maintain four large zones and rotate the GW to each zone for only a single day. Observation and experience is the best guide.

Drip Irrigation with Greywater

Assuming you want to plunge into the troubled waters of drip irrigation with GW, here's the scoop.

More than with regular city water or well water, your drip-irrigation system needs to be protected from clogging. A 200-mesh screen or filter bag is a must. (See Figure 109.) By avoiding the washing machine, you're already reducing your risks. (Remember all the lint you have to clean off the dryer filter? It's an indicator of how much "stuff" is likely to show up in the filter bag.)

It seems—but has not been proven—that some GW particles in suspension glob up together after passing through the filter, so a drip-irrigation emitter with the shortest internal path and the largest exit hole is required. This would suggest an in-line emitter such as the Geoflow Wasteflow® PC 1/2- and 1-gallon-per-hour (gph) products with Treflan® (trifluralin, a pre-emergent herbicide for the control of certain annual grasses and weeds) embedded in the emitters. This tubing is designed for effluent that has more particulates than potable water. Also, consider the Netafim Bioline®, which is impregnated with Vinyzene® to prevent the growth of microbial slime. (Neither of these examples should be used in an organic garden or landscape.)

The best carrier tubing for dirty water, even iron-filled rusty water from a well, is my favorite of the three: Netafim's older pressure non-compensating emitter tubing with a one-gallon-per-hour flow rate. Some manufacturers offer 2 gph emitters that add another level of safety against clogging, and now there's even a new kind of tubing designed just for GW. It's called irriGREY™, and is designed with 2 gph emitters that leave a large path for the water and are designed to clog less than conventional emitters. This tubing has been used for decades in Australia, and was introduced in America in 2013. (See the irriGREY handbook for the Netafim products at the following website: www.netafim.com.au.) Keep all tubing on top of the soil, but cover it with an attractive mulch to protect it from the sun and avoid problems with gophers.

If you still have a problem with clogging, the next step is to add a liquid fertilizer/proportioner to the system to flush the tubing and emitters with a mild acid during or after a rainy day.

For a much more detailed discussion of GW and how to use GW with drip irrigation, see *Drip Irrigation for Every Landscape and All Climates; 2nd Edition*; (Metamorphic Press, 2009).

For Greater Yields, the Dirtier the Better

The "dirt" in GW is also nutrients. In reality, filtering GW for a drip irrigation system is counter-productive, because unfiltered GW can increase the yields of plants.

In a study in South Africa, studies of carrots, "beetroot," spinach, and peppers grown separately with GW, with nutrient solution, and with tap water, always showed a considerable increase in yields with GW, as compared to tap water. The nutrient-solution plants always produced the greatest yields, but the GW plants had "...yields similar to that obtained when using chemical fertilizers". (Furthermore: "Greywater irrigated produce is likely to be safe for human consumption (based on microbiological analyses.)." Notice that two of the crops were root crops.) (*Greywater Use for Agricultural Irrigation in Urban*

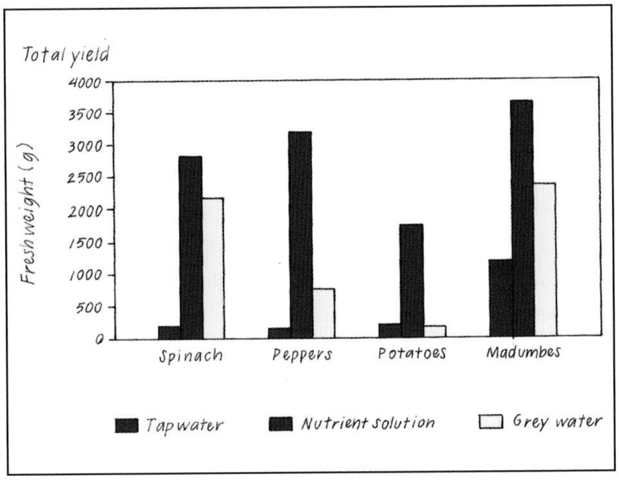

Figure 111. The "dirt" in greywater can act as a fertilizer, as shown here. Greywater is usually superior to tap water for yields. Madumbes are tubers of the taro or elephant-ear plant.

and Peri-Urban Areas, Lumka Salukazana, S. Jackson, N. Rodda, M. Smith, T. Gounden, N. Macleod, University of KwaZulu-Natal; Thekwini Municipalit, South Africa.)

Figure 111 shows the results of another study, at the University of KwaZulu-Natal. Madumbes are tubers of the elephant ear plant (Taro)

Yet another study, at Texas A &M in El Paso, Texas, surveyed the yields of vegetables (tomatoes, bell peppers, and jalapeño peppers) and cotton grown with GW compared to brackish water. The quantity of the vegetables grown with GW increased by 20%. Cotton that was mulched and irrigated with GW produced 70% more cotton fiber (lint), compared to conventional plantings with no mulch and irrigated with brackish water. In the end, they found that: "Cotton plants irrigated with graywater grow better than those irrigated with freshwater." (*Use of Graywater and Brackish Water for Cotton and Vegetable Production*, ZhupingSheng, Ph.D., P.E., Thesis, 2010)

(From "Effects of Graywater on Soil Chemical Composition and Yield Of Tomato Plants," 2008. Yasin Al-Zu'bi * and Fadhil Al-Mohamadi, Food, Agriculture and Environment [JFAE], Vol. 6, Issue 2, p 408-410.)

"...Tomato plants irrigated with ablution water [from the washing of the hands and body] gave significant higher yield than those irrigated with other sources of graywater or those irrigated with tap water."

Detergents for Safe Irrigation & Effective Washing

The single most important aspect of a washing-machine GWS is the choice of detergent for use with the laundry. Some detergents contain chemicals and compounds that are quite caustic to soil structure or plant roots. Other ingredients are actually benign, and some even beneficial. All three categories often don't occur in the same detergent or soap.

The short course of detergent ingredients is simple:

Salts (mostly sodium chloride, or simple "table salt") are the biggest threat to plants irrigated with GW, especially in arid climates where the rains aren't sufficient to leach away salts accumulated during the dry season. Sodium chloride destroys soil structure and kills soil bacteria, beneficial soil life, root hairs and plants. Many compounds in detergents are sodium-based, but their technical names obscure the sodium content. See below for a complete review of these damaging compounds.

Definitions of the Words in Detergent Labels

Aluminosilicates - Detergent builders (a "builder" improves or maintains the efficiency of the detergent's surfactants—the chemicals which make water wetter and thereby improve the cleaning process). Some also make the water "softer."

Anionic Surfactants - Cleaning agents, most commonly some form of sodium salt. Usually found in high sudsing detergents.

Antiredeposition Agents - Prevent loosened soil from resettling on the fabrics. The most common one is a sodium-based chemical—sodium carboxymethylcellulose (CMC).

Enzymes - A group of "complex proteinaceous molecules." They act as catalysts to help break down dirt and stains. Their activity is destroyed by liquid chlorine bleach.

Nonionic Surfactants - Not charged chemically, but very good at removing oily soil. Low sudsing.

Phosphates - Often used to help "soften" laundry water. This is the chemical so famous for "ruining" lakes and rivers, and the focus of many environmentalists and regulatory entities in the 1970s. Phosphorus chemically inactivates calcium, magnesium, iron, and manganese without forming a precipitate. It's a great ingredient for a GWS because plant roots utilize the phosphorus like a fertilizer. Since a GWS is managed for the improved growth of the plants, the more phosphates the better, and with a GWS, it won't be leaching off your property to turn creeks or rivers green with algae bloom.

In the mid-1990s the detergent industry voluntarily banned phosphates from laundry detergents. In 2010, sixteen states banned phosphates from dishwasher detergents. California's governor vetoed the ban, making it the only state to reject it. The other states were Illinois, Indiana, Maryland, Massachusetts, Michigan, Minnesota, Montana, New Hampshire, Ohio, Oregon, Pennsylvania, Utah, Vermont, Virginia, Washington and Wisconsin.

Sodium Carbonate - A strong (harsh, caustic) alkaline salt used in powdered detergents to soften water by precipitating the hard minerals (calcium, magnesium, iron, and manganese) out of solution. Also called soda ash or washing soda.

Sodium Citrate - Used mostly in liquid detergents without phosphates to soften the water and improve the effectiveness of the detergent.

Sodium Perborate - A non-chlorine bleach which is gentler than traditional chlorine bleaches. In powdered detergents, sodium carbonate is often added, to help perborate work better. Releases active oxygen while it's cleaning. Affected by temperature. Hot laundry water shortens the time required for bleaching.

Sodium Phosphate (actually, sodium tripolyphosphate) - The most important form of phosphorus and the first builder (cleansing enhancer) used in modern detergent fabrication and manufacturing. See **Phosphates**.

Sodium Silicate - Used in powdered and some liquid detergents. Helps with the manufacturing of powdered detergents. Also acts as a buffering agent with hard water, inhibits corrosion of the washing machine, and keeps dirt suspended in the wash-water.

Sodium Sulphate - Found in most granular detergents. Mostly neutral as a cleaning agent, but essential in the manufacturing process to increase the powder's pourability, and to help form the granules and keep all the separate detergent compounds evenly mixed.

Soil-Suspending Agents - Keeps dirt suspended in the laundry water and prevents filth from re-depositing onto the clothes. Works much like an anti-redeposition agent.

Water Softeners – These are usually sodium-based compounds to adjust water "hardness" caused by calcium, magnesium, iron, and manganese content. The minerals are removed from the water and replaced chemically by sodium. Works like a household water softener.

You needn't be overly concerned about eliminating *all* sodium if rains in your area are normal. A rudimentary study of GWS, entitled *The City of Los Angeles Office of Water Reclamation Graywater Pilot Project,* was done in Los Angeles during a drought in 1991-1992. The study looked at the sodium buildup from a range of typical detergents in the garden soils of eight different sites. The preliminary conclusion drawn from this one-year study was that "The one site that has consistently used regular powder detergents available on the market exhibited acceptable values for sodium (and) chloride..."

Later in the study the author concludes: "It is not possible to observe any important horticultural effects from a one-year test of gray water application on the landscape plants...no symptoms of harm to the plants have been observed, even on those sites where regular detergents are being used." Off the record, the best guess from this study is that sodium was building up in the soil slowly enough so as to only require a deep flushing with potable water low in salts during normal rain years in Los Angeles—every 6 years or so.

For recycling GW from the laundry, there are four basic guidelines:

First, avoid all powdered detergents. Most of the compounds used to add bulk to these detergents are sodium-based. Instead, use a concentrated liquid detergent, which probably contains much less total sodium.

Second, switch to those laundry products made specifically for GW use, i.e. manufactured with virtually no sodium (one product has only .25% sodium) so as to virtually eliminate this caustic chemical. A sodium-free example is Oasis Biocompatible Super-Concentrated Laundry Liquid, which actually breaks down into plant nutrients.

Third, stop using boron-based products. Boron, once it has been added to the soil, doesn't easily leach away, and plants react negatively to it even at very low levels—irrigation water with only .3-to-2 parts-per-million (ppm) boron will harm many agricultural crops. While boron seldom actually kills plants, it does cause a very noticeable and ugly blackening or browning of the outer edge of their leaves.

Fourth, reduce or eliminate chlorine bleach and detergents with chlorine bleach additives. Chlorine bleach is quite caustic to all life—that's why it's used as cleanser for the coronavirus (but *not* injected). Non-chlorine bleaching agents and hydrogen peroxide can be used to get those whiter whites. (Some GW with bleach can be recycled if other GW with no chlorine bleach—such as shower water—is blended with bleached laundry water or if the two sources are rotated to the same planting zones.)

Greater Garden Yields with Drip Irrigation

(It's not just for droughts anymore.)

Now let's conserve the rest of your garden's *fresh* or *well* water.

Most people think of drip irrigation as primarily useful in times of drought, for saving money on the water bill, or for maintaining container plants. However, as the following examples will demonstrate, drip irrigation can also be used in any gardening situation to *increase* vegetable and plant yields—even in climates that get periodic summer rain or are very humid.

Many studies reveal the overall horticultural efficacy of the slow dripping of water. The rest of this chapter presents some examples of research, all of which demonstrate and verify the fact that plants watered with drip irrigation produce more vegetables (it increases the production of cotton, too!) while using *less* water. (In my experience, this type of irrigation produces the same effect with ornamental growth and bloom, but as the focus of most studies is on economic food crops, and on finding results that farmers can use for increased production, its effect on ornamentals hasn't been widely studied.)

I'll also present a detailed review of how it's possible to simultaneously reduce water use *and* increase yields. This apparent contradiction is due to the intimate relationship of properly applied water with the health of the irrigated soil.

The Healthiest Soil

The first step is promoting healthy soil for healthy irrigation. Healthy soil contains enormous numbers of beneficial critters: worms, "bugs," microbes, algae, mycelia, etc., each with its own small but vital role in making soil fertile and workable. Most of these helpful little organisms need oxygen; they also need to be able to exhaust slightly toxic gases—sort of like miniscule farts—resulting from their metabolic activity. To allow this normal venting of gases, the soil needs healthy pore spaces.

Pore spaces are minute channels within the soil, kind of like distorted and contorted mini-tunnels (see Figure 112) that allow oxygen to enter the soil and vented gases to escape. This underground network is essentially the result of humic acids, humus, and organic matter binding particles of the soil together.

Humic acid (found in humus, which is broken-down organic stuff) is a byproduct of the decomposition of organic matter. It's a relatively stable material that binds to inorganic particles (sand, clay, etc.) and forms, in soils containing these particles, a sort of suspension or web. See Figure 112 for a close-up view of humic aggregates that create minute voids between

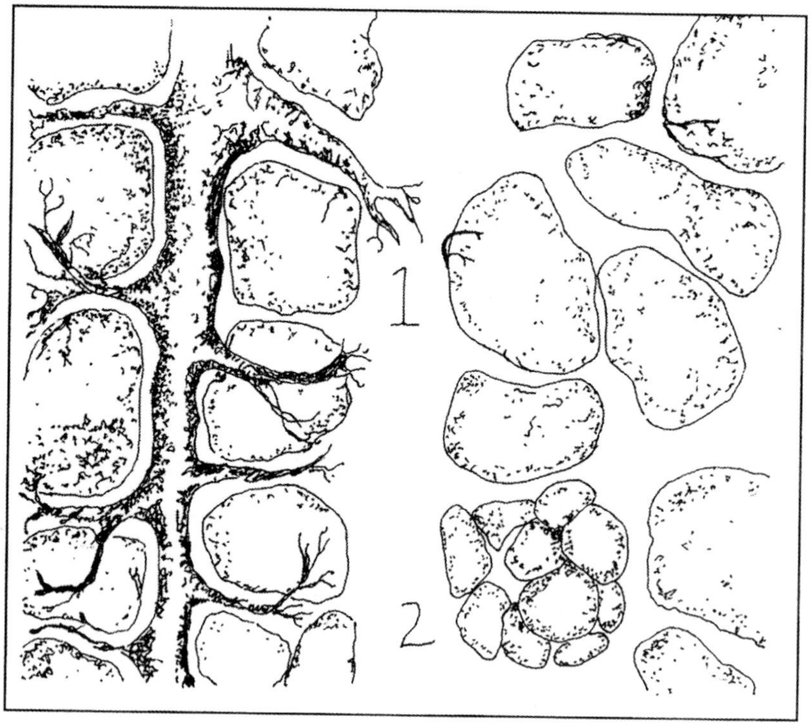

Figure 112. Number 1 in the diagram shows the roots utilizing the pore space in the soil and wrapping around mineral particles. Number 2 shows the continuous pore spaces.

soil particles; these voids are the pore spaces, and they are what makes good soil feel friable (crumbly) to a gardener.

Once created, pore space requires air in order to survive; too much moisture in the soil (and it doesn't take much) will fill and clog pore spaces and prevent the exchange of gases, causing the former aerobic (air-charged) environment of the pore spaces to become anaerobic (lacking beneficial oxygen).

Anaerobic soils are unhealthy soils. Much of the beneficial life in anaerobic soil dies or goes into hibernation, and is thus not available to convert inorganic material (such as minerals) into organic forms that are soluble and can be absorbed in minute amounts of moisture, (as opposed to pore-clogging water) by root hairs. Root hairs are the single-cell extensions of tiny roots that do most of the absorption of nutrients.

Each root hair has one nucleus and lives for only one day. A root hair requires a friable soil for maximum absorption of nutrients. The epidermis is the single-celled outer layer of a tiny root.

Once the source of over-watering is removed, it takes time for the pores to drain and the aerobic activity to begin again. Pore space is easily "drowned," and the heavier the soil, the more slowly the pores drain. Clayey soils have the fewest pores, while sandy soils have the most (see Figure 113).

The dark spots at the top of the soil indicate the difference in drainage width between clayey and sandy soil. The narrow dark top color of the sandy soil (on the right) indicates a relatively quick rate of water drainage, while the much wider dark, gray and white areas of the clay show how slowly clay soil absorbs moisture. Clay soil becomes saturated much more quickly than sandy soil because it drains much more slowly. (The number of minutes and hours will be discussed in the upcoming discussion of emitters and soil.)

Irrigation and Pore Space

Sprinklers, while relatively easy to use, are likely

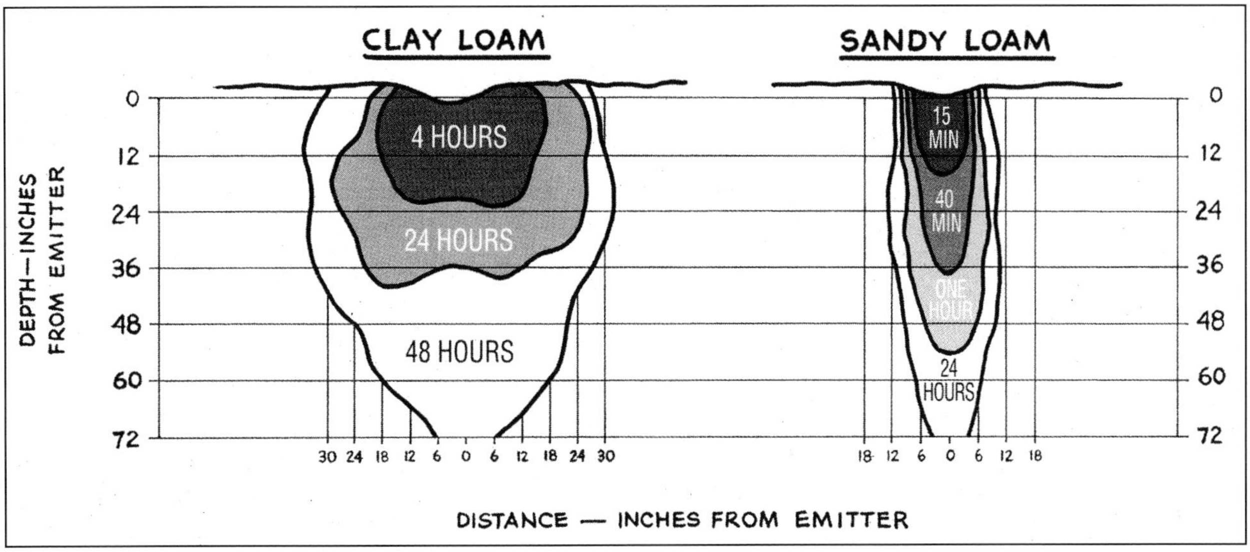

Figure 113. The dry area seen here between the emitters is very deceiving. Down below it's a different story, as the wet spots under the emitters will spread sideways underneath the surface. In sandy soil the wet spot is elongated, like a carrot shape. In clayey soil the shape is wide, more like that of a turnip.

to flood the pore spaces of the soil by completely saturating its upper layers. This over-saturation, if continuous or repeated regularly, will kill or "maim" some or most of the beneficial soil life. If the soil is allowed to drain and dry, allowing unimpeded pore space to re-form, the remaining soil life can repopulate the soil. On the other hand, if the soil is allowed to dry out too much, the pore spaces will also dry out enough to kill off some of the soil life due to oxidation—with the extra air in the pores, the oxygen encourages a sort of slow burn of the soil's microbes, algae, bacteria, etc. They don't flame, but die nonetheless.

Going back and forth between soil that is over-saturated and then allowed to drain/dry can create a too-wet/too-dry yo-yo scenario that can stress plants enough to slow their growth. The plants will still

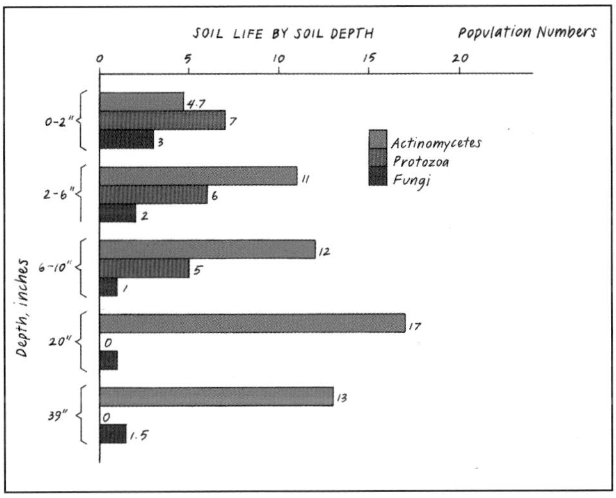

Figure 114. Most soil life prospers near the surface of the soil—the more aerobic the better. Here the protozoa don't even show up 10 inches under the surface. The exceptions to this rule are actinomycetes, which prefer a deeper soil. Note that the fungi still grow 40 inches below the surface.

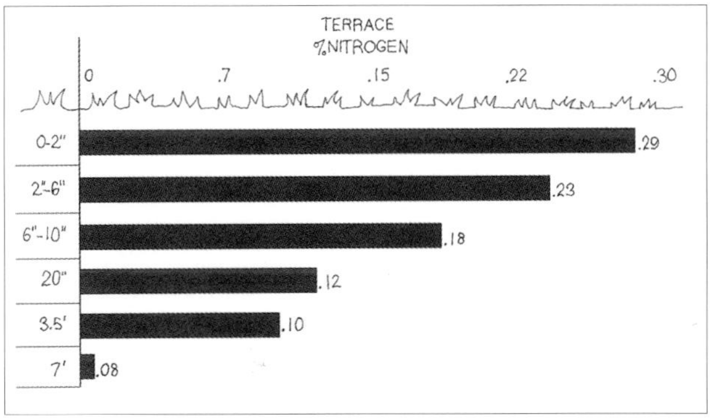

Figure 115. Deeper soils are less hospitable to the microbes, bacteria, and other soil life that are responsible for converting minerals into soluble nutrients; this is especially the case with nitrogen conversion. Notice here the changes of N levels with changes in soil depth. The top two inches of soil contain more soluble nitrogen than the next four inches below, and the two-to-six inch level holds 21 percent less nitrogen than in the top two inches. This applies to most soil life. In one study, the amount of blue-green algae found six inches down was one-fifth of that found in of the top three inches of soil.

grow, but not at their ideal rate.

The only way to use an oscillating sprinkler properly is to turn it on for very brief periods of time, then off for a while, then on again for a short time. (Only experience and careful observation will teach you the proper lengths of time.) The short periods of irrigation allow the water to sink through the pore space. The off period makes sure that the soil remains somewhat aerobic and the soil life doesn't drown. Automatic systems that can be programmed for doing this are rare, and manually turning oscillating sprinklers on and off or moving an arching oscillating sprinkler around is way too much work for most people.

Micro-sprinklers, which emit a finer spray that distributes less water than typical lawn sprinklers, are somewhat less harmful to growth (even though a considerable amount of the misty water often blows away in the wind), but are also often left on long enough to clog the surface pore spaces. Drip systems, which can be easily adjusted to deal with all these situations, provide the healthiest form of irrigation.

Drip is Mighty Bountiful

Here are some articles and quotations, both from working farmers and from scholarly experimenters, on drip-irrigation use and efficacy; all of them reflect actual in-the-field experience.

From: ***Advantages of Drip Irrigation — Micro-irrigation for Crop Production: Design, Operation, and Management***, Freddie R. Lamm, James E. Ayars, Francis S. Nakayama.

1. With 18% less water than sprinklers or furrow irrigation, long gourd yields were 45% higher.

2. With six percent less water than sprinklers or furrow irrigation, the yields for watermelons were 10-32% higher.

3. Micro-irrigation increased production in tomatoes by 39%, 68% for papaya, 14% for eggplants, grapes by 23%, and an amazing 98% greater yield of pomegranates.

4. In another study, yields for banana and grapes were increased by 29% and 19% respectively. And water conservation was, respectively, 29% and 37%. (The cost of equipping these growing areas with drip irrigation could be recovered in one year.)

5. Lettuce grew 7% more than with the use of sprinklers. The proportion of marketable heads, and the volume and the weight per head, were significantly higher with subsurface drip irrigation (SSDI) than with furrow irrigation.

6. Muskmelons had 54-74% greater yields than with no irrigation. The marketable yields were nearly double where sprinklers are used.

7. In Brazil, center-pivot sprinklers yielded 50 tons per hectare compared with 70 tons with drip irrigation.

8. While drip irrigation didn't always increase the yields of melons when compared to furrow irrigation, water use was reduced by 50%. In a study of early melons, the drip irrigated early melons outperformed those irrigated by micro-sprinklers by 2.4 times.

9. In India, production of cabbage was increased by 16-19% with only 37-61% of normal water use. Eggplant yields increased by 8-17%. Drip irrigation was 95% more efficient in enhancing the productivity of nitrogen compared to check basins.

10. Tomato yields were 22-36% greater while water use was reduced by 25%, compared to using surface irrigation.

11. In South Carolina, yields of tomatoes were 21-47% greater and extra-large fruit harvests increased by 33-57%.

"Drip irrigation is popular because it can increase yields and decrease both water requirements and labor. The concept behind drip irrigation is to provide the plant with continuous, near-optimal soil moisture. This is accomplished by conducting water directly to individual plants instead of providing water to the entire garden area, as with flood or sprinkler irrigation. Drip irrigation saves water because only the plant's root zone receives moisture. Little water is lost to deep percolation if the proper amount is applied."

Drip Irrigation for Home Gardens no. 4.702 by I. Broner.

"Drip-irrigation technology has gone a long way since it was first developed in England in the 1940s. Widespread use of this technology began in the 1960s after polyethylene plastics were used to make the drip tubes. Drip irrigation is especially appropriate for the production of capital intensive crops such as vegetables, fruits, and ornamentals. However, the use of drip irrigation requires a high initial capital investment as well as greater management skills than for more conventional irrigation systems. The incorporation of drip irrigation with plastic mulch culture of vegetables result in greater water and fertilizer use efficiency, and has resulted in increased yields of muskmelon, cucumber, eggplant, pepper, squash, tomato, watermelon, among other vegetable crops. For example, bell pepper yields in Puerto Rico were 12 MT/Acre (one metric ton = 2,2005 pounds) with drip and no mulch, vs. 19 MT/Acre with drip and plastic mulch culture. Irrigation efficiency with drip systems ranges from 75-95% compared to 25-50% for surface (furrow); 70-80% solid set sprinklers and 65-75% for portable sprinkler systems." Hector Valenzuela, HITAHR, University of Hawaii

Art Gaus, an extension horticulture specialist with the University of Missouri at Columbia, has had a drip system in his personal garden for nine years. One summer, his bush watermelons with plastic mulch and a drip system produced 32 pounds in a four-by-four-foot area, compared with 9-16 pounds in the same area with conventional irrigation. Gaus reckons that a well-timed drip system "could mean a 100% increase in yields; during the droughts of 1980, '83, and '84, it meant the difference between having a crop or no crop at all."

Farmer Ed Reynolds says the extra investment he makes on his drip cotton may seem excessive until he looks at the added yield.

Reynolds put in his first system, 30 acres worth, four years ago. "The year before I put in that first block, I made 20 bales off that 30 acres," Reynolds says. "The first year with drip, I made 66. The next year I made 88 bales, and the last year that block produced 103 bales."

Drip Irrigation Yield Increase Overshadows Higher Costs —Ron Smith

Michigan State University has documented a 30% yield increase in vegetable crops with drip irrigation, even in its humid, summer-rain climate.

Mr. Van Viet Uan is a 60-year-old farmer in Thua Thien Hue province in Vietnam. [In] one month [after starting drip irrigation], Mr. Uan states that he could already see the benefits—the young plants seemed to be growing very well. When he harvested his crop, Mr. Uan discovered that not only did he have a healthy yield of peanuts in the dry season, but it was also 25% higher than what he achieved during the regular growing season.

I could cite many more examples of the effect of drip-irrigation use on crops, but I think I've made the point: drip irrigation works.

Emitters for the Best Application of Water

Drip-irrigation emitters or "drippers" are the heart of any drip-irrigation system, delivering tiny amounts of water directly to plant roots, thus optimizing soil moisture with less water lost to evaporation, runoff and wind. Their distribution rate is measured in gallons-per-hour (gph).

Figure 116. This emitter is constructed with what's called a "torturous path" to control its flow of water. Every time the water makes a turn thorough it, any silt or dirt present is suspended in a kind of sideways tornado and kept in suspension until it leaves the emitter through a large orifice (not shown here). The emitter shown here is non-compensating, which means that as the run of tubing gets longer, friction causes the amount of water from each emitter to dwindle. Because of this, it's best to keep the length of the tubing running off of each valve at under 100 feet. Since it doesn't clog easily, this is the best emitter for well water and water with iron. Available from: DripWorks, Willits, CA.

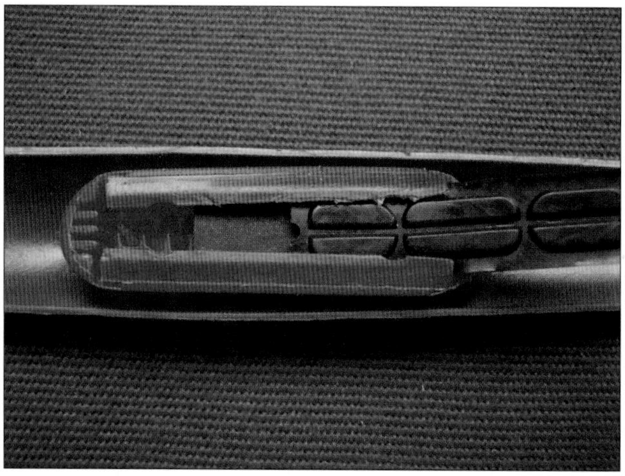

Figure 117. This is a pressure-compensating emitter. The two black flaps revealed here open up slightly when the water is flowing slowly, and squeeze together when the water is flowing swiftly and at greater pressure. This keeps the pressure even for a considerable length (compared to the non-compensating emitter in Figure 116). You can run 250 feet or more of this type of emitter off of one valve or one TEE fitting in a supply line. The tortuous path is much shorter here than in the non-compensating emitters. This factor, along with the flappers, does increase the chance of clogging. This is now the most commonly sold type of in-line emitter tubing.

Emitters are rated at various flow rates. All ½-, 1-, and 2-gph emitters can be punched into solid drip tubing or bought encased inside in-line tubing at regular intervals— usually six, nine, 12, 18, or 24 inches apart—with no external emitters. Because the encased emitters are self-cleaning, the exit hole for water is larger than that of many punched-in emitters.

The most important aspect of drip emitters is that the water is released very slowly. On the surface, a drip-irrigated plot looks like a series of wet spots with dry soil in between, an indicator of how slowly the water is emitted.

The pattern of the merging of wet spots is also shown in Figure 119. This is an important consideration when planning a drip system, as it demonstrates how drip irrigation can irrigate an entire root system without the continuous saturation of pore spaces that occurs with sprinkler use.

Do the following test to determine the emitter spacing for your particular soil:

- Lay down a short length of in-line drip hose, say six or eight feet, connect it to the system, and turn on the system.
- Leave it on for 30 minutes.
- After turning off the water, dig a trench parallel to one or two emitters and close to the tubing.
- Observe the width of the wet spot.
- Turn the system on for another 30 minutes.
- Extend the trench and again observe the wet spot.
- Do this again for an hour; this will amount to

Figure 118. This is the outside of the pressure-compensating in-line emitter tubing. Notice how large the exit hole is.

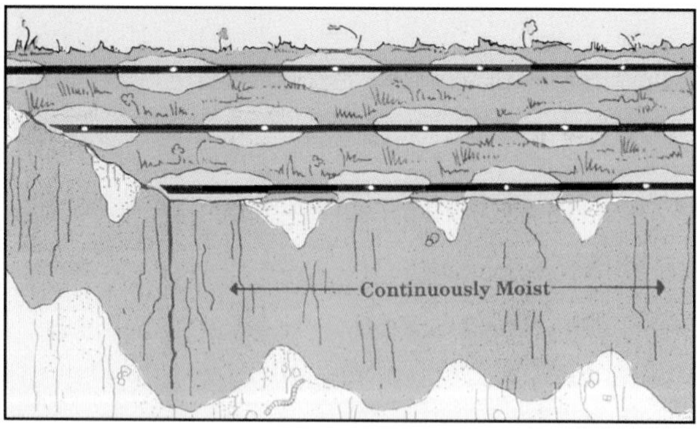

Figure 119. If emitters are placed close together, the wet spots will merge below the soil surface. The desired closeness depends on the soil type: if your soil is very sandy, you may have to either custom-order emitters that are built in every nine inches, or punch in external emitters at six- or nine-inch intervals. In heavy clay soil, you may only need emitters spaced 24 inches apart.

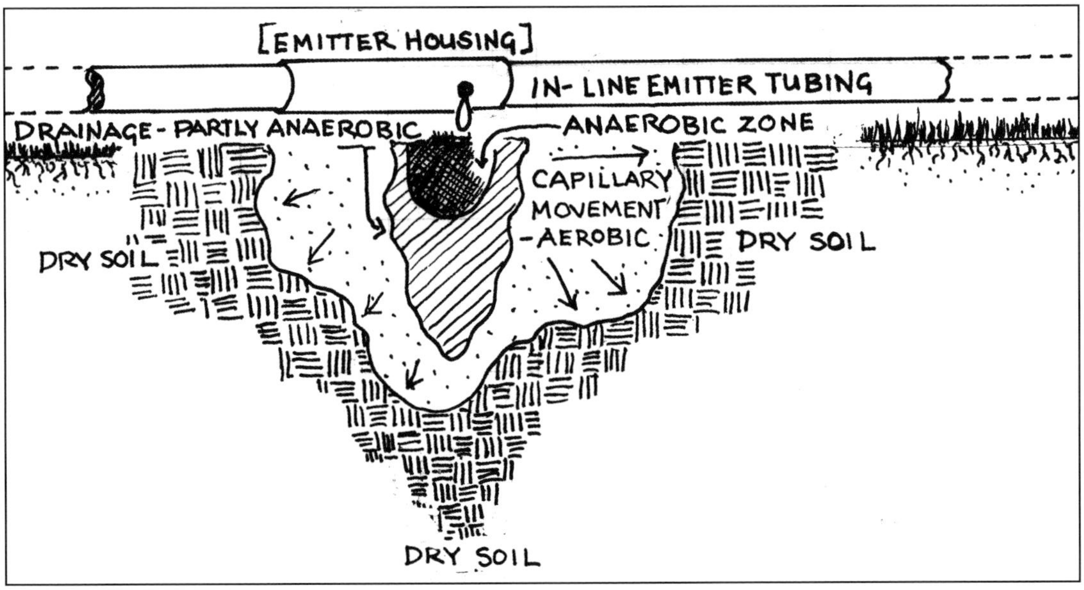

Figure 120. Any emitter, whether punched in or in-line, makes for the best growth, because it results in little anaerobic soil to slow down root development. Aerobic soil is the healthiest and most productive.

a total of two hours of irrigation.

Now you'll know how widely the water will spread below the surface, and how quickly. This will allow you to calculate the interval for which you'll need to program in order for the emitters to water the entire soil volume where the roots will be growing.

How Emitters Increase Yields

This is an important revelation: unlike sprinklers, properly placed emitters create only minute amounts of anaerobic soil, very close to the emitter, as seen in the above illustration. Very little of the pore space around an emitter is flooded; soil beyond the small anaerobic wet spot is moist, but the pore spaces, due to the capillary movement of moisture, are not saturated, and the soil life remains active and viable. This may explain why plant yields can increase with drip irrigation while water is being conserved.

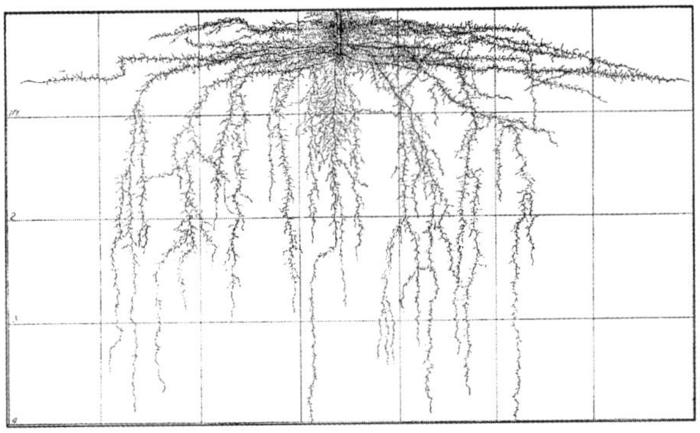

Figure 121. Tomato roots can grow as deep as four feet into the earth. Here, the first inch or so at the very top of the root system is missing as a result of the practice of hoeing or dust mulching (using shallow tillage to cut back weeds and reveal the soil). In this illustration, notice that the top 12 inches of the roots take up about 40-60% of the entire root system, with the greatest portion of the tomato roots exploring the top six inches of the soil. (Each square = one foot.) (From my book *Roots Demystified: Change Your Gardening Habits to Help Roots Thrive)*

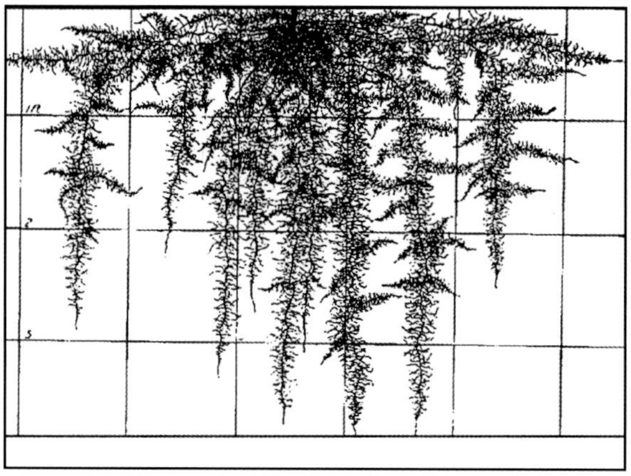

Figure 122. You'll notice that no more than 30-40% of the roots of this pepper plant grow in the top foot of soil, but plenty of them are still contained in the top six inches.
(From my book: *Roots Demystified: Change Your Gardening Habits to Help Roots Thrive)*

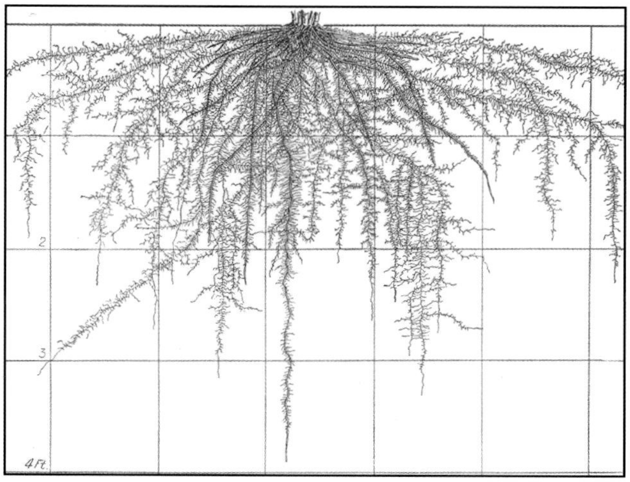

Figure 123. This is the root system of a mature corn plant, showing roots growing down to four feet.
(From my book: *Roots Demystified: Change Your Gardening Habits to Help Roots Thrive*)

Root Depths

Now let's look at where the feeding roots like to live.

The following drawings were done by Dr. John Weaver in the 1920's in Nebraska. It's amazing to see how deep the roots are after only five weeks' growth.

Here are several more of Dr. Weaver's amazing vegetable-excavation drawings, which I show to emphasize the importance of the top two to six inches of soil.

I'll repeat the following information several times in the roots chapter since it's so different from what many believe, and is important and insightful. A 2002 study at the Rutgers University Cooperative Extension found that: "Withholding water does *not* (my emphasis) send peach roots deeper into the soil, nor do light, frequent irrigations encourage shallow rooting." ("Best Management Practices for Irrigating Peach Trees," Jerome Frecon.)

The Value of In-Line Emitter Tubing

As Figures 119 and 124 demonstrate, drip emitters

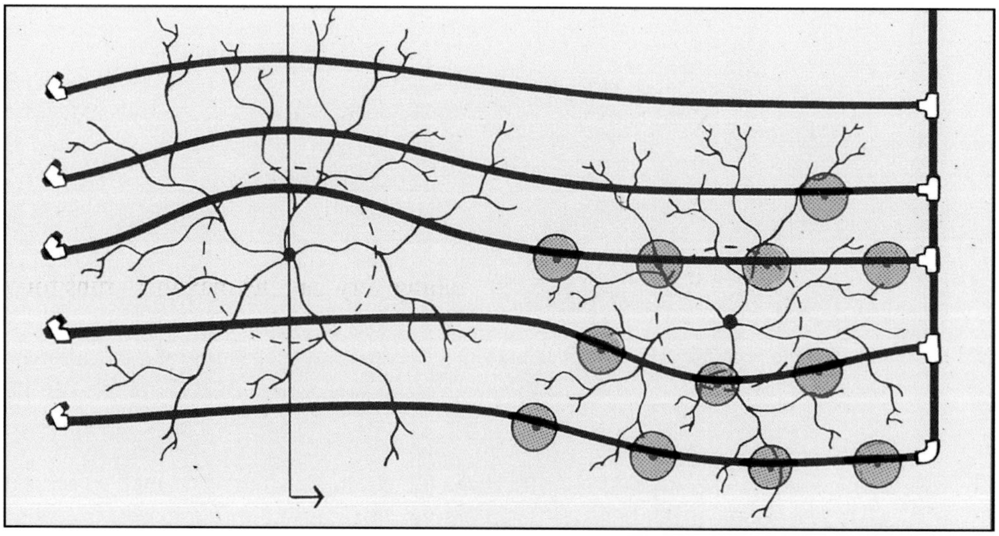

Figure 124. Here, some in-line emitter tubing has been installed among existing trees, with the spots joining together below the surface (as in Figure __) to water the entire root system. The blue circles represent the wet spots on the surface. To prevent root rot, avoid placing emitters close to tree trunks.

Figure 125. Here the laterals to the right supply water topographically; the spacing between the rows is designed to irrigate the entire root system.

need to be spaced out in a regular grid pattern to feed all the roots of a plant efficiently; in-line emitter tubing is the best way to ensure this. Important features of this kind of in-line drip irrigation tubing are:

- There's nothing to break off (a problem with external, punched-in emitters) because the emitter is inside the tubing.
- In-line emitters are engineered to prevent clogging.
- The emitters don't get brittle from sunlight. (The tubing is rated at 10 years in full sunlight. I had some tubing laid down under mulch, and it was still intact after more than 20 years.)
- Eliminates the tedious and painful task of punching emitters into drip hose.

Elegance

In-line tubing is the most elegant system to install. Simply roll out the tubing, flush it out, close the end, and you're ready to go. No messing around on your knees for hours punching in those arm-twisting, hand-stressing old-fashioned emitters. No punching in of emitters at all.

When you first price the in-line tubing, you might think it's expensive, but it's actually about *half* the cost of punched-in emitters and solid tubing. In 2020 pricing, a 100-foot roll of in-line tubing, with 1/2-gph emitters installed every foot, costs about $33. A 100-foot roll of solid ½-inch tubing costs around $12. A 100 punched emitters cost up to $50 dollars. So, in-line = $33, while punched-in emitters = $62! (That's one of several reasons that hardware stores really push those punched-in emitters.)

With in-line tubing, you get no clogging, along with the benefits of easy-to-calculate flow rates; tubing just as sturdy as solid drip hose; no emitter repairs; and nothing to crack off while weeding. It's also easy to calculate the time to leave the system on for each irrigation.

Rain-Barrel Drip

Most pressure-compensating emitters prefer 10 psi (pound-force per square inch) to even begin to operate properly. A barrel on a platform would have to be about 23 feet aboveground to produce 10 psi, because it's the height of the water that builds pressure. A lot of drums strung together may equal a lot of stored water, but the pressure is only as high as the water-line inside the tanks.

Given this information, barrels and drip irrigation don't seem to mix. Enter BlueBarrel Rainwater Catchment Systems, page 128, provider of barrels and kits to store water in one or more 50-gallon drums. They also sell a device that can control the on/off cycle of a barrel-drip system without requiring the special pressure tubing made for use with low-gravity barrel systems. The catch is that you still can't run the tubing very far. The maximum runs for ¼-inch spaghetti tubing off of a ½-inch solid-drip-hose supply line is 30 feet. (Solid ½-inch supply pressure-compensating tubing can run over 250 feet at 45 psi, off of each connection.)

How to Calculate Water & Timing for the Best Growth

Knowing how much water to give your plants is the best way to optimize growth and yields, but how do you figure out how much water your vegetables need?

Sq. Ft. of Plant Cover	ET RATE (IN INCHES/MONTH)									
	1"	2"	3"	4"	5"	6"	7"	8"	9"	10"
1 sq. ft.	0.0187	0.0374	0.062	0.083	0.104	0.125	0.145	0.166	0.187	0.208
4 sq. ft.	0.075	0.15	0.248	0.332	0.416	0.50	0.58	0.664	0.75	0.832
10 sq. ft.	0.187	0.374	0.62	0.83	1.04	1.25	1.45	1.66	1.87	2.08
75 sq. ft.	1.403	2.805	4.65	6.225	7.8	9.4	10.87	12.45	14.	15.6
100 sq. ft.	1.87	3.74	6.2	8.3	10.4	12.5	14.5	16.6	18.7	20.8
200 sq. ft.	3.74	7.480	12.4	16.6	20.8	25.0	29.0	33.2	37 4	41.6
300 sq. ft.	5.61	11.22	18.6	24.9	32.2	37.5	43.5	49.8	56.1	62.4
1 acre solid cover	815	1629	2701	3615	4530	5445	6316	7231	8146	9060

DAILY WATER USE (IN GALLONS PER DAY)
BASED ON VARIOUS EVAPOTRANSPIRATION RATES

Figure 126. To use this chart, find out the estimated ET rate for the coming day or week; select the inches-per-month column; figure out the square footage of the planting; and slide your finger down the monthly ET column to the intersection of the square-footage row. This will give you the total number of gallons you'll need for one day's optimal watering. The yellow represents the average monthly *summer* ET rate for my city (Santa Rosa, California).

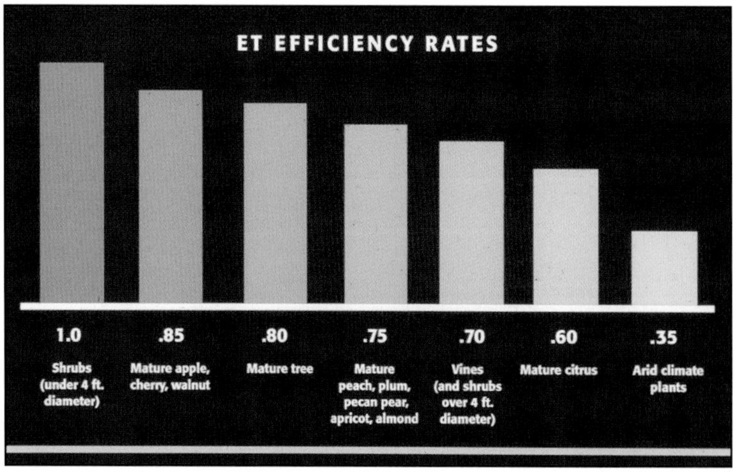

Figure 127. Not all plants absorb the same amount of water for growth; some are more efficient than others. Here, a low-water plant like lavender uses only 35 percent of the water that would keep a lawn happy. To reduce the amount of water as compensation for this efficiency factor, multiply the gallons from the ET chart by the efficiency number.

4' X 10' = 40 sq. ft.

ET = 6"

10 sq.ft. = 1.25 gallons
X 4 = 5 gallons

40 ft. = 40 emitters

5 gal. ÷ 40 - 1gph emitters
= .125 hours

60 X .125 = 7.5 minutes

With 1/2gph emitters – **15 minutes/day**

(once/week = 7 X 15 = 105 min. = 1.75 hrs.)

Figure 128. Vegetables have an efficiency rate equal to that of a lawn —1.0. In this step-by-step calculation, I show how to figure out irrigation for a 40-foot-square vegetable bed. The result shows that daily irrigation (I prefer this rate and will discuss it soon) should be 15 minutes per day or 1.75 hours once a week. Carefully observe your garden's growth and adjust the time accordingly. The same math applies to ornamentals, as long as you're watering the entire root system.

Evapotranspiration (ET) is the amount of water lost to *evaporation* from the soil combined with what is lost in *transpiration*, the natural release of moisture from the leaves. The baseline ET rate is based on that of four inches of cool-season turf grass. While not a perfect stand-in ET for the ground covers, shrubs, vegetables, and trees we might want to grow, this baseline is a good place to start. Any ET rate above zero represents a plant that uses more water than a lawn. Below zero makes it a water-conserving plant.

The problem with most ET charts that show how much water to use is that they tend to list the amount of water in inches. With only this information provided, gardeners who use sprinklers may struggle to figure out to how long to run them, but with persistence it's possible to calculate the timing. For those who prefer drip systems, however, it's virtually impossible to translate inches into the gallons-per-hour (gph) drip-irrigation mode.

The above chart solves that problem by converting the water needed to gallons per square foot. (This is based on the premise that one inch of water per 100 square feet = 62 gallons.)

The ET rate per month at the top of the chart can be obtained from a local Master Gardener hotline; a local farm advisor (based on many years of averaged water needs); agricultural websites (the vintner's page in Sonoma County); or, in evolved cities, the daily newspaper. The ET rate shows how much water should be used per month, based on local conditions. An ET rate of two or three means very little supplemental water is needed because of cool and/or wet weather. During a string of warm to hot days, the ET will be six inches per month. You don't want to be around for a ten-inch-per-month day! (August in Palm Springs, California.)

For example: to find out how much water you'll need for your snow-pea crop with an ET rate of six inches per month, (the average summer ET rate here in Santa Rosa, CA) you'll measure the square-footage of a block of snow-pea plants, say 5 x 20 feet or 100 square feet. Tracing down the six-inch monthly rate, the total number of gallons used in one day by 100 square feet of planting is 12.5 *gallons*. If you water once each weekend, you multiply 12.5 times seven to get: 87.5 *gallons*. (In Palm Springs it would be 146 gallons once per week.) See Figure __ for another example. So how often should you actually water?

Daily Versus Infrequent Watering

Perhaps the biggest debate about irrigation is how often to water. The choices range from daily, to weekly, to monthly. In my opinion, daily irrigation is the best and promotes the greatest growth and yields. This is considered radical by many gardeners. Once the soil is at a proper moisture level (*not* wet) a daily irrigation in summer-dry areas like the one in which I live "keeps the tank topped off." Where there are periodic rains, the system would only be turned on once the soil has returned to an ideal moisture level. (Some irrigation controllers include a cup to turn the system off when it rains.)

This means the drip system is turned on for a *very* short time daily. Daily irrigation uses no more water than irrigating every Saturday, because you've figured out the rate per day, using the ET chart.

In a study of apple orchards, it was determined that the size and storage potential was increased with daily compared to weekly irrigation. (Use Of Drip Irrigation In Orchards, Jan./March 1993. B. Bravdo, E. L. Proebsting, *HortTechnology*, 3 (1).)

One landscape I planted was done in the following eight steps:

1. I prepped and raked out the soil to the contours I desired.

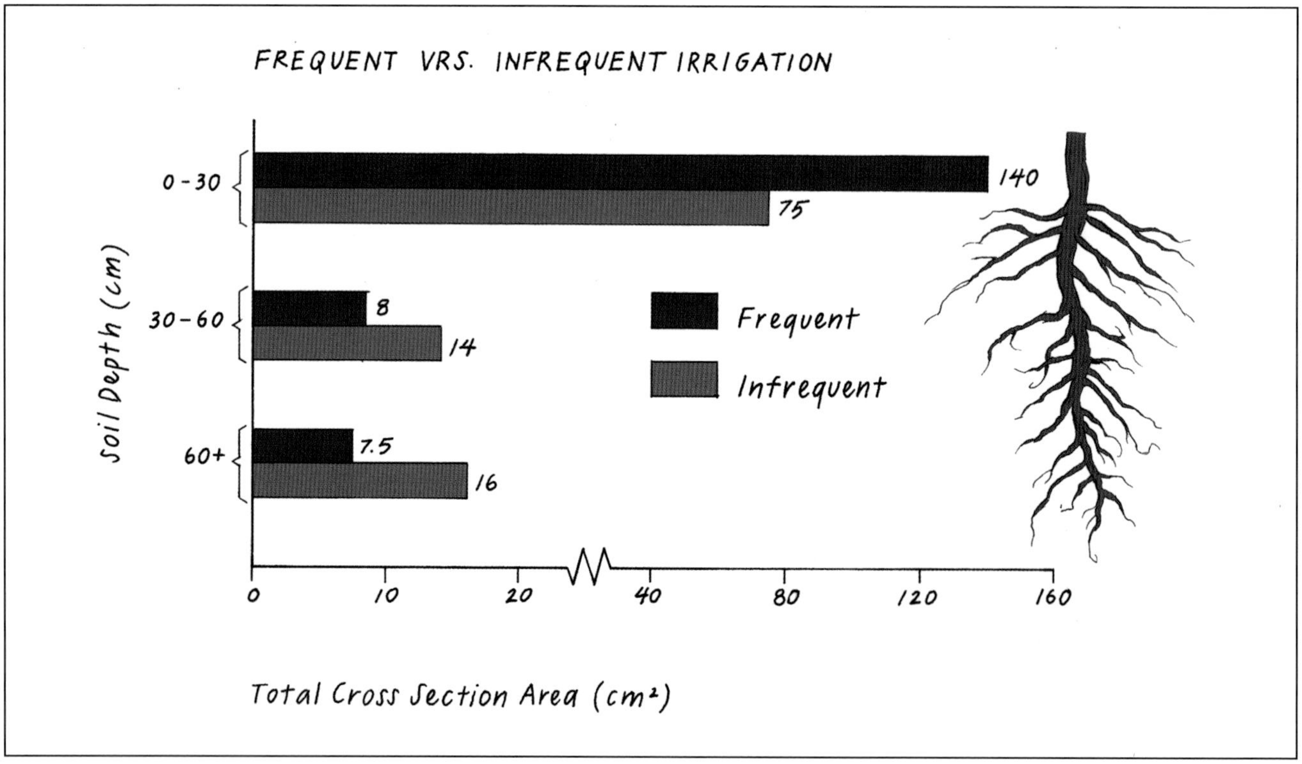

Figure 129. This example of how tree roots grow relative to frequency of watering shows why I recommend daily irrigation. The frequently watered roots (measured as the square centimeters of the cross-section) grew mostly in the soil's top 30cm (12 inches). The roots of an infrequently watered tree grew 20 percent less root mass in the second foot, when compared to the first foot, but slightly more of a cross-section than the frequently watered trees. (This is from: "Impacts of Irrigation and Container Type, 20003. Edward Gilman et al., *Journal of Aboriculture*, 29(4).)

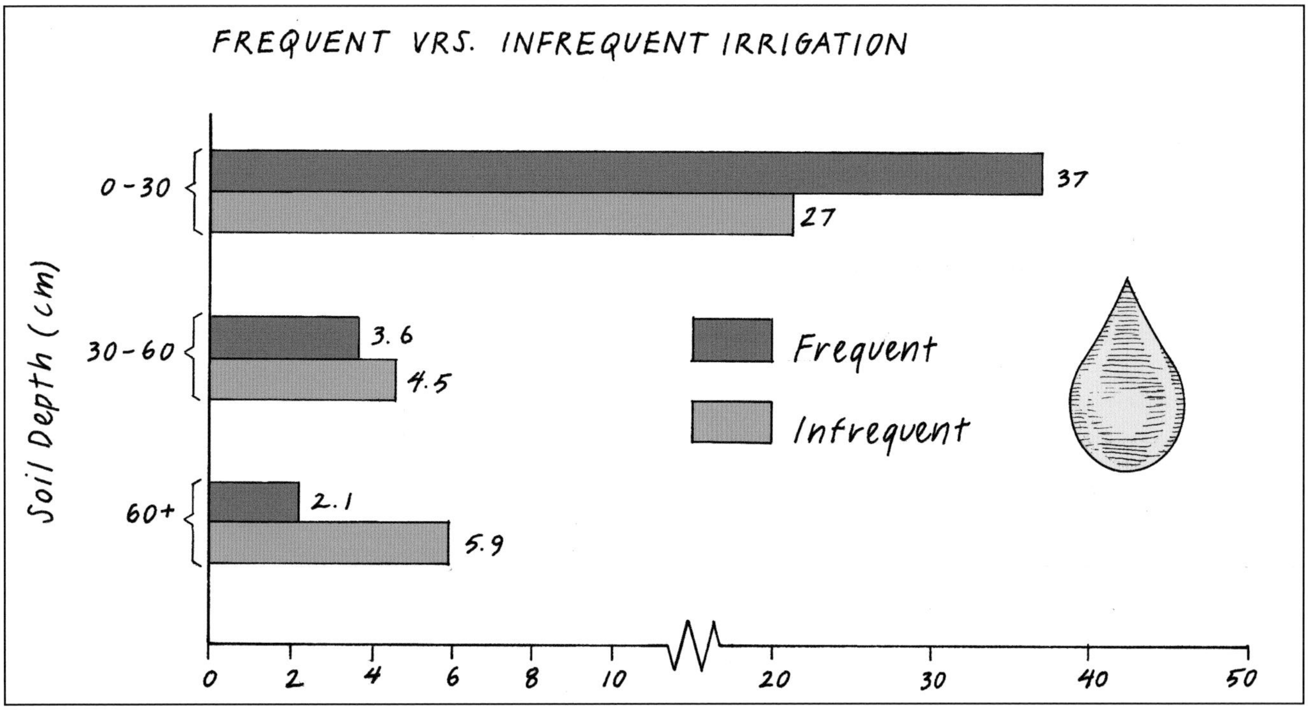

Figure 130. This graph shows the total number of roots resulting from frequent versus infrequent irrigation. This shows the same tree that appears in the previous graph, and again, the top foot is better at root production than the next foot down by a factor of very nearly ten.

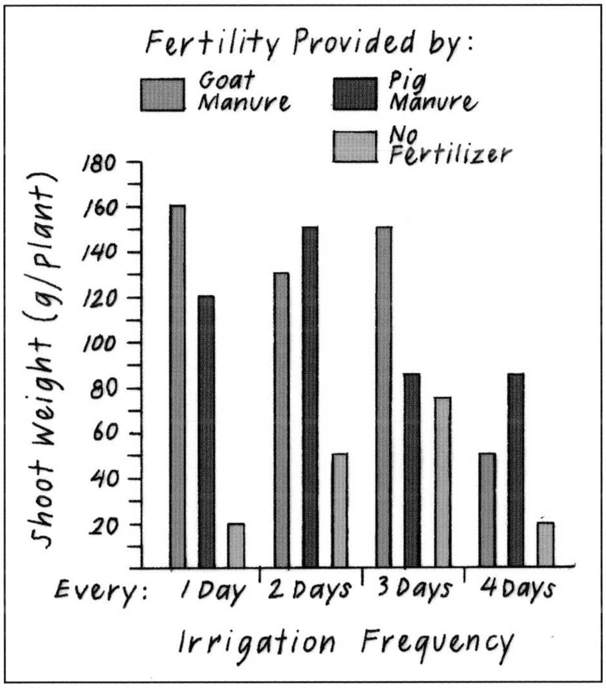

Figure 131. "The results of this study of corn reveal that total dry matter is significantly increased with increasing drip-irrigation frequency. Daily or once every 2 days drip-irrigation frequency is recommended." (Effects of drip irrigation frequency, fertilizer sources and their interaction on the dry matter and yield components of sweet corn, 2014. *Australian Journal of Crop Science.*)

2. I laid out the in-line drip system based on topography (following the lines of the contour—not up and down a hill) and staked it in place.
3. I watered the entire landscape with an overhead oscillating sprinkler until the water had penetrated to about six to eight inches deep in the soil, and returned to plant after the soil had dried enough to be moist, but *not* still wet.
4. As I was using drought-resistant Mediterranean plants, I transplanted them into the soil, with each plant placed halfway between the emitters in each row.
5. Using a hose, I hand-watered the transplants well.
6. I mulched the plants heavily. (This also hides the ugly black tubing and protects the polypropylene.)
7. I turned on the drip system of one-half-gph emitters on day one, to compensate for the daily ET rate.
8. Because the soil was pre-moistened and the plants watered in, the drip-irrigation time that I set on the controller on this jobsite was two minutes per day—mere tablespoons of water.

As mentioned above, if you want to water, say, once a week, just apply water equal to seven times the daily rate.

This is so important that I'll repeat it: Rutgers University Cooperative Extension found that: "Withholding water does *not* (my emphasis) send peach roots deeper into the soil, nor do light, frequent irrigations encourage shallow rooting." (*Best Management Practices for Irrigating Peach Trees*, Jerome Frecon. 2002.) My theory is that the growth of roots is controlled less by available water than by a plant's genetics and preference for an aerobic soil.

To conserve even more water and suppress weeds, I lay newspaper or cardboard over the tubing and make holes in the cardboard to transplant. I then cover the cardboard with a weed-free mulch like wood chips or rice hulls.

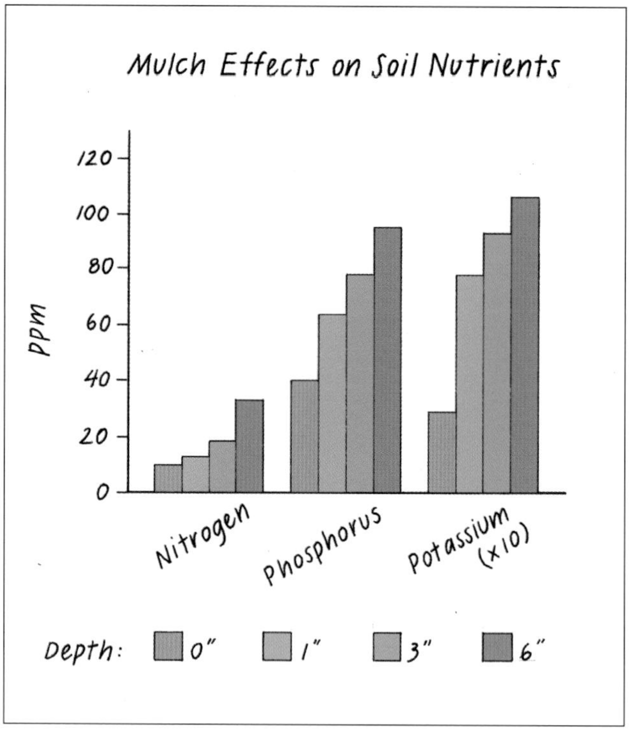

Figure 132. Contrary to my discussion that four inches is all the mulch you need, this graph shows that six-inch deep mulch is even better than three inches of mulch. Ideal of four inches deep is for water conservation. The parts-per-million (ppm) of each nutrient is more available the deeper the mulch gets.

Mulch

The golden rule of sustainable gardening seems to be—mulch, mulch, mulch. Mulch can:

- Save water.
- Control erosion.
- Protect soil from desiccating winds.
- Eliminate the "dust-bowl effect" (soil so dry and loose that it blows away).
- Keep soils a bit cooler during hot summers.
- Add a minute amount of fertility. The more it rots, the more it fertilizes.
- Slow grass-fires a bit (but only the correct mulch, to be discussed later).
- Suppress weeds.
- Increase yields—especially with the correct reflective mulches.
- Prevent mud from sticking to your shoes.
- Promote the growth of beneficial fungi with wood chips.

As with any gardening practice, there's the flip side of the coin. Mulches can have deleterious effects. They can:

- Block warming sunlight in the late winter and early spring.
- Hold too much moisture in the soil and kill plants by encouraging root rot.
- Some mulches (stone and gravel mulches for example) allow weeds to grow through them
- Steal a bit of fertility if the wrong mulch is used.
- Speed up fires if the wrong mulch is used; they can also combust and smolder for days and turn into a spot fire.
- Can harbor termites, slugs, snails, earwigs, and "roly-poly" bugs.
- Hide the movement of moles, voles, and gophers.
- Slide down slopes and expose the soil (if you use the wrong mulch—rice hulls are an example).
- Can blow away to expose soil to wind and erosion—rice hulls again.

I'm not going to provide an exhaustive boring list of possible mulches. You know what materials are available to you in your community. But here are 12 helpful factoids about mulches:

1. If you have a limited amount of mulch or buy expensive mulch, you'll want to use the most effective depth. A 2013 thesis study by Mi-

chael McMillen showed that mulch spread four inches deep was the best trade-off of depth, effect, and cost. Beyond four inches of settled semi-chunky mulch (not straw) the amount of extra moisture conserved as a percentage of the entire mulch is not worth the additional cost or effort. Piling your mulch beyond four inches deep has high diminishing returns.

2. In another study of mulches, "Establishment rates of difficult-to-transplant tree species such as beech can be improved from 10% to 70% [by mulching], while fruit yields of young trees can be increased by 100%." Be sure to mulch well beyond the dripline (leaf canopy edge) of a newly planted tree. As you'll see in Chapter __, the roots of trees grow at least 1.5 to three times wider than their foliage. ("Influence of Pure Mulches on Survival, Growth and Vitality of Containerized and Field Planted Trees," by G.C. Percival, Evangelos Gklavakis and Kelly Noviss; December 2009. *Journal of Environmental Horticulture* #27(4 p 200-206.)

3. Arbor mulch, a mixture of fresh-off-the-truck shredded leaves and wood (it's possible to get it for free if tree work is being done near you) can suppress weeds, but this often requires a six-inch-deep application.

4. According to a study done in Lubbock, Texas, "Three depths of mulch: 0, 7.5, and 15 cm (0, 3, and 6 inches), were applied to 3.8 liter (1 gallon) [container] plants at time of planting. Height, width, stem diameter and foliage fill were recorded for each plant after 18 months. Results indicated no treatment differences within species in their growth or mortality after 18 months. The benefit of pine bark mulching is not apparent in the crown growth." ("Mulch Effects on Crown Growth of Five Southwestern Shrub Species," A. L. Hild and D. L. Morgan, March 1993. *Journal of Environmental Horticulture*, 11(1) p 41-43.)

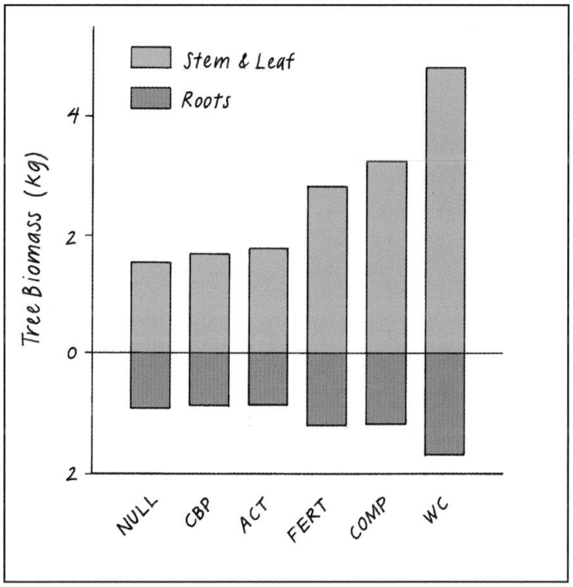

Figure 133. Null = bare, CBP = mychorrhizae, ACT = aerated compost tea, Fert = fertilizer, Comp = compost, and WC = woodchips. Woodchips (WC) used as fertilizer are the superior way to increase the biomass of a tree. Aerated compost tea (ACT) has less than one-half the fertility value of woodchips. (Wood Chips and Compost Improve Soil Quality and Increase Growth of *Acer rubrum* and *Betula nigra* in Compacted Urban Soil, 2014. *Arboriculture and Urban Forestry*, 40(6): 319-331.

5. In a study of five organic and inorganic mulches, black plastic and hay proved to be the two most effective mulches for various impacts on a green ash tree. Wood chips were not as effective as the others. ("Influence of Mulch Materials on Growth of Green Ash," 1983; Margaret Litzow and Harold Pellett, *Journal of Arboriculture*, 9 (1).)

6. In another study, "A research plot with 60 *Acer rubrum* and 60 *Betula nigra* trees was created, mimicking the typical urban landscape disturbance. Wood-chip mulch (WC), compost (COMP), inorganic fertilizer (FERT), aerated compost tea (ACT), a commercial biological product (CBP), and a water control (NULL) were assessed for their impacts on soil quality and tree growth after five years." They found that "This research shows strong evidence that COMP top-dressings and WC mulches are effective and also cost-efficient methods for improving soil quality and stimulating tree growth in compacted urban landscape soils."

7. Note: No significant changes in soil proper ties or tree growth were observed with ACT. For more, see page __. ("Wood Chips and Compost Improve Soil Qual ity and Increase Growth of *Acer rubrum* and *Betula nigra* in Compacted Urban Soil," November

2014; Bryant C. Scharenbroch and Gary W. Watson, *Arboriculture & Urban Forestry;* 40(6), pp 319–331.)

8."It's well known that composted yard waste serves as a high quality, mild organic fertilizer that can meet the nutrient requirements of both microbes and plants as it decomposes. On the other hand, mulching with high C:N (a carbon to nitrogen ratio of 125:l) wood pallets stimulated microbial and fungal growth while adding little nitrogen to the soil. Microbes immobilized a high proportion of the limited nitrogen pool, thereby slowing the growth of river birch. These results are consistent with the hypothesis that soil microbes out-compete plants for nitrogen, and that addition of organic matter with high C:N ratios can induce nutrient efficiencies in plants by stimulating microbial growth." ("Effects of Organic Mulches and Fertilization on Soil Microbial Activity, Nutrient Availability, and Growth of River Birch, 2002," Daniel A. Herms, John E. Lloyd, and Benjamin R. Stinner.)

9. You definitely have to water through the mulch to get good growth: "Negative effects of deep [15 mm (6 in.)] mulch 2 weeks after transplanting occurred for the mixed particle–sized material only. Mulch placed over the root ball intercepted water, causing a drier root ball and resulting in greater tree stress and reduced survival following light applications of water than for trees with no mulch over the root ball." ("Mulch And Planting Depth Affect Live Oak (*Quercus virginiana* Mill) Establishment." Edward F. Gilman and Jason Grabosky, *International Society of Arboriculture*, 204.)

10. Results with mulches can be very different than expected: "Over a 6-year period, blueberries mulched with sawdust produced more than double the yield from similar plots mulched with oak-leaf mold, three times the yield from plots mulched with straw, and almost six times the harvest from plots clean-cultivated. All plots received the same water and fertilization." (*Sawdust Mulches Used For Larger Crops, Better Soils* edited by J. L. Overholser.)

11. When compared to plastic sheathing, cardboard blocks only half the amount of outgassing of carbon dioxide. Wood-chip mulch lets more of this debilitating gas out of the soil than cardboard. The irony is that, if global warming is an issue for you (note: it should be), plastic over all your soil is the way to go, because plastic "mulch" blocks more out-gassing of carbon dioxide than other options.

12. Forest fires are certainly more common now, and they may spread to more and more areas with more frequency. No mulch can stop a forest fire, but it can slow down a small grass fire in the right circumstances. I've also seen mulch acting like a wick taking the fire *toward* the house. Wood-based mulches also smolder very much longer than burnt branches, and can flare up later to restart the fire. Many houses are mulched right up to their foundations. If the house is mostly made of wood, that's a *big* problem.

Here's a chart that will tell you the rates and intensities at which various mulches burn. It turns out that the best mulch is *composted* woodchips, and yes, shredded rubber is a common mulch for children's playgrounds. *The Combustibility of Landscape Mulches*, Stephen Quarles, Wood Performance and Durability Advisor, University of California Cooperative Extension, and Ed Smith, Natural Resource Specialist, University of Nevada Cooperative Extension SP–11–04, 2011.) [www.livingwithfire.com]

Paper and Cardboard - "Organic Herbicides"

My garden requires at least some weeding every spring. By using newspaper and cardboard as a mulch, I usually don't have to weed again until the next spring. To lay down a cardboard mulch, look for appliance stores where you can get huge amounts of cardboard in large pieces. (Make sure you strip any plastic tape off the cardboard or else it will "float" up through the mulch.) Be sure to overlap the cardboard pieces by four to six inches to thwart "crawling" weeds like wild blackberries, poison oak, etc.

When mulching with newspaper, I use 5-10 sheets, overlapped about two to four inches at the edges.

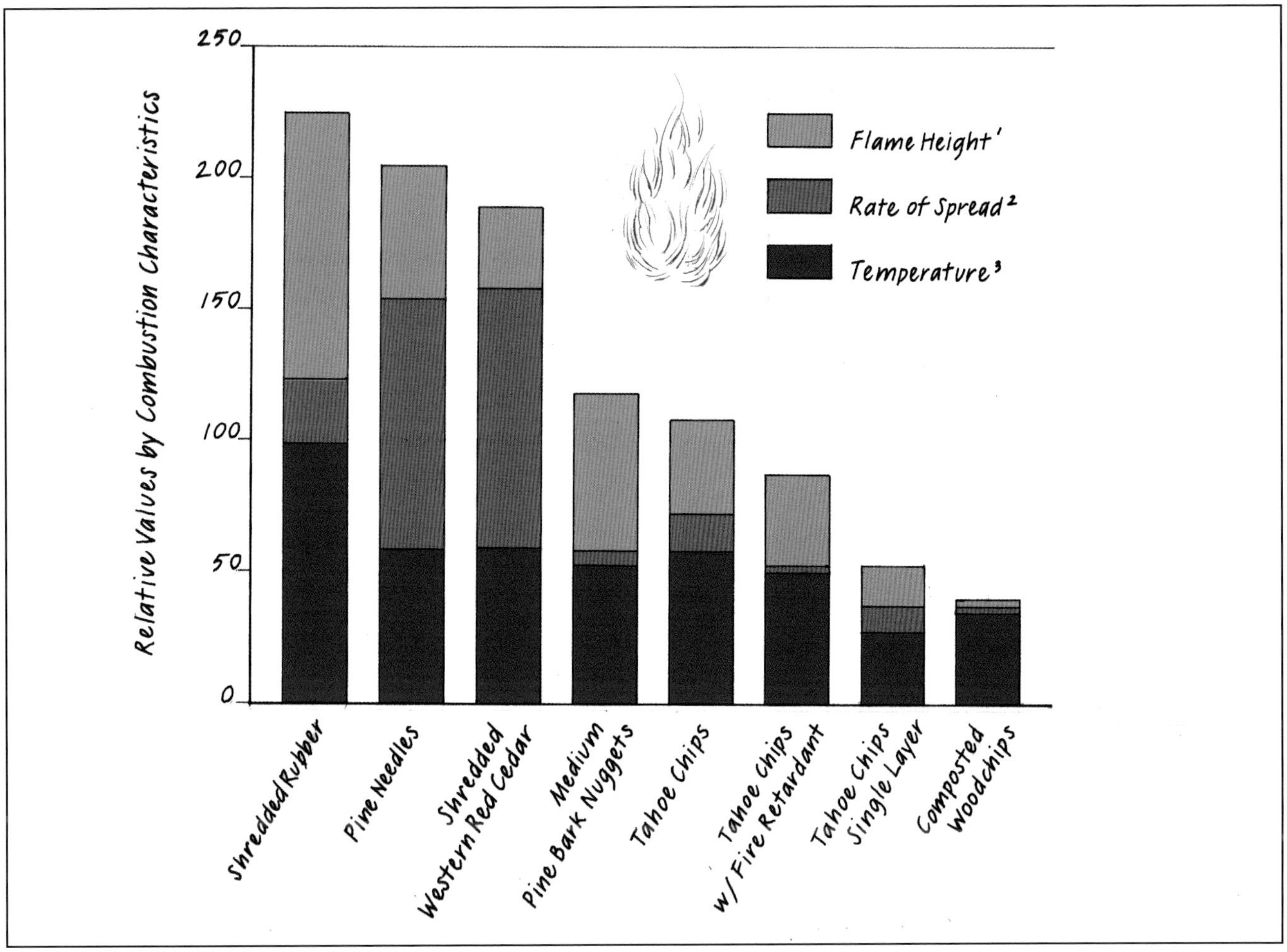

Figure 134. I live in Sonoma County, California, where the last big wildfire burned to within six feet of our house, so I'm very relieved that I found this research. Rubber mulches are most often seen in children's playgrounds. (From: The Combustibilty of Landscape Mulches, 2011. Stephen Quarles, Ed Smith, University of Nevada Cooperative Extension. SP-11-04)

Figure 135. The mulch-and-fire test didn't include shredded-redwood bark, a common material used for mulch in Sonoma County. The 2020 fire completely destroyed the house behind mine, but left its shredded-wood mulch intact. This mulch is hard to work with, as it's very scratchy. Also, when replanting an area, it's hard to keep all the fiber out of the planting hole, and plants can lose nitrogen to the buried mulch.

I use only the B&W paper, not the slick color ads. Cover the cardboard and newspaper with a weed-free mulch like rice hulls, properly made compost, straw, composted leaves, or arborist wood chips.

As I've noted before, gardening is like a double-sided coin. In her blog "*The Myth of Paper-Based Sheet Mulch,*" Dr. Linda Chalker-Scott of Washington State University exposes the flip side of newspaper and cardboard mulching:

- When wet, the layers of paper or cardboard can create an impermeable barrier to water and air exchange.

- Garden pests, moles and gophers can hide under newspaper and cardboard sheet mulches. In her experience, termites preferred to inhabit cardboard over wood chips.

- If the cardboard or newspaper is allowed to dry out, irrigation water may sheet away rather than seep through the layers, especially in arid climates or with well-drained soils. (When I use cardboard or even newspaper for perennials and need to irrigate, I lay out the drip tubing in place *before* putting the cardboard down.)

Organic mulches work as well as cardboard and newspaper to prevent weed growth and kill turf, but it usually takes at least six inches of mulch to do the job—two to three times more material than using cardboard or newspaper as an "organic herbicide." While not ideal, I choose newspaper or cardboard with two inches of mulch because it helps cut down on the cost of mulch. For me, the practicality of using an "organic herbicide" outweighs the slight negative influence it may temporarily have on soil life, and any negative effects probably dissipate after the paper rots. The ease of installation trumps the ideal.

Research

I do like to use paper as biodegradable mulch, but it was difficult to find any data about its use. Then I found a six-year study on shredded-paper mulch conducted by David Granatstein of the Washington State University at Wenatchee, Washington. See Figure __ to realize just how viable sprayed-on paper mulch is as a mulching option. The Wenatchee study found that by combining paper mulch with composted biosolids you get the greatest yields. This result is far above that of a geotextile-fabric mulch barrier and much superior to leaving a bare strip of soil down the row.

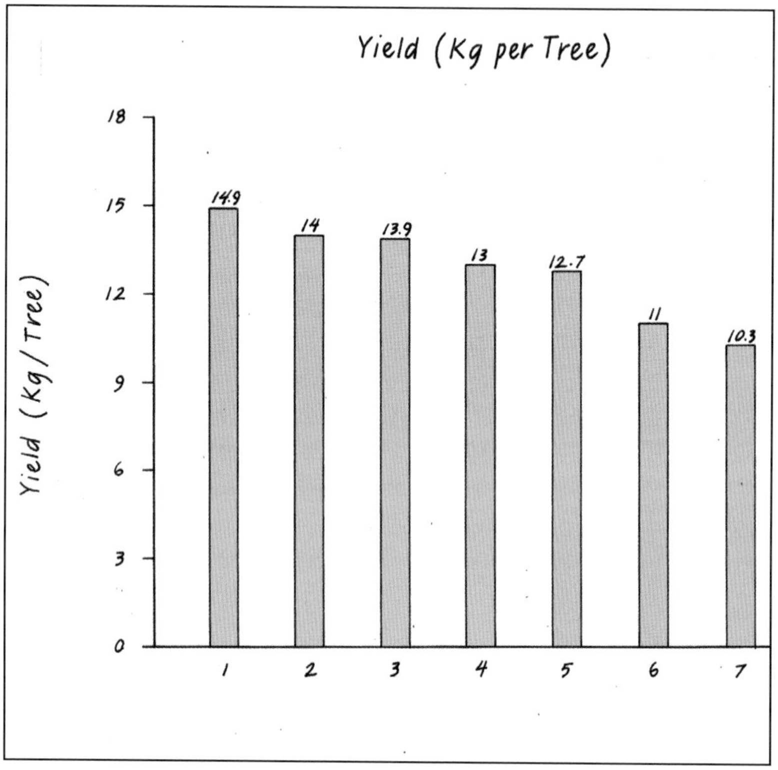

Figure 136. A five-year study in Washington State of dwarf Red Delicious apple trees (M9 rootstock). The code for the numbers is:

1. Composted biosolids with paper (14.9 kg/tree)
2. Alfalfa hay (14 kg/tree)
3. Paper and biosolids (13.9 kg/tree)
4. Paper mulch (13 kg/tree)
5. Geotextile (12.7 kg/tree)
6. Biosolids (11 kg/tree)
7. Check (glyphosate, unmulched with herbicide) (10.3 kg/tree)

In addition, a seven-year study by G. H. Neilsen et al in British Columbia, Canada, demonstrated that shredded paper mulch by itself produced a greater yield than any other mulch or combination—at 81 percent greater yields than a strip on which herbicide had been used.

Not all studies, however, show paper as the superior mulch. A comparison of six treatments (control, newspaper mulch, black plastic mulch, wheat-straw dust, pine- needle mulch, and sawdust) determined that strawberry crops mulched by wheat- straw dust gave maximum fruit size and yield. ("The Effect of Different Mulching Material on the Fruit Size and Yield of the Strawberry under the Agro-Climatic Condition of Mansehra," 2015; Adil Rehman, Muhammad Shireen khan Zakaria, Dr. Ahmad, Abbas Malik, Dr. Asghar ali Muhammad Shahid, *Journal of Natural Sciences Research*; Vol. 5 No.11, p 106-108.)

The Limits of Paper & Cardboard Mulch Used With Drip-Irrigation Tubing

I started using newspaper mulch in the early 1970s when the organic-garden-

Figure 137. Newspaper mulch is not initially ideal for cultivating bacteria and other soil life, but it is pragmatic, as it saves time and kills off some weeds. As the newspaper rots, bacteria will repopulate the soil surface. *Don't* use slick-stock colored paper.

ing debate was hot about whether it contained lead and thus was not organic. Since I discovered that it probably did contain lead, used in the manufacture of the black ink, I stopped. Sometime in the late '70s or early '80s, newspaper printers switched to soy-based inks in the black-and-white sections in order to save money. (Not only did these inks cost less, but also weighed enough less than the lead-based inks to actually have a positive impact on trucking costs.) So I switched back to using newspaper mulch with attractive "natural" mulches on top.

By this time, I was using drip irrigation in my garden, with newspapers (avoiding the toxic slick colored sections) layered on top, followed by an attractive rice-hull or mushroom-compost cover. I used this combination for decades, until I paused to consider some new information.

It turns out that the newspaper or cardboard can make for an anaerobic condition that is slightly detrimental to the exchanges of gases between the soil and the atmosphere. According to Dr. Linda Chalker-Scott, cardboard and newspaper mulches may be useful and not detrimental in annual gardens, but perennial plantings are a different story. Her experience (and her site—www.theinformedgardener.com—is my source of the most fascinating truths and myths about gardening:

- The obvious: newspaper layers are not attractive when the covering mulch blows away or moves; so, keep them well mulched.
- She believes cardboard and newspaper are not as effective as thick layers of organic mulches in keeping down weeds.
- The cardboard /newspaper layers also harbor earwigs, snails, ants, termites, voles, and gophers. (However, a gardener friend found beneficial gopher snakes each spring under large sheets of cardboard.)
- The most fascinating observation is her view that if used on poorly drained, wet soils, the cardboard /newspaper layers cause an anaerobic condition. This applies, as mentioned above, primarily to perennial plantings, as yearly cultivation in tilled gardens awakens the exchange of gases.
- The dry cardboard and newspaper can attract moisture away from the soil in dry-summer areas or where the soil is well drained.

Figure 138. Steep slopes (up to a three-to-one angle) can be protected by jute netting. A thin layer of mulch can be placed on top of the jute—its scratchy texture keeps the mulch from sliding. Use six-inch U-shaped staples every four feet to secure the netting.

Definitely food for thought, especially if you're using the cardboard /newspaper mulch in a no-till garden.

Dry Farming

You can even get away from any watering at all by the practice of dry farming.

Dry farming is a way to conserve soil moisture during long dry periods by using a system of tillage, surface protection, and the planting of drought-resistant varieties. It's not a strategy for yield maximization, but instead allows nature to determine the sustainability of agricultural production.

In a home garden, dry farming usually means watering the transplant thoroughly only at the time of planting, or watering vegetable seed until germination. Steve Solomon, author of *Water-Wise Vegetables: For the Maritime Northwest Gardener,* is an expert at this kind of approach. In one of his experiments he recounts "I sowed a bit of kale, savoy cabbage, purple sprouting broccoli, carrots, beets, parsnips, parsley, endive, dry beans, potatoes, French sorrel, and a couple of field cornstalks. I also tested one compact bush (determinate) and one sprawling (indeterminate) tomato plant. Many of these vegetables grew surprisingly well. I ate unwatered tomatoes July through September; kale, cabbages, parsley, and root crops fed us during the winter. The Purple Sprouting broccoli bloomed abundantly until the next March.

"In terms of quality, all the harvest was acceptable. The root vegetables were far larger but only a little bit tougher and quite a bit sweeter than usual. The potatoes yielded less than I'd been used to and had thicker than usual skin, but also had a better flavor and kept well through the winter."

The trade-off with dry farming is lower yields. As an example, with squash, dry farming produces 3-pound fruits on unirrigated clay, as compared to 7 pounds of fruits on well-irrigated soil. As the following illustration shows, after 10 weeks, the growth of un-irrigated trees with no pruning (or summer or dormant pruning) is incredibly low compared to irrigated trees—five inches versus 70 inches, respectively.

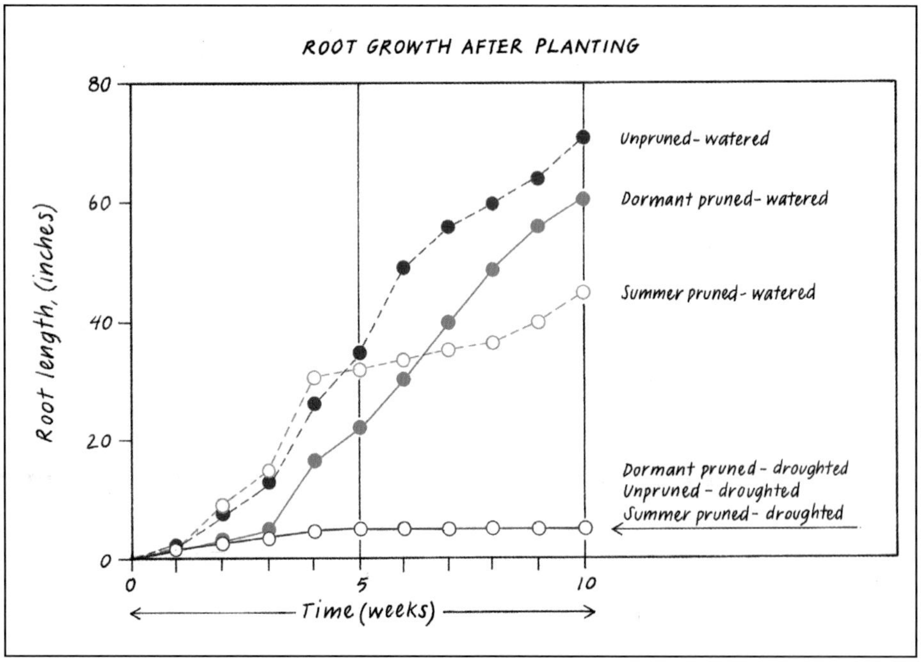

Figure 139. This graph was used before on page __. There, it was used to show how roots can grow as much as five or more feet away from the trunk in just 10 weeks. You can see the difference between irrigated and unirrigated trees. Irrigated trees' roots can grow 12 times further than unirrigated trees.

Bonus Photographs and Graphs

Figure 140. While it doesn't look it, the newspaper is four to six inches away from the stem of the lavender. Bet you can't put the newspaper down without reading it—you can catch up on old news.

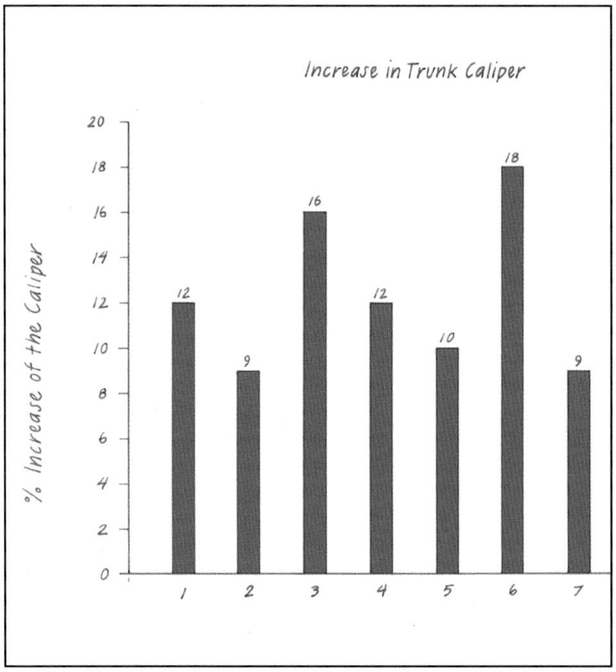

Figure 141. A comparison of various mulches as measured by the increase in the tree's trunk diameter.

1 - "Weed-Chek" fiberglass landscape mat
2 - Wood chips, two inches deep
3 - Hay, six inches deep
4 - Calcined clay, two inches over 4-mil black plastic
5 - Redwood bark, two inches deep over 4-mil plastic
6 - 4-mil black plastic
7 - No mulch

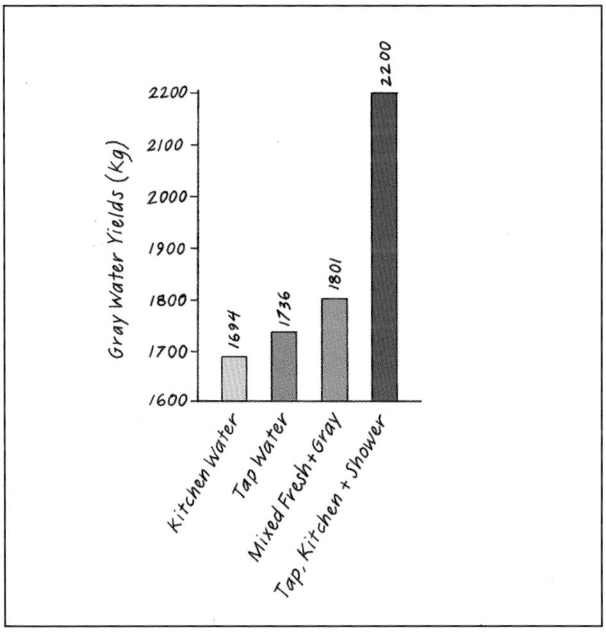

Figure 142. Late-breaking greywater news. A mixture of various "dirty" water with fresh tap water gives the "greytest" yields.

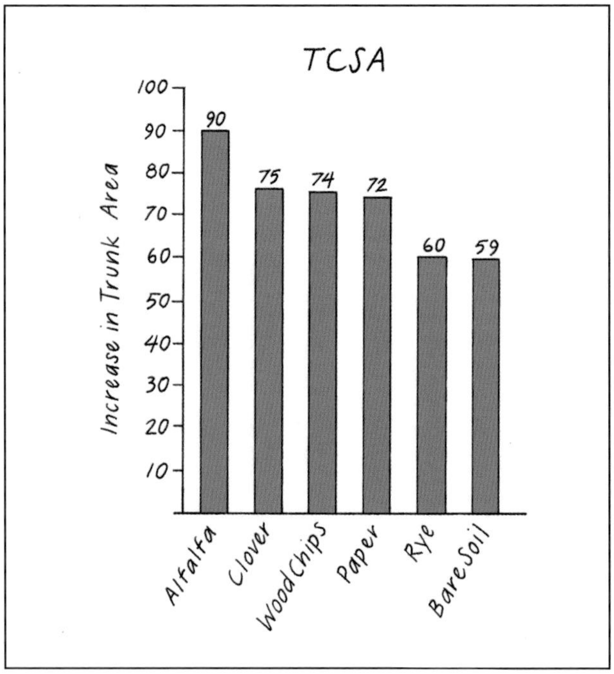

Figure 143. Another study included paper mulch. The paper mulch was better than rye straw, but slightly less than wood chips and clover. TCSA = trunk cross-sectional area and is measured in mm^2.

Figure 144. A new windbreak has been planted to replace the old, ailing windbreak. This windbreak protects a handful of houses on the leeward side at The Sea Ranch, California.

Figure 145. Esalen Institute's young windbreak (Big Sur, California) is composed of *Metrosideros excelsa* trees (New Zealand Christmas tree, USDA Zones 10-11) and *Rhaphiolepis indica* shrubs (USDA Zones 7b-10). Pick your trees carefully.

Figure 146. Esalen Institute's windbreak to protect new frost-sensitive young lemon shrubs. It's 50-percent open so wind is allowed through to jostle the plants to help build stronger trunks.

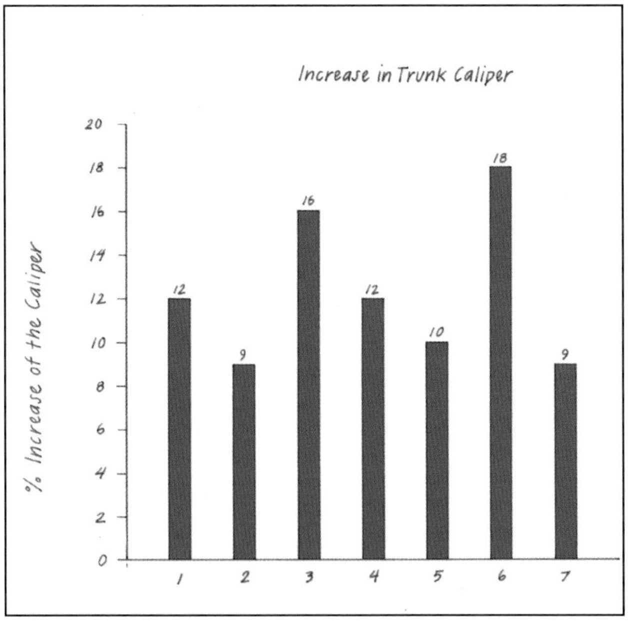

Figure 147. Another study of mulches showing an increase by percentage of the trunk's diameter of a green ash.
1 - Fiberglass landscape mat
2 - Wood chips two inches deep
3 - Hay six inches deep
4 - Clay over 4 ml black plastic
5 - Redwood bark two inches deep over 4 mil black plastic
6 - Four mil black plastic
7 - No mulch
(From: *The Journal of Aboriculture*, 1983. January 9 (1).)

Figure 150. The Sea Ranch community uses sheep to graze down grasses to reduce the fire hazard. Thus, they are left with lots of wool. It becomes a useful mulch in the community garden to protect the soil from winter rains.

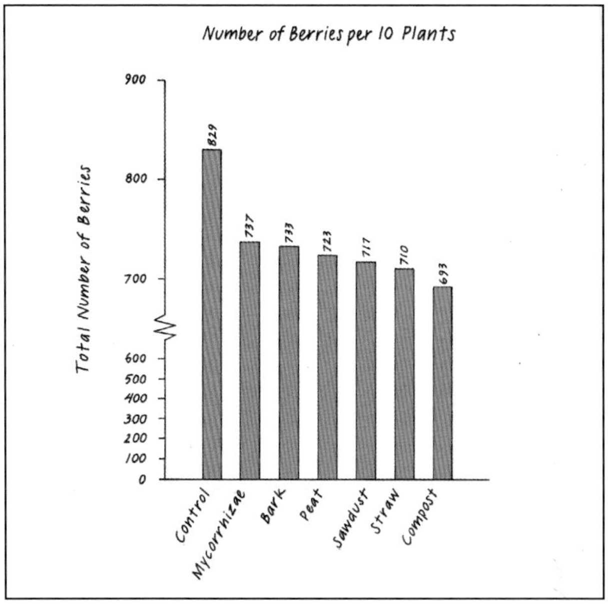

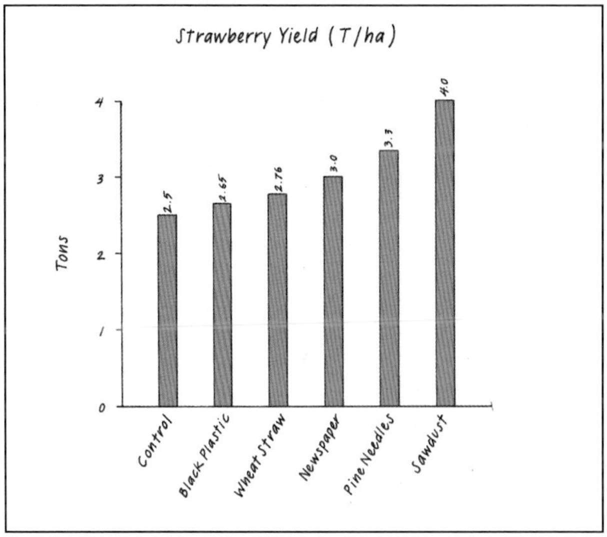

Figures 148, 149. It's very hard to find studies that include newspaper or cardboard used as a mulch. This one from Pakistan (of all places) is fascinating, as two trials were conducted, each with a different combination of mulches. Newspaper gave higher yields in tons than wheat-straw, black plastic, and the unmulched control.

(The Effect of Different Mulching Material on the Fruit Size and Yield of the Strawberry under the Agro-Climatic Condition of Mansehra, 2015. Adil Reham et al, *Journal of Natural Sciences Research*, Vol. 5, No. 11.)

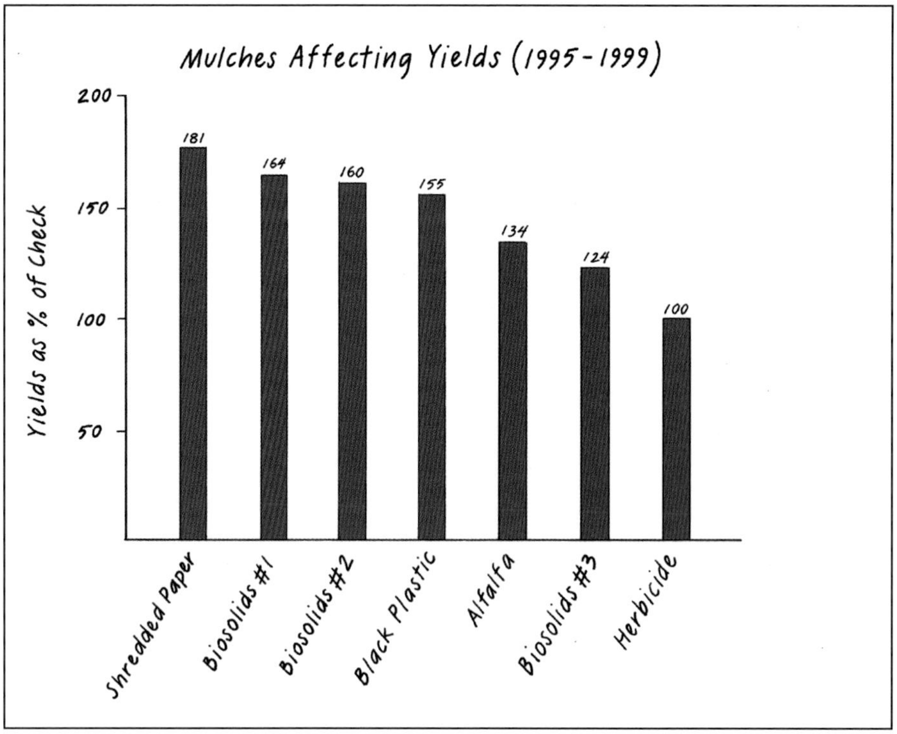

Figure 151. This study used *shredded* paper as a mulch. The paper mulch out-performed all other combinations. The study used three different manufacturer's biosolids. Since the study was done in Canada, I won't mention the product's name because you probably wouldn't be able to buy it.

(Surface aplication of mulches and biosolids affect orchard soil properties after 7 years, 2002. G. H. Neilson, et al, *Canadian Journal of Soil Science*, 83: 131-137.)

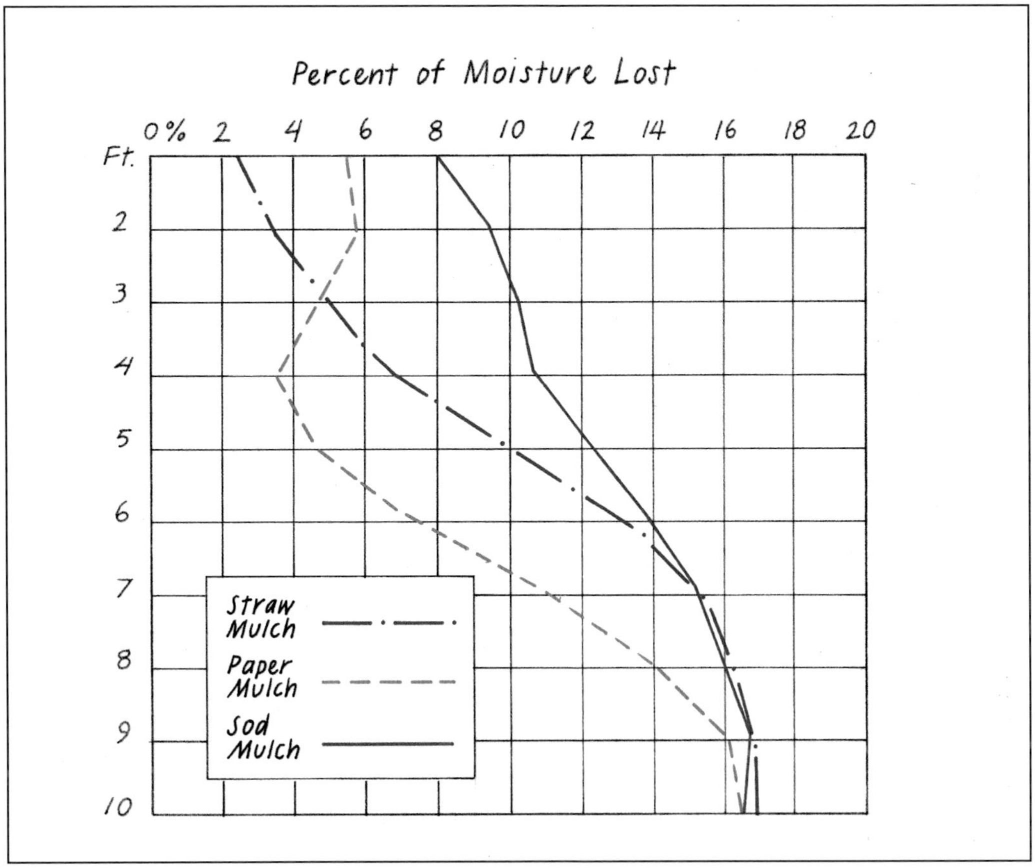

Figure 152. This data is based on a study done by the Nebraska Agricultural Experimental Station in Lincoln, Nebraska. Another graph reveals that paper mulch can be close to as effective as straw mulch in conserving moisture. At the soil's surface, straw mulch is three percent more effective than paper mulch. But at the three-foot depth they conserve the same amount of moisture. At the nine-foot depth all mulches conserve about the same amount of water. The moisture lost at nine feet is due to that absorbed by the roots.

(*Root Development of Young Apple Trees*, 1937. W. W. Yocum, University of Nebraska-Lincoln Extension, Paper 910.)

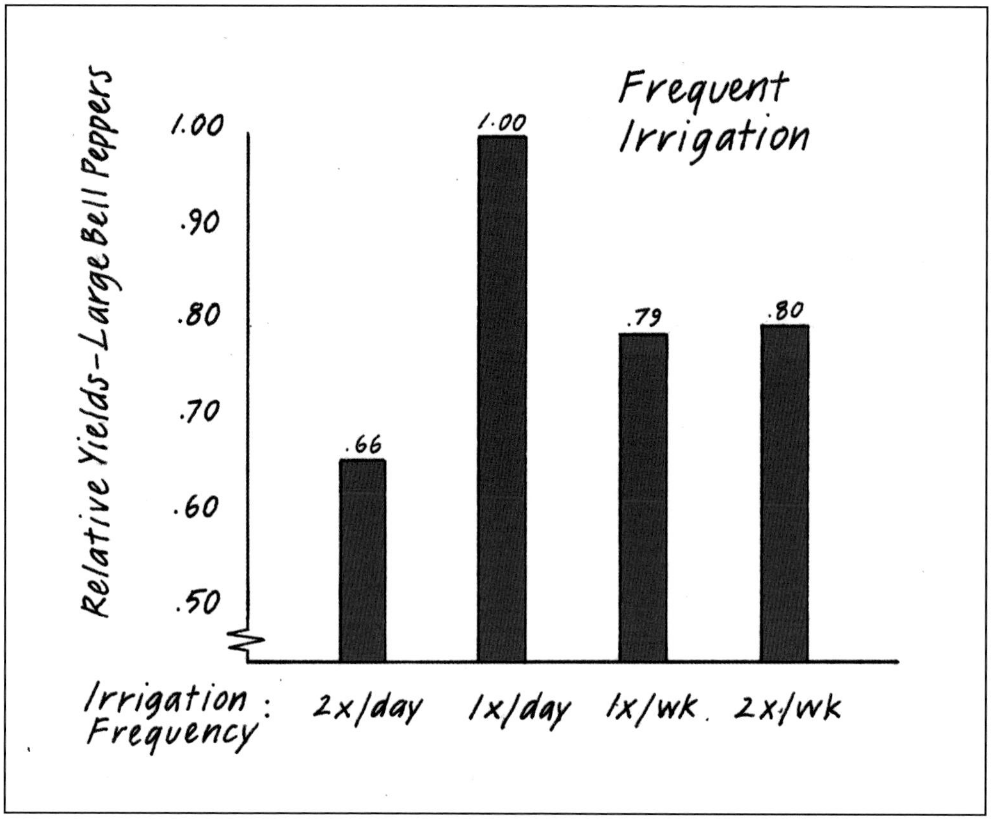

Figure 153. Another study that showed daily irrigation is superior. The same amount of water was applied with each irrigation. I find it fascinating that watering twice a day produces nearly 50 percent less yields of large bell peppers. The slight difference between once a week versus twice a week irrigation still shows more frequent irrigations increase yields. But twice a week is still 20 percent lower yields compared to daily. And twice a day is 40 percent lower yields.
(From, *Horticulture Technology*, January - March 2033, 13(1).)

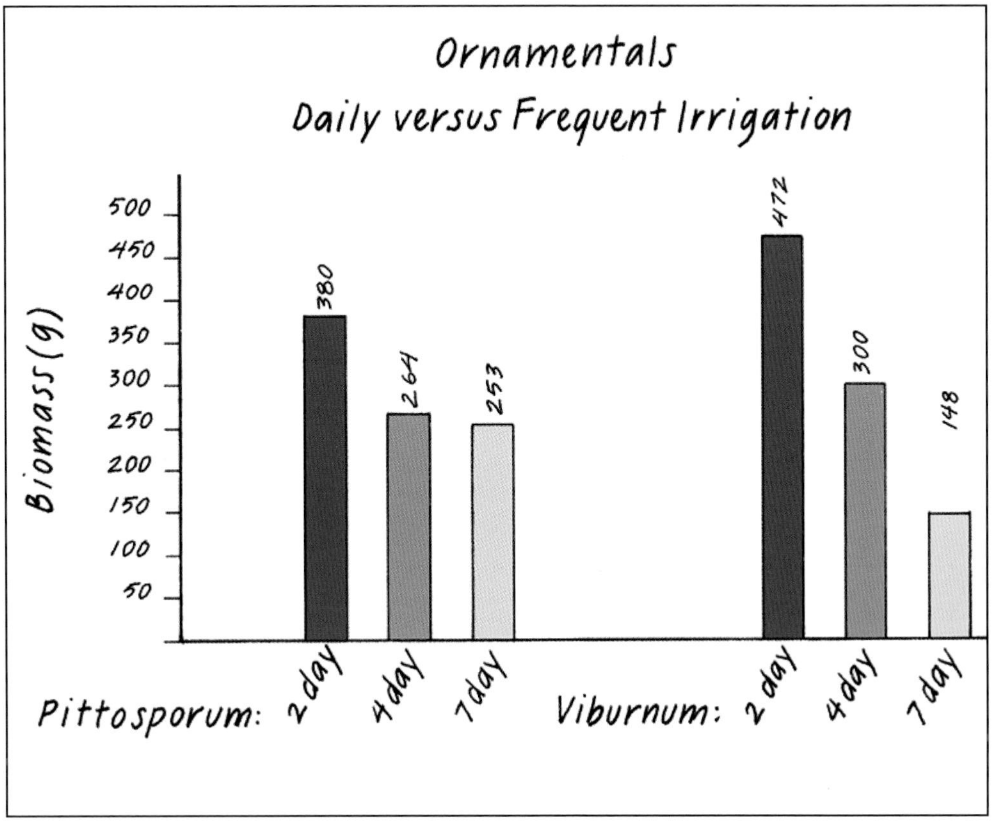

Figure 154. This is the only research I've ever found that measures frequent irrigation versus longer intervals for ornamental plants—not edible crops. Most research is focused on edible crops so the farmer can save money. Here we see that watering two ornamental shrubs every two days is superior to irrigating every four days or once a week. The shrub on the left it *Pittosporum tobira* 'Variegata' and on the right is *Viburnum odoratissumum*. The study was done over a two-month period in Florida.

Chapter 6
Forest Gardens

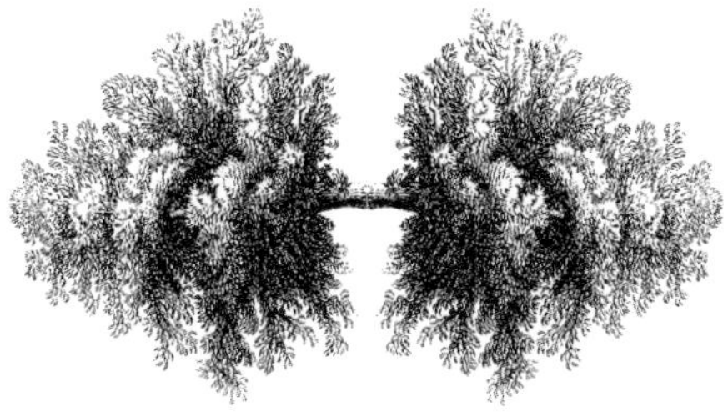

Forest Farming - Robert Hart (aka Robert A. de J. Hart)

One of my earliest reasons for getting into edible landscaping was the exciting work of Robert A. de J. Hart, who (along with coauthor James Sholto Douglas) wrote the book *Forest Farming* back in 1973. It piqued my interest, to say the least—the idea of a mixed forest that could support grazing animals like cattle and pigs, as well as people. This was also the general theme of J. Russell Smith's c.1929 book *Tree Crops: A Permanent Agriculture*. After all these decades, we're just beginning to see the cultural impact of these concepts.

At first Hart's book proposed "free" food sources that would produce plenty of nuts and fruits to be "harvested" by animals living below. In his own words: "I developed my own '3-D' system, which I called OPS—Organic Perennial Subsistence farming. That involved 'cultivating' my hedgerows by encouraging the growth of plants that contained substances particularly nourishing for cattle, such as the elder, wild rose and hazel and sowing some perennial pasture herbs."

As things evolved after the first book, he added: "But my primary aim was self-sufficiency, so I extended my system beyond livestock farming to include trees and other plants—mainly perennial—which would contribute to the health and welfare of human beings. In time, I had adopted a vegan diet." (A strict vegan lifestyle excludes all animal food products, manure fertilizers, leather clothing, and leather shoes.)

Over the years, as chronicled in *Forest Gardening - Cultivating an Edible Landscape* (Chelsea Green Publishing; 1996), Hart refined his concept to that of a layered forest-like "garden," with the following elements:

"An old orchard, unless the trees are severely diseased. My forest garden was planted in a twenty-five-year-old small orchard of apples and pears, some of which were in a pretty poor condition. But the abundant aromatic herbs that have been planted beneath them seem to have rejuvenated them. A decrepit-looking 'Red Ellison' apple was given a new lease of life when Garnet grafted three young 'King of the Pippins' shoots onto it—a trick that was known to the ancient Romans. These old trees constitute the 'canopy' of the forest architecture.

"If one is starting a forest garden from scratch, the best way to form a canopy is by planting standard apples, plums, or pears at the recommended spacing; twenty feet each way. Then fruit or nut trees on dwarfing rootstocks can be planted halfway between the standards, to form the 'low-tree layer,' and fruit bushes between all the trees to form the 'shrub layer.' Herbs and perennial vegetables will constitute the 'herbaceous layer,' and horizontally spreading plants

Tree care can be modeled after the forest—sometimes. The use of guilds is suspect in my opinion. See pages 190-198.

like dewberries and other *Rubus* species, as well as creeping herbs such as buckler-leaved sorrel (*Rumex scutatus*) and the ornamental lady's mantle (*Alchemilla mollis*), will form the 'ground-cover layer.'

"For the root vegetables, mainly radishes and Hamburg parsley, occupying the 'rhizosphere,' a low mound can be raised, so that they will not be swamped by the herbs. As for the climbers that constitute the 'vertical layer': grapevines, nasturtiums and runner beans can be trained up the trees, while raspberries and hybrid berries, such as boysenberries and tayberries, can be trained over a trellis fence, forming a boundary to the garden." (Notice that no traditional annual vegetables except beans, parsley, and radishes are mentioned.)

When I visited Robert Hart in 1998, he was elderly and somewhat feeble (he died in 2000). He lived in a very humble single room. What remained of his food forest was a thicket of perennial edibles. Near the house were mostly cane fruits and an enormous kiwi vine. (How many kiwis can one person eat?) Some local permaculturists had opened up parts of the garden to plant traditional vegetables.

Perhaps the most impressive act of Hart's life was his lifelong care of an elder brother who was too mentally challenged to care for himself. Sadly and ironically just as I left, a "Meals-on-Wheels" type group delivered his evening meal.

Figure 155. Robert Hart, 1998.

While I was touring gardens in England in 1998, Patrick Whitefield, author of some of the most popular books on forest farming, including *How to Make a Forest Garden* (Permanent Publications, UK 2002), took me to see his edible forest. At 15 years old (in 1998), the food forest, modeled after the work of Robert Hart and Bill Mollison, was the most mature edible-food forest in England.

On the way to visit his edible forest, Whitefield remarked: "It's really just a jam and jelly forest." Most of the wild fruit trees had matured and the bulk of the crops were well above easy-harvesting range. This drove home the point for me that for any type of forest "farming" to really work, some pruning is in order, especially to keep the trees low enough to beat the birds to the harvest. (This doesn't apply to squirrels. They just take what they want!)

When I arrived, "meadows" had been recently opened up to plant apple trees, other more traditional fruit trees, and conventional vegetables. Interestingly, Whitefield (who died in 2015 at age 66) did not live in his forest, but in a house some miles away, with a typical vegetable garden and greenhouse.

"Climax" Ecosystems

According to Robert Whittaker, author of the 1953 book *Communities and Ecosystems*, "climax" is "a steady state of ecological community productivity and population, with the dynamic balance of its populations determined in relation to its site." The final or stable community was considered the *climax community* or *climatic vegetation*. It's self-perpetuating and in equilibrium with the physical habitat. Although this was the traditional definition of a climax ecosystem, things have changed, to say the least.

Climax Community

In scientific ecology, *climax community*, or *climatic climax community*, is an historical term for a boreal-forest community of plants, animals, and fungi that, through the process of ecological succession in the development of vegetation in an area over time, have reached a steady state. This equilibrium was thought to occur because the climax community is composed of species best adapted to average conditions in that area.

Perhaps the strongest criticism of the view held by permaculturists that New England's deciduous forests (considered a climax forest by some) were more highly productive than farmland is to be found in a review of Toby Hemenway's book *Gaia's Garden*, published in the Winter 2001 edition of the *Whole Earth Review*. In it, Greg Williams asserted that Hemenway's view was based on a theory of ecological succession that postulates that net productivity declines as ecosystems mature.

Hemenway's response in the same magazine, citing data from ecologist Robert Whittaker's book *Communities and Ecosystems,* disputes Williams's claim on productivity as focusing on climax conditions rather than on maturing forests. Hemenway was also critical of Williams' characterization of permaculture as "simply forest gardening."

Succession and "climax" theories have since become less monolithic and more complex.

According to Dan Botkin (Professor of Biology [Emeritus], University of California, Santa Barbara) "The older concept, the end point of succession called a climax, was assumed to be self-perpetuating in both species persistence and their relative abundance. This final, stage would therefore remain in equilibrium with the physical habitat. There would be no net annual accumulation of organic matter in a climax community. In addition, it would remain relatively unchanged until destroyed by an event such as fire, storms or introduction of other species that were toxic to the "climax" species. In a diluted form a phrase used to refer to what might otherwise be called mature or old-growth communities.

Recent research on actual ecological succession shows that even this "final" stage can be subjected to many changes, because small areas within a mature forest would now and again, be subjected to local disturbances that would reset the process of succession within that small part of the forest (or whatever ecosystem was present, from marine, freshwater, to solid land). In summary, the current idea is that a large area would approach a final "climax" ecosystem, formed by species adapted to this "mature" stage but with variations in the relative abundance of each species and even variation in the kinds of species that were adapted to and persisted long term." (Copyright 2020, from Botkin's forthcoming book, *Discordant Harmonies for the 21st Century*.)

Figure 156 on page 172 (from *Bob's Honest-to-Goodness Newsletter*, Vol. 2/ #2, 1996) shows the extremes and the mean productivity for seven ecosystems. The darker horizontal bar within each vertical bar is the mean productivity. A label lists the mean productivity for each ecosystem. The top of the vertical bar indicates the highest productivity. The bottom of each vertical bar indicates the lowest productivity.

Notice that the mean for tropical rainforests exceeds that of estuaries, and the *maximum* productivity of cultivated land can only equal that of estuaries. Note that the mean productivity for cultivated land (based on agriculture, but may also represent our gardens) is lower than all other ecosystems. But its highest productivity is also about equal to the highest productivity of a tropical rainforest.

Ecocentric Gardens

Forest gardening (a smaller version of forest farming, and the phrase I'll use hereafter) in America is mostly ecologically egocentric. Many of its proponents live in or close to primarily deciduous forests that are often thought of as climax ecosystems. The most heavily forested region in the US is New England. In the mid-1880s, settlers had reduced the forest cover to 30-40 percent of its original size, but now 80 percent of New England is covered by second-growth forest or by thick woods. (Deer were down to several hundred in Massachusetts at the beginning of the 20th century; but today, there are about 85,000 white-tailed deer in the state. The predator-free suburbs seem to help maintain the deer population.)

All the forests in America once totaled around one billion acres, but are now reduced to 747 million acres. Thus there are large ecosystems where trees are *not* the predominate "climax."

Consider the Great Plains (see map, on page 173). Prairies are defined (by Wikipedia) as "ecosystems considered part of the temperate grasslands, savannas, and shrublands biome by ecologists, based on similar temperate climates, moderate rainfall, and a composi-

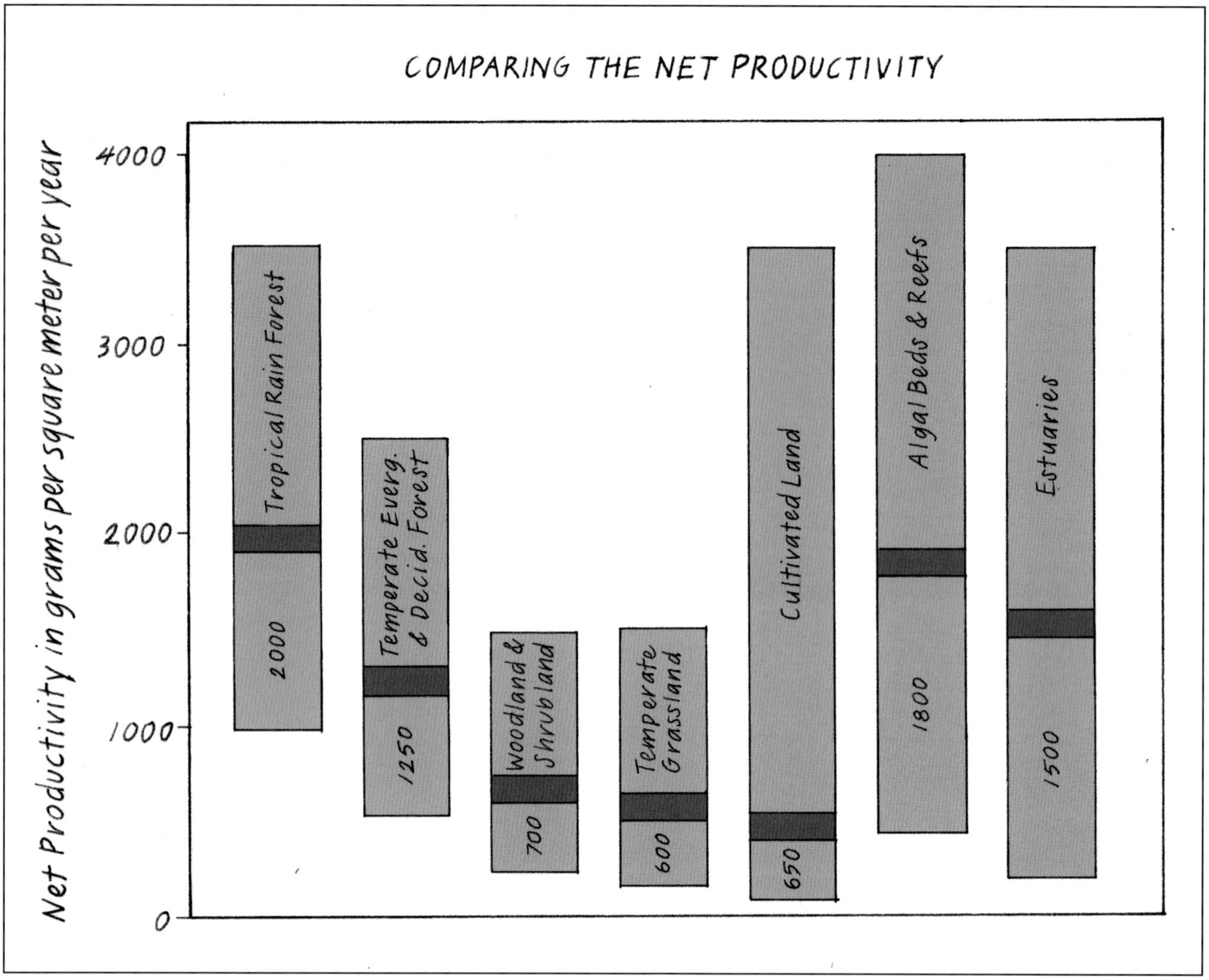

Figure 156. The dark blue bar indicates the median between the two extremes. For example, cultivated land has a low productivity of around 100-200 grams per square meter per year and a high of about 3500. (The median is 650 grams, compared to the tropics with a median of 2000.) This reveals that the highest productivity of cultivated land can just about equal the highest productivity of a tropical rainforest.

tion of grasses, herbs, and shrubs, rather than trees, as the dominant vegetation type." Some sources say the Great Plains grassland ecosystem used to cover 320 to 367 million acres. It has now shrunk (due in great part to agriculture) to 76,000,000 acres, a 79 percent reduction.

We don't hear as much from permaculturists about grasslands gardens as about temperate forest gardens, but Sepp Holzer and a few others are working on the growing of grains in Salzburg, Austria. Josef "Sepp" Holzer is a farmer, an author, and an international consultant for natural agriculture. "Holzer Permaculture" is a branch of permaculture that he developed independently from the mainstream permaculture in Austria. It's particularly noteworthy because it grew out of practical application, relatively detached from the scientific community.
This not to say that research is lacking. Wes Jackson's Land Institute in Salina, Kansas has been working on developing a commercial strain of perennial wheat for more than 40 years.

Two important facts to know about the current status of grain crops is that annual crops are 50 times more susceptible to soil erosion than are perennial pasture crops, and commercial annual wheat crops currently use 30 to 50 times more nitrogen than perennial grains. Annual grains like cereals, oilseeds, and legumes are planted on almost 70 percent of the world's croplands.

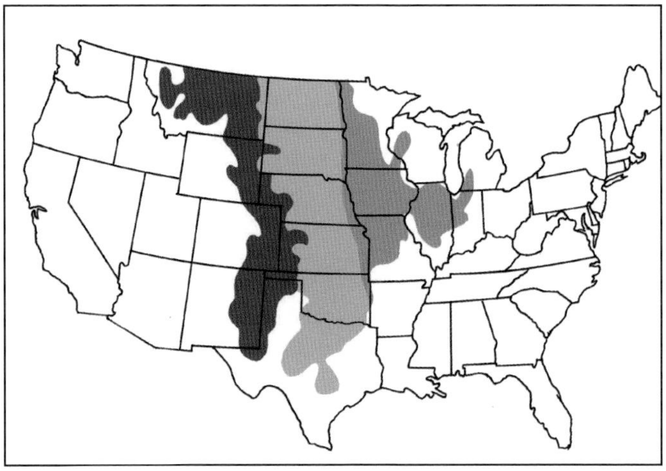

Figure 157. The three major prairie zones in the US are, from left to right: dark brown = shortgrass prairie; tan = mixed grass prairie; and green = tallgrass prairie. All together they equal about 76 million acres, down from 320 million acres before the area was settled.

In "Increased Food and Ecosystem Security via Perennial Grains" (J.D. Glover et al, *Science,* 25 June 2010; Vol. 328 no. 5986, p 1638-1639), we read: "Annual grain crops can lose five times as much water and 35 times as much nitrate [a form of nitrogen—ed.] as perennial crops. Perennial crops require fewer passes of farm equipment and less fertilizer and herbicide, important attributes in regions most needing agricultural advancement." Also, greater root mass maintains more soil carbon in comparison with annual crops. Perennial plants help to protect soil against erosion in ways that annuals, with their smaller root systems, can't.

(The perennial "wheat" (called Kernza®) that the Land Institute has developed is a cousin of annual wheat. It's a domesticated perennial grain originating from a forage grasses known as intermediate wheatgrass (*Thinopyrum intermedium*). Kernza® is in the early stages of commercialization, and is not a plant for home gardens.)

"Climax" forests are found from New England to the California Sierras. The California chaparral, however, is another large area (it covers about 4.6 million acres, or 14 percent of California) that doesn't always rapidly evolve to a climax situation. According to "*Chaparral in Southern and Central Coastal California in the Mid-1990s: Area, Ownership, Condition, and Change,*" by Jeremy S. Fried, Charles L. Bolsinger, and Debby Beardsley (Forest Service Resource Bulletin PNW-RB-240, 2004):

"'Chaparral' [is] derived from the Spanish term *chaparral,* meaning scrub oak (McMinn 1939), and refers to a group of biotic communities or vegetation types comprising several genera of shrubs, some of which exhibit treelike growth forms under optimal conditions. Chaparral is often referred to simply as 'brush,' although in the past it was also called 'elfin forest,' a poetic term apparently no longer in use. "In California, chaparral occurs within the Mediterranean-type climate zone, which is characterized by hot, dry summers and mild, moist winters. Chaparral grades into conifer forest (*Pinus* spp., *Pseudotsuga macrocarpa, P. menziesii,* and *Sequoia sempervirens*) or hardwood woodland (*Quercus* spp.) on moister sites at higher elevations, or on north-facing slopes, and in canyons."

The most arid regions of North America are the Chihuahuan, Sonoran, and Mojave deserts, which receive less than 10 inches of rain per year, and along with other much smaller deserts, equal about 12,160,000 acres. In such areas, large trees do cluster together in riparian habitats, but small trees, shrubs, and xeric plants predominate.

Every one of the above regions was used for agriculture for hundreds of years by indigenous peoples, though in some areas it was harder than in others to support a family or clan. Certainly a "forest garden" in the Mojave Desert is rare indeed. Such climates tend to support far fewer people than New England's forests, but there are plenty of models to mimic for a desert garden. Internationally celebrated nature writer, agrarian activist and ethnobiologist Gary Nabhan has written extensively and lyrically about native foods in the southwest, and how they were and are managed and cultivated. Two important examples of Nabhan's books are: *Growing Food in a Hotter, Drier Land – Lessons from Desert Farmers on Adapting to Climate Uncertainty*, 2013, Chelsea Green Publications, and *Desert Terroir, Exploring the Unique Flavors and Sundry Places of the Borderlands*; 2012, University of Texas Press.

PermaEdge

Proponents of forest gardening believe that it's the interfaces, the edges between two ecosystems (for example, where trees meet shrubs and grasslands) that are the most productive. Where one type of large or macroscopic ecosystem meets another, there's a mixing of plants and animals indigenous to both. This produces a unique environment with some of the virtues and diversity of both ecosystems, as well as of plants and animals that thrive only in the mixed zone.

In *Permaculture: A Designers' Manual* (1988), Bill Mollison defines this mixed zone, which he terms the "edge," as "...normally richer in species than the contributing media themselves." He also states that "...complex systems of land/ocean [edge], such as estuaries and coral reefs—show the highest production per unit area of any of the major ecosystems."

But it's not that simple. For instance, the assumption that estuaries show the highest net productivity of any major ecosystem is refuted, as described in this Chapter by respected ecologists. In the book *Communities and Ecosystems*, by Cornell University ecologist Robert Whittaker, an informative table ranks the net primary productivity (a gauge of fertility) of various ecosystems. (See Figure 156.) Once again, looking at the mean amounts, estuaries (at 1500 grams per square meter per year) are *less* fertile than tropical rainforests and only 250 g/sq. m/yr. *more* productive than our temperate evergreen forests.

Since fertility is not a fixed number, but a fluctuating range of figures, it's misleading to state unequivocally that estuaries show the highest production of any of the major ecosystems. In addition, most of the accumulated fertility or productivity of an edge must stay *within* an ecosystem in order for it to persist. We can only "extract" a small portion of the productivity each year as food, timber, or fiber if the ecosystem is to remain self-sustaining.

From the assumption that all "edges" are the most fertile productive environments, permaculture proponents focus on creating more "edge" to increase the dynamic of the environment. In *Permaculture: A Designers' Manual,* we read, "...a complex edge... can be considered the basis of the art of productive landscape design. And most certainly, increased edge makes for a more stimulating landscape. As designers we can also *create* a harmonic with plants, water, or buildings."

At this point, some perma-enthusiasts attempt to construct an intellectually unifying pattern or theory based upon the many forms of a sine wave, the spiral form represented by the Fibonacci series (as in a nautilus shell), and the various curves manifest in the turbulence of water and smoke (see *Sensitive Chaos: The Creation of Flowing Forms in Water and Air*, by Theodor Schwenk; 1996, Rudolf Steiner Press).

These curves and spirals are offered as virtually universal forms that can be used in landscape design, as represented in a herbal spiral—a common feature of a classic permaculture garden. In most designs, the spiral is planted with herbs in an attractive pattern, but as they mature, the spiral usually becomes too crowded, and the plants must compete for sunlight, moisture, and nutrition.

As a guideline, I'd say don't waste time with spiral herbal plantings unless you're *very* tight on garden space. Not to mention that in many curved gardens, all those lovely winding paths make it difficult to get quickly from point A to point B on garden errands. If you use a hose, it's a nightmare to drag it along and around circular vegetable beds. At best, you'll need a lot more hose-bibs (faucets), and will need to couple and uncouple the hose frequently. A garden that looks wonderfully symmetrical on a garden plan set down on paper while inside a comfortable house, often looks far from symmetrical when one is standing in the actual garden.

Designs in permaculture books also tend to be full of aesthetically pleasing parabolic curves. This recurring design element is, in my opinion, more of an intellectual extraction than an actual frequent element in the natural landscape. Permaculturists have chosen the mathematical form of a parabolic edge and then inferred the biological implications.

While any curved line can lend artistic beauty to a design, it's a stretch to say each curve represents an optimal biological design, or that the ecological dynamic has been enhanced, especially in a suburban backyard. The ecology of an edge is a macroscopic dynamic between large plant and animal communities, not one that can necessarily be reduced to a few acres or a 1000-square-foot lot. The recommendations presented in many permaculture tomes are an artful approach to design, but not one that necessarily enhances the ecology of the landscape. In short, unless you want to make a ton of extra work for yourself, do *not* shape your paths or vegetable beds in circular, mandala, or spiral patterns.

Somehow, the abstract discussion of "edge" takes on a very literal interpretation in many permaculture books. The cross-sections of various planting schemes in many permaculture books feature a polite progression of small plants to large trees forming a convenient diagonal line, although a natural edge seldom takes this form. Arranging the tallest plants on the north side of a landscape and graduating each shorter plant toward the sunny south has, by the way, little to do with the lofty and complicated theories of "edge harmonics"—it's just good-old common sense.

Layers—"Stacking"

In permaculture and forest gardening, seven layers of the food forest are identified (see below):

1. The canopy
2. Low tree layer (dwarf fruit trees)
3. Shrubs
4. Herbaceous plants
5. Soil surface (cover crops)
6. Rhizosphere (root crops)
7. Vertical layer (climbers, vines)

The 8th layer, or Mycosphere (fungi), is often included in stacking.

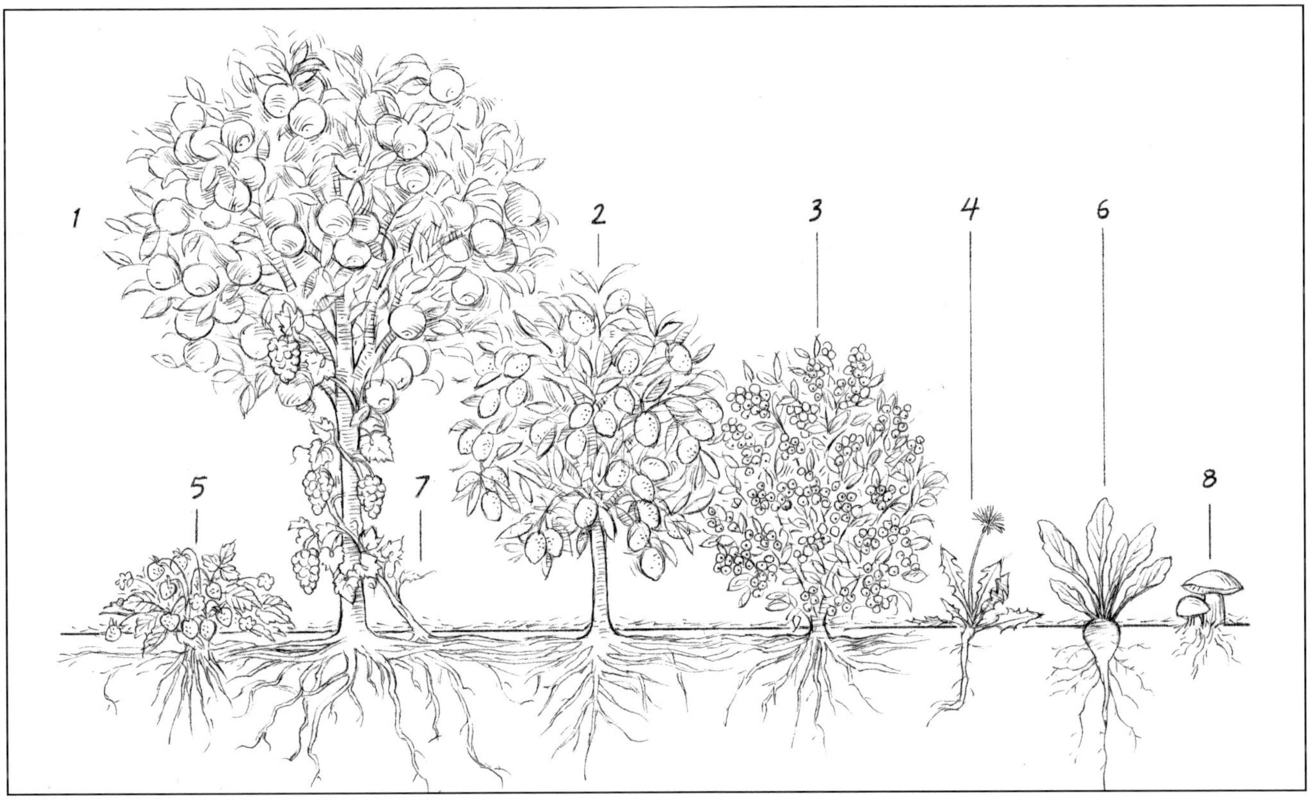

Figure 158. The main concept of a forest garden is to layer all aspects of it to replicate the structure of a natural forest. Here are the eight layers of an "ideal" forest garden (see above for the key to the numbers). The idea is to stack all components in the same garden, some literally, as with grapevines trained to grow up a fruit tree. As the vine matures, however, the fruit grows further from the gardener, and pruning vine and tree for ideal fruiting becomes more difficult. Some "cover crops" are literally crops, such as strawberries grown as an edible ground cover. Other examples of this include: chives, society garlic, and garlic chives. If they're prolific, these ground covers can make the pruning and harvesting of the other layers troublesome. Fallen fruits, with their potential for harboring pests and/or diseases, are harder to find and clean up. I call this "peasant-based agriculture."

The permaculture thinking about forest garden goes something like this: a mature ecosystem such as a woodland, has many relationships between its components: trees, understory, ground covers, soil, fungi, insects, and animals. The various plants grow at different heights, which allows for a diverse community of life to thrive in a relatively small space. This layering reflects that found in a natural forest. To mimic the forest is to have a more vibrant, natural ecology. The assumption is that mimicking the structure of a forest will result in the most sustainable, if not the most productive, food forest possible.

(Though the general principles seem to be sensible, early texts promoted the use of non-native, invasive species or environmental weeds, including many *Acacia* spp. trees, the autumn olive shrub (*Elaeagnus umbellata*), honeysuckle shrubs and vines [*Lonicera* spp.], and kudzu [*Pueraria montana* var. *lobata*]. Be sure to read Chapter 3 of this book.)

Perennial Forest Gardens

Many proponents of forest gardens, including Robert Hart, favor all-perennial plantings. This is based on the concept that trees yield a more stable ecosystem, which in turn is based more on the tropical model with its continuous cover of trees than on forests in temperate climates. The argument is that most of the trees are edible or medicinal; this is mostly absurd, since most people don't eat only perennial fruits and nuts. This would require a major shift in diet—something that is culturally very difficult from the popular soda-pop, doughnut, and hamburger diet.

One simplistic rationale, stated by Angelo Eliades of Melbourne, Australia, is: "Annuals are in the minority in nature and the majority in our food." They obviously haven't lived in my Northern California chaparral ecosystem, or near the flower meadows of New England. Eliades also states that: "Forest succession is an ecological process which Nature uses to turn bare soil into a forest over a period of 50-150 years." Fifty years is certainly more the exception than the actuality; 150 years is perhaps possible, but both goals are out of reach of most gardeners.

Agroforestry

Agroforestry is one approach to perennial-based cropping, and is different from permaculture in that there's some physical separation between trees, animals, and annual crops. Agroforestry has its roots in the land management of Native Americans, who managed forests to further the acorn harvest, while mammals they could hunt roamed below. The meadows and forest were routinely burned so as to kill young trees, shrubbery, and grasses and sustain the growth of oak trees to enhance the production of acorns for food and further the deer population. The native peoples effectively put the natural evolution of the forest on hold to manage the forest for increased yields. When Anglos first arrived in Yosemite Valley, for example, they didn't recognize this management as agriculture.

Nowadays, an agroforestry farm often grows linear rows of one or more practical trees that produce, for example, edible crops, lumber, or firewood. There's usually more space left between trees than in a natural forest. In this country, agroforestry is well suited to temperate ecosystems, and provides the closest examples to temperate permaculture forest farms and gardens.

From *An Introduction to Agroforestry*, by Chloe Johnson at the Kerr Center for Sustainable Agriculture; USDA Risk Management Agency; Oklahoma Department of Agriculture, Food and Forestry; Summer 2006: "Multistory Cropping is an agroforestry design where farmers and ranchers harvest trees, crops and livestock to make the most of their land. This method is especially popular outside the U.S. and is also known as 'forest farming.' In many countries native forests are managed in order to produce crops or forage. The system yields wood or tree products [such as pecans] in addition to agronomic crops or forage, improves crop or forage quality and improves soil nutrient recycling for crop or forage use. It also reduces excess subsurface water and provides favorable habitat for beneficial plant and animal species. Trees should be planted or native forests thinned to allow adequate light into the understory crops or forage. Tree canopies range from 5 to 40 percent crown cover."

"Alley cropping," as practiced for many decades, is a common approach to agroforestry. It often involves planting one or two rows of a commercial tree crop with wide spaces between the rows for cash crops of

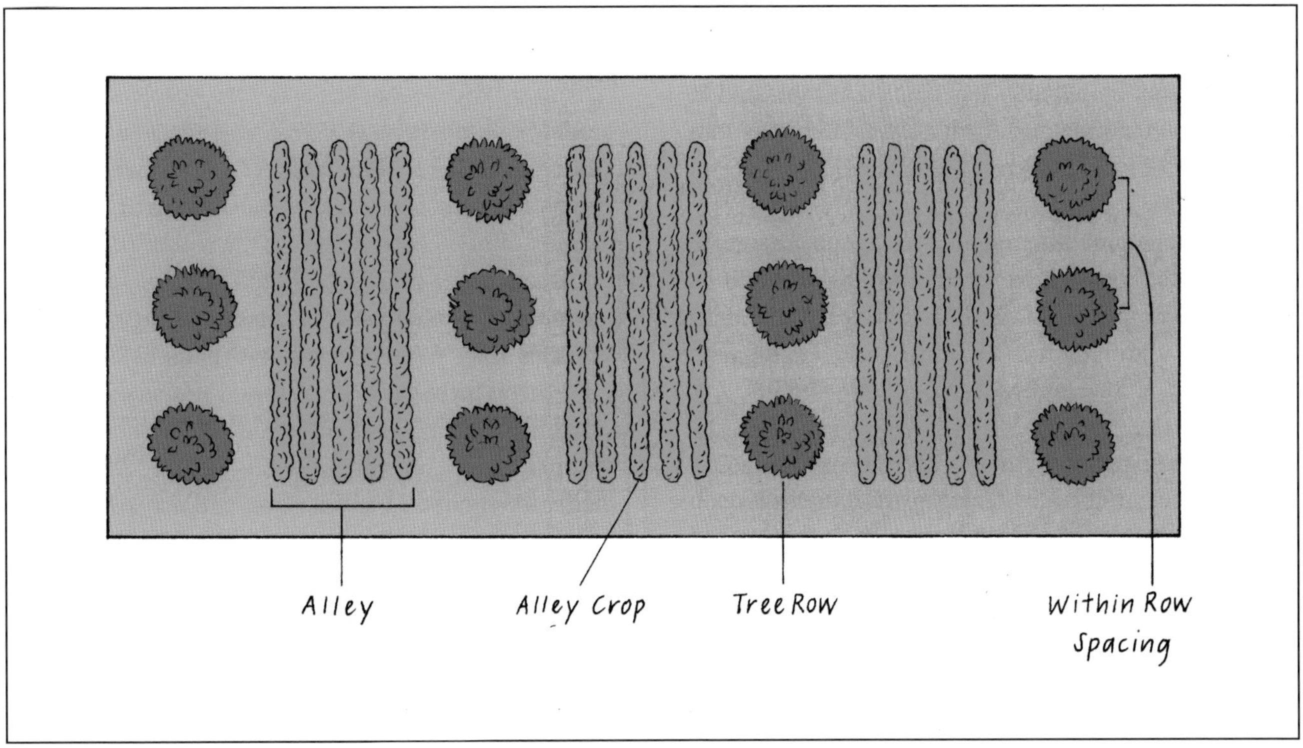

Figure 159. In the United States, agroforestry came along before permaculture. The alley crop designated here may be annual vegetables or grains or another smaller tree such as a filbert. As well as nut trees, the tree row may contain trees for firewood or veneer.

annuals, or to provide perennial pasture for livestock. The temperate-zone alleys are planted with the trees further apart than in tropical alleys, where the practice originated.

One common example of American agroforestry is the planting of black walnut trees (*Juglans nigra*). The black walnut is a very desirable tree. both for its wood and for the nut harvest. A single tree to make veneer wood for furniture can sell for as much as $30,000. (Veneer is a very thin layer of particularly fine wood, such as walnut, that has been glued on to inferior wood to produce a smooth and attractive finish.)

In an agroforestry black-walnut orchard, trees are planted much further apart than the 18-foot spacing of an ordinary orchard, leaving as much as 36 or more feet between the rows. With the wider spacing, corn, winter wheat, barley, sorghum, or soybeans can often be grown between the alleys for up to 10

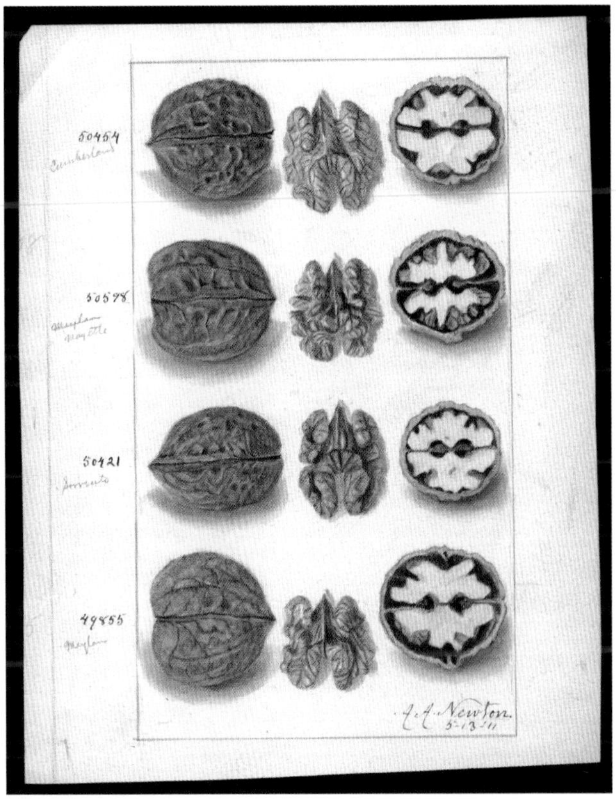

Figure 160. Walnuts are best known for the nuts. But they make superior furniture wood.
(From the U.S. Department of Agriculture Pomological Watercolor Collection. Rare and Special Collections, National Agricultural Library, Beltsville, MD 20705, circa 5-3-1911.)

years, after which root competition and shade from the trees begins to crowd out the other crops. Pecans are another nut crop that is also planted alley-fashion; often with cotton, milo, wheat, oats, potatoes, or other row crops in between.

To learn about alley planting in the US, consider contacting the Association for Temperate Agroforestry (AFTA), School of Natural Resources, 1-30 Agriculture Bldg., University of Missouri, Columbia, MO 65211. A membership includes a subscription to a quarterly newsletter, The Temperate Agroforester, as well as discounts on association events and on many agroforestry books ordered through the association.

In *An Introduction to Agroforestry*, we read: "Most of the [tropical] systems are dominated by perennial rather than annual crops, resulting in a relatively high ratio of nutrients stored in the vegetation to that stored in the soil." (From *An Introduction to Agroforestry* by P.K. Ramachandran Nair, 1993, Kluwer Academic Publishers, 101 Philip Drive, Norwell, MA 02061, USA.) In a temperate zone; this is reversed, with more of the nutrients, because of the extensive planting of row crops, stored in the soil.

Obviously, the alley concept is too large to use in most suburban gardens, and too difficult to scale down to a typical yard. Therefore, I propose growing a mixture of annual and perennial edible plants (favoring the annuals), and gradually shifting your diet toward the products of some fruit and nut trees. The non-commercial growing of fruit and nut trees should be in moderation, as a fruitarian diet is highly out of balance and many nut trees require a lot of space and take many years to come into bearing.

Figure 161. The photograph at left shows deciduous peach trees along a path in the California winter. Since most of the plants below are perennials that die back or cut back, it's fairly easy to get to the trees for annual pruning and spraying for peach-leaf curl. In the summer, as shown on the right, it's a different story. This design by Maile Arnold for Two Acre Wood in Sebastopol, CA, is very attractive in the spring and summer, and the members of the co-housing community harvest nearly every peach, but it does take extra time and coordination to step around all the perennials. The public lands in the development are so small that a mulched orchard system is less likely—the beauty of perennials prevails.

Peasant-Based Forest Gardens

As mentioned earlier, greatly due to the influence of Bill Mollison, Robert Hart, and many other permaculturists, forest gardens have become one of the big horticultural trends. I learned my lesson the hard way in the mid-1970s with my own forest-garden experiment.

My mixing of layered plants to attempt to create a forest garden with and under fruit trees was a royal pain in the @#s. I was forever looking for a place to step without squishing the lower edible plants. In my experience, it greatly increased the time I spent thinning fruit, picking fruit, and pruning. Another problem was finding all the fallen fruit, so as to not leave any pests or diseases around to infect the tree the next year. (Chickens can help with the removal of fallen fruit.)

Figure 162. The perennials along this path are lovely, but restrict access to the fruit trees and slow down the picking of the ripe fruit. As a result, fallen fruit may get lost and promote pests and diseases the following year. It's a small orchard, however, so "peasant-style" intensive care fits into the community's (CoHousing) gardening time.

The entire concept of a mixture of layers of bearing crops is really based on peasant culture, especially in the tropics. If you read the permaculture books carefully, you'll find a lot of the "models" are actually based on non-temperate-zone (tropical and sub-tropical) plantings. All the forest gardening-centered permaculture workshops and events that I've attended spend a lot of time, whether consciously or unwittingly, on presenting tropical examples. The tropical ecosystem, combined with skilled knowledge of techniques for working in it, often makes for lush, multi-layered plantings, but home gardeners in the tropics generally learn via many generations of cultivators, in areas where tried and trusted cultivation techniques are common. (For more, see Chapter 1.)

According to Food & Nutrition Bulletin (Vol. 7,No. 3, page 4): "Layering is not as definite a feature…in the cooler, sun-poor regions of the world. In temperate gardens, trees and bushes are well spaced, leaving ground-covering plants without shade. Annual seed culture is prominent…defined crop areas are the rule…with reduced inter-cropping and mixed cropping resulting in a more open vegetation canopy that takes full advantage of solar radiation."

As mentioned earlier, one great inspiration to start my career as an environmentally sound horticulturist/organic landscape gardener working to reverse the paving of America with lawns was the "Dump Heap and the Origin of Agriculture" chapter of Edgar Anderson's book *Plants, Man, and Life*. In this chapter, we read of many plants grown by a farmer in Santa Lucia, Guatemala in what at first appeared to be chaos. Further study by Anderson over the years revealed that the garden was a collection of mostly tropical and some temperate-zone plants (squashes and pumpkins) grown in what appeared to be a random fashion. In fact, many years of trial and error had taught this farmer, probably through oral history, a number of lessons: how to plant for the greatest ease of planting and care; the use of perennial food crops; "random" plantings to help thwart pests; and planting of chayotes, as a horticultural crop and for shade. Many of these plants were grown in a vertical layering to make good use of limited space.

In contemporary America, most families have two working adults. What to do in their "spare time" is

Figure 163. Peach trees at David Ulmer's are planted three to a hole to increase variety in a small space.

critical. A densely layered planting can take so much extra time that I don't recommend it for the novice gardener. For example, rather than growing peach trees with roses, irises, and dozens of other plants set beneath and beyond the dripline of the tree, I prefer clustering fruit trees in a zone with continuous mulch, much like David Ulmer's yard in Sebastopol, CA (Figure 163) With this approach, the time saved for other recreational or gardening activities is greatly increased. David's planting of beneficial-insect-attracting plants, well beyond the current and future diameter of the fruit tree, works fine, as most beneficial insects can easily fly into the fruit-tree zone to prey upon pests.

Another excellent approach is to plant two, three, four, or more trees in the same planting area, as David Ulmer does. The photograph showing fruit trees without lower-zone plants and with a (mostly) continuous mulch beneath, exemplifies Ulmer's wise use of fruit trees, and demonstrates how to increase the variety of fruit grown while making the planting

Figure 164. More fruit trees planted three to a hole. For specifics, go to Dave Wilson Nursery's website and view the video on this topic.

look less regimented than a typical orchard. An even more compact diversity can be achieved by grafting more than one variety of fruit onto one tree.

It's a matter of choosing between an aesthetic bias, a preference, an intellectual concept, and the convenience of zoning the trees together. This doesn't mean we're required to plant trees in linear rows like those in a commercial orchard. On the contrary, I often plant fruit trees close together with irregular spacing to look more grove-like.

Yields

"Permaculture in the tropics, as expressed in Bill Mollison's *Permaculture: A Designers' Manual*, did not produce significant amounts of food or fruit when applied in northern New South Wales and Queensland, largely since the closed canopy is not conducive to fruit production. The system, while very healthy in and of itself, yielded very little produce." (From: a March 17, 2013 entry on the blog by Autonopedia).

Meanwhile, in Melbourne Australia, Angelo Eliades is growing a backyard forest garden. He first cleared and leveled his backyard, and then replanted and mulched many of the existing plants. From this he created 915 sq. feet (85 sq. meters) of garden beds. He now grows 30 fruit trees, 20 types of berries, 50 medicinal herbs, and some vegetables.

The annual production of edibles from Eliades' garden was:
First year: 133 kg. (292 pounds)
Second year: 204 kg (450 pounds)
Third year: 195 kg (as a consequence of a minor disaster) (430 pounds)
Fourth year: 234 kg (516 pounds)
This is equivalent to 32,628 pounds per acre, per year or 75 pounds for each 100 sq. ft. for a year-round garden.
His average monthly amount of produce is 19.5 kg. (43 pounds), or 551 pounds for 12 months. And it only takes him two hours per week to maintain.

For comparison, look at Graham Bell's 26-year-old organic permaculture food- forest garden close to the Scottish Border near Coldstream in England. It's larger than average at 8,712 square feet, and he has documented harvests of 551 pounds of fruits and vegetables—about 63 pounds per 100 sq.ft. per year, not too different than Angelo's yields.

In the USA, the "average" person spends two to 4.5 hours per week tending an average 600-square-foot garden. Food gardens that are 100 percent vegetables, take as much as 20 hours of tending per season per 100 square feet. A typical annual vegetable garden with wide rows may yield 45 to 60 pounds per 100 sq. ft. per season.

Bruce Butterfield was the research director for the National Gardening Association for 25 years. His surveys have shown that a well-maintained garden can produce a half-pound of fresh vegetables for every square foot of garden space (50 pounds per 100 square feet) at an average market value of about one dollar per square foot.
Butterfield's surveys also found that the median size

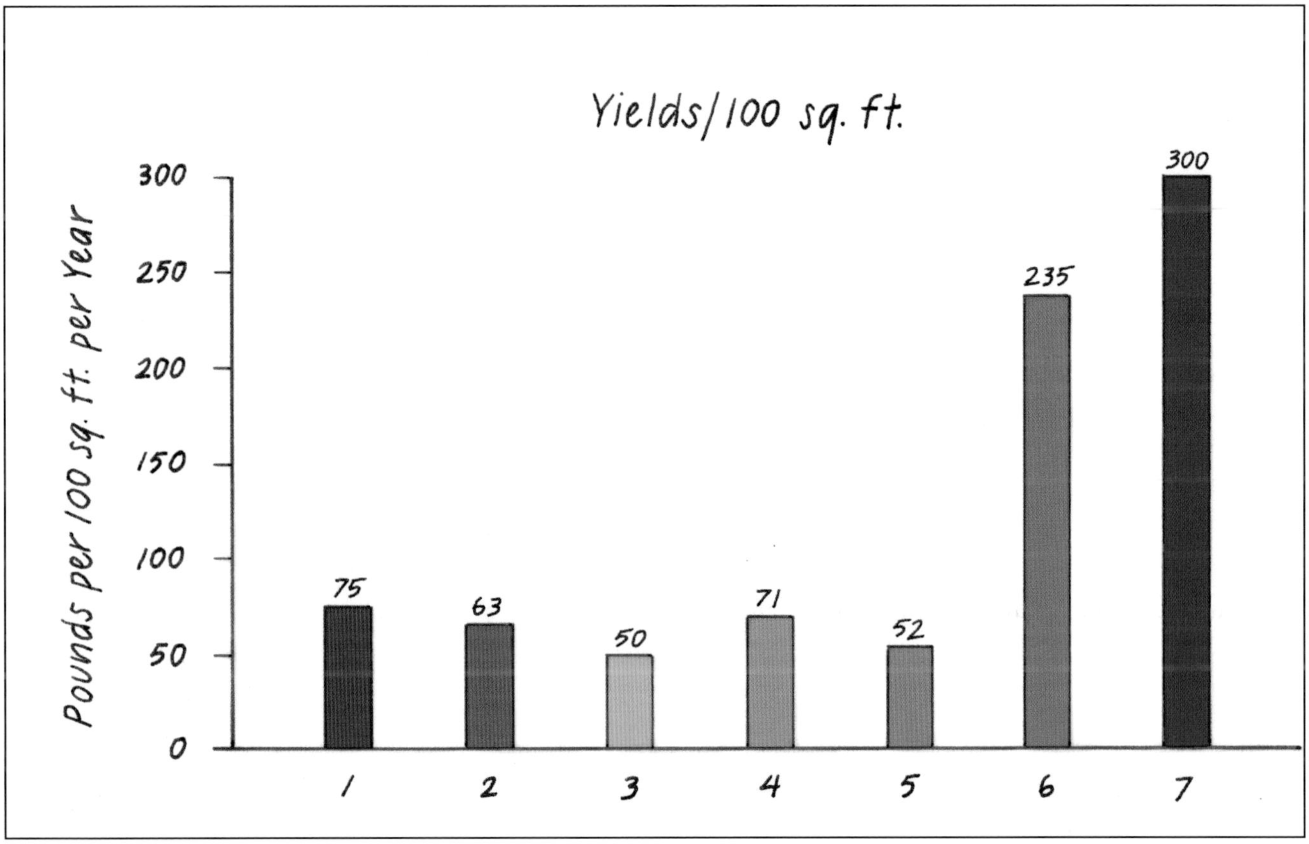

Figure 165. Key = Crop yields for: 1-Angelo Eliades (in the sub-tropics); 2- Graham Bell; 3- NGA-USA average; 4- Naomi van der Velden; 5- Roger Doiron; 6- Ros Creasy; and 7- John Jeavons. Except for Creasy and Jeavons, the average yield in pounds of produce per 100 square feet per year is just over 62 pounds. (Of course the mixture of vegetables and fruit was completely different for each garden.) All gardens except for Angelo Eliades' were grown in various temperate zones.

of an edible garden is about 100 square feet. For a family of four, a growing space of 200 sq. ft. should keep the family in veggies all summer long. He says one should plan to spend four hours a week tending the garden, with an initial eight to 12 hours for preparing the planting area in spring, shopping for seeds and seedlings and sowing crops.

Dr. Naomi van der Velden of the University of Cumbria coordinated a one-season trial with 50 gardeners participating. Each was assigned a square meter of a high- density polyculture (similar to the vegetable section of a food forest, with crops like peas, radish, perennial spinach and beets, onions, lettuce, sweet corn, coriander, runner beans, kale, marigolds, and arugula) and one square meter of low-density vegetables (peas, radish, perennial spinach, and beets). In the trial both types of plot averaged about 71 pounds per 100 square feet for one May-October season. (I was amazed that the polycultures were able to match the low-density crops, given all the complexity and competition for water, light, and nutrients.)

But the effort involved is another story. "On average, 3.4 kilograms were harvested from a square metre of the low-diversity plots for every hour of effort put in, and the high-diversity plots yielded 2.3 kilograms." In other words, the high-diversity plots yielded about 40 percent less food per hour of effort. (And contrary to many assumptions about polycultures, "...the high-diversity plot suffered the most damage. It may be that plants were more stressed from competition for space and nutrients and less able to cope with additional pressures.")

Roger Doiron of Kitchen Gardeners International grew a 1600 square-foot garden in Maine and harvested 52 pounds per 100 square feet. Or, "If you use the $2400 figure and consider that our garden is roughly 1/25th of an acre, it means that home gardens like ours can gross $60,000/acre. You can also calculate it on a square foot basis, which in our case works out to be roughly $1.50 per square foot. That would mean that a smaller garden of, say, 400 square feet would produce $600 of produce." (From "What's a Garden Worth? Roger and Jacqueline Doiron Do the Math and Calculate the Cost of Their Garden Vegetables for One Growing Season." Fine Gardening Magazine, March, 2009.)

In one six-month season, a well-designed and very intensive edible landscape (in this case, Rosalind Creasy's) yielded about $700 of food in only 100 square feet of gardening space.

In Creasy's words: "From April to September, this little organic garden produced 77.5 pounds of tomatoes, 15.5 pounds of bell peppers, 14.3 pounds of lettuce, and 2.5 pounds of basil — plus a whopping 126 pounds of zucchini! Next time I won't feel bad about pulling out those extra plants.

"I (Ros) figured the total value of my 2008 summer trial garden harvest was $746.52. In order to get a fair picture, I also needed to subtract the cost of seeds, plants and compost (I can't make enough to keep up with my garden), which added up to $63.09. That leaves $683.43 in savings on fresh vegetables. Of course, I'd have a hard time eating 126 pounds of zukes."

John Jeavons, of double-digging fame, maintains that very intensive planting can yield 300 pounds yearly in a 100 sq. ft. bed, but to achieve these yields requires a very strong back, lots of skill and knowledge, plenty of compost, and years of practice.

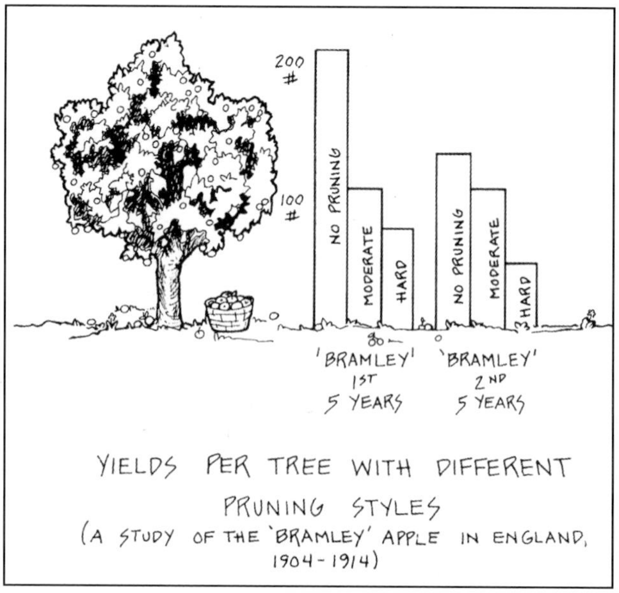

Figure 166. Pruning reduces the number of fruits and total weight in pounds of fruit. Unpruned apple trees produce plentifully, but the fruit is very small (good for cider, though).

Except for these last two special garden situations, the numbers are similar, noting that the Australian garden is heavily weighted towards fruit crops.

Pruning Fruit Trees for Good Yields

Unfortunately some of the proponents of perennial-food gardens prefer not to prune. They think you can walk away from a fruit tree after planting and return to harvest. Not so.

Back in 1986, my book on edible landscaping (*Designing and Maintaining Your Edible Landscape - Naturally*, Metamorphic Press) devoted four chapters to what were then the most modern—meaning innovative—pruning guidelines for fruit trees.

What follows is a review of the 1986 recommendations (labeled **Then...**) as compared with what I've learned since (labeled **Now...**). Let's have a look.

Pruning Proverbs Probed

Then..."No pruning at all gives the biggest harvest." The actual 1986 text begins: "Studies have shown that the yields of unpruned trees, in total pounds, are often higher than the yields for pruned trees." Then I qualified this point with the comments that such trees have better flavor (if unirrigated), require less work, don't have as much fruit near the ground, often get uneven skin color, make smaller-sized fruit, may slip into alternate bearing (a good harvest one year followed by a poor one, etc.), are kind of ugly, *and* are prone to broken limbs from the weight of crop.

Now... This series of points has stood up quite well. Basically, the research I've seen still indicates that the harder you prune a tree in the early part of its life, the longer you'll delay the time at which it begins to fruit. *But*, while unpruned trees make more total pounds of fruit than a pruned tree, the fruit usually is far from what most people consider good for eating fresh out of the hand—it's best used to make juice or cider.

Then..."Prune to the branch collar."
The text and illustration were a virtual duplication of the work of Dr. Alex Shigo—formerly a forest pathologist with US Forest Service in New Hampshire (1).

Now... Just as valid as 40 years ago.

Fortunately, the concept of not cutting flush to the trunk or branch is beginning to seep into mainstream gardening. While many who attend my pruning workshops still haven't heard of keeping the branch collar intact, more and more gardeners, arborists and landscapers have taken Shigo's wisdom to heart.

Shigo did more to reshape our understanding of tree growth and pruning than perhaps any other single person. He has dissected over 15,000 trees with a chainsaw to determine how injury, pathogens and pruning affected growth, rot and disease. One of his early studies involved planting walnut trees, pruning them in various ways and waiting 12 years to cut open the trees and analyze the long-term effects of pruning.

Then..."You don't need to use pruning tars and paints."

Now... This till holds true. After thousands of cuts I've made on my own trees, the properly made ones have callused over quite nicely. However, to this guideline I'd add an important amendment: I've seen so many young tree trunks damaged from sunburn, that I want to remind you to use paint (50 percent water and 50 percent interior or exterior *latex-only* paint) to cover the *entire* trunk up past the first few limbs on the very day the tree is planted. Also, trees that are radically pruned—summer or winter—may need the paint applied to the tops of newly exposed limbs.

Another good use of paint is in conjunction with the removal of large limbs, or when too much of the canopy is removed. Most deciduous fruit and nut trees can benefit from this protective painting when large amounts of foliage are pruned off (although sunburn is actually more a sign of having pruned too radically). The need to paint can be avoided by simply making judicious thinning cuts over the course of several years. While a proper pruning cut doesn't need paint to form a healthy callus, some plants (such as tropical and subtropical fruit trees like avocado and citrus) are extra-sensitive to sunburn and a quick coat of the 50/50 paint-and-water mixture may prevent sun damage.

Then..."A branch at a 45-60^0 angle promotes the formation of flower buds and fruit."

Now... While this holds true for fruiting trees, ornamentals are another story. From the book on pruning

I wrote for Smith & Hawken: "As a *general* guideline, the best position for a shoot to naturally favor both lateral growth and flower budding is between 60° and 45° from horizontal. With this placement, some of the apical (tip) bud's influence is lost [this is known as diminished apical dominance] so that flowering and branching are encouraged and a reasonable amount of tip growth remains so the plant doesn't become too narrow or spindly.

Pruning or shaping new growth to this ideal zone of position will promote:

- More laterals without extra pruning.

- A wider angle of attachment of the younger lateral to the trunk or primary scaffold. (This is good for fruit trees, because it lets more light into the interior of the canopy for better fruit color and flavor.)

- More flower buds for greater ornamental beauty and, with food crops, more to harvest."

The important overriding concept is that there are *numerous* exceptions to pruning "rules," depending on a tree's natural growth habit. Nature made many plants that flower profusely on limbs in an upright position (forsythia, matilija poppy and rugosa rose shrubs, for instance); on those that cascade down (weeping cherry, weeping hemlock, weeping mulberry and other weeping species of trees); or on limbs at just about any angle (manzanita, rockrose, winter daphne, tobira shrubs, strawberry tree (*Arbutus unedo*), Russian olive [now known as an invasive exotic], holly, and oak trees). Follow the natural archetypes, not pruning dogma (even mine), especially with ornamental trees and shrubs.

Then..."Summer pinching and pruning promotes flower buds."

Now... Alas, this time-honored method from Europe is not standing up well to the scrutiny of modern science. In the Smith & Hawken book I wrote: "This classic approach to training espalier fruit trees involves repeated pinching back of the new growth to induce flower buds by late summer. When trying to force flower fruit buds by pinching, you must start *early* in the summer. After three to five pinchings, depending on the season, the amount of growth and the climate, only a whorl of three to five leaves remains—each with fat flower buds at the base of the leaf stems. (See the illustration below.) One theory says that the repeated removal of the apical bud and the partial de-

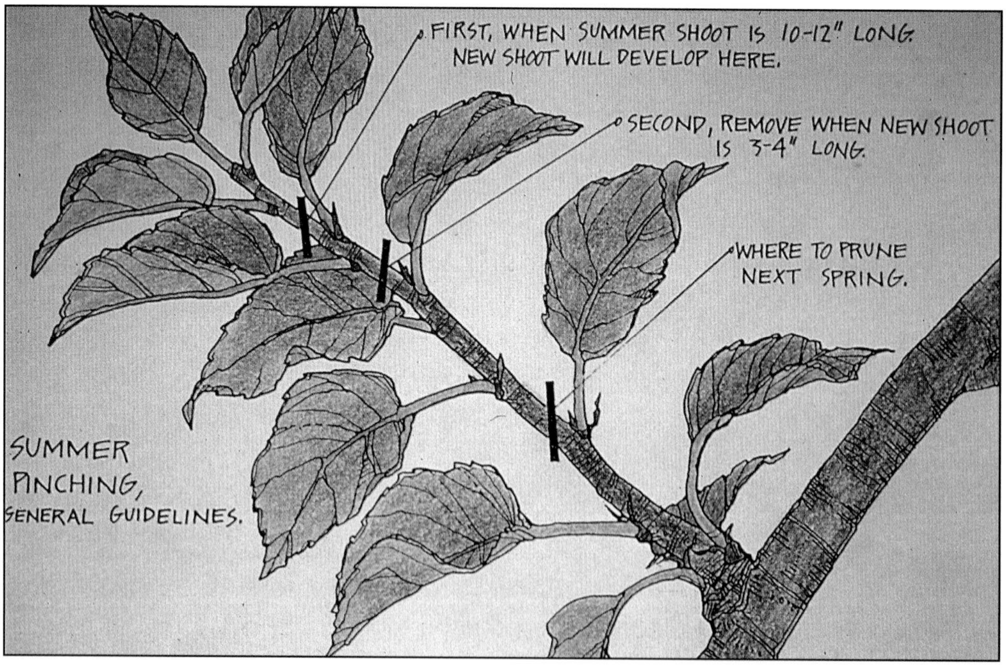

Figure 167. The French developed the Lorette System of pruning for fruit production, in which a series of pinchings are made on the shoot through the summer. Each pinching causes the foliage to store more carbohydrates in the leaves near the base of the shoot. A final cut in late winter or early spring encourages flower buds in the places that have stored up a relatively high amount of carbohydrates. This method takes a lot of time throughout the summer, and using clothespins to stimulate fruiting is much faster, as it only needs to be done once in late spring when shoots are about eight-12-inches long.

foliation act to force more carbohydrates to be stored in the dormant buds. These "carbo-loaded" buds turn into flower buds by fall, and flower the next spring."

Here I cheated a bit by using the words "this classic approach" and "one theory." However, these pinchings can't hurt the tree and are often required to maintain an espalier tree's architectural form. Whether or not this technique increases flower formation (initiation) is still highly debated. From my favorite book on fruit culture, *Temperate-Zone Pomology, Physiology and Culture* (Melvin Neal Westwood, Timber Press, 1978): "With apple [trees], increased photosynthesis and flower bud formation inside summer pruned trees was reported by Stembridge and Lakso but Greene and Lord found no increase in flower buds. [According to] Marini and Barden, pruning vigorous apple shoots severely in June or July increases [flower bud] initiation more than late-summer pruning, which misses the natural [flower bud] initiation period. Many reports show that apple fruit size is smaller with summer pruning."

Then..."A 45° to 90° angle of attachment makes the strongest branch."

Now... This "dictum" is frequently repeated in fruit-tree literature. To my chagrin, I could have found the correction for this in the *1983* edition of *Arboriculture: Care of Trees, Shrubs, and Vines in the Landscape* (Prentice-Hall; First Edition, 1983). It wasn't until recently that I discovered (on page 401-402) Richard Harris's reminder that "When Miller subjected 60 apple branches of the same variety, age and size to stress until they split, there was no correlation between the angle of attachment and the strength of branch attachments, but a branch will grow more upright and vigorously when the angle of attachment is more acute." (The paper was published in *1959*, for crying out loud!) In fact, with both fruiting *and* ornamental trees "For a strong attachment, a branch must be *smaller* than the trunk or limb from which it arises." However, any angle of attachment is weakened when there's a condition known as embedded bark (also called included or invaginated bark). To spot embedded bark, look for a prominent invaginated branch ridge where the cambium of the bark has folded in on itself. (See the illustration on page 374.) Also according to Harris (page 33): "Angle-of-branch attachment is primarily influenced by the genetic makeup of the plant." So, I propose: "Trees come a variety of natural branching patterns. If a narrow angle of attachment is intrinsic to a tree's branches, you'd have to spend a lot of time to try and "correct" this tendency (for instance, by using "spreaders"— see below.). But you needn't bother; chances are that the tree has evolved tissues which compensate for the narrow angles."

Then..."Summer pruning is the best way to remove water sprouts, suckers and control a tree's growth."

Now... Guess what? Summer pruning doesn't control tree growth (reckless or well-behaved) any better than winter pruning.

The clue to debunking this myth is tucked away on page 216 of *Temperate-Zone Pomology, Physiology and Culture*, in which Westwood writes: "Marini and Barden have summarized the literature on summer pruning and conclude that studies often do not support many of the long-held views on the practice. Many studies indicate that summer pruning is no more effective than dormant [winter or early spring] pruning for limiting size and suppressing shoot growth, but the timing of growth suppression can be useful." After reading this I called Westwood and he was generous enough to spend a lengthy amount of time discussing the details.

After our conversation, this is how I choose to present the topic: "A time-honored tradition (especially in Europe) holds that the removal of foliage during the growing season reduces the amount of leaves available to make food for storage in the tree or shrub's mass. This reduction in food has a gentle debilitating effect on the subsequent spring's growth. This amounts to a mild form of dwarfing. Under this theory, more ambitious ("harder") summer pruning equals more control over growth.

The European tradition that summer is the *only* time to prune trees to keep their size down doesn't stand up to modern research, but now we know that summer pruning is *part* of the solution. Controlling the size of a tree or shrub usually requires summer pruning simply because more than one pruning is required to keep the size of a tree or shrub within desired bounds.

If the plant in question isn't very vigorous and is somewhat small to begin with, then one pruning

Figure168 . Using clothespins in early spring just as a new shoot starts to grow is a good way to "train" it into a 45-degree angle of attachment, which promotes fruiting. The clothespins are removed in late summer.

a year might be enough to limit its height. This single pruning could be done in the winter, spring or summer. When trying to restrain a vigorous tree or shrub, pruning is best done in winter, spring *and* summer. Some pruning all season long is best for controlling overgrown trees and shrubs.

I suppose I can comfort myself a bit with the knowledge that Marini and Barden didn't release their paper until 1987, a year after my first edible-landscaping book was published, but I was amazed how long it took me to update this well-worn "pruning commandment."

Then..."Toothpicks and clothespins can be used to position a shoot for more fruit."

Now... This concept remains true, but I learned some smart ways to use other "tools" as well. Read on.

Shaping Fruit Trees

You don't need pruning clippers (shears) to get an abundant crop. Here are some tricks to make it even easier to shape a tree, avoid pruning, and harvest plenty of fruit. These techniques are for apple, pear, and Asian pear trees only.

When to Summer Prune

Doing some careful pruning all summer long is a balanced approach to caring for trees, shrubs, and perennials. Pruning in the summer is also helpful when trying to restrain a vigorous tree or shrub, in order to continue thinning out unwanted growth. This doesn't mean I've cast aside all winter pruning; there's almost always a place for it in a Mediterranean climate. Winter pruning invigorates a tree and stimulates new growth, so I use it to increase the growth of side shoots where there are bare areas in the canopy. I also cut back tall, unbranched young trees to help them form side branches exactly where I want them—I can force a new side shoot to grow in any desired direction by cutting a branch back just above a bud that faces the preferred orientation.

Toothpicks

If a branch growing at a 45-degree angle is ideal for a balance of growth and fruiting, why not start early? A toothpick can be used to carefully spread a young four-to-10-inch shoot to the ideal angle. Stick one end of the toothpick into the branch of origin, and the other end into the new shoot, so that it's held out and away at a 45-60-degree angle. At the end of the summer you can flick the toothpick away with your finger, and the newer shoot will remain at the new angle. The more you use toothpicks, the less pruning for fruit you'll need to do in the future.

Clothespins

Clothespins (the old-fashioned kind with a coiled spring in the middle —see Figure 169) are the next size up in spreading devices. I hang them on tender young spring shoots (about six to 12 inches long) to weigh them down to form the angle I want. This slows down the tip growth and encourages fruiting buds (as shown in Figure 169). Bending down the tip of a shoot means that the apical (tip) dominance hormone, which favors vertical shoot growth, doesn't get through to the tip as much, with the result that the tip stops growing or slows down considerably. With young short growth it only takes one clothespin to get

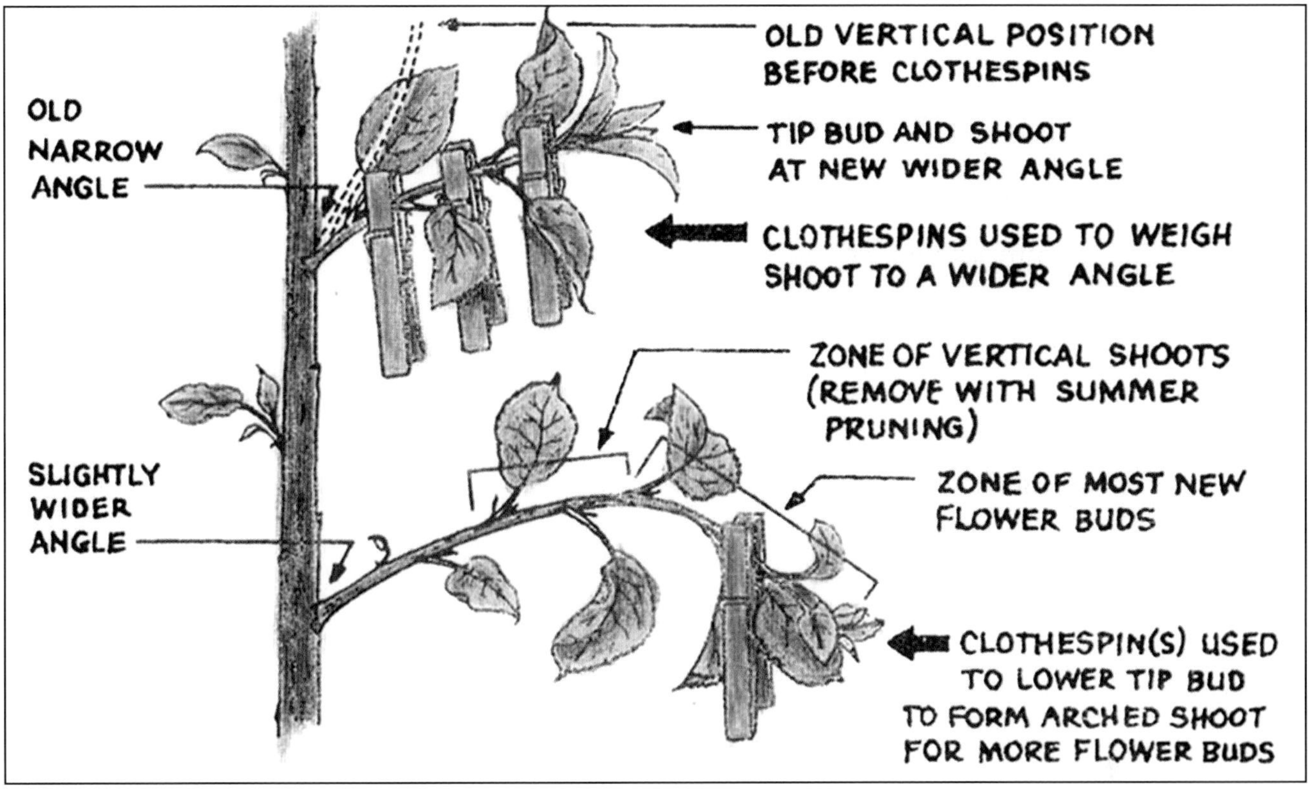

Figure 169. I experimented with two approaches in my Asian pear orchard (shown below), pruning half of the trees with the Lorette system and using clothespins to form angles on the other half. After two years I gave up on the Lorette system because it took so much extra time, was very tedious, and didn't work as well. The trick is to get the clothespins clipped on before the new shoot gets too stiff. These were oblique-cordon trees, and they made more fruit because a forty-degree-angled branch was more fruitful than other angles. So, I figured, why not plant the whole tree at this angle? It's the easiest espalier method and takes less work, but the trees on which I used clothespins were about three to four feet wide—not a classic espalier.

the tip to dangle downwards. If the shoot is longer and stiffer, you can clip more clothespins onto its outer curve. I've used as many as four (or more) clothespins to achieve the goal of greatly reducing tip growth.

You can also use one or more clip-on clothespins to weigh down a slightly older shoot to a curve or to a desired angle. (See the illustration above.) If tone clothespin doesn't lower the shoot enough, keep adding more until the ideal angle is achieved. Don't clip the clothespins too close to the end of the shoot or you'll bend the tip over and thus stop it from lengthening at all. When fall comes, you can remove the clothespins, because the shoots will have become more rigid and will hold their position. You can re-use the same clothespins each year.

Instead of summer pinching, there's a simpler, less time-consuming way to form fruit, but it doesn't keep the canopy architecturally manicured in the classic espalier pattern, in which the branches of a fruit tree or ornamental shrub are trained to grow flat against a wall, or supported on a lattice or a framework of stakes.

I practiced this technique with 40 Asian pear trees planted at 45° (known as the oblique-cordon espalier form) and tied to a wire trellis. When each lateral shoot grew to eight or more inches long, I'd weigh it down with one or more clothespins. This bent the shoot into a miniature Belgium arch (a lesser-known, but naturally productive espalier form) and caused heavy flowering on the outside of the arch the next spring.

Thereafter, the weight of the fruit continued to hold the limb in a fruitful arch and I could move the clothespins to new shoots. Although my Asian pear trees were four or more feet deep (instead of less than 24 inches deep as with a classic oblique cordon), the overall effort was *considerably* reduced when compared to the work required by typical summer-pinching and clipping. I think of this as an Americanized form of the European ideal.

Figure 170. These boards have a sharp nail-point on each end (drive the nail in and clip it off at an angle). This method can be used on branches that are too stiff to use clothespins. Spread the limbs wide and stick a nail into each. After a few years the sticks can be removed and the branch will hold its angle. The later you wait to do this, the longer it will take the branch to stiffen up.

Once the shoot is bent over, buds "germinate" into fruiting buds on the curve nearest the outside branch. The following year some of these buds will make fruit. A bud may sprout at the very top of the curved shoot due to apical dominance; such growth needs to be rubbed off in the spring or clipped once in late summer.

This approach to making fruit is so simple and so easy that it only takes 20 to 60 minutes at several different times to shape 40 trees—the number of trees in my test orchard. (And once you've shaped the curve it stays in place and you can unclip the shoot and use the clothespins over again.) Pruning the trees using traditional espalier methods takes many hours for the same effect. If you use clothespins with trees that have been wired to a trellis, you can easily clip the shoots to a wire to direct them as you choose. The clothespin approach means the tree will be much wider than traditional espalier fruit trees, but I'll take it any day. Saving many hours of pruning is my cup of tea, wider trees notwithstanding. I used only 100 clothespins over and over to tend my 40 trees. If it's so easy for that many trees, think how much less time

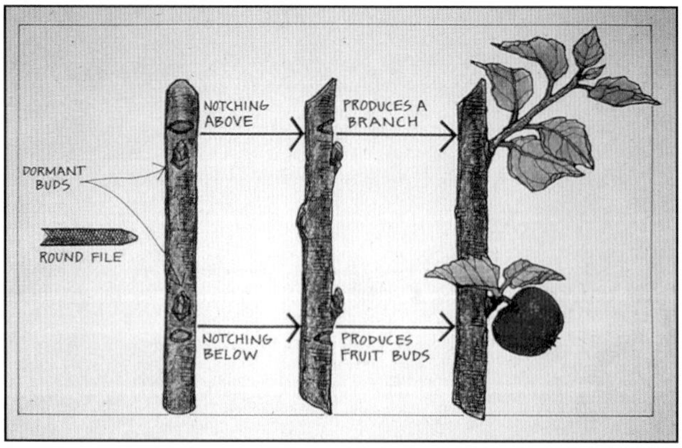

Figure 171. This is the most interesting way to get a fruiting spur or shoot to form exactly where you want or need one (it works only on one- or two-year-old shoots). The article showing how to do it appeared in *The Journal of the American Society of Horticultural Science*, 1994, 119 (4), p. 678-682.

and effort it'll take you with one or two.

Spreaders

Spreaders are used for older limbs. If you didn't use a toothpick or clothespin in the youth of a sprout, you have to catch up by spreading more mature limbs with spreader sticks. To make them, take one-by-one wooden sticks of different lengths, hammer a nail in each end, and snip the head of the nail off at an angle, creating a sharp point to stick into the wood of more mature limbs in the same way you used the toothpicks on new shoots. See Figure __ for how spreaders look after being inserted into place.

Notching

There's an awesome way to produce a flower bud or shoot growth, because the hormones of a tree move around in such way as to allow you to place a shoot or fruit bud *exactly* where you want one. There's a hormone that sends a signal from the tip of the branch (called an apical bud) down to the buds below, "telling" them not to sprout. By scouring a notch *above* a dormant bud, the bud doesn't receive this hormonal signal from the apical bud, and as a result, the dormant bud will sprout into a shoot.

The carbohydrates of a tree must be stored in a bud for it to form a flower. When carbohydrates are generated, they're often sent down the limb into the trunk and roots. Notching *below* the dormant bud stops the carbohydrates from traveling down to the roots and stores them in the bud, causing it to become a flower bud.

Notching works best on shoots that are one or two

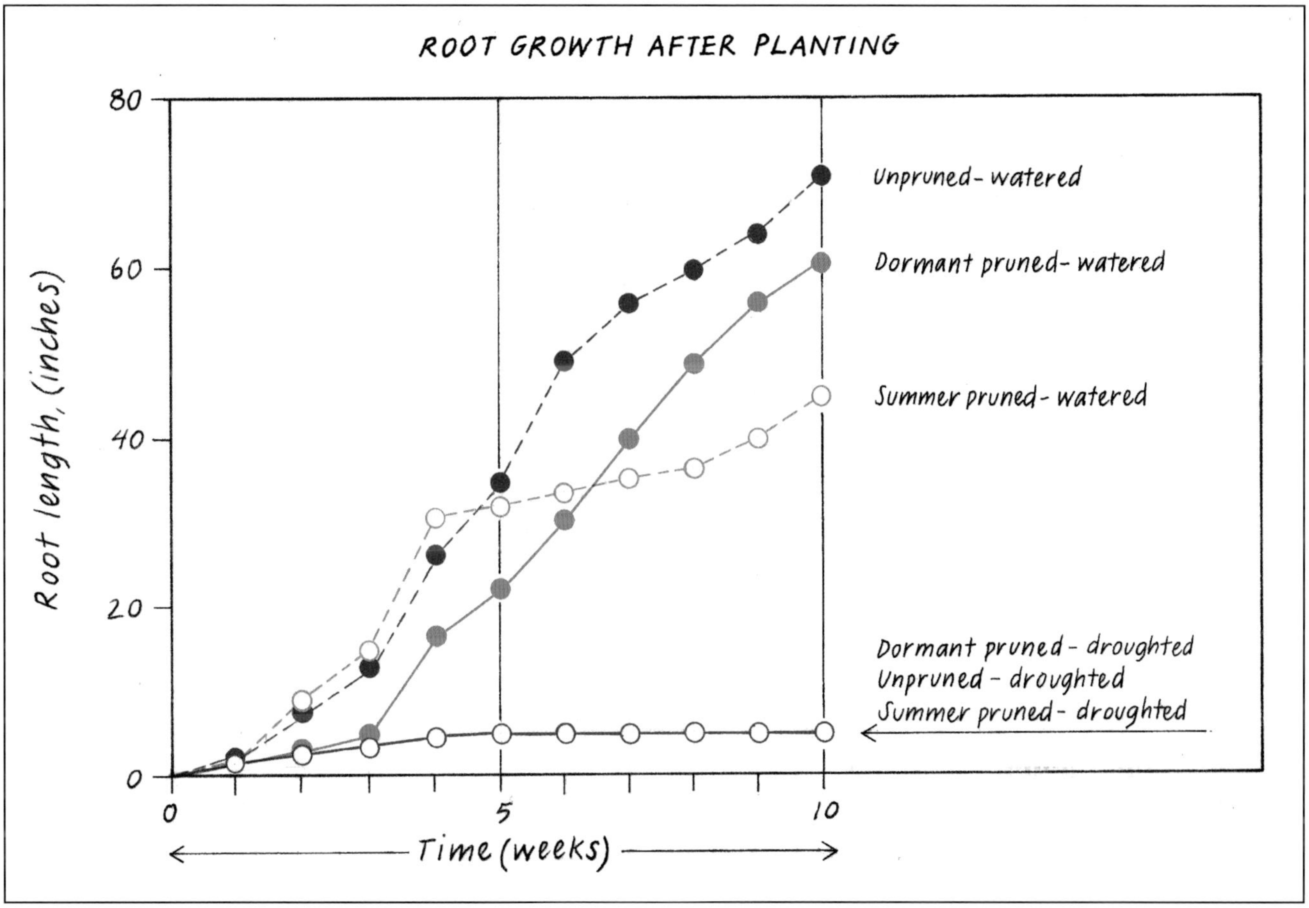

Figure 172. This important chart shows how irrigation affects growth, and is also intended to show the impact of pruning. Un-irrigated trees, at ten weeks after planting, showed less than five inches of root growth. Un-pruned trees registered about 70 inches of root growth. Summer pruning had the greatest effect on reducing root growth, and un-pruned trees showed the most new growth—about six feet. This chart also shows that, if properly planted, the roots of a tree can easily grow past its planting hole in just four to five weeks. This means that you should set your irrigation at a wider distance from the new tree as early as one month. The only reason to prune a new tree is if it has serious scaffold (branching) problems.

years old and is best done at two to four weeks before full bloom. To notch, take a 3/8-inch round rat-tail file and use it (like you're playing the violin) to score above or below the bud just deep enough to remove the green tissue (the cambium) under the bark. Be sure to score at least halfway around the shoot. One twist of the wrist and you've got a new shoot or flower bud, a nice result from so little effort.

I learned this technique from Emil Linquist in 1980. It wasn't until 2018 that I finally found some research to substantiate the technique. In a study by the Department of Plant and Soil Science, 1994, I found out the best time to notch is two to four weeks before full bloom. And notching above a bud increases shoot production by around 600 percent.

Little or No Pruning

A native forest doesn't prune itself, so sometimes we should likewise refrain, depending on circumstances. I shear my lavender heavily and religiously each year to encourage new growth. I prune my fruit trees every summer (and sometimes in the winter). I shape, not shear, my Jerusalem sage (*Phlomis fruticosa*) once a year, otherwise it will take up ever-increasing space in the garden. But I've never pruned the strawberry tree (*Arbutus unedo)* that I planted 25 years ago. It's lopsided, but so are some trees in the forest just 50 feet away. It has edible fruit that I have to gather from a ladder because I don't prune it, a compromise I'm willing to make. All too often we prune more than we need to, and heavily clipped trees and shrubs can look ridiculous or horrifying to a sustainable gardener.

Limbs do fall in the forest, from trees that happened to grow with a weakness. We should always prune a young tree if it needs some judicious snipping to help it get off to a sturdy start. When two trunks share the same spot, the two shoots compete for tissue. The trunk cannot share its circumference in this situation, so one of the shoots will inevitably fall off or weaken. Pruning off one of the parallel shoots when the tree is very young provides a strong start to a tree that may never have to be pruned again.

You also might want to consider refraining from picking up, chain-sawing, or composting some of those limbs that fall in the forest, as leaving them makes habitats for lizards, snakes, beneficial beetles, insects that abet decomposition, and other small critters. (I know one woman who made large piles of brush, 10 to 25 feet in diameter and three to six feet tall, to create an animal habitat. Tacky-looking to most gardeners, but even some foxes took up residence there.)

So go ahead, keep those hand pruners and saws sharp, oiled, and ready to go. Just think twice before shearing in a "natural" landscape.

Now, on to a classic component of most forest gardens.

Guilds

A major design element of many permaculture gardens is that of "guilds" planted around fruit trees. The original concept for guilds (in the 1970s) by ecologists goes way back, and their definitions are varied and complex. A more recent definition is "... a group of species that exploit the same class of environmental resources in a similar way." (From: "The Guild Concept and the Structure of Ecological Communities," 1991; *Annual Review of Ecological Systems*, 22: 115-143.)
This is a far cry from the "modern" concept of guilds as proposed by today's permaculturists. Here's a definition that's closer to their mistaken current use of term. It focuses on insectaries, and not on other dynamics, such as how legumes improve the soil's nitrogen.

"The term "guild" has been used botanically to describe groups of plants in some way dependent on other plants, such as the epiphytes, saprophytes, parasites, or climbing vines. More recently the term has been used in a broader sense to characterize ecologically unified, functional groups of organisms. Our usage denotes individuals that are functionally dependent or interdependent with respect to their herbivores, and does not necessarily imply spatial association. Guild members function as antiherbivore resources in three major ways: (i) as insectary plants that aid in the maintenance of herbivore predators and parasites; (ii) as repellent plants, either directly or indirectly causing the herbivore to fail to locate or reject its normal prey; and (iii) as attractant-decoy plants causing the herbivore to feed on alternative prey." (From "Plant Dense Guilds; Many Plants Are Functionally Interdependent With Respect to Their Herbivores," Peter R. Atsatt and Dennis J. O'Dowd, *Science*, New Series, Vol. 193, No. 4247, p 24-29; July 1976.)

I want to spend some time discussing guilds, in order to provide a thorough look at their ecological appropriateness. (Spoiler alert: I seldom see the need for "classic" fruit- or nut-tree guilds.)

Toby Hemenway, of *Gaia's Garden* fame, offers a simpler definition: "A harmoniously interwoven group of plants and animals, often centered around one major species, that benefits humans while creating habitat."

Bill Mollison says, "A plant guild is a harmonious assembly clustered around central element. This assembly acts in relationship to the element to assist its health, aid our work in management, or buffer adverse environmental effects."

Another permaculture definition is: "…a planting arrangement that incorporates several different species of plant, each of which performs a function that has a positive impact upon the growing system as a whole, giving all the members of the guild the best conditions in which to survive and thrive and, in turn, maximizing the potential for yield for the permaculture gardener."

I would pay attention to the word "potential." Most proponents say the guilds will increase the crop yields of the central tree, but there's no data that shows one way or the other if this is the case.

(There's an amazing permaculture group in Bulgaria called the Balkan Ecology Project. They've embarked on a study of polyculture versus monoculture yields with their Polyculture Project, coordinated by Paul Alfrey and Sophie Roberts.

Their summary is: "Little data exists showing the productive capacity of polyculture systems and the economic viability of them. We believe there is a need to fill this gap and provide data and concise coherent models that can be replicated easily. Our project intends to go some of the way in filling that gap. We make no claim that polyculture can feed the world but we do claim with 100% conviction that anyone can grow polyculture and by doing so they are taking a small but significant step in reversing a negative trend and we are pretty certain they will have a great time doing it." This is the only group I know of that plans to run a detailed review of what's basically a study of fruit-tree guilds. The goals of the overall experiment will include records of the following:

- Time and costs taken to establish and plant the beds
- Soil analysis of each bed from 2018 - 2020
- Growth rates of nitrogen-fixing trees and shrubs in biomass beds from 2018 - 2020
- Biomass harvested from pruning for each species from 2018 - 2020
- Time taken to harvest biomass
- Nitrogen fixed by each species from 2018 - 2020 - Sofia University Assoc. Prof Ganka Chaneva /Assoc. Prof. Miroslava Zhiponova / Assist. Prof. Detelina Petrova
- Records of invertebrate diversity in the gardens
- Quantity of harvest

According to some permaculturists, here are the main components of an enlightened guild:

• Nitrogen-fixing plants (sometimes referred to as just "the fixers.") to supply nitrogen.
• Mineral-accumulator plants, to help supply nutrients other than nitrogen. This category often refers to plants that have a deep taproot. These are sometimes called "diggers."
• Beneficial-insect plants to minimize pest problems. Also called insectary plants.
• Bee plants, for wild bees, bumblebees, and honeybees.
• Plants with differing root system depths, to exploit the soil layers efficiently. (Under the false assumption that deeper roots acquire an important amount of nutrients that shallower roots can't absorb. See pages

__.)
• Aromatic plants, to confuse pests and increase system health—so-called "companion plants."
• Ground cover plants, to cover the soil surface densely.

When I read about guilds, I see lists of plants referred to as one person's idea of a guild, as compared to another person's choices. Often I wonder how these choices were made. Here's an example: each guild is supposed to have at least one beneficial-insect attractor, but no science is offered as to why a particular plant has been chosen.

As seen in the long lists on pages 361 and 362, there are specific plants that attract specific insects. (Even this list is anecdotal; read the rest of the chapter for scientific studies of specific predators and specific plants.) To put it simply, the predatory (beneficial) insect needs to match the pest to be preyed on. For example, if you're dealing with apple codling moths, it does no good to plant species that attract insects that prey on aphids or tomato hornworms.

As to mineral-accumulators, see my critique on page 241-245 of my own original 1986 lists of dynamic accumulators (I even see mistakes like referring to comfrey as a nitrogen fixer. This is far from the truth, as it's not in the legume family.)

Let's look at a few proposed fruit- and nut-tree guilds.

Apple Tree (*Malus domestica*) Guild.

1. Daffodils, irises, or other early-flowering bulbs. They provide beauty for us and attract pollinators to the tree.

2. Dill and onions, planted under the dripline of the apple tree, repel pests that are otherwise attracted to the tree.

3. Beans, peas, or purple clovers fix soil nitrogen, improving the quality of the soil in which the tree is growing.

4. Comfrey and borage are herbs with many human uses. In addition, they grow quickly,

U.S. Department of Agriculture Pomological Watercolor Collection. Rare and Special Collections, National Agricultural Library, Beltsville, MD 20705,

shade the ground, and can be cut back and used as mulch around the tree.

Let's look at each of these recommended plants:

1. Yes, daffodils provide wonderful beauty in early spring, and often continue into mid- and late spring but, according to Rusty Burlew (see her amazing blog – honeybeesuite.com), "...daffodils are not what we assume when it comes to attracting bees. Daffodil breeders hand-pollinate the flowers.... Like most flowers that have been highly manipulated by plant breeders, daffodils are no longer particularly attractive to insects. The reason for this is simple. When it comes to ornamental flowers, plant breeders select for beauty. Daffodils, in particular, are selected for the color of the flower, the symmetry of the flower, the angle of the flower (up or down), its size, and other arbitrary characteristics that daffodil judges look for. Rarely will these highly bred flowers cross without human intervention."

   Other enthusiasts say that a circle of daffodils will stop grasses in their tracks— "When planted in a circle at the drip line of a fruit or nut tree, daffodils keep grasses from creeping under the tree, preventing the grass from competing with the tree for water and soil nutrients." See the adjacent photograph of a daffodil nearly overwhelmed by grasses; I rest my case.

And yet, according to Toby Hemenway in *Gaia's Garden*, "Planting one circle of daffodils around the trunk and a second thick circle of them at the dripline will temper the depredations of browsers, burrowers, and bark-chewers." This is *not* true, as I know, from 33 years of living with plenty of deer in my garden, of no plant whose smell deters deer from eating other nrearby plants. As one example, in my garden, rosemary went uneaten by deer (although one deer took a nap on top of it!), but even plants leaning up against it were eaten if they weren't in the deer-resistant category. The deer also won't eat the poisonous daffodil or narcissus (nor will gophers), but any non-deer-resistant plant next to either will be eagerly consumed. Gophers will actually move daffodil bulbs around in their

Figure 174. Here, shown in the early spring, is my favorite example of how useless daffodils and narcissus are in stopping grass growth. Don't bother using them in a guild.

eagerness to pillage adjacent plants—I've seen individual bulbs pop up and bloom over 20 feet from where they were originally planted.

So, let's strike daffodils from the guild list for bug-attracting and deer-repelling purposes, but they're still a great treat for the eyes, and I always plant plenty.

2. Companion-planting recommendations are often based only on anecdotal information, but it's hard to find scientific studies that confirm those anecdotes. I found one research test (in the *Annual Review of Entomology*, 51: p45-66.) that showed that dill repelled pests that prey on the Brassica family. In another study, (*Advances in Horticultural Science* 11(4): 175-181) dill attracted predators of the Colorado potato beetle.

Since these studies have nothing to do with apples. I wouldn't make the leap from brassicas and Colorado potato beetles to fruit trees, mostly because fruit-tree pests are totally different critters. I've also found nothing about the efficacy of onions, which are sometimes recommended for guilds as attracting or repelling companions for any tree crop. (See the chapter on beneficial insects for much more on this topic.)

3. These plants fix nitrogen: Beans (50 pounds per acre), peas (40 pounds per acre), Dutch white clover, which produces .0034 pounds of nitrogen per 100 square feet, fixes 100-150 pounds per acre. But our current subject, apple trees, very rarely need extra nitrogen if their new growth each year is at least six or 10 inches long. (See Figure 8 in Chapter 1.) In all of my 40-plus years as a consultant, I've seldom seen growth of less than six inches on an unfertilized apple tree. If additional nitrogen is needed, use 0.02 pound actual nitrogen for every 100 square feet, but be careful, as too much nitrogen can cause outbreaks of aphids (see page 15). I suggest that you keep an eye on your apple trees, and wait

to see if their new growth shows the need for any extra nitrogen.

4. Many people think the "chop-and-drop" approach to mulching with comfrey, as a frequent example, will help fertilize an apple tree. After the comfrey leaves are chopped into mulch, it's thought that they'll feed the tree as they rot. Well, first, in arid or summer-dry areas, the leaves are likely to desiccate before they have a chance to decay, "burning" off much of the nitrogen. (This happens in most climates to some degree.) Secondly, comfrey is a *very* heavy feeder and, as with all rapid growers, needs a lot of nitrogen. Comfrey planted around an apple tree will actually be competing with it for nitrogen.

   In the book *Comfrey, Past, Present, and Future* by Lawrence D. Hills we read: "... It needs very good land, plenty of dung and attention." At 100 tons of leaf production per acre (about five pounds per 100 square feet) each year, I assume comfrey has to be a heavy feeder to produce so much foliage. Because of this, I plant it where it will be well nourished, like around compost piles or poultry fencing, *not* around trees.
   Older comfrey plantings with leaves showing a lighter-green color will usually require nitrogen fertilizer. Recommended rates vary from 40 to 100 pounds of nitrogen per acre (this could be supplied by Dutch white clover) depending on your soil's organic matter. While it will grow in many types of soil, comfrey prefers the moist, fertile kind.

   In summary, keep comfrey from competing with your trees for nutrients and moisture by planting it in patches far away from them. To fertilize your apple trees adequately, I suggest that you simply use four to six inches of arbor mulch (the mixed woody- and green-leaf chippings straight from a tree service truck) to suppress weeds and eventually rot, improving the soil from the top down.

(For more on growing and using comfrey, see Appendix I.)

Rhubarb is often suggested as another cut-and-drop foliage, the idea being that the gardener can cut the foliage and use it as mulch within the guild. It's even supposed to add minerals drawn up from deep soils, but there is no scientific literature or peer-reviewed papers that show rhubarb acts as a dynamic accumulator. Also, rhubarb prefers a very rich soil. On her farm in southern Missouri, my aunt always grew the rhubarb next to the chicken coop so it could soak up the nitrogen it prefers. Also, as discussed earlier, fruit trees seldom need extra nitrogen, and certainly not the amount that rhubarb requires.

Another proposed guild:
Walnut Trees (*Juglans* spp.) as the Central Element of the Guild.

1. Hackberries (grass-suppressing shrub that tolerates walnut's juglone toxins)
2. Currants (fruit-bearing shrub)
3. Wolfberry (Goji Berry) (fruit-bearing shrub)
4. Tomato, pepper or other members of the nightshade family (annual vegetable)
5. Russian olive (N-fixer)

Walnut trees can grow to be very large (up to 30- 60 feet high), and take seven to 10 years to produce a significant crop, so they're not a major candidate for many suburban yards. Besides the excessive shade, the big problem with walnut trees is that they produce a special chemical called *juglone* that stunts or kills some plants growing around them. The chemical is washed off the foliage and fallen leaves and is also exuded by the roots of the tree. (By the way, it's the black walnut—*Juglans nigra* —that we're concerned about here. The grafted English walnut tree doesn't produce juglone, and thus doesn't cause much of a problem.)

All members of the *Solanaceae* (nightshade) family—tomatoes, potatoes, and peppers are examples—appear on the not-tolerant list, as they do on many such lists concerning the Midwest and New England where the black walnut is native.

1. Hackberries (*Celtis occidentalis*) *may* suppress grasses, but they're shrubs. A continu-

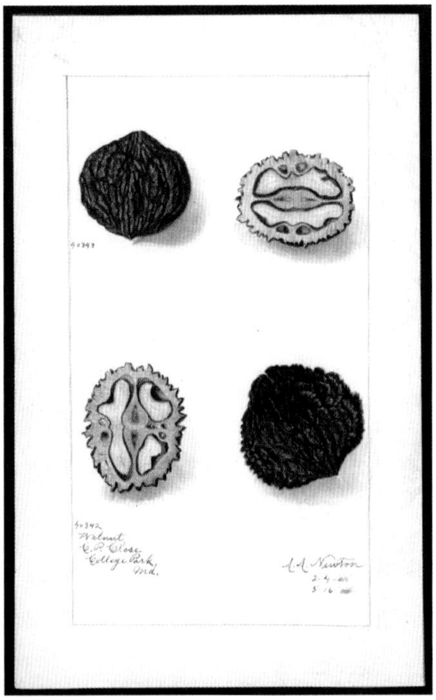

U.S. Department of Agriculture Pomological Watercolor Collection. Rare and Special Collections, National Agricultural Library, Beltsville, MD 20705,

ous planting of hackberries under the foliage would really slow up the harvesting and pruning of the tree, and would also make it difficult to plant annuals. To prevent disease and pests of the walnut in the following year, all fallen nuts that aren't eaten should be picked up at the end of each season. This would be very difficult if one had to work through shrubbery to pick up the wayward nuts. (There are actually 65 species of insects and diseases affecting the genus *Juglans* and more specifically black walnuts. A major pest, the black walnut curculio, develops after the female curculio deposits an egg within a developing nut. The larva hatches, then feeds on the nut within the husk.) Again, I suggest simply weeding by hand to control wayward grasses.

2. Currants (*Ribes* spp.) do bear tasty fruit, but present the same difficulties as hackberries with regards to shrubbery getting in the way of harvesting and pruning. They can also get *Anthracnose spp*., a fungal disease similar to leaf spot that attacks red and white currant bushes and causes tan spots with dark purple outlines to appear on their foliage. *Anthracnose* also affects tomatoes, so be wary: you're planting an attractant for anthracnose in the same suggested planting with nightshade plants—not a good combo in my book. I suggest planting the currants in a separate bed for more accurate control of fertilizing, watering, and preventing disease.

3. Goji berries or wolfberries (*Lycium barbarum*) are also in the *Solanaceae* (nightshade), family, and present the same problems as tomatoes. Otherwise, they're a very adaptable drought-resistant fruit that thrives in USDA Zones 3-10. A potential problem with guild use may be that full sun exposure is required to produce goji berries. The solution is to plant on the south or west side of the tree, near the dripline of what will become the mature tree. Again, this three-to-10-foot-tall shrub/tree would interfere with caring for and harvesting the walnuts. I would also suggest that other berries be planted near (but outside of) the dripline.
According to herbal lore, the berry contains betaine, which is a known liver protectant, but can also be used to induce menstruation and abortion, so its use should be avoided by pregnant women. Chinese herbalists suggest abstaining from using wolfberry during times of cold and flu. Wolfberry has an estrogen-mimicking effect, so it should not be used by people who are pregnant, or who have diseases that are sensitive to estrogen. I suggest growing other fruit trees if you plan to raise tomato plants near a planting of goji berries.

4. I think the concept of planting *Solanaceae* plants with walnuts may have originated with Toby Hemenway in his book *Gaia's Garden*, where he writes: "The lofty walnut, together with its guild of hackberries, currants, and solanaceous annuals such as peppers and tomatoes, will go at the north end of the yard." (He doesn't say if it's a black or English walnut.) I never got to ask him which kind he

meant before he died in 2019, but so many lists from Midwest Farm Advisors say not to plant solanums near or under black walnut trees that I would skip its use as a central guild tree. By planting tomatoes in clusters elsewhere in the yard you can also control late-summer watering to get the best flavor.

In general, I don't suggest planting annuals under the foliage of trees, as the annual tillage of the soil destroys tree roots. In actuality, as many as 50 percent of tree roots are in the top four to 12 inches of the soil. Since tillage easily reaches nearly that deep and can destroy a lot of feeding roots, I suggest keeping tomatoes that aren't cared for with no-till techniques in their own plot in the vegetable garden.

4. Russian olive (*Elaeagnus angustifolia*) may be a nitrogen fixer, but it's also considered an invasive plant, especially in New England. There are many other nitrogen-fixing plants that aren't as invasive. Again, I wouldn't suggest shrubby plants, as they hinder tree care and hide fallen nuts. Stick with perennial Dutch white clover.

A Suggested Guild for Peach Trees.

1. Siberian pea shrub (N-fixing shrub)
2. Sunflowers (insectary plant)
3. Cabbage (annual vegetable)
4. Clover (N-fixing ground cover)
5. Comfrey (nutrient accumulator)
6. Horseradish (nutrient accumulator)
7. Calendula (insectary plant)

Here are my thoughts on it:

1. Siberian pea shrub (*Caragana arborescens*) is another invasive perennial shrub/small tree. I wouldn't suggest it because it hides fallen peaches. Go with another clover, perhaps the dramatic red clover.

2. Sunflowers are in the *Asteraceae* family, (used to be called the composite family) and are known as good insectary plants.

U.S. Department of Agriculture Pomological Watercolor Collection. Rare and Special Collections, National Agricultural Library, Beltsville, MD 20705

I would, however, choose a *perennial* insectary plant so that the soil isn't disturbed, and the peach-tree roots aren't damaged by annual tilling. Osentowski may be referring in his list to the Maximilian sunflower (*Helianthus maximiliani*), another invasive plant with flower spikes that grow to a height of six feet. I suggest using low-growing perennials like thymes (*Thymus vulgaris*); chicory (*Cichorium intybus*) with its deep roots; yarrows (*Achillea* spp.) that can be kept within the garden; or some deer- resistant artemisias (*Artemisia* spp.). Again, I would plant all these carefully, so the soil isn't greatly disturbed and the peach-tree roots damaged. I suggest keeping all annual vegetables in beds near the insectary plants, but not necessarily under the trees. No-till gardening is a good option.

3. Peach trees are more likely to need more nitrogen than apple trees, but don't need extra nitrogen unless the new growth

is less than six to 12 inches. Consider a perennial Dutch white clover as a nitrogen-fixer here, but just make sure it doesn't spread out of the garden. Also, don't plant it if you're allergic to bee-stings, as bees love Dutch white clover.

4. Again, keep comfrey from competing with the tree for nutrients and moisture by planting it in patches far away from the tree. I suggest planting comfrey far away from *everything*, perhaps in your neighbor's yard!

5. With foliage that grows up to to four feet high, and roots that can reach as much as fourteen feet deep, horseradish, once established, is a very hard plant to eradicate if you need to. (Whether it accumulates any special nutrients remains to be seen.) I suggest that you treat horseradish like comfrey and plant it alone and away from all other plants. You can always cut the foliage and move it elsewhere to be used as a mulch or added to compost.

6. Calendulas are good insectary plants, and even though they're annuals, they'll easily self-sow, so you only have to plant them once. They can spread outside of the garden if left unchecked.

An Oak (*Quercus* spp.) Guild.

Oak (the central element)

1. Pawpaw (shade-tolerant fruit tree)
2. Hazel (nut-bearing shrub)

Figure 175. An old lithograph shows how a small piece of a comfrey root can sprout as a new plant. Comfrey foliage has a high NPK rating (1.8-0.5-5.3), and is high in calcium. Its foliage is very useful to assist decomposition in a compost pile, but it should certainly not be planted where it will be tilled into the soil. It works best as a mulch where there are summer rains to help it rot.

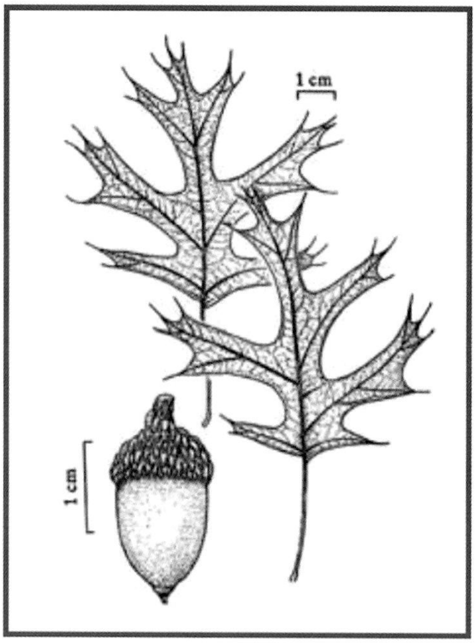

3. Gooseberries or currants (fruit-bearing shrubs)
4. Comfrey (nutrient accumulator)
5. Horseradish (nutrient accumulator)
6. Borage (bee attractant)
7. Strawberries (groundcover)

1. Oak trees for acorns, I presume. Not a very common food in suburbia. Try English walnuts or pecans instead.
2. Pawpaws - a good choice, unless your tree is a native California-coast live oak that prefers no summer water. Pawpaws love the summer rains in Missouri and other mid-western states where they grow.
3. Hazelnut - OK choice, but prefers a bit more sun than under the oak.
4. Gooseberries - same as hazelnut.
5. See #4 under apple tree guild recommendations above. I don't think it's needed.
6. Horseradish - must be a permanent placement, as it's very hard to get rid of.
7. Good choice for bees and other beneficial insects.
8. Strawberries – dreadful choice; you'd be stepping on the strawberry plants while pruning and harvesting. The acorns might get lost in the foliage. And there's too much difference in water needs, at least with native evergreen California-coast live oaks (*Quercus agrifolia*). (The image to the left: red oak *Q. rubra - Field Guide to Native Oak Species of Eastern North America*, USDA, 2003.)

Guilds come under the heading of polyculture, but far too often, as described above, the mixture of plants actually works *against* the intent of the gardener.
In some cases, just a deep and extensive mulching beneath the tree may be sufficient to harvest plenty of fruit and nuts. Suggestions for a productive and lower-maintenance approach to polycultures can be seen in the final chapter of this book.

Hügelkultur

This strange German word refers to a special technique that allows for the decomposition of woody plant material into soil while growing vegetables. *Hügel* in German means "hill," and *Hügelkultur* refers to the practice of growing on mounds. I acquired a booklet called *Mound Cultivation: the Gardening Method of the Future*, by Hans Beba and Hermann Andra, printed in 1979. When I discovered it, there were already 50,000 copies of it in print after five years. Figure 176 from the handbook shows the cross-sections of the mound with suggested dimensions.

This method, as presented here, is used to recycle twigs and small branches into soil beneath growing vegetables. The idea is that mound settles and rots down quite a bit in one season, and this approach is renewed each year, maintaining the same dimensions, to improve the soil and slowly build up its level. This technique is a good way to decompose small woody materials that wouldn't work well in a compost pile, especially in cool (mesophilic) piles that don't heat up.

This approach has fascinated many gardeners, and it has evolved somewhat from the original. It now often involves digging pits or troughs (swales) and filling them with large branches and logs before adding the layers above the soil surface. The important difference of this system is the use of logs, firewood, and big limbs and branches. Much of this approach stems from the work of Sepp Holzer. The use of logs—and nowadays there seems to be a contest among permaculturists to use the biggest logs—radically changes the dynamic. Sepp's approach is meant to create three- to four-foot-high steep slopes, for easy harvesting without bending over.

Here are some ideas from proponents of Hügelkultur, with some comments from me.

Advantages:

1. You add a massive amount of carbon, the building block for all organic matter, deep in the soil, and it eventually becomes humus over a very long period of time.
2. This constitutes a long-term and slow-release fertilizer.
3. Logs supposedly rot from the inside out.
4. The logs are said to act like sponges and release water on demand to the soil.
5. The method eliminates anaerobic conditions. All pathogens are anaerobes. (This is *not* the case, a pervasive myth—Ed.)
6. Some think heat is given off as logs compost. (I think otherwise. There's too much carbon in the area around the logs to create a hot [thermophilic – 106° to 252°F] zone. Yes, the above layers can be arranged so that a very mild heat is formed by mesophilic (moderate-temperature) action, creating something like a slightly warm compost heap.)
7. This heat supposedly reduces the effects of ground frost. (Where I live, there's almost no frost, so I don't have a clue if this helps.
8. As the logs rot, they take up and release water and therefore expand and contract. This heaving constantly supposedly moves and loosens the topsoil above it.

Disadvantages:

1. The big concern in my mind is that the bigger the logs, the more the nitrogen is locked up in them. The material that is not yet rotted will "rob" the nitrogen from the adjacent soil to aid in the decomposition process.
2. Organic matter is often not incorporated into the wood layers, and the large logs are too woody to easily decompose. Nitrogen lock-up may occur in the immediate vicinity of the logs but not in the surrounding soil. The nitrogen is locked up and composting or rotting only occurs in the zone between the soil and the undecomposed organic matter. All this leads to a slow rate of decomposition.
3. The wood and all-organic matter will rot if there is enough moisture, but this is much less likely in arid climates. Any logs used have to be fresh or very moist.
4. It takes a *lot* of soil and wood to make a tall, sloped Hügelkultur.

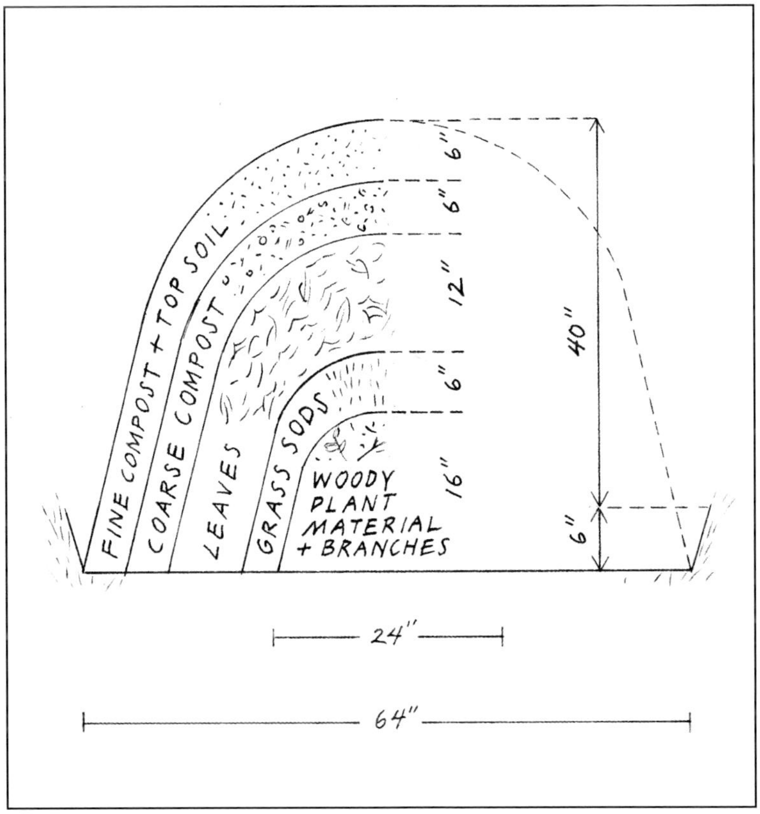

Figure 176. Here's an illustration based on the handbook *Mound Cultivation: the Gardening Method of the Future*, by Hans Beba and Hermann Andra, printed in 1979.

5. According to the author of *The King's Table* blog how won't give his/her name), "Mice love the hügel bed. There are crevices between the logs that dirt didn't fill in, so I find new mice holes in the side of the bed every year. There are fewer crevices as the dirt settles and the logs decompose, so I've only seen one hole this year. The mice haven't caused a problem except the time they dug a new hole near a plant and uprooted a couple of radishes."
6. It also takes a *lot* of time to construct. See below.

In my mind hügel beds this don't qualify as a universal concept or process. There are plenty of arid climates where they're less than ideal, as mixing warmth and moisture makes for the best conditions for rotting. Many of the proponents of this method (Sepp Holzer is the most renowned) of gardening live where there are summer rains—in Sepp's case 29.6 inches of rain, mainly from April to September. In contrast, Tucson, AZ receives 12 inches, mostly between hot July and October. Los Angeles, CA averages 14.7 inches only in the cool winter when things rot less easily.

Many of the books or articles also come from rainy-summer climates with seasonal rain. For example, noted permaculturist Geoff Lawton's farm in New South Wales, Australia, gets 60 inches during the summer alone. The combination of warmth and moisture is the best for the rotting of the buried logs.

James Paris, who lives in temperate, humid England, wrote: *HUGELKULTUR - Raised Bed Vegetable Gardening With Hügelkultur: an Introduction to Growing Vegetables in Tree Cuttings and Turf Heaps*.

Evaluation

But where's the science? Nowhere that I can find in permaculture literature, and certainly not in any peer-reviewed journals.

One insight is provided by a 2016 thesis by Megan Laffon. "A Quantitative Analysis of Hügelkultur and Its Potential Application on Karst Rocky Desertified Areas in China" was done in Bowling Green, Kentucky, where it rains 45 inches in the summer. She found that "Over the course of three months, the water- concentration levels in the hügels stayed consistently high; fluctuations (in) moisture reflected precipitation levels..."

Another Study

One important informal study (in that it's the only one of two I could find) was done at the University of Wisconsin in 2013. The real goal was to recycle as many wood chips as possible so they didn't go to the municipal landfill (dump). The title of the finished paper is: "*Hügelkultur Gardening Technique Does Not Result in Plant Nutrient Deficiencies and is a Potential Source Reduction Strategy for Yard Trimmings*." In the cross-section, notice how the soil "cap" is compared to the more traditional method illustrated above. At 16 inches thick, this is plenty of soil to satisfy the growth of most vegetable roots for at least one year, and probably more, as compared to the six inches recommended in the *Mound Cultivation* booklet.

After only one season of the Wisconsin study, the summary was: "Due to the limited parameters of this study, it is impossible to say how beneficial the method of hügelkulture is over conventional techniques in organic gardening. This is in part due to site selection and inability to obtain reliable yield information. This being said, the treatment did not result in major nutrient deficiencies, and did represent a sizeable diversion of yard-trimmings wastes." They also commented: "[It] may be a suitable form of yard trimming waste minimization in the short term." So Hügelkulture is good for keeping the dump free of woody debris, but not that important for growing vegetables.

Finally, installing a Hügelkultur takes a large quantity of imported elements—they're resource-intense, not to mention a lot of work. (The study in Wisconsin found that their simple 25-by-50-foot raised-bed Hügelkultur layering took 200 person hours—25 *eight-hour-days*—to construct!) Most people's gardens don't produce a lot of logs or huge branches; in many cases even small twigs and branches may have to be saved up for a while. I've seen Hügelkulturs that were started off with such deep and wide troughs that they required a small backhoe. Not an appropriate tool or process for many suburban yards.

Even growing enough green matter for the layering can be difficult, and many gardens don't contain any grass sods to dig up. As described in the *Mound Cultivation* book, a ten-foot-long bed would need well over 10 cubic *yards* (3000 pounds) of leaves alone. In St. Louis, MO, with its many deciduous trees, my dad could easily rake up that many leaves, and the trees were also constantly sloughing off twigs and small branches. In my mostly evergreen Mediterranean forest, however, gathering this amount of leaves would hardly be possible. Gathering and compiling Hügelkulture material takes many hours. Years ago, circa 1981, when I constructed a simple mound, using lots of straw, some soil, and compost, a fellow gardener observed: "I could have double-dug that bed in less time." (We were in our mid-30s and still had strong backs.)

The goal of a truly sustainable garden is to limit as many "imports" as possible. A reasonable approach, instead of hauling in a lot of wood for a Hügelkultur, is to grow a green manure instead of buying all the coarse compost and green matter required by the mound method. Till in the foliage or place it in a compost heap to make coarse compost—you'll improve the soil and raise a crop with no imports other than the seed. (When you start out, leave a few plants to bloom. and save some seed for future green manuring.) For tips on using green manure, see Chapter 8.

An Alternative to Hügelkultur

In 1988, before I found the Hügelkultur book, I'd evolved a method of using inexpensive soil amendments to form large planting mounds—curvilinear berms that gently sweep through a landscape. I stockpile wood chips from local tree-trimming services—not sawdust, but large, chunky chips from an "industrial-strength" shredder, mixed with the shredded tree foliage. I call this "arbor mulch." I try to avoid any shredded material that might easily resprout—such as willows, alders, and various vines.

Another type of chipped material to watch out for is that made from highly *allelopathic* plants whose foliage stunts or kills the growth of nearby plants. Examples of allelopathic plants include: black walnut (*Juglans nigra*), sagebrush (*Artemisia californica*), and mesquite (*Prosopis glandulosa torreyana*). Conifer trees, pine trees, and Douglas fir (*Pseudotsuga menziesii*) trees are rather acidic, and more resistant to decay, and I think the terpenes and resins in conifers, firs, and pines are a bit rough when added to the soil as an amendment or in mounds, so I reserve conifer, pine, and fir chips for mulch in my rustic pathways. The best chips I've used for making berms are hardwoods, my favorite being oak.
Because arbor chips are so high in carbon, I layer or mix them with manure. I start the mound with the largest, chunkiest pieces; this keeps the bottom of the pile from settling too much and helps keep the whole mound slightly aerobic. For all of the subsequent layers I use a blend of smaller chips (1/2-inch to 6) inches long, and manure. You can use other types of high-nitrogen materials to help decompose the woody chips—fresh grass clippings and cover crops such as buckwheat, vetch, bell beans, and clover. (Don't use kitchen scraps unless you're fond of raccoons.) The more nitrogen you add, the faster the mound will decompose. Pile this mixture of high-carbon and high-nitrogen materials at least 50 percent higher than you want your final soil level to be. Depending upon the types of materials used, the mound may settle as much as 25 to 50 percent. Experimentation and experience are the only good teachers.

To the top of the chip mound, add a soil cap at least four inches thick—the thicker, the better. Once the mound of chips starts rotting, the root hairs of the plants will follow right behind the decomposition, taking advantage of the newly available nutrients. The roots will naturally avoid areas that are too hot due to thermophilic decomposition (hot composting), but the plants do get some benefit from the subtle amounts of heat that the mound may generate—much like the enhanced growth of bottom-heated greenhouse flats.

Eventually the chips fully decompose, the whole mound settles to a lower level, the shrubs fully-root into the native soil, and the berm becomes a wonderful, curvilinear feature for the landscape. For me, mounded plantings seemed to be a heap of contradiction at first, but soon they became one of my preferred techniques for quick soil development; a cheap way to create "instant" soil drainage; and an aesthetically pleasing way to enhance the topography of my landscape.

I made the work of mound-building even simpler the next year by finding a source of arbor mulch (chips mixed with shredded leaves). To create my mound, I first rocked a garden fork back and forth on the mound site to crack the soil underneath, then mounded up the arbor mulch in a form that made the garden look more attractive. On top of this, I added a cap of four to six inches of topsoil taken from the area next to the mound when I dug a long swale there to help with drainage in a wet spot in the garden.

I build mounds *only* in November, just as our winter rains are beginning, and mulch heavily in the winter and spring, and I plant only very drought-resistant ground covers, perennials, and shrubs—mostly Mediterranean varieties. (The native *Salvia* spp. worked nicely and weren't eaten by the deer.) I maintained this garden for over 30 years with no summer irrigation. (See the photo on page __ to see how attractive such a garden can be.) In the process, I learned that there are three tradeoffs for never having to water again:

1. The plants grow more slowly.
2. They don't grow quite as wide in shape, so you need to plant them more closely together.
3. Most importantly, they don't bloom nearly as long, but there's plenty of bloom nonetheless. The lavenders and Jerusalem sage are the best bloomers.

I like that this is a gardening technique that only a home gardener can do, since landscape contractors have to plant year-round to make a living. Pretty cool.

Temperate Zone "Forest Gardens"—Russian Inspiration

So, are there *any* good temperate-zone examples of

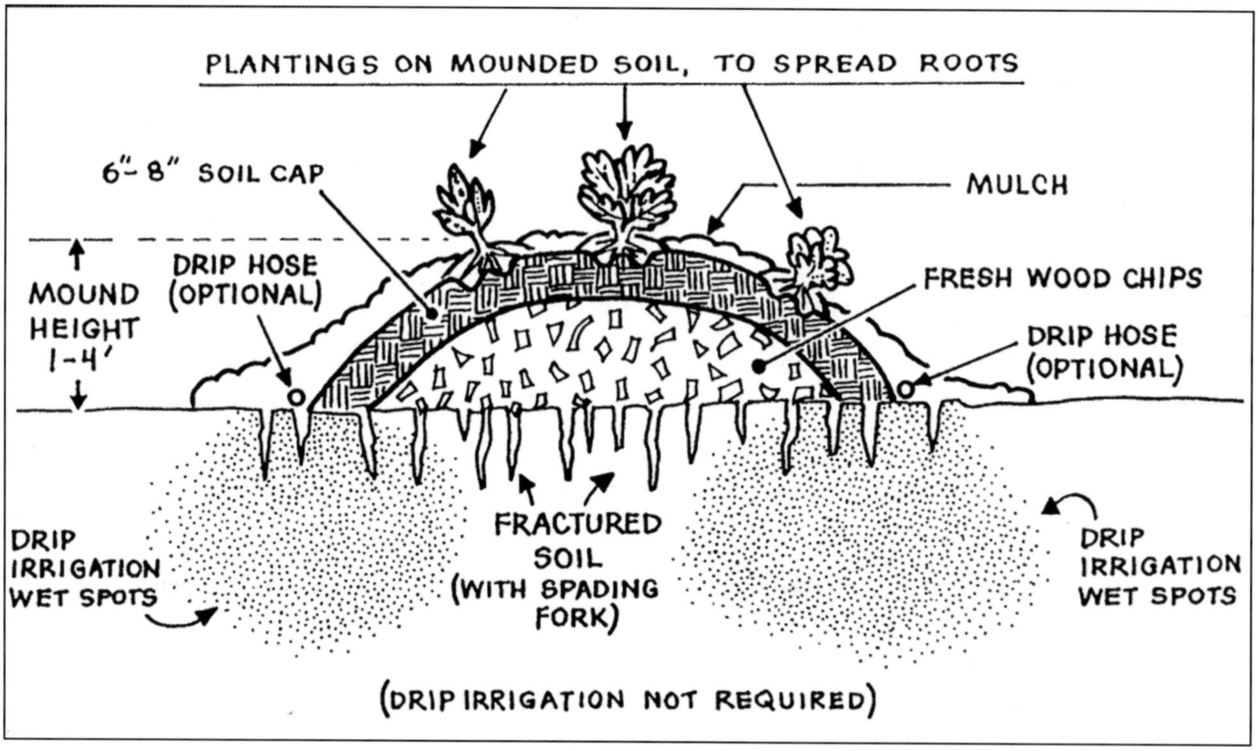

Figure 177. This is the technique I used to grow a garden with no irrigation. I often plant on mounds, building them in the only fall before our California rains begin (in this climate, the roots grow all winter). I start by fracturing the area that will be under the mound with a sturdy spading fork. The plants are transplanted into a six- to eight-inch-deep soil cap layered on top of a mound of arbor chips. A very mild composting process encourages root growth into the composting chips. As the soil dries out in the spring, there will be a wide root system to absorb moisture. For mild-winter fall planting only.

sustainable food gardens containing fruit and nut trees, vegetables, and animals? For the answer, we turn our attention to a place not often considered by US gardeners—Russia. Some of the personal gardens around Russian dachas (second homes or summer retreats in rural areas near a city, often surrounded by gardens) provide great models for a temperate-zone forest garden.

As detailed in an intriguing paper entitled *Thirty Million Agroforestries: Russia's Family Gardens* by Leonid Sharashkin, Michael Gold. (Deep Snow Press, 2009), Russian agroforestry *does* exist in the country's many rural second homes, summer homes (called *dachas*) and urban plots. In fact, Russia may be home to the best existing models for a temperate-climate forest garden.
A surprising 80-95 percent of the Russian population is comprised of gardeners who grow 65 percent of the perennial crops in the country. These plots also

Figure 178. Here my good friend Chester Aaron (1923- 2019) plants lavender on top of a wood-chip mound made by piling wood-shreddings and leaves two to four feet high, and capping them with six to eight inches of topsoil.

provide 90 percent of the potatoes, 80 percent of the vegetables and melons, 65 percent of the perennial fruits and nuts, 52 percent of the milk, and 52 percent of the meat for the Russian diet. These are gardens with enough variety to shelter animals along with diversified annual and perennial foods, and on the average they grow 13 different vegetable crops and seven fruit, berry, and nut crops on the same plot.

"According to official government statistics in 2000, over 35 million families (approximately 105 million people or 71 percent of the population) were engaged in *dacha* gardening." These people are known as gardener-inhabitants, or *datchniki*." (From *Russian Dacha Gardening – Homescale Agriculture Feeding Everyone,* Underwood Gardens, 2013. (Leonid Sharashkin, Michael Gold. Deep Snow Press, 2009.) In the Vladimir district 115 miles northeast of Moscow, 95 percent of all households rely on homegrown fruits and vegetables for part of what they eat.

In 1998, six percent of Russia's farmable land was found to be around rural second homes or in urban areas. The *dachas* represent 49 percent of the country's garden plots. A 2008 survey by the Public Moscow Opinion Fund found that 56 percent of urban Russian families or individuals possess a *dacha,* and this six percent of urban-area "kitchen gardens" yielded over 50 percent of Russia's food.

"The law pertaining to *dachas* (prior to 2006 when they were officially recognized as private property) even enforced a certain number of apple, pear, cherry, and plum trees, and raspberry and blackcurrant bushes, as well as defining a minimum area for strawberry plants and vegetable patches. The majority of plots have a surface area of 600m^2 (a rectangle of 20 meters by 30 meters) and are called the "six-one hundred square meters" in Russian. In general, they surround or adjoin a *dacha*, which serves as temporary living space, as well as for storing gardening tools and harvested fruits.

"The mixture of fruit trees, vegetable patch plants and ornamental crops creates a varied island of

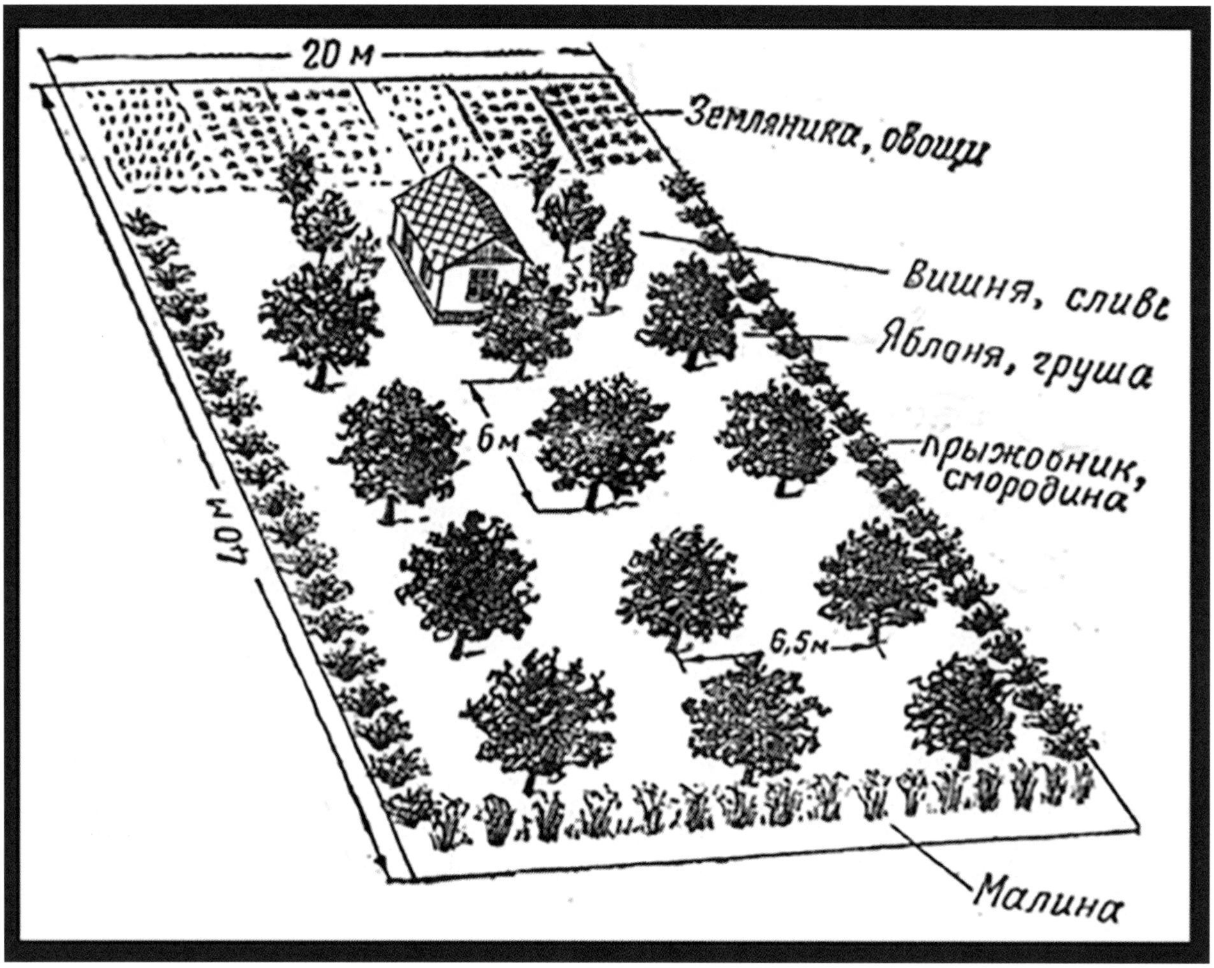

Figure 179. Here is the Russian government's suggestion for planting a 600 square-meter (65' X 135') dacha garden, From top to bottom, to the right, it recommends: wild strawberries (what we call Alpine strawberries or fraises des bois), vegetables, sour cherries, plums, apples, pears, gooseberries, black currants, and raspberries. Here we see more perennial fruits than on most modern dachas. On larger dachas, people keep chickens and even larger livestock. Because the growing season is so short—110 days—most modern dacha properties include cold-frames and/or greenhouses.

greenery. The gardens host great biodiversity, which decreases vulnerability to disease, pests and adverse weather conditions." (*A "Dacha" for Everyone? Community Gardens and Food Security in Russia*, Arthur Grimonpont, 2020.)

According to Sharashkin and Gold, "...Russian family garden-plots can be seen as miniature agroforestry systems, since their combination of annual and woody perennial plantings is *intentional, intensive, integrated, and interactive*. (The four "I"s are being considered by more and more current ecologists as the definition of an agroforest farm.) This diversity is improved due to the fact that 37 percent of urban and 74 percent of rural gardeners respectively use chiefly organic methods and 86 percent use manure. Although 53 percent of all Russian gardeners are using chemical pesticides."

Most Russian *dacha* gardens are larger than most American suburban gardens, ranging from 26,900 square feet (.25 acres) to 6,500 square feet (600 square meters or .13 acres) in rural and urban areas respectively. The average for both rural and urban gardens is 19,200 (.44 acres) square feet to 4,000 square feet (.09 acres). These figures include a mixture of vegetables, fruit and nut trees, and animals in each plot. All of this is accomplished while using only 2.5 percent of the country's land and just 5.8 percent of all its agricultural lands.

From what I've been able to find in photographs (see pages 206 and 207), the main distinction between *dacha* gardens and permaculture forest gardens is that the former often have their vines, vegetables, fruit trees, and animals growing in separate areas. The sides of vegetable beds can be lined, but they're not necessarily raised beds.

By and large, Russian *dacha* gardens don't often combine everything into the vertical integration promoted for forest gardens by permaculturists like Eric Toensmeier, Patrick Whitfield, and Toby Hemenway. The *dacha* plantings look more like large productive community gardens or, as they're called in England, allotments. Because of the short growing season—about 110 days near Moscow— each *dacha* almost always incorporates a greenhouse and perhaps cold frames.

There are actually some models of agroforestry on commercial farms in America and Canada. In *An Introduction to Agroforestry* by P.K. Ramachandran Nair, 1993, we read: "For example, leeks, corn, and strawberries are grown in peach orchards in Ontario, Canada; oats are grown in some New York apple orchards; and potatoes, grains, soybeans, squash, and peaches have been planted in pecan (*Carya illinoensis*) orchards in the southern United States. Approximately 10 percent of all fruit and nut orchards in Washington State (USA) are intercropped with vegetables for home use; and in another 25 percent of the orchards, cattle or sheep are grazed during part of the year."

A Russian Garden Book for All

Russia is a *big* country, and one of the most popular gardening books in the world—at over one million copies sold—was published there. Virtually unknown in America, its delightful title is *Growing Vegetables with a Smile: Bring your Backyard Garden to the Front of your Mind*, by Nikolay Kurdyumov (Deep Snow Press, Ava, MO, 65608.) This 420-page easy-to-read-through gardening manual favors organic methods and as little tillage as possible. Most US gardeners will learn a lot from this book, and I recommend that you get a copy.

Russia's climate ranges from that of Siberia, with lows easily below -45°F, to areas in Azerbaijan, next to the Caspian Sea, where temperatures seldom drop below 15°F. Kurdyumov's book shows the numbered zones for different techniques and crops, but has one severe fault—it doesn't provide a map of the zones, or even a chart of the range of temperatures for each zone, and figuring these out requires a complicated Google search. Nonetheless, I found the book to be an amazing insight into Russian gardening, with a lot of good information that's applicable to much of the temperate United States.

The author, Nikolay Kurdyumov, is in his sixties, and spent 10 years "mingling closely with gardeners at their dachas." He attended the Tirmiriazev Agricultural Academy (founded in 1865), and follows in the footsteps of the Russian agriculturist/agronomist Ivan Evgenievich Ovsinsky (sort of the Robert Rodale of Russia), who published treatises on natural, no-till farming in 1900. Ovsinsky claims that: "you need no fertilizers at all if you handle the soil correctly" and "it is possible to grow plants without rains at all." He developed a no-till method of farming and shunned all chemical fertilizers, yet his yields of rye were five to 10 times higher than his neighbors'.

Ovsinsky began to conduct practical experiments in his fields in 1871, using organic mulch without prior plowing. He described the experience gained thereby in *The New Agriculture System*, which was reprinted three times. Semen Antonets, with the scientific support of Dr Mykola Shykula, established the first Ukrainian organic ("agro-ecological") farm in 1977, based on Ovsinsky's principles, and employing more modern agricultural techniques.

As described in the 2020 article "Zero-Till Technology, Or "No-Till" Technology As A Complex Technological Model," by Satyr R. Yu and Ukrainian Yanchenko N.:

"When using the No-Till technology, the stubble remaining on the field is not burnt and does not burrow into the ground, and all organic residues are crushed to a certain size and uniformly distributed over the field in the form of mulch, so the main requirement when processing the land with this method is its flat surface. Scattered by a rather thick layer, the mulch creates a powerful protective coating in the field

Figure 180. Here's a typical *dacha* garden. With a growing season near Moscow of only 110 days, most dacha gardens contain one or more greenhouses and/or coldframes. Berries are growing in the lower right of the photo near the fence. The lower left plot might be potatoes. (© by Wolfgang Kaehler, Kirkland, WA.)

preserving and restoring the upper layer of fertile soil, which allows you to save moisture. Mulch also excellently protects the soil from wind erosion, prevents weeds from growing and promotes the formation of active microflora with an abundance of micro and macro elements that provide high crop yields."

Kurdyumov was also inspired by the work of Edward Faulkner (*Plowman's Folly*) and Masanobu Fukuoka (*The One-Straw Revolution*). His summary of his approach is:

The Code of Intelligent Gardening:

1. *Do not dig the soil without a definite need.* For loosening and creating structure, use organic wastes, mulch, compost, and green manures.
2. *Do not till the soil deeper than 2 inches (5cm).* Deep tilling ruins soil structure. Create a friable upper layer, which produces the proper conditions for underground condensation.
3. *As with your favorite pet, feed and coddle the creatures of the soil.* Give back to the soil more organic matter than you take from it. Create compost directly on beds. Utilize everything that will rot. Raise anything that will grow.
4. *Never leave soil exposed.* Sow green manures! Time succession plantings so that when the crop is harvested, a second is already emerging.
5. *Consider things seven times over before believing recommendations urging you to expend more labor or money.* Everything you need is in nature. It is just that we often do not know how to utilize it!
6. *Do not hurry!* Organic matter is not like a shot of medicine, but a way of life on earth.

Figure 181. This *dacha* garden contains both crops and ornamental flowers. (Poppies and cosmos in the lower right.) I think the white or pale-yellow flowers may be the blossoms of potatoes, as they're a common vegetable in such gardens. Dacha gardens grow 90 percent of Russia's potatoes. (© by Wolfgang Kaeh, Kirkland, WA)

> Soil that has been ravished for decades will not be repaired in a year. It needs time to remake itself. Help it do this!"

Kurdyumov also wrote a bestseller on fruit cultivation—*Growing Fruit with a Smile, Planting Seeds of Laughter, Reaping Fruits of Joy*. (Deep Snow Press, Ava, MO, 65608.)

In Praise of the Suburbs: Village Homes

Change the suburbs and you change much of America; about 56 percent of the US population lives in the suburbs and in small "metros," as opposed to about 27 percent in the urban core. I'm focusing on suburbia in this book, since 1) that's where most of us live, and 2) it's much more difficult to garden sustainably on a suburban plot than on a farm or large property.

Village Homes is answer to the quandary of many suburbs—housing versus the environment, and lawns versus food. Village Homes is a 70-acre private and environmentally sound community started by Mike and Judy Corbett in 1975 in Davis, CA. This innovative community provides many examples of symbiosis for a synergetic effect in a temperate zone. For all intents and purposes, Village Homes is the closest thing to a mature permacultural development in the US. The Village includes 23 acres in greenbelts; orchards with 30 varieties of fruits and nuts; vineyards; vegetable gardens; and edible landscapes for 225 homes. Tony Lembke (now the head of the Village Homeowner's Association landscape crew) says Village Home gardeners currently grow about 25 percent of the produce they consume.

The community-owned lands include small areas of lawn, one large sports field, and intensively planted perennial food crops—almonds, grapes, plums, jujubes, apricots, pineapple guavas, figs and pears, to name a few. (See Figure 191.) Most of the homes are

Figure 182. A peach tree grows near this Village Homes house, along with a variety of colorful ornamentals—including sweet peas on the left. Vegetables grow in the half-barrel, and a grapevine shades the south-facing wall to help protect it from late-summer sun and heat.

augmented by vegetable plots or edible landscapes. The average lot size is 4000 square feet. There's also a community garden.

Catchments (berms and swales) for surface storage and groundwater recharging *are* possible on a community-wide basis. At Village Homes, a series of catchments were built on previously flat land prior to the construction of the homes. The swales meander throughout each neighborhood and add a dynamic complexity to the landscape. They also fill up with heavy winter rains, trapping any soil that might be washing away, and allowing time for the water to percolate deep into the subsoil. While most suburbs just dump the rain into a storm-drain or sewer, the swales at Village Homes keep the winter rains within the suburb and store 90 percent of the water in the 10 feet of topsoil for the landscape to use into the dry summer.

While this doesn't eliminate summer irrigation, it does delay the time at which watering begins. Since the community-owned lands and most, if not all, of the private landscapes are tended with organic methods, the community's groundwater isn't threatened by herbicides and pesticides. Bill Mollison would like to see swales and ponds and lakes proliferate throughout America.

At Village Homes, soil erosion is virtually non-existent, due to a number of factors: The nearly complete coverage of the soil with lawn, clover, ornamental and edible ground covers; the water-slowing swales; plenty of tree canopies to diffuse hard rains; and the frequent use of mulch on recently cultivated soil. Each homeowner has done their share to control destructive erosion. Virtually every square foot of each yard (except for short narrow driveways) is either covered with living foliage or well mulched.

A popular soil-knitting ground cover is Dutch white clover, which usually stays below 12 inches without mowing, though an occasional mowing makes it look much like a verdant, green lawn. Being a legume, it fixes atmospheric nitrogen, at the rate of 100 pounds per acre per year. Other homeowners use decorative ground covers such as Peruvian verbena, prostrate rosemary, iceplant (an invasive plant outside the urban or suburban garden), low-growing coyote bush, and *Myoporum parvifolium*.

Nearly every Village Home uses passively solar-heated water, and passive solar design provides a good percentage of their heating and cooling in many of the homes. Deciduous fruit trees abound near southern and western walls of many of the 225 homes. During the full leafing of summer, the trees help keep the houses cooler. Come winter, the deciduous branches of the trees let *some* light through to warm the home. (The branches can shade from 30 to 85 percent of the total light of the house in the winter. See Figure 29 and 30, pages 39 and 40, respectively.)

The chickens that many people keep in their yards are often located beneath ornamental shade trees—the tree provides welcome shade in a climate that's overly hot and dry for chickens, and the chickens fertilize the tree to further the growth of its protective shade. In addition, the chickens convert vegetable garden and kitchen scraps into eggs with richly colored yolks.

Most noticeable to the passing visitor is the extraordinary amount and diversity of beneficial insects. Ladybugs, praying mantises and predatory wasps abound—you practically have to sweep them off the mailbox to get to the mail. The increased population of helpful insects is due, in great part, to the incredible variety of flower blooms that unfold throughout the year. The pollen and nectar in the rich complex of native and exotic plants provides the food energy that many of these beneficial insects require for searching out other insects to eat or parasitize.

While this doesn't mean that the Village Homes'

Figure 183. Bill Mollison (1928-2016) standing next to the Village Homes community's orange orchard. As the co-founder of Permaculture, with a great interest in polyculture and perennial plantings, Bill was intrigued by the orchards and vineyards in the common areas of the village.

homeowners can abandon their bug-spray tanks entirely, it does mean that nature is more likely to balance things out without the intervention of a gardener. Most homes use organic techniques for the management of their gardens, which can be somewhat easier on the beneficial insects.

This is perhaps, after more than 45 years, still the largest and most intelligently planned, most solar-powered, and most edibly-landscaped suburb in America.

Bill Mollison in an American Suburb

In 1981, I figured Village Homes would be an appropriate place to discuss a concept set out in Bill Mollison's *Permaculture - A Designers' Manual*, to wit: "assembling conceptual material, and strategic components in a pattern which functions to benefit life in all its forms." According to Bill, "This is the only place in America that I like to visit. Mark Rubald [the head of the Village Homeowner's Association landscape crew for about 30 years] told me at that time Village Homes grew 86 percent of its produce. Village Homes is the best suburban example of Permaculture in America." In a previous interview, by Jamie Jobb for the *CoEvolution Quarterly*, Bill's comment was: "Suburban family farmers are the salvation of the world!"

Talking With Bill

While touring Village Homes with Bill, a native of Australia, I began by asking him about his observations of the American landscape. He began with large, more cultural or political problems, pointing out three major areas of concern: the lack of clean water, the great amount of soil erosion, and the lack of trees.

Water

The lack of clean water is, according to Bill, due in part to the lowering of aquifers due to excessive use of water in agriculture. "The American lawn," in Bill's view, "is the world's largest 'agriculture' [meaning a cultivated

Figure 184 . This south-facing solar home has a clover "lawn" with a variety of fruit trees. An almond tree grows in the foreground.

Figure 185. This garden contains a mixture of ornamentals (the lavender in the background) and vegetables (onions in the foreground.)

"crop" requiring water and fertilizer]. Flying over Denver, Colorado, what you see is a lawn desert—an ecological void due to the proliferation of lawns. Lawns in the dry lands are completely inappropriate. In the future, I hope to see lawns heavily taxed because they are such a waste of resources and the chemicals that are used to keep them up are toxic pollutants to the groundwater."

Instead, Bill proposes food crops to replace lawns because "it's easier to grow food than a lawn." (I disagree with this last point. But food instead of lawns is *very* important.) Most houses at Village Homes have very small lawns, if they have one at all, most homeowners choosing to plant their small landscapes to food, flowers, or ornamental trees and shrubs. Throughout the community there are small patches of lawn, much less than 1/5 acre each, which serve as playgrounds for children. Bill despises lawns and is much more excited to see the Dutch white clover ground cover that's used as an alternative to lawn under some of the groves of fruit trees in the community.

Part of a clean-water supply involves a community that captures and stores as much of its own water as possible. In looking over the American landscape, Bill deplores the lack of "overland catchments" —small ponds and lakes. The system of studying the lay of the land, carving a series of catchments, berms and swales, and controlling the dispersal of the water to both the soil surface and the deeper water storage in the aquifer is known in Australia as the Keyline system, developed by P. A. Yeomans.

Figure 186. A small raised bed for vegetables flourishes on the south side of this solar house. A grape arbor shades the lower south-facing windows.

Figure 187. The west-facing wall of this house, as seen from rose-bordered community vineyard, will soon be covered by a growing vine to cool the house.

The Keyline system shows up throughout Bill's books. "A good ecological design," Bill says, "incorporates resiliency, flex, and adaptability. Swales incorporate a natural flex that allows part of the stream to build up a bank and another part to destroy a bank. But the power of the water is moderated within this natural dynamic." (The cost of bulldozing a pond or building a dam is very different in Australia. A single six-acre-foot dam built in 1980 at the Farallones Institute Rural Center in Occidental, California cost $12,000 in 1980 dollars; an Australian contractor can bulldoze more than two dams of the same size for that amount.)

Another strategy that is common in Australia is the extensive use of cisterns, and Bill said: "Village Homes needs to have cisterns." Because every home uses some form of passive solar heating and cooling, Bill was, during our walk together, musing out loud about integrating cisterns into the water-storage heat sinks built into some of the solar-heated homes. Since summer rains don't fall to replenish the cisterns, they aren't as appropriate in Village Homes as in the rest of the country where many homes, during the summer of the 1988 drought and since, would have benefited from a large rain-filled cistern.

Figure 188. These rocks line part of the swale of the Village's public winter drainage. Here a straw-covered mound protects rows of vegetables from winter rains.

Figure 189. This is the entry to a row of houses via the public pathway at Village Homes. A clover ground cover grows underneath one of the trees community apple orchards, a situation that may lead to mildew.

Figure 190. Not all gardens have vegetables. This is an attractive ornamental garden with a lawn. The public path separates two rows of houses. The lots are quite small. So, it's easy to see into the houses from the public pathway—thus a lot of foliage (and curtains).

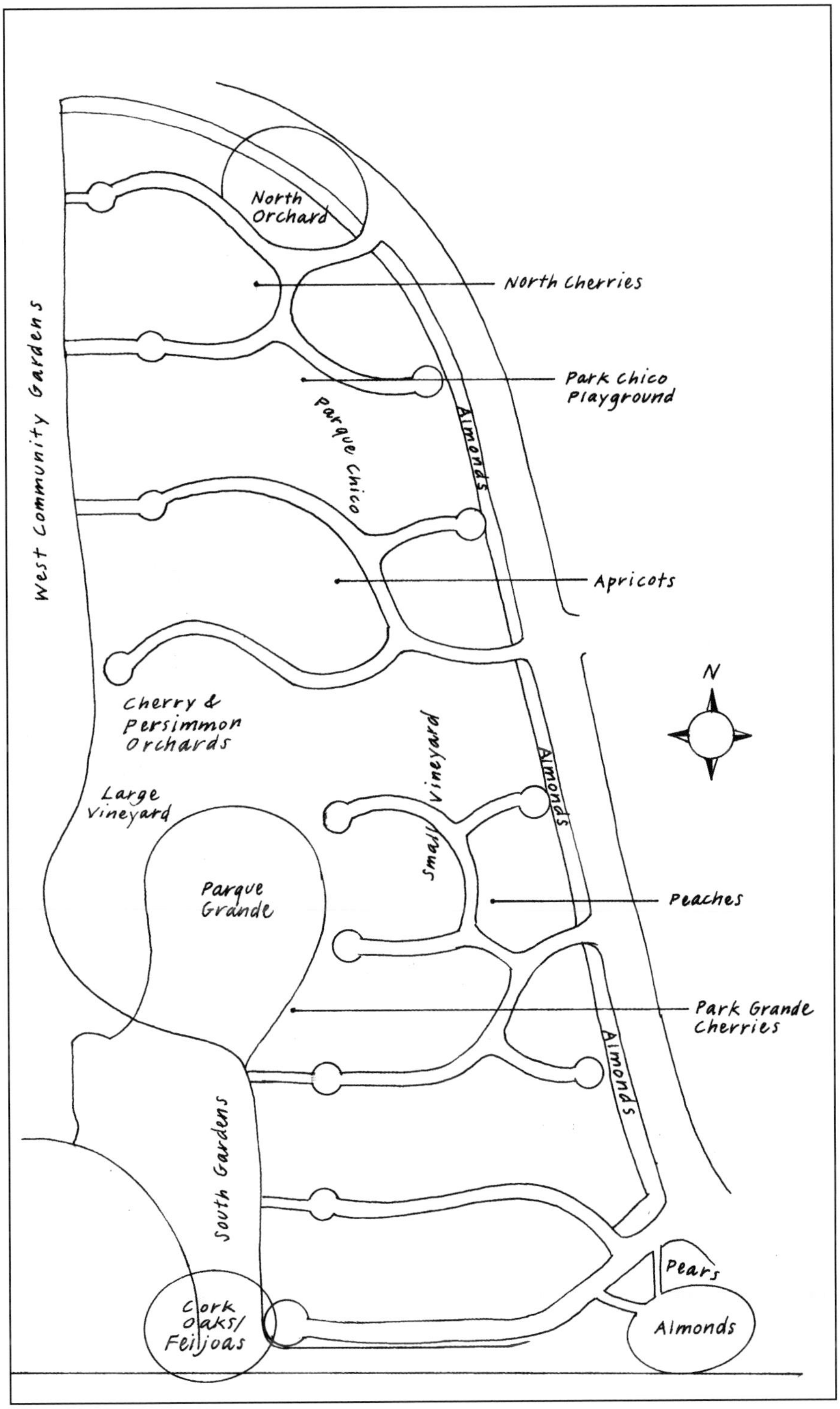

Figure 191. This map of Village Homes shows only the community orchards and vineyards. The streets run east to west, insuring that each home can be passively heated in the winter and use solar water heaters. Solar-electric panels were too expensive to use in the late '70s and early '80s when these homes were being built.

Bonus Charts and Photos:

Figure 192. A common process in food forests is known as "chop-and-drop." This means that a variety of cover crops are grown just so they can be cut and left in place like a mulch. Besides conserving moisture, such an approach is thought to add nutrients to the upper soil. It's also thought that the deep roots of the chop-and-drop plants may bring up nutrients that shallower roots can't grab.

Two common plants recommended for chop-and-drop programs are comfrey and rhubarb, the idea being that their deep roots gather minerals, and that chopping/dropping the large fertile leaves adds nutrients to the surface soil. One problem I have with this method is that these two plants in particular are heavy feeders, which means they require a lot of nitrogen to prosper and to support their prolific foliage. (My aunt grew all her rhubarb next to the chicken or hog pens.) The chop-and-drop technique is often used in guilds, but the amount of nitrogen that keeps the rhubarb and comfrey "happy" can over-stimulate other plants—fruit trees in particular.

As explained in a handbook available on my website (*Aerobic Soils are Fertile Soils; Deep Soils are not as Fertile*), deep roots don't gather more nutrients than shallow roots; neither do they gather minerals that shallower roots can't. The myth of deep roots being very important in gathering nutrients is common in the literature, but aerobic soil that's found near the surface almost always provides more fertility than deeper soil regions do. (See the 76-page handbook for the details.) Deep roots are important during times of drought and in keeping a large plant or tree from blowing over.

Figure 193. This cherry tree has been planted on a mound. I prefer planting on mounds, as it protects the upper root system from wet soil and root rot, to which cherry trees in particular are susceptible. As the soil begins to dry out, the roots will spread further from the mound, encouraging a wider root system, making for a sturdy tree, and allowing the roots to explore more soil for nutrients and moisture. Here in this commercial planting, the drip-irrigation tubing is on top of the mound because cover crops will be planted and mowed between the mounds. Homeowners with mulch can place the tubing between the trees on level ground to irrigate more of the root system, and cover the tubing with an attractive mulch.

Figure 194. Here we see the white blossoms of almonds that have been grafted onto a peach tree with its pale-pink blossoms. Multiple grafting allows for more diversity in a smaller space. The peach trees are self-fertile, so having only one variety of fruit is fine. With other trees, like almond and apple, you need at least two varieties for good pollination and a bountiful crop. (Apple trees are the easiest to graft.) With the exception of cherries, all stone fruits can be grafted to a single tree with varying degrees of success. Almonds are more difficult to graft successfully than plums.

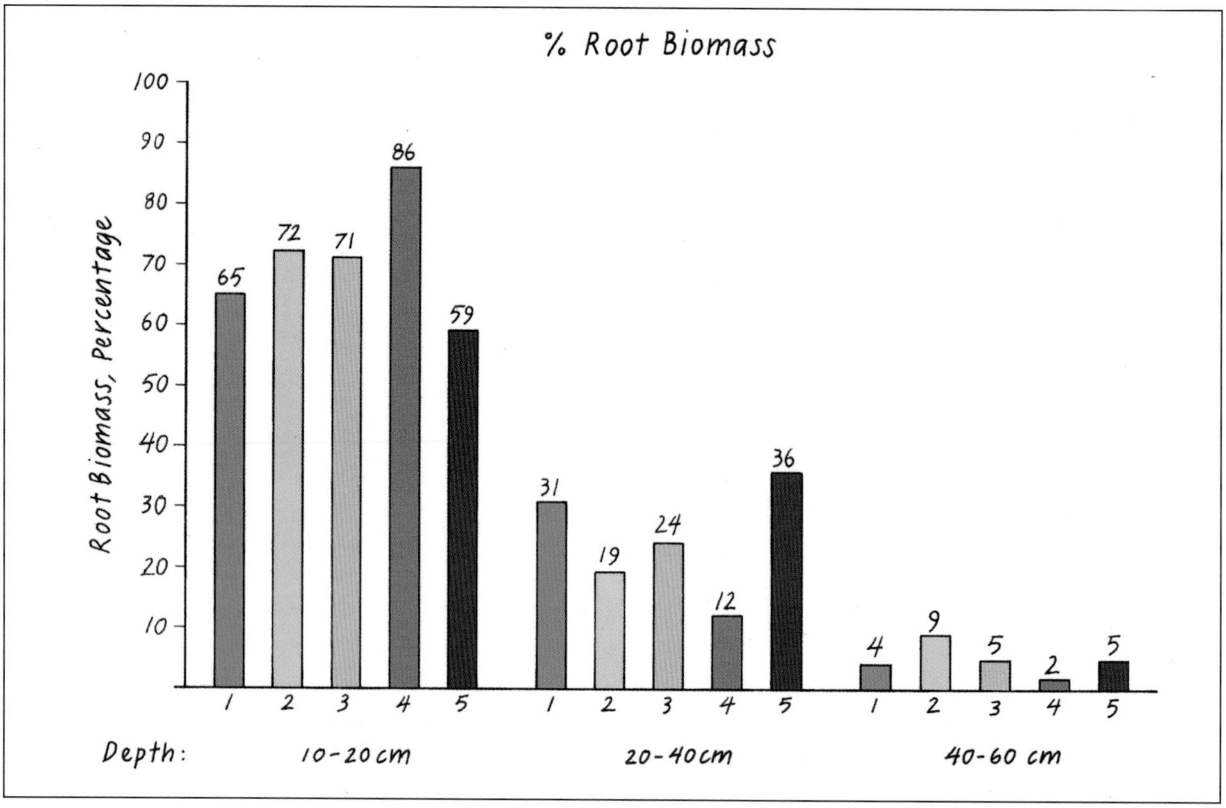

Figure 195. This is a very revealing graph. Most people assume that California native plants from the chaparral ecosystem grow deep roots to gather moisture and nutrients because they often need to survive poor soils and summer droughts. Here we see that, contrary to myth, more roots grow near the surface than in soil lower down. (30 cm = one foot.) Thus, *Haplopappus pinifolius* has produced over 10 times more roots in the top 10-20 cm of soil than at just 60 cm (two feet) lower down. *Adenostoma fascicula-tum* produces 86 percent of its roots above the one-foot-deep level. As the "White Paper" for sale on my website states: deep roots no; shallow roots, yes. Here we see that even the uptake of moisture takes place almost completely above the two-foot level. (This is so important that I'll repeat this graph in the chapter on roots.) It's another case for using mulch and no tillage. (Note: irrigation should be kept far away from the base of these types of plants to avoid root rot.)

1 -*Arctostaphylos pungens* - Pointleaf Manzanitas

2 - *Ceanothus greggii* - Desert ceanothus

3 - *Erigonum fasiculatum* - California buckwheat (not at all related to the edible grain)

4 - *Adenostoma fasciculatum* - Chamise

5 - *Haplopappus pinifolius* - Pine Bush

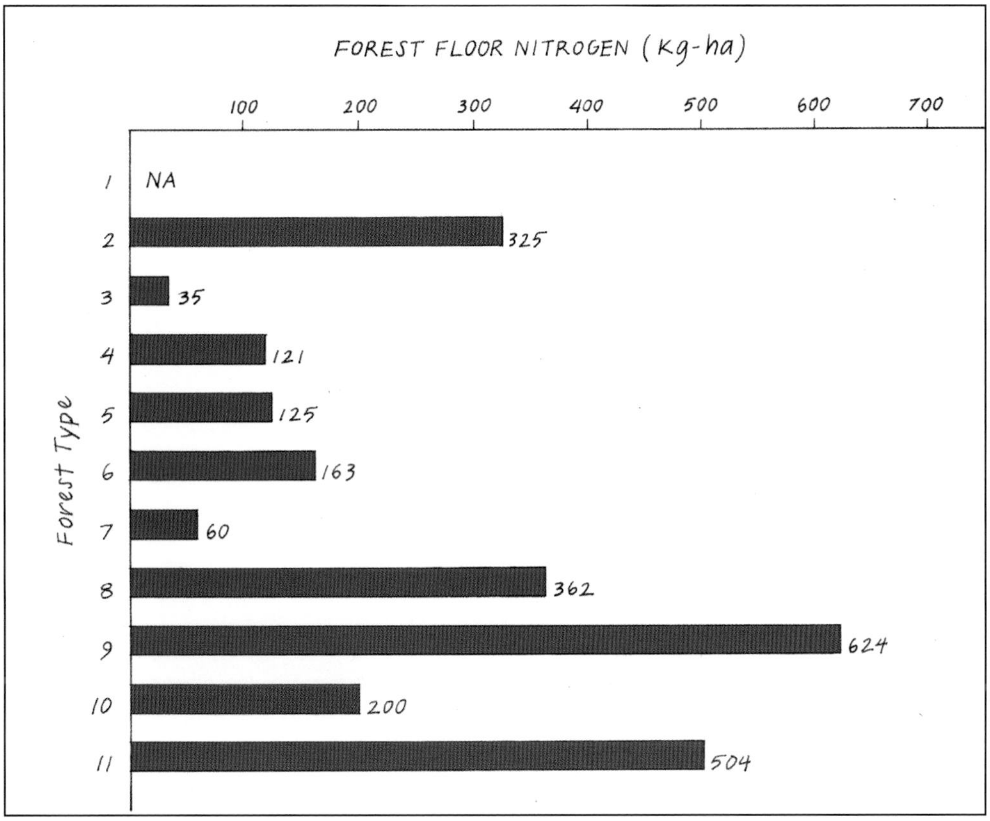

Figure 196. Another comparison of the tropics versus other ecosystems.

1 - Tropical broadleaf deciduous
2 - Tropical broadleaf evergreen
3 - Tropical broadleaf semideciduous
4 - Subtropical broadleaf evergreen
5 - Mediterranean broadleaf evergreen
6 - Warm temperature broadleaf deciduous
7 - Warm temperature broadleaf evergreen
8 - Warm temperature needleleaf evergreen
9 - Cold temperate broadleaf deciduous
10 - Cold temperate needleleaf deciduous
11 - Cold temperature needleleaf evergreen

Figure 197. Mounds for planting trees don't need to be very high to work. Here is a fruit tree planted on a five-foot wide mound that is eight inches high at the trunk. The water still drains away from the crown of the root system. Unlike Figure 177, mound plantings don't have to be "lumpy." Make them wide and gentle.

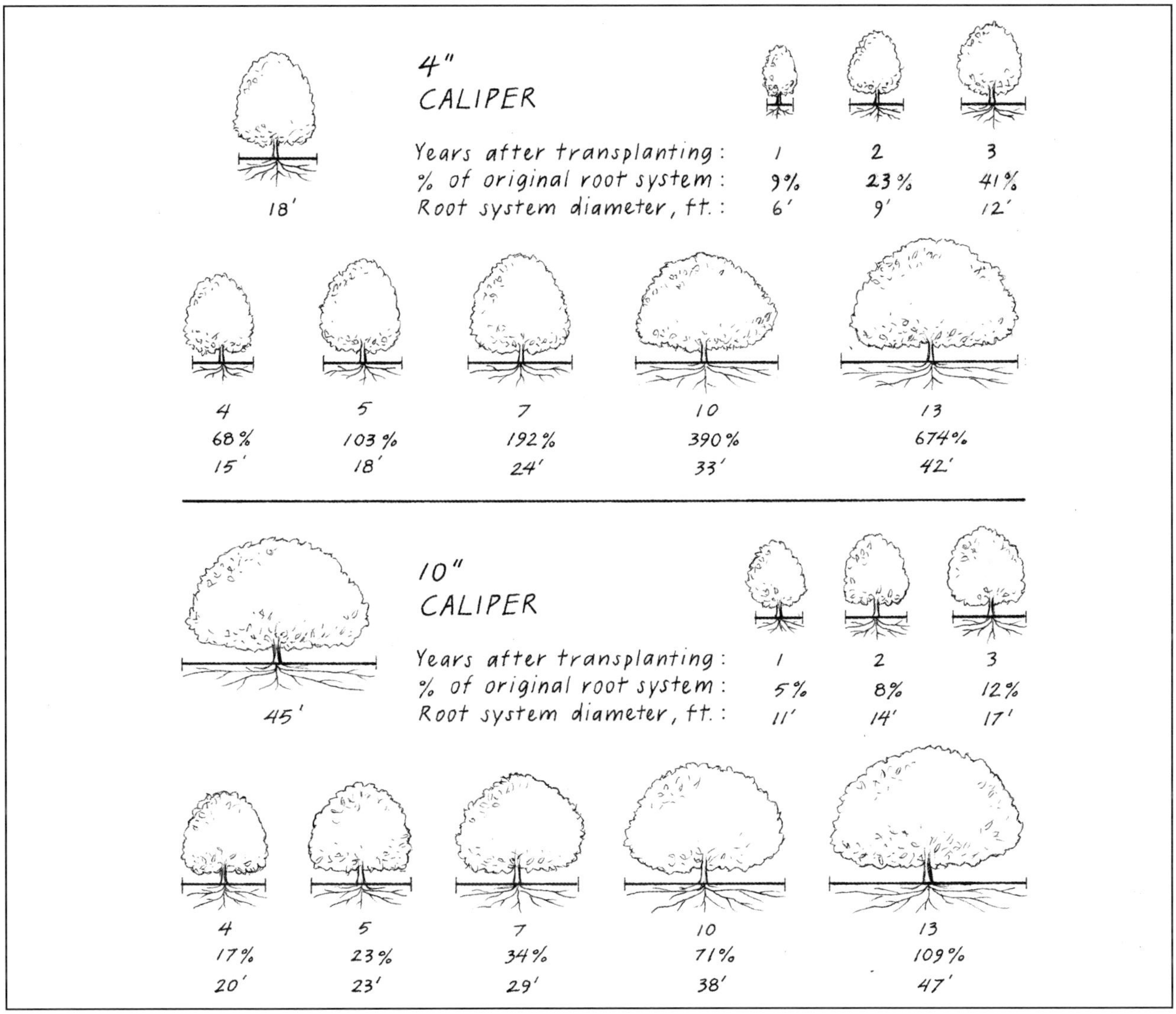

Figure 198. This is a bit complicated or "dense." Please take the time to review the drawing. It shows one example where planting smaller plants/trees produces a bigger mature plant/tree. The four inch caliper (trunk diameter) tree is dug up with a six-foot rootball. After five years the diameter of the root system is equal to the roots' diameter before transplant. A tree with a ten-inch caliper takes nearly 13 years to do the same and get back to the size of the roots prior to transplanting. At 13 years, the four-inch caliper tree is six times the percentage wider than its original transplant size and the 10-inch caliper tree is only the same as the original root diameter. Smaller is better. This also applies to the four-inch pots I buy instead of one-gallon "cans." (They really were metal cans when I first started landscaping in 1974.)

Chapter 7
Container Gardening

For many urban gardeners, container plants are their only option for access to soothing touches of foliage, color, and fragrance, fresh herbs or vegetables, or tree-ripened fruit. Ornasmental and edible plants displayed in attractive pots are a wonderfully easy way to provide colorful seasonal highlights and fresh produce—and you can even move them to follow the sunlight.

Planting Containers

Although I seldom see this done, every clay, cement, or (especially) terracotta pot should have its inside sealed so the water isn't constantly being wicked out of the soil by the container's porous interior. To prevent wicking, I brush an inexpensive, nontoxic sealer (swimming-pool and cement sealers are cheaper than other kinds) on the inside surfaces of clay pots. Sometimes the clay soaks up so much sealer that two coats are required.

Figure 199. Here, in the large blue pot, we see the edible flowers of marigolds and a dwarf lime tree. Most of these containers are glazed ceramic, which won't wick water away from the plants.

To start your container garden, choose pots much bigger and deeper than those in which your plants arrived, so that you don't have to repot right away. Be sure to wash the soil off of all of the roots of the container plants and trim off all broken, kinked, and circling roots. Dip the remaining roots into water into which you've mixed a mycorrhizal inoculant—this is one of the few times you'll need to use one.

Spread the roots wide on a cone of potting soil. (I recommend using a soil mixture with no peat moss, as this is basically a non-renewable resource and much of it comes all the way from Quebec in Canada.) As much as possible, avoid potting soils that contain a lot of fiber or chips, as they'll rob nitrogen from the soil. *Never* use a potting soil that contains "nitrolized redwood bark," because redwood is so slow to rot that it ties up a lot of nitrogen for a longer time than compost or most potting soil does. (Redwood bark is also a byproduct of milling, and I choose not to support the lumber industry.)

Irrigation

Compared with their down-to-earth cousins, container plants are quite unforgiving of neglect, especially when it comes to watering—even properly sealed pots will dry out quickly. Periodic and/or infrequent drenching harms the plant by inflicting intermittent periods of drought-stress and, when you *do* get around to watering, by leaching valuable and limited nutrients from the soil. During a "dry spell," the soil will pull away from the sides of the pot, and the

Above, pomegranate trees are one of many fruit trees that, with care, can be grown in large containers. See page234.

next time you water, a lot of what you pour in can spill down through this gap and run out of the pot's drainage hole. Daily watering with tiny amounts of water, just enough to keep the soil moist, is essential to maintaining both consistent moisture and nutrient availability in the soil. Daily watering also means the soil won't pull away from the side of the container.

Drip Irrigation

Enter drip irrigation. With a well-planned drip system you can add a very small and measured amount of water to each pot every day. Consistent watering of container plants will reward the gardener with better growth and a longer period of bloom (and fruit, if you're lucky enough to have a large enough deck or patio to accommodate fruit trees). Furthermore, drip irrigation can eliminate the mildew and fungus problems that are encouraged and exacerbated by overhead watering.

A drip system works much better if the pots on each zone or program are the same size. If one or more of the pots is smaller than the rest, extra water will drain through it, washing out valuable nutrients.

Drip irrigation of pots calls for a variety of hardware seldom used in landscapes. This is the only situation in which I mess with so-called "spaghetti," the trade name for 1/8- or 1/4-inch tubing. Its small diameter and low visual profile make it easy to hide among plant foliage. When carefully installed in small- to medium-sized containers, spaghetti in-line tubing is manageable and doesn't end up in the tangled mess so accurately suggested by its name.

Trickle Your Pot's Fancy

Remember, before adding soil, drip irrigation, and planting your pots, be sure to seal their inside surfaces. Also, don't fill the bottom of the pot with gravel or packing pellets, as the water will then just settle in a puddle there, which is *not* good for the roots. A small piece of broken pottery is enough to keep soil from falling through the hole.

Small Pots

For small pots, up to six inches in diameter, it's best to use a single emitter, no matter how many plants are squeezed in. To avoid overwatering small pots, I find it best to use 1/2-gph emitters—if more water is needed, I can always set the timer to stay on a little

Figure 200. From above, clockwise: A single 1-gph (gallon-per-hour) emitter, held in place by a U-shaped metal stake; a petunia s with 1/4-inch in-line tubing with a "goof plug" closing the end; and another 1-gph emitter held in place by a plastic stake. The emitters are placed at the end of 1/4-inch tubing, which is linked by a transfer barb to1/2-inch supply tubing.

longer. I always use pressure-compensating emitters for all my containers. Be sure to buy the right spaghetti tubing for your emitter's barbed post (usually ¼-inch spaghetti tubing), and *do not* punch the emitter into the supply 1/2-inch tubing then add a length of solid 1/4-inch tubing to the pot. Instead, punch a transfer barb into the 1/2-inch supply tubing, add the proper length of ¼-inch tubing, and insert the emitter at the end.

Medium Pots

For medium-sized pots [six to 18 inches in diameter], the rule is to use one emitter for every one to two plants. In a multi-plant container, you might get away with a single emitter with a higher flow rate, but if the water spreads unevenly through the potting soil, some areas may go unwatered. Again, before adding the spaghetti tubing and planting your pot, be sure to seal the inside. Remember that terracotta pots are especially prone to wicking moisture through their walls.

Installing Drip Irrigation In Medium Pots

The most invisible drip-irrigation systems for pots are those that are installed as the containers are planted. To begin, run a length of 1/2-inch solid drip hose from the timer or controller to a spot near the base of the pot. At each pot, punch a hole in the solid drip hose and insert a transfer barb with a barbed post at each end—one to go into the drip hose and the other to insert into the spaghetti tubing. (Be sure to use a barb that's the same size as the spaghetti tubing.) Run the spaghetti from the drip hose to the pot, and attach a 1/4-inch barbed 90° "spaghetti elbow."

Add a length of 1/4-inch in-line spaghetti tubing to the other end of the elbow, cutting it four to six inches longer than the projected soil depth in the pot, and thread the tubing through the hole in the bottom of the pot *before* you add any potting soil. After planting, clip the spaghetti off just above the soil, add another elbow, and attach about a four-inch piece of spaghetti in-line tubing to lie flat on the soil.

Again, to keep potting soil from falling out, use a single piece of broken pottery or a flat pebble to close off any space around the spaghetti where it comes through the drainage hole. If your pots are on a flat surface like a patio or rooftop, you'll need to use blocks to lift each one off the surface a bit so the spaghetti can come up through its bottom without getting flattened. On a wooden deck, simply set the drainage hole of each pot over the space between two boards, setting the pot up on blocks to prevent the decking from rotting.

Now, fill the container with a rich, well-drained potting soil with lots of compost. The spaghetti in-line tubing coming through the bottom drainage hole should be lifted up while you add the soil, and should wind up projecting four to six inches above the finished surface. Use a transfer barb to add another elbow, and circle the loose spaghetti on the surface of the soil. Leave this tubing open, and flush it for a minute or so to clear out any soil inside. A little bit of mulch will further disguise the tubing and emitter. Even on the day you plant, your container drip system should be practically invisible.

If a pot in your system should become empty, you can save water by clipping the emitter off the spaghetti tubing and closing it off by inserting a barbed "goof plug."

Large Pots

In pots over 18 inches in diameter, you *could* use a number of punched-in emitters, but it's preferable to

Figure 201. A single black ½-gph emitter dripping on *Armeria maritima* that's growing in a metal container.

Figure 202. A spiral of 1/4-inch spaghetti tubing irrigates sugar pea seedlings planted in half of a wine barrel.

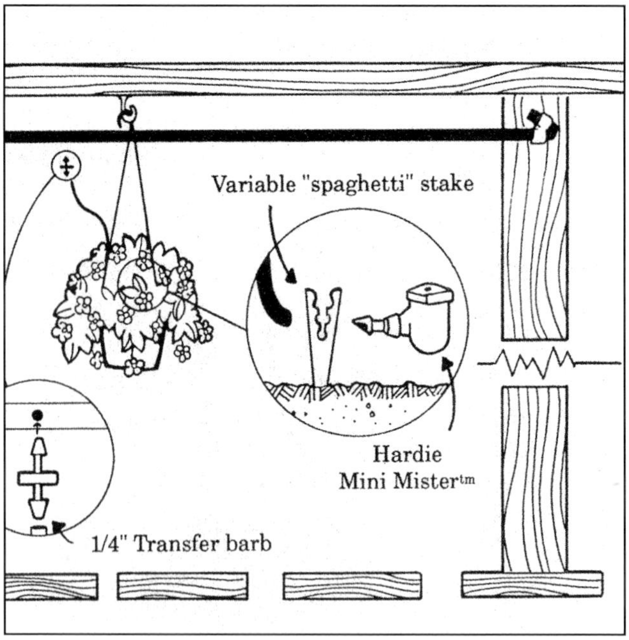

Figure 203. The best drip irrigation for hanging pots is the Hardie Mini-Mister™. It's attached with 1/4-inch spaghetti tubing to a 1/2-inch supply line (shown here as solid black) with the use of a transfer barb. The mister is turned on briefly a few times a day.

use one or more circles of the 1/4-inch in-line tubing that has emitters built in every six inches, and passes 1/2 gph. Install the tubing in your pots as described above, and fill the pot with soil, leaving a few inches of solid 1/4-inch spaghetti tubing projecting above the potting soil. [See Figure __.] After potting the plant, flush the system. Next, clip the spaghetti off just above the soil surface and attach a 1/4-inch barbed spaghetti elbow (90°). Add a length of 1/4-inch in-line tubing to the elbow and circle it around the pot just a few inches in from the perimeter. With large pots you can coil the tubing in a spiral pattern, but make sure you won't be watering directly on top of the crown of the root system. Flush the tubing again and close the end of it with a goof plug. You can hold the tubing in place with metal J-stakes every foot or so.

In pots bigger than three feet in circumference, you can use regular ½-inch in-line emitter tubing. The pot or container should be big enough that the tubing won't kink shut when you circle it on the soil surface.

Hanging Up Drip Emitters

You can use regular emitters in hanging pots, but these pots usually need more water than the same-sized pot on the ground. If you're using, for example, a single 1/2-gph emitter in a 10-inch pot on your patio, you might want to try at least two emitters or one 1-gph emitter in a hanging pot of the same size. "Spot spitters," as seen in Figure __, put out a good amount of water over a large area, but the flow rate is so great that the water may run off the surface or down the sides of the pot faster than it can percolate into the potting mix. I don't recommend using spot spitters in most hanging plantings.

On the other hand, many hanging plants are water and humidity lovers, and like the fog-like vapors of a "micro-mister." While the misters pass more water than some of the lowest-flow emitters, they use only three gallons per hour to generate a delightfully cool cloud. When the the mister is placed beneath the foliage of the plant, the moist fog condenses on the leaves and drip-irrigates much of the soil surface below, although some water will dribble from any foliage that hangs outside the pot circumference. Two name brands for 3-gph misters are Hit ProMist™ and the Hardie Mini-Mister.™ Each comes with a special stake to hold the mister upright so the spray will billow up into the leaves. The Hardie Mini-Mister costs less; its stake and mister are considerably less bulky and therefore not as noticeable as those of the Hit ProMist. The use of either requires a 200-mesh filter screen on the drip-system filter to prevent the tiny misting hole from clogging. (Talk to your supplier.)

Hanging plants are sometimes located so high off the ground that 1/4-inch spaghetti, with its 15-foot limit, can't reach them. In this case, 1/2-inch solid drip hose can be laid out for a length of up to 100 feet, and you can plug solid 1/4-inch tubing into the drip hose to get as close to each hanging plant as possible. [See Figure 57.] It's important not to fasten the tubing securely to the post, and not to stretch it taut, because the tubing will shrink significantly in cold weather; this may damage fittings. Paint the tubing brown (which actually shows up less than most green paints) to protect it from ultraviolet rays and help it blend in with its surroundings. Use a 1/4-inch transfer barb to attach a length of spaghetti tubing to each hanging pot, and run the spaghetti down one of the wires that holds the pot, tying the tubing to the wire for a low visual profile. Flush the system and close the 1/4-inch tubing with a goof plug.

Because their water needs are so special, hanging plants should also always be plumbed onto their own valve with a timer. If you live in an especially hot or windy area, get a timer which allows you to turn on the system automatically several times during the same day to keep hanging plants well moistened and healthy. Slight amounts of daily water will keep the soil mixture moist, plump, and in full contact with the walls of the container. The daily amount of water added should equal what is lost by evaporation from

Figure 204. The upper photo of a romantic Italian alleyway looks idyllic, but I wouldn't want to have to repot these shrubs! *Never* put a plant into a pot with the opening smaller in diameter than the width of the pot halfway down (the "hips"). The bulge looks attractive, but you'll do a lot of root damage trying to cut out the roots from the hip section (and it takes a *lot* of extra time).

the soil and container and by transpiration from the plant's leaves (see Chapter 5). By adding just enough water to equal these losses, you will also prevent leaching of limited and valuable nutrients in the potting mixture.

Only experience and experimentation can provide the specific answer to "How long do I leave the drip system on?" If you're going to automate your container irrigation, start with very short times— only a minute or two. But beware; not all irrigation timers or controllers can be set for times that short. Check with your supplier to make sure you purchase a timer that permits a very short "run" time.

Growing Food Plants in Containers

While it's often a bit harder to manage than tending ornamentals, growing food in pots is definitely possible. It's the perfect solution for people who have only small gardening spaces—patios, sidewalks, porches, roofs, boats (I've seen it!), and decks. Of course, the bigger the pot, the better, and always buy bush-sized vegetable plants for easier growing. When buying tomato plants, get a determinate variety that

Figure 205. Here are some suggestions as to the kinds and sizes of edible plants and herbs to grow in different-sized pots. (The bigger the pot you can afford, the better the growth.) All of these suggestions assume you're using a very fertile soil with daily irrigation and frequent fertilization. When growing perennials, repot the plant every two to three years, completely replacing the potting soil and adding compost and mycorrhizae.

Figure 206. This is a bird's-eye view of a planter made from a livestock water trough. The 1/4-inch spaghetti tubing irrigates both vegetables and edible flowers.

stays bushy. (For ornamentals, look for the Latin species name '*nana*' or '*compacta*'.)

Rooftop and Deck Gardening

Rooftop and deck gardens sound great, but they need some important analysis upfront. First, check with the building department to see if it's even legal to create a rooftop garden, or (if you're a renter) whether the homeowner is willing to build the deck you need. If so, research the guidelines (height and style of railings, for instance) needed to keep it up to code. Find out if you need a permit and what fees might be involved.

If you're growing on a roof or deck, be *very* sure to check its capability to support the weight of pots, people, and soil. Damp soil is very heavy—a mere cubic foot of it can weigh 75-125 pounds. Be sure to determine whether the roof can bear the additional weight of people, and if so, how many. You *must* get a licensed, structural engineer to inspect the support of a roof or deck on which you want to establish a large garden.

Soil

Use the lightest possible potting soil, and vermiculite or pumice instead of sand for drainage. It may be best to use plastic pots that look like terracotta from a distance, instead of actual terracotta or ceramic. Attractive flooring is nice, but only adds weight. Keep it simple.

Since you'll be growing plants in an essentially unnatural environment, you can justify (and may even need) the use of mycorrhizal inoculants. (See pages 79-87 in Chapter 4.) You can buy a powdered inoculant, mix it with water, and apply it according to the label instructions, or buy potting-soil mixes that are already inoculated. Be sure to treat bean and pea seeds with the correct nitrogen-fixing inoculant before you plant—peas and beans require different types of inoculant. This will ensure that the roots will form nitrogen-fixing nodules to feed the plant. Use foliar fertilizer sprays, not soil drenches, to compensate for any nutritional deficiencies.

When growing food crops in containers, I've found it is best to blend 25-50% organic compost with a high-quality organic potting soil, which I replenish with new compost every year. As mentioned above, avoid as much as possible potting soil with a lot of fiber or chips as they rob nitrogen from both the compost and the soil.

You can avoid using much soil by growing on top of

Figure 207. A rooftop garden in New York City, with lots of evergreen shrubs and seasonal color. The fiber pots, being much lighter than terracotta pots reduce the load on the roof. To mitigate wind damage, most plants grown are no taller than the side railings of the garden.

Figure 208. Another NYC rooftop garden, by Barbara Britton, is equipped for gatherings, with chairs, benches, and a BBQ grill.

Figure 209. Lonnie Zamora tends the salad garden he grows on the sixth-floor roof of his NYC building. Drip irrigation keeps his lettuces well watered.

straw bales (See Chapter 4), but they may be just as heavy as containers with soil, and it may be too difficult to get them to the roof without an elevator. If looks are not a big factor, you might also consider growing directly in sacks of potting soil or compost as described on page 231.

The proper soil blend is critical. For more information, contact local community gardeners, other rooftop gardeners, landscape architects with rooftop experience, or a book called *The Rooftop Growing Guide: How to Transform your Roof Into a Vegetable Garden or Farm*, Annie Novak, Ten Speed Press, 2016.

Even with a good soil mix, the plants require periodic fertilizing. Check the label of the fertilizing product to gauge how much to apply and how often. If the label calls for diluting a powder, it's best to use distilled or filtered water to keep chlorine and slightly caustic minerals out of the soil in the pot. When you're growing food plants in containers, it can be hard to dilute some organic fertilizers properly, although pellets with the proper NPK amounts are now available.

But, and it's a *big* but, the water must flow *though* the pellet to dissolve the nutrients. This is difficult to achieve with drip emitters. It's better to make a solution with the pellets and periodically add it to the plants with a sprinkling can. Kelp in solution or powder is fairly worthless as a fertilizer because it's only good for correcting small mineral deficiencies. The NPK of kelp is only 1-0-4, so if it helps, it's only very slightly.

You also need to be aware that some fertilizers are based on manure or urea, and can really stink up your deck or rooftop. It's possible to add a device called a proportioner to apply a dilute amount of soluble fertilizer every time you drip-irrigate, but it's complicated to install, and for the proportioner to work properly, you need at least 60 1-gph (gallon-per-hour) emitters on one valve/zone to produce

Figure 210. Lonnie Zamora's company built the ornamental flower boxes shown on the 16th floor of a building in downtown Manhattan. The window boxes in the photo below are at Ladew Topiary Gardens, Monkton, Maryland, and contain some herbs.

the minimum flow rate. (If you're using 1/2-gph emitters, this means you need to be irrigating 120 plants. That's a *lot* of container plants.) For more details, see my book *Drip Irrigation for Every Landscape and All Climates; Second Edition*. (2009, Metamorphic Press.)

Water

You'll need to figure out how to supply water to your roof or deck garden, but be careful, as you definitely want to avoid leaks. If you can, it's best to install a faucet, as using sprinkling cans for daily watering can be tedious and the cans heavy to carry. A rooftop garden is the perfect place to install an automated drip-irrigation system, using the same guidelines as described for irrigating containers on the ground. In addition, you *must* have a backflow preventer on every hose bib or faucet, so you don't siphon fertilizer or compost into your home's drinking supply. A simple brass backflow preventer will cost you less than 12 dollars.

Leaks

A critical aspect of rooftop gardening is preventing water leaking through the roof to the floors below. See Figure 211 for how to layer materials for wate-proofing. I recommend that you include a very thick rubberized sheeting of EPDM (ethylene propylene diene terpolymer), a material made for roofing. It's also fire-resistant, almost impossible to ignite, and can actually impede the progress of a fire. In addition, it's *very* hard to puncture. (I once saw some 12-inch rocks roll 20 feet down a slope and hit an EPDM lining, without resulting in a single puncture.)

Wind

Most roofs are exposed to wind at least some of the time. Create windbreaks as needed with walls and trellises. If you live in an especially windy location, you may not be able to grow hanging plants or raise trees, even small ones, on your roof or deck.

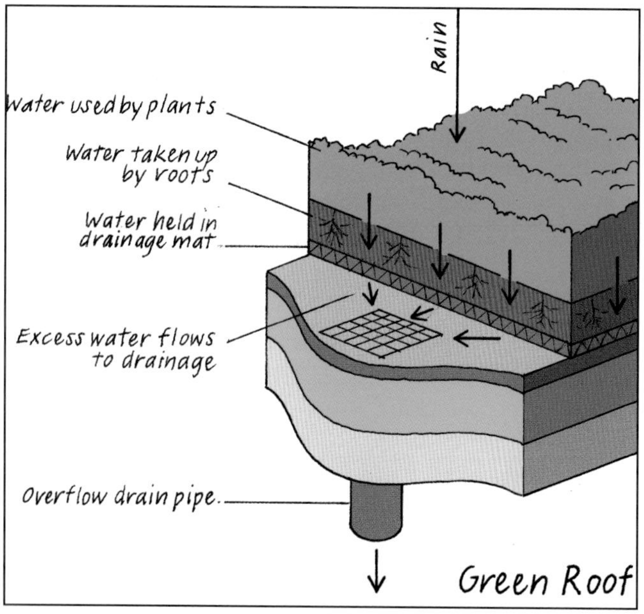

Figure 211. This is a diagram of all the layers of a waterproof roof garden; the drainage mat is *very* important. Unless you have previous experience with this detailed layering of elements, it should probably be done by a professional contractor.

Ornamental Plants

When selecting plants for a roof garden, use smaller species or selections. Consider a shrub known as 'Mock Orange.' (It's not at all related to citrus; the blossoms just happen to have a similar fragrance.) Its botanical name is *Pittosporum tobira*, and it can grow up to be 33 feet tall and 10 feet wide. There's a smaller variety, called *Pittosporum tobira 'Compacta'*, that grows10 feet tall and wide, and another smaller selection comes in a compact form called *Pittosporum tobira* 'Nanum' or 'Nana', a compact plant that grows only two feet high and wide. Yet another selection is *Pittosporum tobira* 'Wheeler's Dwarf'—three feet tall by five feet wide. So, you do have choices. (There's even a variegated selection of this plant two to two-and-one-half feet tall and wide—the Cream de Mint™ Dwarf Pittosporum, also known as *Pittosporum tobira* 'Shima'.)

Planting a lot of bulbs, summer potting plants, and annual vegetables makes for a barren deck or roof in the winter. Be sure to add some low evergreen shrubbery for a bit of winter interest. If you live in

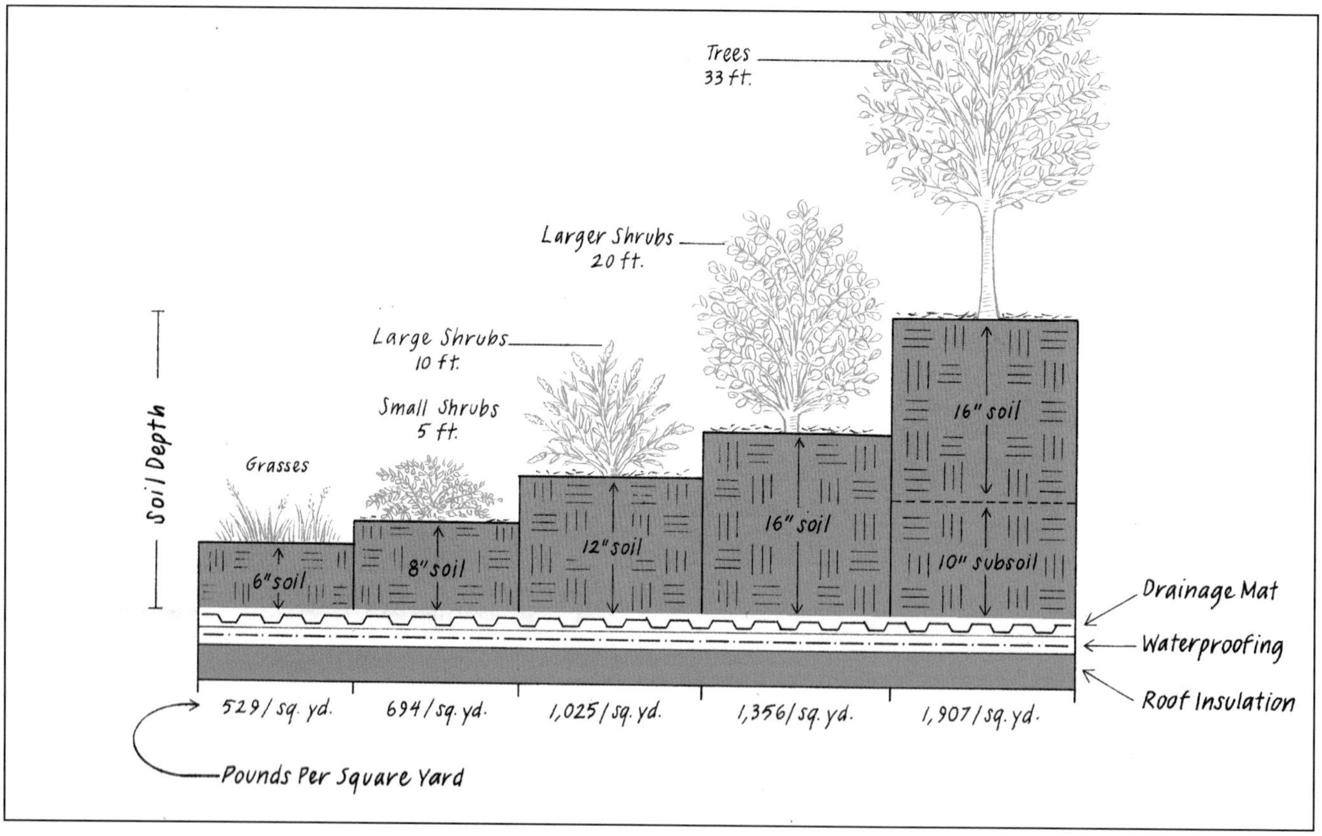

Figure 212. A healthy rooftop garden needs plenty of soil; this chart shows the depth required to keep different-sized plants happy. Please notice how heavy moist soil can be, even without people standing on it. Have a structural engineer inspect the supporting structure of your roof. This chart also shows the ideal depth of pots for various plants—as a rule, the wider they are, the better.

a climate with winter snow, consider growing only deciduous trees to keep piled snow from breaking limbs. "Weeping" trees, such as ornamental weeping mulberries (not very edible) and Japanese ornamental weeping cherries present less of a profile to wind.

If you'll be looking out of a window at your rooftop garden with any frequency, be sure to scale up the pots from smallest, near the window, to larger ones in the back. Consider keeping the larger pots near the edges or railings of the garden, so there's more support for the extra weight.

Herbs

Fresh herbs cost a *lot* of money from the grocery store, but they're the most cost-effective plant for containers. For example, rosemary (mostly in USDA Zones 8-10.) is very well adapted to container culture in terracotta pots, window boxes, wooden troughs, and tubs. Rosemary is a plant that requires good drainage, so use a potting soil with plenty of vermiculite or perlite. Container-grown rosemary can flourish on a sunny south-facing kitchen window for easy harvest and a pungently sweet aroma.

Vertical gardens are becoming popular, but can be expensive. As a shortcut, you can use a cloth shoe rack to grow plants such as perennial herbs. This usually requires frequent watering by hand.

Salad Greens

As your next starter crop, you can grow mixed salad greens. Even though it's sold all year round, a mix of several leafy greens is sold as "Spring Mix" (also called mesclun) in my area—go figure. To grow it, you can use a "cut-and-come-again" gardening technique that's more productive than growing individual heads of lettuce, and is a great way to harvest food from small containers. Here a slight amount of sweat will save you a healthy amount of money.

If you plant your pots full of just one kind of seedlings, repetitive harvesting will be convenient, relaxed and productive. If you plant all the salad varieties, including lettuces, together, you'll only be able to harvest everything a few times at best. By separating the varieties into separate pots, you'll be able to harvest most leafy green ingredients six to eight times as they re-grow—just cut off the foliage two to three inches above the soil surface. Some of the greens that can be harvested this frequently include: arugula (*Eruca vesicaria*), 'Mizuna' (*Brassica japonica*), red Russian kale (*Brassica oleracea*, Acephala Group), white mustard (*Brassica hirta*), and garden cress (*Lepidium sativum*). **In Figure 209 Lonnie Zamora tends a small productive salad garden many stories above the ground.**

It doesn't take a very big pot or trough to grow enough plants for frequent salad harvesting. Containers are also a great way to foil gophers, creeping weeds, voles and other critters, including—with a copper-foil band around the pot circumference—snails.

Vegetables

The quick-and-dirtiest way to grow plants like tomatoes on a deck or driveway is to buy a sack of potting soil or compost, lay it flat on its widest side, slit it open, and plant it with tomato or pepper seedlings. I've seen this done many times in greenhouses in England. You can also experiment with planting other vegetables like broccoli.

Got an old rotten wheelbarrow? Use it for growing small veggies, fruits like strawberries, and flowers

Figure 213. These peppers are growing in a greenhouse at the RHS Wisley Garden in Woking, UK, in sacks of potting soil made just for this process. You can also use sacks of compost inoculated with mycorrhizae.

and move it around to follow the sun.

Figure 214. Jan Mettler's old wheelbarrow makes a great portable planter to follow the sun. (Cat not required.)

Figure 215. This fig tree in Sheila Larsen's garden was purchased at Thomas Jefferson's home in Monticello, and was propagated from the 'Brown Turkey' fig trees grown there. The 20-year-old tree produces 24 or more figs per year, and Sheila repots it every few years. She gardens along with her husband and two sons in Atlanta, Georgia, in Zone 7, and you can see her excellent blog at: redheadgarden.com.

Containers for Shade-Tolerant Vegetables

Another very nice advantage to containers is that smaller pots located on small decks, ledges, or patios can be moved from shady to sunny spots as the season progresses. See Figure 217 for ideas about how much sun leafy greens need. You'll be surprised to learn that many vegetables don't need eight hours of sun to be happy and productive. For example, the mesclun salad mix described above only needs two to three hours a day of direct sunlight to make a decent salad.

Perennial Food Crops

Some gardeners are now promoting perennial-food-crop diets for easier gardening, as the crops only have to be planted once. Examples of perennial foods include: Egyptian top-set onions, prickly-pear cactus, sea kale, bamboo shoots (grown *only* in containers, as they can be wicked invasive), bunching onions, radicchio, a perennial nasturtium called mashua (*Tropaeolum tuberosum*) and society garlic. In mild climates, you can grow "tree kale" (*Brassica oleracea* var. *acephala,* USDA Zones 7-10), also called "tree collards" and "walking-stick kale," as a semi-perennial, as it survives temperatures down to 18F°.

Figure 216. This espalier by the Natural Gardening Company is a hard kind to grow, as the limbs are not at 45° angles. (See Figure 223.)

Shade Tolerant Veggies

2 Hours - Asian Greens, Mesculun Mix

3 Hours - Culinary Herbs, Mustard Greens

3-4 Hours - Arugula, Kale, Lettuce, Mizuna, Spinach, Swiss Chard

4-5 Hours - Beans, Beets, Carrots, Celery, Collards, Pak Choy, Peas, Radishes, Rutabagas

5 Hours - Broccoli, Brussels Sprouts, Cabbage, Cauliflower

Figure 217. Some examples of half-shade edibles are: Alpine strawberry, day lily, huckleberry, American persimmon, pawpaw, rhubarb, mallow, peppermint, currant, thimbleberry, and French sorrel.

Urtica urens.

Figure 218. Here's *Urtica urens*, the annual form of stinging nettle (*U. dioica* is the perennial form). Both are *very* nutritious, richly flavored, and need only to be steamed to eat. Both tolerate full shade. Wear gloves to harvest them!

You can gather the young leaves from their long spindly stem—which can grow four to 12 feet tall—for up to four or five years.

Cautious people living in a city should wash all produce.

Fruit Trees

One of the beauties of container gardening is that the pots can be equipped with stands on wheels and moved around with the seasons. By doing this, many people get away with growing citrus and figs in areas with freezing temperatures. Although it will be heavy, a large pot can support special, dwarf varieties of these trees. Use a fertile, loamy potting soil, not garden soil.

Deciduous trees, once they've lost their leaves, can be moved to any location that doesn't freeze. For citrus trees, it's best to find a sunny spot indoors, or, in moderate climates, under an eave. There are many varieties of citrus that can be grown on dwarfing rootstock. A common one with lots of fruit is the type 'Meyer Lemon'. If you're moving pots that aren't equipped with wheels, you'll need a good strong dolly to make it easy on your back.

Fig trees are a bit tricky. If you want to grow them, look for the semi-dwarf varieties 'Miss Little Figgy' or 'Black Jack'. These can be kept very small with radical pruning and sacrificing the mid-summer fruit. You'll only get fruit in the second crop in the very late summer or early fall. If fall comes too soon to ripen the late crop, you must harvest what is called the summer "breba" crop. A **breba** is a fig that develops in the spring on previous shoot growth. The main fig crop develops on the current shoot growth and ripens in late summer or fall. Leave some branches unpruned each year to favor the breba crop.

There is a relatively new type of fruit tree called a "columnar" tree. These have no branches, and fruit only along the trunk; as a result of this trait, they can fit into very small spaces. The choice of columnar trees currently consists of a few kinds of apple trees and one variety of peach ('Crimson Rocket'). Your choice of apple trees includes:

- 'Northpole' will remind you of a McIntosh apple.
- 'Golden Sentinel' is like a Golden Delicious apple in flavor.

Figure 219. To grow a columnar apple in a pot, use one that's 24 inches wide and deep. Smaller containers also work if you fertilize often. Water daily with tiny amounts of water.

Figure 220. The 'Tasty Red'tm Urban® columnar apple tree. Sometimes a side branch or shoot will show up; simply prune it off. Developed by Dr. Jaroslav Tupy of the Czech Republic. Needs a pollinator apple tree. (Photo courtesy of James Owen.)

Figure 221. This dwarf Mandarin orange grows in a small pot. (Notice, no "hips") For best results with citrus, be sure to keep the soil pH slightly acidic. (Photo courtesy of Four Winds Growers, CA.)

- 'Scarlet Sentinel' produces green-yellow apples with a red blush.
- 'Tasty Red'tm Urban® is a bright-red sweet and juicy apple that ripens mid-September.
- 'Golden Treat'tm is a beautiful golden apple with a sweet delicious flavor.
- 'Blushing Delight'® has a sweet flavor and a pretty red blush over a green background. It ripens in mid-September.

(The tm symbol means the tree is registered by the breeder and can't be produced without paying a fee. These tm trees are from the "Urban® Apples" series developed in the Czech Republic.)

Columnar trees, which are hardy in USDA Zones 4-9, can be as grown as close as two feet apart, and can be kept under a height of eight feet. They'll need regular watering and an occasional boost of fertilizer, and be sure to plant at least two different trees for good pollination. (Two good sources of columnar and EMLA-27 trees are Raintree Nursery and Stark Brothers Nursery.) Frankly, the flavor of some columnar apples is not yet as good as that of the best of heirloom apple varieties, but they're pretty much your only choice for very small decks, roofs, patios, and the like.

You can also plant dwarf fruit trees in bigger pots.

If you have space for foliage up to four to five feet across, you can use a dwarfing apple rootstock called EMLA 27 (only 20-30 percent of the height of a normal apple tree), and graft on any tasty heirloom-apple fruit stock. This rootstock will allow you to produce a "mini-dwarf" tree that will grow up to five feet tall and wide. To encourage low, wide branching, be sure to cut the trunk back to about one foot tall when planting.

If your rooftop has a space that's at least six to eight feet wide, you can plant one or more espaliers. Buy only trees that have already been trained as espalier

Figure 223. Seen at the United Nations Plaza in NYC, an espaliered tree with each limb trained to its most productive angle (45^0) for good fruiting with less pruning. This is similar to an oblique-cordon tree, and each limb here could be a different variety of apple. This can also be done with pears an Asian pears.

Figure 222. Early research on columnar apples at RHS Garden Wisley, in Woking, UK. This photo was taken in the spring of 1998.

plants, ideally with several varieties of fruit already grafted onto the tree. They're not cheap, and I only recommend espaliers if you're an experienced pruner, as they take very special care to keep their form and bear fruit at the same time.

To keep the cost down you could learn grafting so you can have more than one variety on each tree. Grafting isn't really that difficult, but it's a good idea to practice on standard-sized trees before attempting grafts onto a dwarf rootstock. The only type of espalier I recommend is called an "oblique cordon"—usually only done with apple or pear trees, the easiest kinds to graft. Another nice thing about espalier apples is you can graft a number of tasty heirloom varieties onto their dwarfing rootstock.

Spurring on Fruit

If you're determined to grow dwarf or espalier apples, here's another tip: apple trees are pruned or shaped in part to encourage fruiting areas called spurs. Once a flowering spur is formed on an apple tree, it may go on fruiting at the same place for 25 or more years. But what if you could nearly eliminate pruning for the sake of flower formation? Meet the spur-type tree, developed from rare trees that were seen to make more spurs (flowering buds) naturally with no regard as to pruning. The flower spurs grow closely together on the limb, making the trees even a bit more dwarfed and even more prolific. Look for the word "spur" or "spur-type" on the label. Examples include:

- 'Arkansas Black Spur'
- 'Cox's Orange Pippin Spur Type'
- 'Compspur Granny Smith'
- 'Compspur Rome Beauty'
- 'Empress Spur Golden Delicious'
- Spur 'Gala-go-Red'
- Spur 'Red Gala #42'
- 'Golden Delicious, Empress Spur'
- 'Granny Smith, Spur'
- 'Granny Smith, Starkspur'
- 'Iwamatsa Spur' ('Fuji' spur type)
- 'McIntosh, Compact'
- 'McIntosh, Compspur'
- 'Starkspur McIntosh'
- 'McIntosh, MacSpur'
- 'Oregon Spur Delicious'
- 'Red Delicious, Starkspur'
- 'Spur Red Rome Beauty'
- 'Spur Winter Banana'
- 'Spuree Rome Beauty'
- 'Starkspur Winesap'
- 'Winesap, Starkspur'
- 'Winter Banana, spur type'
- 'Yellow Delicious, Spur'

While dwarf fruit trees often start bearing in the second year, some *very* dwarf fruit trees, such as apples grafted onto a rootstock called M27 and M9, (originally from Malling Research Station in England) will fruit the first year. It's not instant gratification however, because ideally the fruit should be stripped off early for a year or two to allow the tree to grow, as fruiting can virtually stop these trees from growing new shoots. For the average gardener dwarf fruit trees in the garden or a container are a must—the care and harvesting of the shorter trees is so much easier than that required for a full-sized tree, and they still provide a reasonable harvest.

Figure 224. An Arkansas Black apple.

U.S. Department of Agriculture Pomological Watercolor Collection. Rare and Special Collections, National Agricultural Library, Beltsville, MD 20705

Bonus Photos:

Figure 225. Here's another way to use bagged potting soil or compost to grow plants such as strawberries. This was also taken in a greenhouse at the RHS Garden Wisley in the UK, but I suspect it would work outdoors as well.

Figure 226. A photograph (© 2016) by Annie Novak, from her great book: *The Rooftop Growing Guide: How to Transform Your Roof Into a Vegetable Garden or Farm* (Ten Speed Press, 2016). This is the Eagle Street Rooftop Farm in Greenpoint, Brooklyn, New York, "... a garden grown out of the astonishing generosity of Tony Argento and his sister Gina Argento."

Figure 227. Elvin Bishop (see pages 35-37) built a two-foot-tall raised bed to grow Japanese mountain yams. The long tubers (top photograph) are difficult to harvest, so Bishop constructed a box with screwed sides, which he can unfasten to expose the soil and harvest the yams. This also works with burdock (gobo in Japanese), daikon radishes, and other large root crops.

Figure 228. In the top photo above, a stone path going past fruit trees (nectarines, apples, and pears) surrounded by a perennial herb garden. Below is a view across a deck toward blueberry and raspberry bushes and a stand of fruit trees—on the 35th floor of a building! Designed by Mark K. Morrison Landscape Architecture, PC, NYC.

Figure 229. When he needed to repair a dilapidated roof on his studio and gallery (LaHaye Studio of Sonoma), sculptor Jim Callahan commissioned this rooftop garden, designed by SYMBIOS ECOTECTURE of Sebastopol, California. The 900-square-foot garden has a wide range of California native plants, as well as herbs and two citrus trees. A rainwater-catchment system captures rain from the roof, stores it in three 1000-gallon tanks, and pumps it up to the garden when needed.

Figure 230. The Callahan rooftop as it has matured. The citrus trees next to the umbrella are bigger and a huge rosemary in the herb section cascades thru the railing. Unlike most jobs, SYMBIOS doesn't do the maintenance—Jim gives the garden all its care. I like the design and I think this garden would also look good on the ground.

Figure 231. This shows the rooftop garden's drainage mat as it's covered with a lava-rock layer before the soil is added to sustain the herbs and lawn.

Figure 232. The lemon trees, herbs, and ornamental plants after some initial growth. The soil for the trees is 20 inches deep.

Figure 233. Ah, Ah, matés here a come a boat-load of 'matters. These gardeners can pluck a tomato for a sandwich while sitting at the nearby table. I didn't know whether-or-not to include this here, or in the eccentric section at the back of the book.

Figure 234. Here's a houseboat floating in the Seatlle area with a view of the Space Needle in the distance. Every square foot is covered with plants. This is the boat of horticulturist Bob Lily—a very approproate name!

Figure 235. Two topiary rosemary plants in contianers at the Missouri Botanic Garden. They are containerized so they can be wheeled into a greenhouse for the winter.

Chapter 8
Free Fertilizer

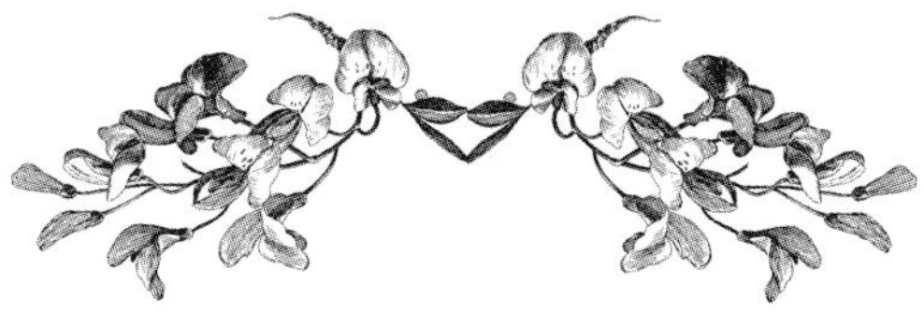

Dynamic Accumulators

I found my first explanation of so-called "dynamic accumulators" in 1978, while reading a book called *The Organic Method Primer*, by Bargyla and Gylver Rateaver, "Some plants" they wrote, "will, through root action, bring into solution and absorb more minerals than other plants." This was my inspiration for the dynamic-accumulator list that I included in my 1986 book *Designing and Maintaining Your Edible Landscape – Naturally*.

I eventually came to wish that I'd never written about dynamic accumulators; my 1986 list was mostly a compilation of unscientific information from six books:

1. *Weeds: Guardians of the Soil;* Joseph Cocannouer
2. *Practical Organic Gardening*; Ben Easy
3. *Stalking the Healthful Herbs*; Euell Gibbons
4. *Weeds as Indicators of Soil Conditions*; Stuart Hill
5. *Weeds and What They Tell*; Ehrenfried Pfeiffer
6. *The Organic Method Primer;* Bargyla and Gylver Rateaver

Very few of the above books included information on the actual content (in ppm) of the nutrients supposedly "accumulated" by some plants as compared to others. As gardening lore, it's as bad as the "oral history" of companion planting (which I had enough sense at the time to disregard). I listed the invasive weeds on my original list with the intent of showing which nutrients would be going into the compost with their foliage.

It's taken me years to admit that almost nothing in that chart is scientific. Even worse, there was never any information as to what the net gain (in ppm or milligrams per some unit) in the mentioned nutrient or micronutrient might be, or any documentation that proved that the plants in question would add anything significant to a compost pile. Unfortunately the chart, in whole or part, keeps popping up on the Internet and in other books like a weed (luckily for me, mostly without citation, so at least the reader doesn't know where the bum information came from).

During my 1986 tour to promote my new edible-landscape book, I gave a slide show that included a section on dynamic accumulators for compost piles. I was arrogant enough to recommend edible dynamic accumulator weeds to places like the Garden Club of the Arnold Arboretum (of Harvard University) in Boston. Sheesh! I wish I could take it all back. I was well intentioned, but I've learned a lot since. When I wrote: "Grow dynamic accumulators in or around your edible landscape to gather nutrients for use in the edible areas," I didn't intend to imply that gardeners should plant weeds for their dynamic-accumulation effect. I was just trying to show, in a poorly written

Legumes have distinctive blossoms like the ones pictured above. Proper use of legumes can provide some practically-free fertilizer.

sentence, that some weeds contributed "added value."

Here are some of the changes to consider applying to that original list: eliminate plants that accumulate copper, fluorine, boron, and cobalt, as soils are very rarely deficient in these micronutrients (and in small amounts, boron is toxic to some plants). Eliminate lemon balm (*Melissa officinalis*) because it is so invasive and should not be planted except in very contained circumstances. Mustards (*Brassica* spp.) should be removed, as there is a possibility of heavy metals accumulating in their roots from contaminated soils. However, if you don't have high levels of lead, cadmium, or zinc in your soil, mustards accumulate sulfur and phosphorus and do make a good addition to the compost pile. "Oak bark" is gone, because it's absurd that any gardener would "harvest" it. Same goes for "Willow bark." In fact, you should eliminate or control willow, as small pieces root so easily and it may become a weed. Nobody should grow tobacco, so it's off the list. Tansy spreads very rapidly, so it should be removed it from the chart. It turns out that mistletoe is parasitic (don't know why I overlooked this), and very invasive, so it sure didn't belong on the 1986 list.

When I visited Dr. James Duke at his USDA office in Beltsville, MD in 1986, I realized that I had stumbled onto a database very useful to the determining the correct function of dynamic accumulators. At the time, Dr. Duke was the only herbalist in the USDA, and he was looking for natural compounds that might fight cancer. In the process, he analyzed thousands of plants and determined the content, in ppm, of many of the chemicals and compounds in their leaves, seeds, stems, and/or fruit. Those plants that were appropriate on my 1986 chart are displayed in **Bold** type.

What I've done, using Dr. Duke's Phytochemical and Ethnobotanical Databases, is to list the top ten plants (referring either to their leaves, or a with a notation that the whole plant or herb is concerned, but not roots, stems, or seeds individually) that accumulate each nutrient as per his research. The

POTASSIUM

Common Name	Botanical Name	low ppm	high ppm
Lettuce	*Lactuca sativa*	2,900	121,800
Endive	*Cichorium endivia*	2,915	96,000
Lambsquarter	*Chenopodium album*	87,100	87,100
Chinese Cabbage	*Brassica pekinensis*	74,100	81,900
Purslane	*Portulaca oleracea* (herb)	4,880	81,200
Pigweed	*Amaranthus spp.*	2,630	73,503
Spinach	*Spinacia oleracea* (plant)	2,060	69,077
Borage	*Borago officinalis*	4,700	67,210
Watercress	*Nasturtium officinale*	3,300	66,000
Garden Beet	*Beta vulgaris*	4,380	65,950

MAGNESIUM

Common Name	Botanical Name	low ppm	high ppm
Purslane	*Portulaca oleracea* (plant)	670	18,700
Oats	*Avena sativa*	300	14,800
Spinach	*Spinacia oleracea* (plant)	420	11,000
Wild Cherry	*Prunus serotina*	435	9,600
Lettuce	*Lactuca sativa*	110	8,700
Stinging Nettle	*Urtica dioica*	860	8,600
Cilantro	*Coriandrum sativum*	6,940	7,488
Borage	*Borago officinalis*	520	7,436
Chives	*Allium schoenoprasum*	550	6,875
Pigweed	*Amaranthus* spp.	550	6,700

Figure 236. Purslane is a common "weed."When you throw it into the compost pile, you get more potassium.

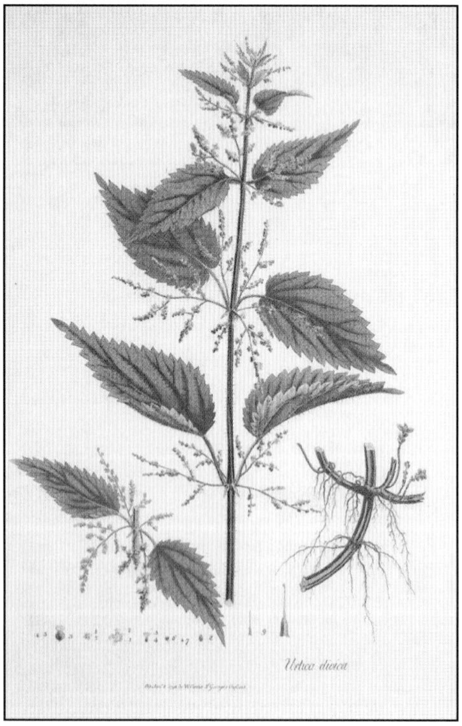

Figure 237. Stinging nettle is a power house. It shows up on many dynamic accumulator lists. See Figures 238, 240, and 242.

SILICON			
Common Name	**Botanical Name**	**low ppm**	**high ppm**
Stinging Nettle	*Urtica dioica*	117	33,330
Parsley	*Petroselinum crispum*	60	1,425
Spinach	*Spinacia oleracea*	1	855
Lettuce	*Lactuca sativa*	10	800
Garden Dill	*A. graveolens* (plant)	30	700
Horsetail	*Equisetum arvense*	386	386
Eyebright	*Euphrasia officinalis*	303	303
Thyme	*Thymus vulgaris*	202	202
Oats	*Avena sativa* (plant)	183	183
European Pennyroyal	*Mentha pulegium*	182	182

Comfrey shows up as SILICON in **root** at:	35	35
Comfrey shows up as SILICIC-ACID in **leaf** at:	40,000	40,000
Horsetail Equisetum arvense shows up as a plant SILICIC-ACID		80,000

silicic-acid (leaf) 40,000 ppm- (Roots is only 35ppm silicon) *Symphytum* is number three on the top plants list for silicic-acid.)

PHOSPHOROUS			
Common Name	**Botanical Name**	**low ppm**	**high ppm**
Malanga	*Xanthosoma sagittifolium*	610	38,416
Lambsquarter	*Chenopodium album*	720	36,833
Bitter Melon	*Momordica charantia*	540	33,467
Horsetail	*Equisetum arvense*	930	14,762
Luffa	*Luffa aegyptiaca*	330	14,141
Sesame	*Sesamum indicum*	14,000	14,000
Lettuce	*Lactuca sativa*	108	13,920
Swamp Cabbage	*Ipomoea aquatica*	420	12,360
Watercress	*Nasturtium officinale*	600	12,000
Pigweed	*Amaranthus* spp.	487	10,082

Figure 238. The leaves of comfrey have a tremendous amount of silicon, more than the roots. Horsetails have less silicon than stinging nettle.

IRON			
Common Name	**Botanical Name**	**low ppm**	**high ppm**
Chickweed	*Stellaria media* (plant)	2530	2530
Mullien	*Verbascum thapsus*	2360	2360
European Pennyroyal	*Mentha pulegium* (plant)	2310	2310
Butterbur	*Petasites japonicus*	2000	2100
Stinging Nettle	*Urtica dioica*	44	2000
Spiny pigweed	*Amaranthus spinosus*	22	1965
Black Tupelo	*Nyssa sylvatica*	8	1820
Pigweed	*Amaranthus spp.*	23	1527
Garden Thyme	*Thymus vulgaris*	1075	1508
Tea	*Camellia sinensis*	189	1500

CALCIUM			
Common Name	**Botanical Name**	**low ppm**	**high ppm**
Buffalo Gourd	*Cucurbita foetidissima*		77,600
Tomato	*Lycopersicon esculentum*	60,800	60,800
Luffa	*Luffa aegyptiaca*	560	55,000
Spiny Pigweed	*Amaranthus spinosus*	4,760	53,335
Pigweed	*Amaranthus* spp.	1,559	53,333
Stinging Nettle	*Urtica dioica*	5,940	41,900
Lambsquarter	*Chenopodium album*	3,090	33,800
Wild Cherry	*Prunus serotina*	4,060	32,640
Anise-Scented Basil	*Ocimum tenuiflorum*	31,500	31,500
Calendula	*Calendula officinalis*	30,400	30,400

Figure 240. Here we find stinging nettle again in this case, for both iron and calcium. It's easy to grow but hard to get rid of. Best gathered from the wild.

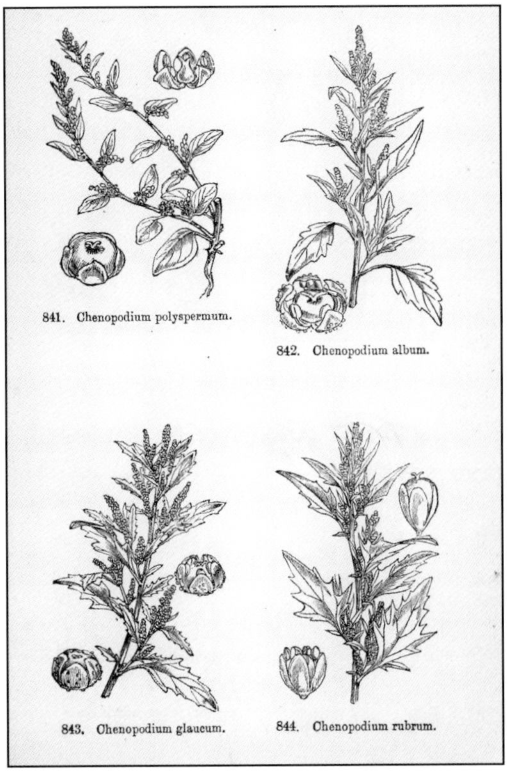

Figure 239. The most common lambsquaters in my garden is *Chenopodium album*. It accumulates calcium, phosphorus, and potassium. Also see Figure 242.

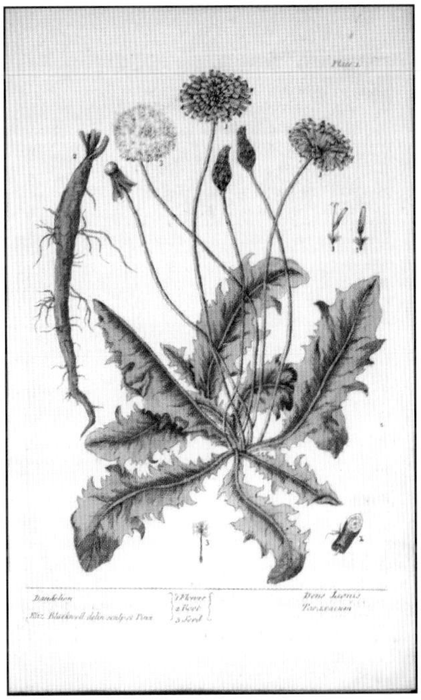

Figure 241. Dandelions have deep roots, but don't show up on any of Dr. Duke's lists. This may dispel the myth that deep-rooted plants are dynamic accumulators.

low parts-per-million (ppm) usually indicates the fresh plant with its moisture. The high ppm is what Dr. Duke calculated after the foliage had been dried. Each plant contains a different level of moisture. Whatever the actual ppm, the charts show the relative ranking of the plants listed. Not all entries in the 1986 chart were entirely wrong.

Weedy Accumulators

Agri-Dynamics of Martins Creek, PA, in the process of trying to find an alternative fodder to alfalfa, tested many weeds and obtained accurate measurements of nutrient content—not just guesses. (Note that the content for comfrey of calcium, phosphate, and potassium is surpassed by at least

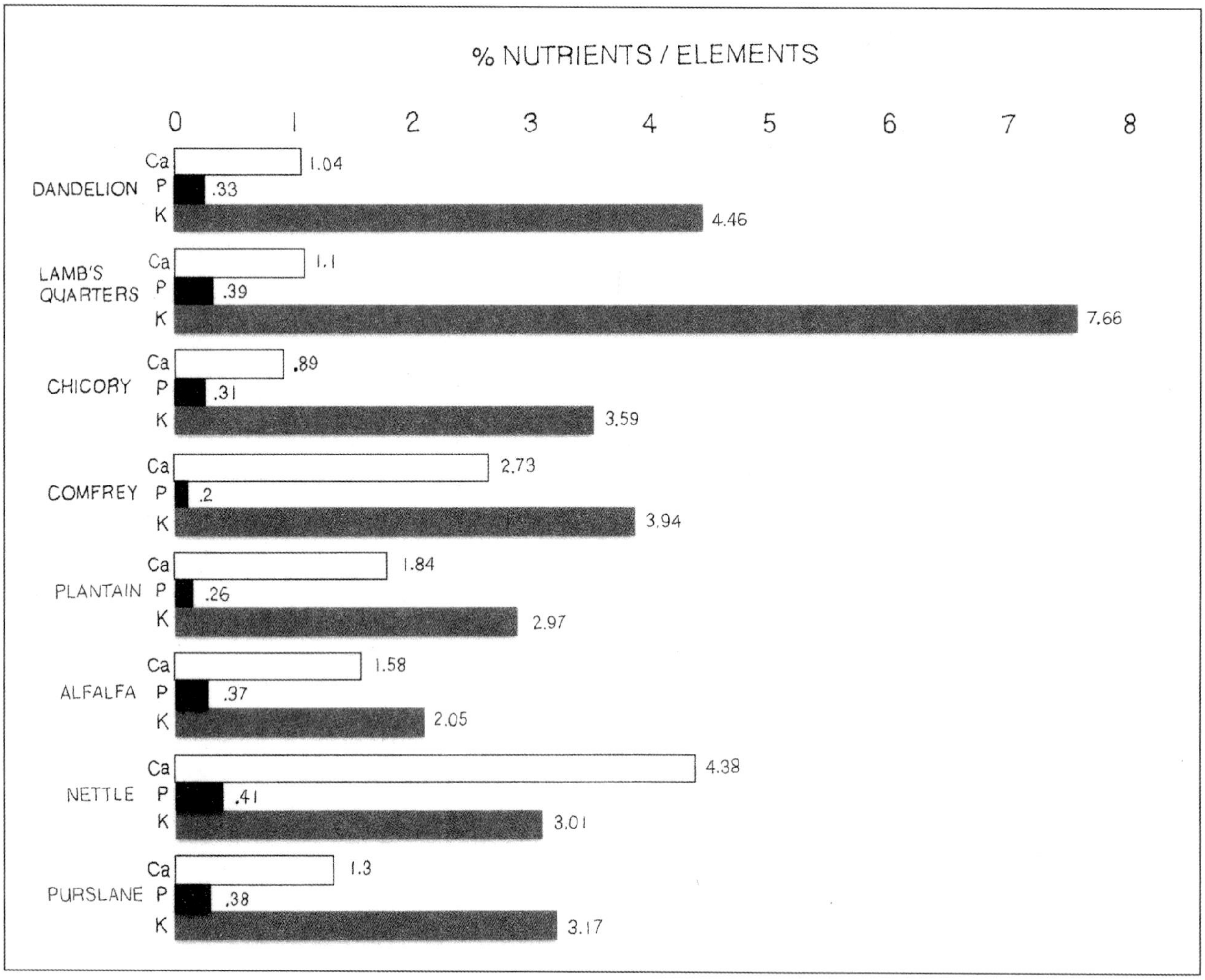

Figure 242. There is much to learn from this chart. This study was done to try and find an alternative to alfalfa as forage for cattle. The difficulty is that the plants that are superior to alfalfa are aslo classifiied as invasives. Otherwise the foremost observation is that comfrey doesn't gather the most calcium, phosphorus or poatsh (potassium) as some other "weeds." Stinging nettle gathers almost twice the amount of calcium as comfrey. While comfrey is relatively higher in its potash content, it is still lower that dandelions and lamb's quarters. With potash comfrey has nearly one-half as much potash as lamb's quarters. But lamb's quarters has less calcium than all examples and just a tiny bit more than dandelions. Only dandelion and chicory have tap-roots like alfalfa. The other "dynamic accumulators" have lateral roots in the top 12 inches or thereabouts of soil. Don't plant these invasive plants. Just know that adding them to a compost pile after weeding is helpful.

one other weed.) Figure 242 is a welcome addition to Dr. Duke's real data on dynamic accumulators. Broadleaf lupine can persist in low-fertility soils because of its ability to fix nitrogen, and may very slowly increase soil fertility. On sites in the Olympic Mountains, the soils directly surrounding the broadleaf lupine plants contained twice the nitrogen, more organic matter, and more phosphorus than did adjacent soils

Nitrogen Accumulators

The references to all the nitrogen accumulators in my 1986 chart are probably accurate. At least there's plenty of non-questionable literature on nitrogen. (See pages 247 - 251.)

Allelopathy and Hyperaccumulation

Compared to its lack of research on dynamic accumulators, science has made great strides in allelopathy, hyperaccumulation, and phytoremediation since my edible-landscape book was published in 1986. These last two areas of research are much more important than my old dynamic-accumulators list. The science of allelopathy, (which, according to the Oxford Dictionary, is "the chemical inhibition of one plant or organism by another due to the release into the environment of substances acting as germination or growth inhibitors") was not well researched in 1986, so I choose not to include it in my book. The terms "hyperaccumulation," which indicates that certain plants can absorb and concentrate extremely high levels of metals in their tissues, and "phytoremediation"— the treatment of pollutants or waste in contaminated soil or groundwater by the use of green plants—weren't even "invented" then. (See pages 310 - 314 in Chapter 9.)

This book focuses on growing stuff in suburbia, since that's where most of us live, and it's much more difficult to be sustainable in a small yard than on a farm or large property. So, let's see how we can manage fertility.

Soil Testing

Not knowing the nitrogen, phosphorus, and potassium, pH, and organic matter of your soil means you're just guessing, so you *must* do a soil test if you don't want to waste unneeded amounts of fertilizers or compost. (In my area, for instance, there are many soils that don't need any extra phosphorus.) Collect a soil sample as instructed by the soil-testing lab you use, and be sure to tell them if you want results and recommendations for particular crops like vegetables or fruit trees.

Make certain that you choose a lab that can both interpret the test *and* tell you how many pounds of *organic* fertilizers to add in order to balance each soil element, if needed. I use Harmony Farm Supply and Nursery (PO Box 460, Graton, CA, 95444 www.harmonyfarm.com) which does mail-order soil tests. You can also use the through-the-mail test from Peaceful Valley Farm Supply (www.groworganic.com; 888-784-1722). Many state Cooperative Farm Extention will do less expensive testing but will not interpret the results. If your garden or landscape is large, be sure to do at least two tests in different areas or in places where soils are noticeably different.

Organic, Sustainable Fertilizers?

Nitrogen and Corn

Gardeners often "import" fertilizers without thinking of their full environmental impact. Consider the following examples involving the cultivation of corn.

Corn needs lots of nitrogen, and leguminous plants can provide it with a large part (or even all) of the N that it requires. In a sustainable garden, the ultimate goal is to eliminate all imported nutrients. Horse manure, cow manure, sacks of bone meal, blood meal, green sand, bat guano, colloidal phosphate, etc., all add additional fertility, and all qualify as "natural," but each comes with various environmental costs attached, such as mining, transportation, energy use, and wasted bulk.

As an example of the energy invested in nitrogen for corn, consider how blood meal is produced. According to the Food and Agriculture Organization's regulations, blood is introduced into a (processing) tank as a coagulated mass previously obtained by a steam-action process. Ideally, as much liquid as possible is squeezed from the coagulum. Heating is initiated at 82°C (180°F), progressively raised to 94°C (200°F) for about three hours, then elevated to

SOIL ANALYSIS COMMENTS & RECOMMENDATIONS

HARMONY
FARM SUPPLY & NURSERY
3244 Hwy 116 N, Sebastopol
707/823-9125 fax 707/823-5586

TEST TYPE:	[x] Basic [ ] Complete [ ] Special
PREPARED BY:	*Kenny Phillips* DATE: 1/14/20
CLIENT NAME:	*Robert Kourik*
LAB #:	*56353*
Add:	[x] per 100 sqft [ ] tons per acre

NOTE: "s" = slightly; m = "moderately"; "v" = very

Organic Matter Level: Optimum (3 - 5%) ***4.5%***	[ ] Low	*Add compost: 1 cubic ft/100 sqft for each % increase.*		
	[x] Adequate	*Continue applying maintenance amounts or cover cropping.*		
	[ ] High	*Stop adding organic matter.*		
Est. Nitrogen Release:	[ ] Low	[x] Adequate	[ ] High	(from organic matter)
Phosphorus (P) ***20*** *ppm* / *ppm*	[ ] Low	[x] Adequate	[ ] High	Optimum 20 - 50 ppm (WB) or 10 - 25 ppm (OM)
Important for reproductive processes and the development of DNA, but high levels can inhibit uptake of Zinc and Iron. Watch for symptoms of deficiencies.				
Potassium (K): ***153*** *ppm*	[ ] Low	[x] Adequate	[ ] High	Optimum 150 - 250 ppm
Potassium is key to the regulation of stomata and the uptake of nutrients. Amend by adding Sulfate of Potash at a rate of 1 lb per 100 sqft.				
Magnesium (Mg): ***351*** *ppm*	[ ] Low	[ ] Adequate	[v] High	Optimum 100 - 250 ppm
Important for sugar synthesis, but high levels can adversely affect availabilty of Calicum and/or Potassium in additon to affecting soil structure.				
Calcium (Ca): ***1399*** *ppm*	[x] Low	[ ] Adequate	[ ] High	Optimum level is relative.
Low relative to Potassium and Magnesium. Amend with Gypsum at the rate of 5 lbs per 100 sqft.				
Sodium (Na): ***15*** *ppm*	[x] Low	[ ] Moderate	[ ] High	Optimum < 80 ppm
Low is okay. High levels can adversely affect plants ability to uptake water.				
Soil pH: ***6.7***	[s] Acidic	[ ] Alkaline	[ ] Neutral	Optimum 6.2 - 6.7 pH
Soil pH is in the optimum range.				
Hydrogen (H): ***0.5***	[ ] Low	[x] Adequate	[ ] High	Optimum < 3.0

Figure 243. This is another soil test. It makes me jealous: as it represents a very decent soil. Because the organic matter is 4.5 percent, *no extra compost* is required. Phosphorus is just at the adequtae level and is not needed as a fertilizer. Only calcium is shy of ideal and requires the addition of five pounds of gypsum per 100 square feet.

100°C (212°F) for seven hours. (That's a LOT of energy.) Drying is complete when the final moisture level in the dried product reaches about 12 percent.

According to Todd M. Kananen, P.E., Corporate Manager of Utilities, ConAgra Beef Co., in *Where's the Beef? ConAgra's Approach to Energy Sources & Efficiency in Meat Processing*, "Blood dryers and air make-up units will have a firing rate of 4.5 to 30 million BTUs per hour. Four and one-half million BTUs equals 1,318,825 watt-hours. Thirty million equals 8,792,167 watt-hours." Compare that to your lightbulbs!

No matter what your source of imported nitrogen, whether for soil preparation or as a summer application for growth, it's most effective to spread it relatively far from the cornstalk itself in order to feed the massive width of the corn-root system most efficiently. One method of doing this would be to fertilize *between* the rows rather than at the base of each plant or on the rows themselves. If you plant

Figure 244. This is a popular fertilizer for nitrogen to promote leafy growth. (NOT popular with vegans! :)

intensively, be sure to add plenty of nutrients for this hungry crop.

In the early 1930s Dr. John Weaver of the University of Nebraska, curious about just how roots grew, dug countless trenches and drew many meticulously measured images of root systems as they appeared underground. One such drawing was that of corn, pictured here as if looking down from above. (This root pattern depicts all the roots in the top six inches of soil.) Dr. Weaver inferred that corn roots absorb nitrates at all levels of the root system.

He cautioned, however, against adding manures to the "hill" (a practice in which corn is planted and a small mound is hoed up onto the roots near the stalk as it grows), as it promotes early growth but provides little benefit at the time of the formation of the corn ears. In his observation, those plants that were fertilized at the hill produced smaller root systems and were more susceptible to drought damage (this was in Nebraska and he didn't irrigate) and reduced yields. Weaver thus deduced that localized application of fertilizers also localizes roots and prevents them from naturally ramifying (exploring) a larger volume of soil.

Figure 245. Alfalfa is a great source of nitrogen due to nitrogen-fixing bacteria.

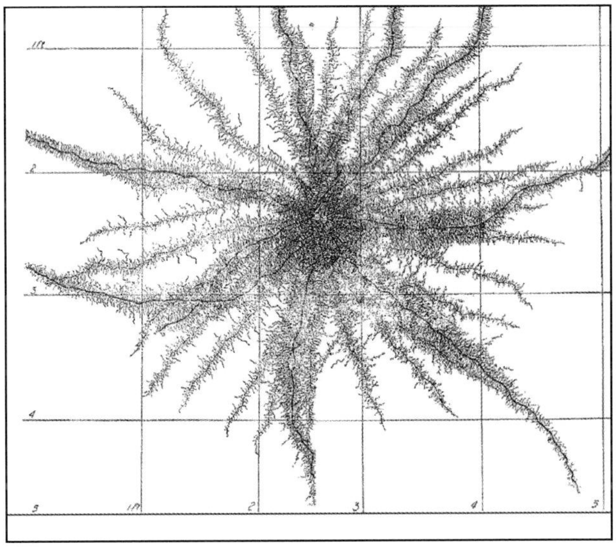

Figure 246. A view looking down on the top six inches of corn roots. As excavated by Dr. John Weaver.

Grow Your Own Nitrogen

Another way to feed corn is to grow a legume crop solely for the nitrogen that it gathers from the atmosphere and accumulates in nodules located on its roots. One of the few legumes that produce enough nitrogen for corn is alfalfa, at 250 pounds of N per acre. Tilling the alfalfa foliage under before planting corn, a practice called "green manuring," would increase the corn yield. This is another illustration of the importance of crop rotations in any vegetable garden. Don't grow corn in the same spot every year, as it will exhaust much of the nitrogen. Instead, alternate corn crops with green manures. See Figure 249 for more on using green-manure crops.

Here are some legumes to consider:

- **Alfalfa** (*Medicago sativa*): Has very deep roots—six or more feet deep. Don't let this perennial legume get established in your garden, as it can become a perennial "weed." It's best to till alfalfa under before it blooms so it can't set seed. If you want to keep it as a perennial, it's best left in pathways and mowed on a regular basis. Watch for spreading roots between the pathways and all planted areas, and cut them off with a straight-backed garden spade. (Alfalfa produces 70-225 pounds of nitrogen per acre, or about two ounces of nitrogen per 100 square feet.)

Figure 247. Fava beans makes a great winter cover crop in warm winter climates. Plant in Sepetember and irrigate until the winter rains begin.

Figure 248. This shows the nodules of beneficial nitrogen-fixing bacteria on a green bean. White nodules with pink centers are the healthiest.

- **Beans** mung, soy, and velvet (*Pisum* spp.): Good for warm soils. Plant late spring or early summer. (Soybeans produce 20-275 pounds of nitrogen per acre, four ounces of nitrogen per 100 square feet.)
- **Bur Clover** (*Medicago polymorpha*): A good plant to seed in the fall in warm-winter climates, and one of the best legumes for all mild climates. (Produces 100 pounds of nitrogen per acre, 1.5 ounces of nitrogen per 100 square feet.)
- **Crimson Clover** (*Trifolium incarnatum*): Will grow well in slightly acid soils. Not winter hardy north of New Jersey, but can be planted in the fall elsewhere. Can become naturalized and invasive. (Produces 50-200 pounds of nitrogen per acre, 1.5 ounces of nitrogen per 100 square feet.)
- **Cow Pea** (*Vigna unguiculata*): Withstands drought and will tolerate some shade. Must be planted after the soils have warmed up. (Produces 100 pounds of nitrogen per acre, four ounces of nitrogen per 100 square feet.)
- **Fava** (*Vicia faba*): Fall planting in warm winter zones. Plant one-quarter pound of seed for every 100 square feet, and be sure to coat the seed with two to three tablespoons of fava bean inoculant (*Rhizobium leguminosarum viciae*) If you

harvest a crop of beans for the table, you sacrifice much of the nitrogen in the soil. Cut the favas in early bloom and compost their tops, then till the stubble under. Yields approximately one-half pound of nitrogen for every 100 pounds of foliage turned under. (Produces 100-200 pounds of nitrogen per acre.)

- **Austrian Winter Peas** (*Pisum sativum* subsp. *arvense*): Spring sowing in cold climates; fall planting in warm-winter zones. (Produces 155-175 pounds of nitrogen per acre, three ounces of nitrogen per 100 square feet.)
- **Lupines** (*Lupinus* spp): Thrive on acid soil low in fertility. (Produces 90-120 pounds of nitrogen per acre, two to three ounces of nitrogen per 100 square feet.)
- **White Clover** (*Trifolium repens*): Very hardy. Fall plantings winter over. Don't let this one go to seed, as it will become invasive. Like alfalfa, it's a perennial and spreads. Best left in pathways and mowed on a regular basis. As with alfalfa, cut any spreading roots off with a straight-backed garden spade. (Produces 100-150 pounds of nitrogen per acre, two ounces of nitrogen per 100 square feet.)
- **Red Clover** (*Trifolium pratense*): Shade- and drought-tolerant. Needs well-drained soil. (Fixes about .27 pounds per 100 square feet or 200 pounds per acre, two ounces of nitrogen per 100 square feet.)
- **Vetch, Hairy** (*Vicia villosa*): Not winter-hardy in severe climates. Till under before seed forms or it will become a pest—it twines around anything it can find, including trellises, plants, and fences. (Produces 110 pounds of nitrogen per acre, 1.5 ounces of nitrogen per 100 square feet.)
- **Birdsfoot Trefoil** (*Lotus corniculata*): Slow-growing. Needs well-drained soil. (Produces 200 pounds of nitrogen per acre or 10-20 pounds per 100 square feet.)

Recommendations: broadcast peas, hairy vetch, wheat, oats, grain rye, and other large-seeded cover crops at the rate of one-quarter pound per 100 square feet. Plant small-seeded cover crops, such as buckwheat, mustard, and ryegrass at the rate of one-sixth of a pound per 100 square feet.

John Jeavons states that: "...Purple and Hairy Vetch each fix about 0.22 lbs. of nitrogen in the soil per 100 square feet...(Before planting, soak two ounces of each kind of vetch seed...), if they are harvested when the vetch is at 50% flower. (If you harvest after bloom, the fixed nitrogen has already been drawn up from the roots into the plant to produce the seed.)

On the other hand, the delightfully named **Woolly Pod Vetch** (*Vicia villosa* ssp. *dasycarpa*) has the capacity to fix up to 0.64 lbs. of nitrogen in the same area—almost three times as much as the other two vetches. While at first this would seem to be ideal, there are both advantages and some potential disadvantages to high-nitrogen levels. Insufficient nitrogen generally means less crop growth, but too much nitrogen usually means fast, weaker tissue growth and a lower dry biomass percentage at harvest."

One study by Greg Hoyt (Department of Soil Science, North Carolina State University) found that Austrian peas, hairy vetch, crimson clover, and subterranean clover yielded 119, 114, 101, and 86 pounds of N per acre respectively.

The difference in the amount of yields when the cover crop is either harvested or tilled in is fascinating, see Figure 249.

Nitrogen-fixing plants in temperate climates didn't evolve for the express purpose of sharing accumulated nitrogen directly with other plants. The ability of legumes to fix free nitrogen "out of thin air," in their root nodules is a competitive evolutionary development that allows leguminous plants to colonize nutrient-poor soils. Legumes are some of the first plants to inhabit disturbed soils such as landslides, cliffs, beaches, sand dunes, fire-scorched areas, and the compacted clay subsoil that's left after most houses are built.

Figure 249. This seven-year study shows that you lose five bushels of corn when you cut the clover cover crop and take it out of the field. In a garden, thjis means yields of corn could be 10 percent less when removing the clover. Growing no legumes in a rotation produces far less corn than the rotaion of clover-corn.
(From *Soils & Men*, USDA Yearbook, 1938.)

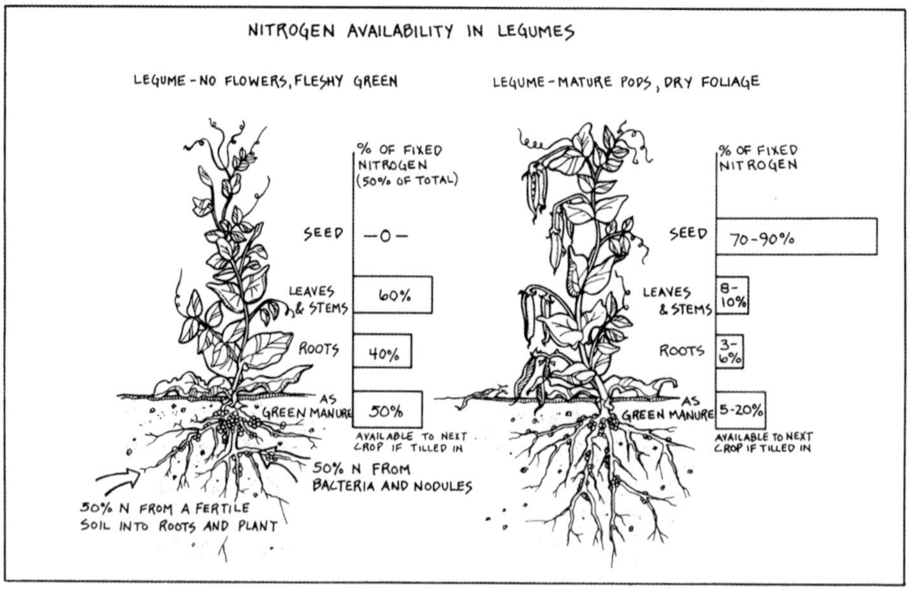

Figure 250. An illustration of the cycle of nitrogen during the life of a legume. After harvest of seed only 5-20% of the nitrogen is in the soil for the next crop.

The legumes use all the nitrogen that they accumulate to grow and make high-nitrogen seed. While the plant is growing, the nitrogen from the nodules on its roots is temporarily "banked," or stored, in the stems and leaves of the plant, providing plenty of nitrogen for rapid assimilation into the blossom once seed begins to form. Consequently, that green-bean plant growing up a cornstalk isn't about to share much nitrogen with the corn during the current growing season, as the mature beans will contain 70 to 90 percent of the entire plant's total nitrogen. (See Figure 250.)

After the bean crop is harvested, the roots and their nodules contain as little as three to six percent of the plant's total nitrogen—not much of a legacy or fertilizer portfolio for the "heirs." So, if you harvest a legume crop like beans, you'll be only getting a very tiny amount of the nitrogen back; because 1) the seed has taken up most of the nitrogen, 2) little nitrogen is left in the stems, roots, and leaves, and 3) the tops are too carbonaceous for easy decomposition. Instead, to harvest maximum nitrogen, you should cut the tops back as the bloom approaches about ten percent. (The Figure 250 is from page 265 of my book *Designing and Maintaining Your Edible Landscape—Naturally.*)

Surplus Nitrogen

Most gardeners overuse nitrogen fertilizer (even in the form of manures) at rates even higher than the worst farmers. Here's an example based on some very nitrogen-hungry crops—celery, cabbage and potatoes. Farmers often apply about 280 pounds

Figure 251. Cannabis growers love bat guano for the flower-stimulating phosphorus. Do NOT buy Jamacian guano as its harvest is killing the bats.

of actual (elemental) nitrogen to each *acre* of these plants in order to replace the nitrogen lost when the crop is harvested. This might seem like a lot of fertilizer, but it calculates out to only 0.006 pounds of elemental nitrogen *per square foot*.

Kate Burroughs, Certified Crop Advisor with the Society of Agronomy in Sebastopol, CA, estimates that most homemade compost contains 0.7%—or less—of actual/elemental nitrogen. It takes 0.0045 pounds of compost per square foot to equal the elemental needs of celery, cabbage, and potatoes. So, each 100 square feet of celery, cabbage, or potatoes actually requires 4.45 pounds of compost—not much at all, and far less than most gardeners would guess I think.

Kate Burroughs recommends raw steer or horse manure or compost (all about 0.7% nitrogen) spread only one-quarter to one-third inch deep over the soil to equal the nitrogen that farmers apply to replace the lost nitrogen (280 pounds per acre). And uses the same guidelines for homegrown sweet corn and lettuce. (Broccoli and pear trees need only 1/12th- to 1/9th-inch of compost to equal farmer applications.) Burroughs also commonly recommends such a "light dusting" for apple trees—"if the apple trees need any additional nitrogen at all." (See page 16 for more on this.) I often observe gardeners applying compost (and often as mulch) far deeper than this, wasting fertilizer, money, and nitrogen. Bottom line? Don't waste valuable compost by guessing. Take a soil sample and base the amount of additional compost on the test's results. More is not better.

Too much of a good thing can be counterproductive and wasteful. Excess nitrogen, whether it's over-applied in a synthetic or "organic" form, quickly leaves our gardens as a gas or leaches away toward fresh water supplies. Applied nitrogen can also set back the productive and effortless activity of vesicular-arbuscular mycorrhizae—beneficial fungi that pump phosphorus, nitrogen and other nutrients into the cells of roots. Surplus nitrogen has been shown to stimulate weaker, succulent growth that's more prone to pest and disease problems, and yields may be reduced due to the nitrogen-induced emphasis on growing foliage, rather than stems, tubers, or fruits.

Total Nitrogen Fixed (compared to)	**Horse Manure**	**20-0-0 Fertilizer**
Red alder (*Alnus rubra*) Fix 25-300 lb./acre/yr. (depending upon the study)	85-414 lbs.	5-18 lbs.
Mountain alder (*Alnusviridis* ssp. *crispa*) fix 55- 362 lbs./acre/yr. (depending upon the study)	185-499 lbs.	7-18.5 lbs.
Alders (*Alnus* spp.) leaf fall adds 55 lbs./acre/yr.	185 lbs.	7 lbs.
Mixed leguminous ann./peren. crops fix an average of 100-200 lbs./acre/yr.	337-674 lbs.	12.5-25 lbs.

Other examples include:

Redstem ceanothus (*Ceanothus sanguineus*)	21-45 pounds/acre/year
Chaparral whitethorn (*Ceanothus leucodermis*)	2 pounds/acre/year
Wedgeleaf ceanothus (*Ceanothus cuneatus* var. *fascicularis*)	54 pounds/acre/year
Mountain buckbrush (*Ceanothus greggii*)	27-53 pounds/acre/year
Deerbrush (*Ceanothus integerrimus*)	< 2.2 pounds/acre/year
Prostrate ceanothus (*Ceanothus prostratus*)	.28 pounds/acre/year
Snowbrush ceanothus (*Ceanothus velutinus*)	6-45 pounds/acre/year
Antlope bitterbrush (*Purshia tridentata*)	.18-.8 pounds/acre/year
Seaberry (*Hippophae rhamnoides*)	62 pounds/acre/year
Sitka alder (*Alnus viridis* ssp. *sinuata*)	1.8 pounds/acre/year
Honey mesquite (*Prosopis glandulosa*)	22-66 pounds/acre/year
Acacia (*Acacia* spp.)	46-59 pounds/acre/year
Australian pine tree (*Casuarina equisetifolia*)	52-180 pounds/acre/year
Black locust (*Robinia pseudoacacia*)	23-217 pounds/acre/year
Sweetgale (*Myrica gale*)	8 pounds/acre/year

(To get pounds-per-100 square feet from an acre, divide by 440. For example, the rate of 22 pounds-per-acre for honey mesquite means you would get 0.05 pounds of nitrogen for 100 square feet—one-half of one percent.)

Figure 252. Growing trees for nitrogen for adjacent crops is getting popular. But this chart shows few of the nitrogen-fixing trees and shrubs produce as much nitrogen as a cover crop like clovers (50-90 pounds per acre). Exceptions are, on the high range, seaberry, honey mesquite, acacia, black locust, and *Casuarina equisetifolia*.

Nitrogen Gathered by Legumes (depending upon the type of legumes and the study)

Common Name (*Botanical Name*)	**Pounds per acre**	**Vegetables with equal or less nitrogen**
Field peas (*Pisum arvensis*)	40	
Hairy vetch (*Vicia villosa*)	80	Broccoli
Dutch white clover (*Trifolium repens*)	100	
Soybeans (*Glycine max*)	100	Asparagus, lettuce, squash, and broccoli
Alsike clover (*Trifolium hybridum*)	140	
Red clover (*Trifolium pratense*)	140	
Sweet white clover (*Melitousa alba*)	150	sweet potatoes and all of the above
Alfalfa (*Medicago sativa*)	250	Corn, beans, tomatoes, cantaloupes and all the above

Figure 253. This chart shows that broccoli needs 80 pounds of nitrogen per acre. All the legumes except field peas will provide that much nitrogen. Also, it shows that only alfalfa as a green manure can provide the nitrogen needed by corn, beans, tomatoes, and and cantaloupes.

Purpose
BD - Bio Diversity
EC - Erosion Control
EN - Excessive Nutrients
NF - Nitrogen Fixation
OM - Organic Matter
SF - Supplemental Forage
SM - Soil Moisture Mgt.
WS - Weed or Pest Suppression

BD	EC	EN	NF	OM	SF	SM	WS	Cover Crop Species	Life Cycle	Nitrogen Value (lb/Ac)
								Legumes		
	x		x	x	x		S	Annual medic	SA	40 - 100
	x		x	x	x		x	Berseem clover	SA	60 - 90
	x		x	x	x			Crimson clover	SA	50 - 60
	x		x	x	x		x	Field peas	SA	30 - 100
	x		x	x				Hairy vetch	WA	60 - 180
x	x		x	x	x		x	Mammoth red clover	B	60 - 70
x		x	x	x	x		x	Sweetclover	B	70 - 90
x		x	x	x	x		x	Alfalfa	P	50 - 150
x	x		x		x			White clover	P	60 - 100
	x	x	x	x	x			Medium red clover	P	60 - 70
			x					Alsike clover	B/P	60 - 70
			x					Birdsfoot trefoil	P	40 - 100
x	x	x	x	x				60/40 mix (RC/SC)	B/P	60 - 90
			x		x			Soybeans	SA	0 - 40
x	x		x	x	x	x		Crownvetch	P	50 - 100
								Non-Legumes		
x		x						Buckwheat	SA	NA
	x	x		x	x	x	x	Corn	SA	NA
	x			x		x	x	Field Bromegrass	SA	NA
	x	x			x		x	Forage turnips	SA	NA
x	x				x	x		Oats	SA	NA
	x	x		x	x	x		Oilseed radish	SA	NA
		x					x	Rape	SA	NA
	x	x		x	x	x	x	Sudan Grass	SA	NA
	x	x		x	x	x	x	Annual ryegrass	AW	NA
	x			x	x			Barley	AW	NA
x	x	x		x	x	x		Cereal Rye	AW	NA
x	x	x		x		x	x	Triticale	AW	NA
x	x	x		x	x		x	Wheat	AW	NA
							x	**Re Forest Areas**		
								Redtop		NA
								Canada Wildrye		NA
								Virginia Wildrye		NA
								Riverbank Wildrye		NA

Figure 254. This chart shows the per-acre yields of legumes in pounds. It is also very informative as to the other benefits cover crops have. Bio diversity also means it attracts beneficial insects.

(Used with the permission of Dana Visalli, of the Methow Biodiversity Project, PO Box 175, Winthrop, WA 98862).

If you need one-half pound of nitrogen for a given area and are using blood meal (12-1-1), divide .5 (= one-half pound of nitrogen) by .12 (the percent of nitrogen in the product). This tells you that 4.2 pounds of blood meal are needed to supply one-half pound of actual nitrogen.

Non-Leguminous Nitrogen Fixers

There are special plants (examples in Figure 252) that are not legumes but also have the ability to fix nitrogen in nodules on their roots. *Frankia* is a genus of non-leguminous nitrogen-fixing soil actinomycetes (a specific group of bacteria that resemble fungi) that also have the ability to fix nitrogen on plants. While *rhizobia* bacteria are responsible for nitrogen fixation in legumes, *Frankia* are capable of infecting a wider range of host plants than most of the other rhizobial bacteria, and also of sharing nitrogen with adjacent plants.

You'd never till in any of the plants in the following list for green manure, as they're all very woody. Examples of temperate-zone plants with non-leguminous nitrogen-fixing *Frankia* bacteria include:

Alders(*Alnus* spp.)
Bayberry (*Myrica pensylvanica*)
Sweetgale (*Myrica gale*)
Sweet fern (*Comptonia peregrina*)
New Jersey tea (*Ceanothus americanus*)
California summer lilac (*Ceanothus* spp.)
Canada buffaloberry (*Shepherdia canadensis*)
Silver buffaloberry (*Shepherdia argentea*)
Round-leaf buffaloberry (*Shepherdia rotundifolia*)
Antelope bitterbrush (*Purshia tridentata*)
Mountain mahoganies (*Cercocarpus* spp.)
Australian pine (*Casurina* spp.) (Note: an invasive exotic.)
Autumn olive (*Elaeagnus umbellata*) (Note: an invasive exotic.)
Russian olive (*Elaeagnus angustifolia*) (Note: an invasive exotic.)

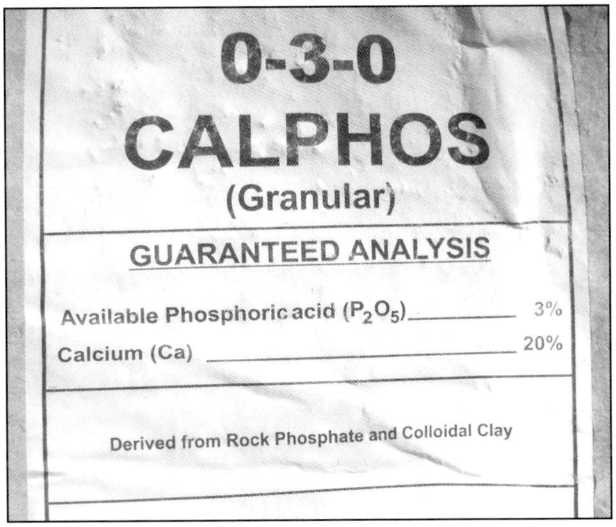

These plants must die, or be grazed, burned or cut down in order for other plants to "gather" a significant amount of nitrogen from them. I doubt if you have cattle, and it's unlikely you'll cut these perennial shrubs down or burn them, especially if you live in the suburbs. Therefore the nitrogen derived from them will be accumulated slowly by the gradual "shedding" of nodules into the soil, as some of the nodules naturally die and release their nitrogen to the roots of adjacent plants during the change of seasons. This process is slow but constant.
Here's an example (Figure 252): nitrogen accumulations of the non-leguminous alder trees can be measured and here compared to horse manure and nitrogen fertilizer.

For more examples of the nitrogen-per-acre amounts, plus more information, see Figure 254 (used with the permission of Dana Visalli, of the Methow Biodiversity Project (PO Box 175, Winthrop, WA 98862).

Bracken fern, another non-leguminous nitrogen accumulator that I listed in my 1986 dynamic-accumulator chart, should still be listed as an accumulator of nitrogen—at 65 to 116 pounds per acre per year. According to research by the Department of Forest Science at Oregon State University, "Bracken fern may have a competitive advantage by sequestering [accumulating] nitrogen to the point that reduced soil mineral nitrogen…limit(s) growth of competing vegetation."

Phosphorus & Corn

Purplish strips at the edges of corn leaves usually indicate a deficiency of phosphorus. Some organic gardeners use colloidal phosphate as the solution to this deficiency, but consider this: colloidal phosphate is often strip-mined in Florida, washed with water

FIG. 1. *Fagopyrum esculentum*. Illustration by Jean Henri Jaume St.-Hilaire (1828–1832).

Figure 255. Buckwheat (*Fagopyrum esculentum*) is tolerant of acid soils, and soluble nitrogen fertilizer will favor vegetative growth. Buckwheat responds well to cover crops, such as rye and hairy vetch used as green manures. Plant three ounces of seed per 100 square feet of seedbed soil. It's best to till the crop under when it's four to six inches tall, as it's easier than if you wait longer. Buckwheat is a fast grower, taking just a bit more than two months to mature. By planting green manures early, you can fit quite a few green-manure crops into one summer. (Although, since buckwheat doesn't like hot, dry weather; spring or fall might be best.)

(and Florida has a *big* problem with supplies of fresh water), loaded on train cars, and shipped to places as far away as California and Washington, where it's sacked up and shipped to your local garden-supply store.

In addition, the total amount of phosphorus (P_2O_5) in the sack is only 16 percent of all the bagged-up bulk, of which a mere two to three percent is available the first gardening season, since the phosphorus is locked up in a mineralized form that requires the activity of soil microbes, soil bacteria, and exudates. (As mentioned on pages __ to __; exudates aid in the nutrient-release process by releasing sugars, organic acids and other compounds to dissolve minerals into a soluble state and stimulate the soil's microbial action.) Unnecessary water use, exploitation of limited resources and wasted energy—all wrapped in a single bag. A carbon footprint as large as Cleveland.

As a side note: Florida mines 75 percent of the phosphorus used by American farmers and about 25 percent of the entire world production. Phosphate ore must be chemically processed with sulfuric acid. When sulfuric acid reacts with the phosphate it produces a slightly radioactive byproduct known as *phosphogypsum*. According to the Florida Department of Environmental Protection, there are a billion tons of phosphogypsum stacked across the state and 30 million tons are generated each year.

Also, while the mining produces radium, the EPA's limit for use is 10 picocuries of radium per gram, well below the levels usually found naturally in the mounds.

A recent statement from the EPA reads: "Only two uses (for the gypsum) are permitted: limited agricultural use and research. Other uses may be proposed, but otherwise the phosphogypsum must be returned to mines or stored in stacks."

While the presence of uranium-decay products makes gypsum a tough sell for the phosphate industry, the uranium has, at various times, presented the industry with a business opportunity of its own.

One of the lesser-known facts about the phosphate industry is that its processing facilities have produced and sold sizeable quantities of uranium.

In 1997, just two phosphate plants in Louisiana produced 950,000 pounds of commercial uranium, which amounted to roughly 16% of the domestically produced uranium in the US.

(George Glasser of the University of Florida states he has conducted extensive research on rock phosphate contaminants and claims there are *no* rock phosphate sources in the world that do not contain contaminants such as selenium, cadmium, fluoride, and radioactive nuclides.) Add to this the fact that there's a limited supply of easily mined colloidal phosphorus. It's much like oil: will we have enough in the future? Will we be able to find enough new supplies if the current mines are exhausted? Some say the U.S. supply will be gone by 2035. The world supply may be depleted in 50-100 years. (Seems everything goes bad in 50-100 years.)

I took a sample of "organic" colloidal phosphate to a lab that measures radioactivity. The clicks spiked up noticeably when the gauge was put near the phosphate; but not, according to the lab, enough to be a hazard.

Grow Your Own Phosphorus

From the 1938 USDA Yearbook, *Soils & Men*: "Phosphorus has sometimes been called the master key to agriculture. Its importance in general farming is indicated by the fact that low crop production is due more often to a lack of phosphorus than to the lack of any other element." So let's grow some phosphorus.

To grow your own phosphorus, use common grain buckwheat (Buckwheat straw: 0-2-0, 0% nitrogen, 2% phosphorus and 0% potassium) as a green manure. From Conservation Systems Research USDA-ARS-NSDL411 S., Donahue Dr., Auburn, we read: "[Buckwheat is] a scavenger of phosphorus and calcium. Grows better on low-fertility soils than cereal grains. Not drought-tolerant."

Another source says: "Buckwheat is known as a "phosphorus pump" because it is assumed the roots solubilize phosphorus from phosphate precipitates in the soil profile and subsoil. The following characteristics make it outstanding for its phosphorus efficiency:

"1) A finely divided root system with a high ratio of root surface to root or shoot length;
2) A high storage capacity for inorganic phosphorus;
3) An increased release of protons and solubilizing substances by phosphorus deficient plants;
4) A favorable ratio of phosphorus uptake to root mass growth, especially as a low phosphorus supply;
5) A high activity of acid phosphates in the rhizosphere and the capability to use phosphorus from organic sources." (Gardner and Boundy. 1983. "The Acquisition of Phosphorus by *Lupinus albus*;" *Plant and Soil*, Vol. 70, pp. 391-402.)

From the *Conservation Fact Sheet No. 04, 2008* (National Soil-Dynamics Laboratory, Auburn, CA) : we read: "[Buckwheat is] A scavenger of phosphorus and calcium. Grows better on low-fertility soils than cereal grains. Not drought-tolerant. Terminate within seven to 10 days after flowering to prevent seed production."

Also, "Buckwheat solubilizes and takes up phosphorus that is otherwise unavailable to crops, then releases these nutrients to later crops as the residue breaks down. The roots of the plants produce mild acids that release nutrients from the soil. These acids also activate slow-releasing, organic fertilizers, such as rock phosphate. Buckwheat's dense, fibrous roots cluster in the top 10 inches of soil, providing a large root surface area for nutrient uptake." (From A. Clark's *Managing Cover Crops Profitably*, 3rd ed.; National SARE Outreach Handbook Series, Book 9, published by the National Agricultural Laboratory, Beltsville, MD, 2007.)

Jean English of the Maine Organic Farmers and Gardeners Association says: "If you time it right, you can fit in a crop of weed-smothering buckwheat (it's a strong weed suppressor, but may be sensitive to herbicide carryover.) this summer and follow it with an overwintered green manure/cover crop of hairy vetch and rye, adding abundant organic matter and nitrogen to the soil to nourish your garden next year.

"Eliot Coleman suggests one possible rotation to prepare a fertile garden bed: Seed winter rye in the fall; undersow it with biennial sweet clover the following spring; mow the rye in midsummer; let the sweet clover grow through the winter; mow the clover the second summer and follow it with buckwheat; mow the buckwheat in the fall and sow a rye-hairy vetch mix; till this in in the spring and plant vegetables. The mowed cover crops can be used to make compost."

When grown as green manures, turning in a 1.2 tons per acre yield of lucerne, red clover, and white clover/grass would return 5, 9 or 2 pounds phosphorus per acre to the soil, respectively, plus the additional phosphorus contained in stubble and roots.

There are some exceptions: a study by the Henry Doubleday Research Association in England found that, when planted in poor soils, green manure crops make very little contribution to the phosphorus and potassium content of the following crop. They also found, however, that if green manures were added for a full year and up to three years they can take up a "substantial" amount of phosphorus and potassium. Their study showed that deep-rooted chicory took up 105 pounds/acre—50% more than the more shallow-rooted rye grass.

A study in Canada (using the metric system) found that some green manures are better than buckwheat at gathering phosphorus. The uptake (in kilogram per hectare) for alfalfa (19) was higher than oats (10.6) and oats were higher than buckwheat (7.9).

Note: Planting clover helps prevent the loss of phosphorus. The main source of phosphorous is in the clover stems. (*Soils & Men*, 1938 USDA Yearbook)

Lupine

Another plant to use for gathering phosphorus is white lupine (*Lupinus alba*). As mentioned in the abstract "Physiological Aspects of Cluster Root Function and Development in Phosphorus-deficient White Lupine (*Lupinus albus*)" in *Better Crops*, Vol. 83, 1999, No. 1: "Phosphorus deficiency-induced adaptations of white lupine, involved in P acquisition and mobilization of sparingly available P sources, are predominantly confined to the cluster roots..."

From the Green Harvest catalog (Australia) we read that lupine, "...adds nitrogen; long taproot opens and aerates the soil; accumulates phosphorus; flowers are attractive to bees and other beneficial insects. Sow in temperate areas autumn and spring..."

From what I'm able to find, all lupines share similar abilities to accumulate phosphorous, which plays a key role in the symbiotic N (nitrogen) fixation process as well:

1. "...[P] improves the density of Rhizobia bacteria in the soil surrounding the root..."
2. "The concentration of P (phosphorus) in the tissue of healthy, active nodules is often two to three times higher than in the roots on which they are formed."
3. "When grown in soils with sparingly available phosphorus, white lupine (*Lupinus albus*) forms special root structures, called cluster roots, which secrete large amounts of organic acids and concomitantly acidify the rhizosphere. Our results demonstrated that over a one-year cultivation period in the absence of an external phosphorus supply, white lupine was able to acquire phosphate from the soil and that the processes leading to this P uptake took place preferentially in the rhizosphere of cluster roots." (R. C. Le Bayon, L. Weisskopf, E. Martinoia, J. Jansa, E. Frossard, F. Keller, K. B. Föllmi, J.-M. Goba, 2006, "Soil Phosphorus Uptake

by Continuously Cropped Lupinus albus: A New Microcosm Design;" *Plant and Soil* Volume: 283, Issue: 1-2, Pages: 309-321.)

In "Phosphorus Mobilization by Buckwheat," Dan Olsgaard and Lee Thomas, wrote: "In 2003, the soil P concentration at the Thomas location was significantly greater where buckwheat was planted two years previously. The trend was similar at the Olsgaard location, but the increase was not statistically significant. This difference may be explained by the difference in buckwheat biomass production, which was significantly greater at the Thomas location."

It seems that deep-rooted legumes, such as a sweet clover, *Melilotus officinalis*) or alfalfa (*Medicago sativa*) green manure do increase phosphorus in the subsoil, perhaps through root decomposition there. Sweet clover is better than other cover crops for accumulating phosphorus.

Grow Your Own Potassium

Potash (its chemical symbol is K, which comes from the Mediaeval Latin *kalium*, which may have derived from the Arabic word *qali*, meaning alkali) controls the movement of water and plant foods from the roots to the foliage as well as the formation of grains, fruits, and flowers. It's essential to balance the activity of other plant foods and to create vigorous, healthy growth with resistance to pests, diseases and cold, wetness, and drought.

In New Mexico, with its significant deposits of potash-bearing rock, the processing of K involves the washing of sodium salts from the rock supply (plants don't require as much sodium as they do potassium). No chemical process is involved. The pipelines for this water come from a distant freshwater aquifer to the potash-processing operation, and the company is proud to say that the freshwater pipeline provides water for local ranchers, reducing reliance on variable rainfall in this arid area.

Wood ash, containing about three percent potash, is the most commonly known form of potassium, but its levels will vary depending on the age of the wood that was burnt. Young wood, such as tree prunings, will have a higher potassium content than will older, thicker branches. Vegetables grow best in soil with a pH of about 6.5; testing your soil's K level before adding wood ash is recommended, so as not to raise the pH too much higher than pH7. But who these days has access to wood ash? Are you going to raid Boy Scout campfires? I think not. So, here are the alternatives:

- **Granite dust** at 3-5% potassium, 67% silica (sand), and micronutrients. Slow release. One application can last 10 years. Improves soil structure. Broadcast 10 lbs. per 100 sq. ft. on soils low in potassium.
- **Greensand** is a sand-based fertilizer mined from dried ocean deposits. An application can last up to 10 years. It contains around 5-7% potassium, trace minerals, and silica. Apply in autumn at 10 lbs. per 100 sq. ft. (The only active greensand mine in the country was in southern New Jersey, until Inversand Mining Co. decided to sell the property. The company now sells a new product called GreensandPlus that's mined in Brazil.)
- **Kelp** (Seaweed) comes as a solid or as a liquid concentrate. It's high in potassium and micronutrients. With the solid concentrate, use about 10 lbs. per 100 sq. ft. Follow label directions for dilution with the liquid form.
- **Rape and mustard** can be used as green manures to collect potash. Brassicas like these can root to depths of six feet or more, scavenging nutrients from below the rooting depth of most crops. This so-called "biodrilling" is most effective when the plants are grown at a time of year when the soil is moist and easier to penetrate. (Their deep rooting also allows these crops to scavenge nutrients from deep in the soil profile. As the large taproots decompose, they leave channels open to the surface. These increase water infiltration and improve the subsequent growth and soil penetration of crop roots. Smaller roots decompose and leave channels through the soil, improving penetration by the roots of subsequent crops.)
- **Rye** increases the concentration of exchangeable potassium near the soil surface

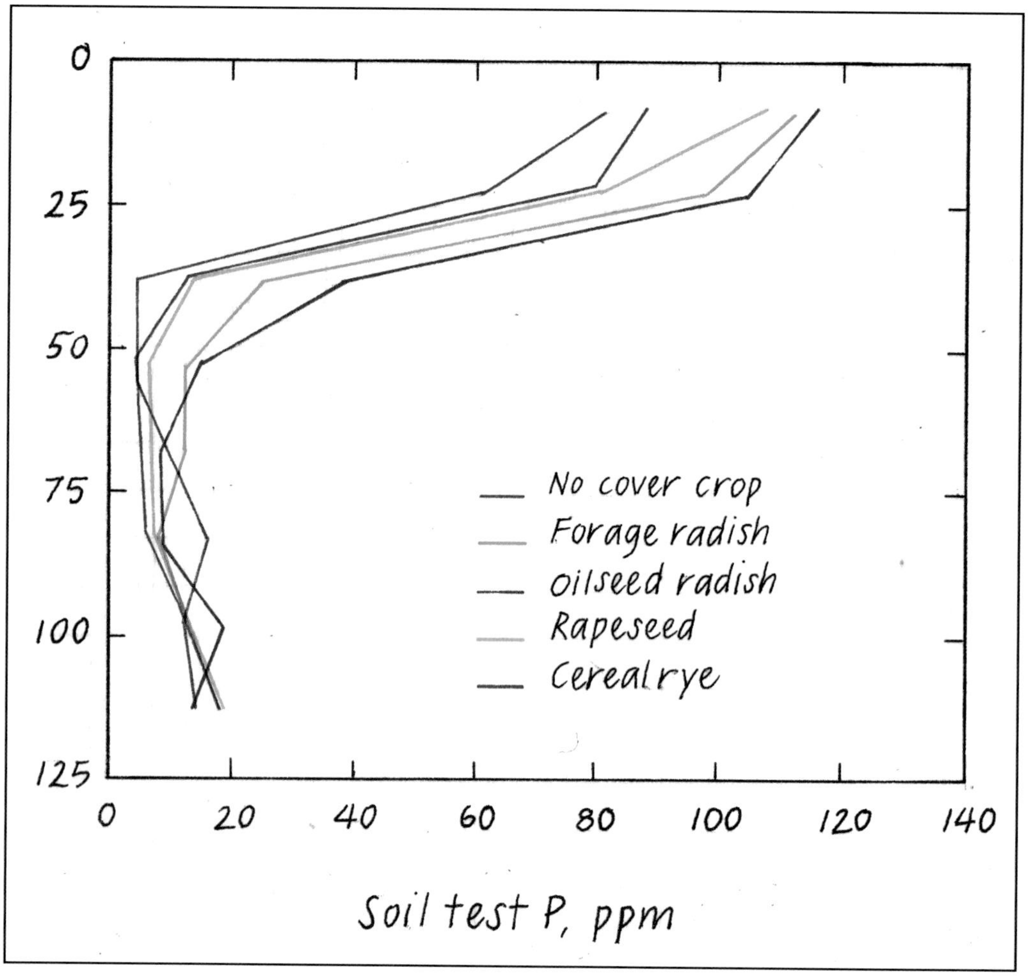

Figure 256. Forage and oilseed radishes provide the greatest amount of phosphorous (115 ppm) in the upper 25 cm of soil (10 inches.) (The left-hand range is in cm.) At around 60 cm (24 inches) all crops have about the same amount of phosphorus—15 ppm. This is also another example as to how important shallow soils can be. Tillage disturbs soil biota, making it more difficult to rellease phosphorus. Another argument for no-till gardening.

by bringing it up from lower levels in the soil. Grass hay mulches contain about 2% potash on a dry-weight basis, and a four-inch mulch may deliver 200 pounds of potash per acre or more.

- **Yellow sweetclover** (*Melilotus officinalis*) and **white sweetclover** (*M. alba*) are the most drought-tolerant of forage legumes, and are quite winter-hardy. They can extract phosphorus, potassium, and other micronutrients that are otherwise unavailable to crops from the soil, then release them for use by other plants. They also appear to have a greater ability to extract soil nutrients from insoluble minerals than most other cover crops. Root branches take in minerals from seldom-disturbed soil horizons; these nutrients become available as the tops and roots decompose.

Nutrient Catch Crop

Wheat enhances cycling of nitrogen, phosphorus and potassium. A heavy N feeder in spring, it takes up the element relatively slowly in autumn. This adds up, however: a Maryland study showed that a September-seeded stand absorbed 40 lb. nitrogen per acre by December. If used as an overwintering cover rather than a grain crop, wheat wouldn't need fertilizer in either fall or spring.

A 50-bushel wheat crop can use up to 20 to 25 lbs. of phosphorus and 60 lbs. of potassium per acre by the time it reaches "boot stage," when the seed is still developing in the sheath. About 80 percent of the potassium is recycled if the stems and leaves aren't removed from the field at harvest. All the nutrients are recycled when wheat is managed as a cover crop and given a role in scavenging excess nitrogen.

In general, according to studies, plant-available phosphorus and potassium additions were significantly greater following velvet-bean incorporation and were determined to be 59 and 260 kg ha respectively. Legumes tend to enhance availability of phosphorus, while grasses, especially rye, enhance availability of potassium.
Sustainability can be modeled from older European farms. I have friends who toured Swiss organic farms in the 1970s. The farmers managed field crops, animals, cover crops, green manures, and meadows, and the rotations of each of these elements of farming could take as much as 25 years to get back to the same place. More incredibly, the farmers shared each others' land to complete the rotations. (Sylvie Blanchet – pers. communication, 2013) That sure doesn't happen here! Most home gardeners don't have animals, although chickens are becoming "legal" in more and more suburbs. In Appendix 2, you can read about the vegan gardeners that don't use manure. Space, expertise, years of practice, and some plants grown as nutrients instead of food can replace animal manures, while at the same time allowing for far fewer imports.

Biodrilling - to Grow Phosphorous

"Biodrilling," also called "biological subsoil tillage," is the buzzword for using deep roots to improve soil texture, open heavy subsoil, deposit organic matter deeply (via the dead root-hairs), and relieve compaction in zones closer to the surface. In the process of biodrilling, the accumulation of phosphorus is one of the few examples of the possible dynamic-accumulator effect discussed earlier.

A detailed 2009 thesis by Guihua Chen offers some great insights: his four-year study found that using three cover crops—forage radish [*Raphanus sativus var. longipinnatus* 'Daikon'], rape [*Brassica napus*] and cereal rye [*Secale cereale*)—produced greater yields with corn than the yields in fields with no cover crops. He attributed this to biodrilling, and found that the beneficial influence of forage radish was greater than that of rape, which was greater than that of rye. In a similar study, forage radish and rapeseed (*Brassica napus*) enhanced access to subsurface soil water for corn by providing deep roots channels in compacted soil. (Root growth and yield of maize as affected by soil compaction and cover crops, 2011. Guihua Chen & Ray Weil; *Soil & Tillage Research*, 117 P 17-27.)

The use of tap-rooted cover crops in crop rotations helps to improve soil compaction up to a depth of slightly more than one meter (about 39 inches). This practice costs far less money and produces fewer negative environmental effects than the common practice of "deep-ripping" or "sub-soiling" tractor tillage that's traditionally been used on farms to reduce soil compaction or to break up hardpan. (In deep ripping, a set of vertical shafts or tines, as much as three or four feet long, are drawn through compacted soil to fracture it.)

It takes about five years of biodrilling to affect and lessen deep soil compaction, but the all-important soil horizons and soil biota are left intact and relatively undisturbed. Of course, you can also use the French-Intensive Biodynamic double-digging process (see Chapter_) to create well-fractured soil that retains its integrity (that is, if your back can handle it).

So which plants are recommended by experts as the most "efficient" for biodrilling? As mentioned, some brassicas (forage radish, rapeseed, turnip) produce large taproots that can penetrate up to three to six feet deep to alleviate soil compaction. This deep rooting also allows these crops to scavenge nutrients from further down in the soil profile.

According to a product study done by the "manufacturer" of a strain called "Tillage Radish®," when fields were prepared by planting its "product," corn yields went up an average of 12 bushels per acre, soybean yields went up eight bushels per acre, and winter-wheat yields went up to five to seven bushels per acre. (I haven't yet found any peer-reviewed studies to back this up, as all of the available research has been funded by the seed companies.)

As mentioned previously, biodrilling is most effective when the plants are growing at a time when the soil is moist and easier for the roots to penetrate. The deep rooting allows the roots to scavenge nutrients from deep in the soil, and as the large taproots decompose, they leave channels open to the surface. These increase water infiltration and improve the subsequent growth and soil penetration for the roots of the following crop. Smaller roots also decompose and leave channels through the hardpan. Unlike those patented biodrilling radishes, most mustards have a more fibrous root system and don't root as deeply.

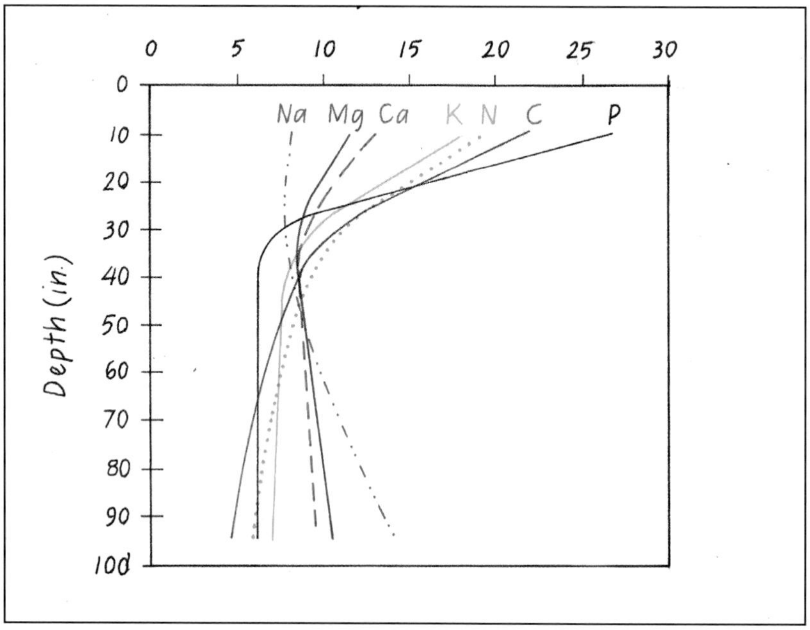

Figure 257. Most nutrients have more availability in the top 10 inches of soil compared to deeper soils. About 28 percent of phosphorus is available in the top10 inches of soil. At 90 inches there is only eight or so percent of available phosphorus. Yet another example of the need for no-till gardneing. (Are you beginning to get the point?)

Now that I've got you excited about the possibilities of biodrilling, I have to mention that a necessary condition for its use is a "hard" fall freeze (into the mid-teens Fahrenheit), to kill off the plants. The death of nearly mature radishes or daikon (*Raphanus sativus* var. *longipinnatus*, a mild-flavored winter radish usually characterized by fast-growing leaves and a long fat white tapering root) opens up large holes in the soil. This allows for air and rain to easily penetrate the soils and assist in increasing friability (looseness). This means that biodrilling isn't really feasible in warm-winter areas without hard frosts. Undaunted, growers in my mild winter area are experimenting with mowing or tillage to kill the fully grown radishes.

Alleviation of Soil Compaction by Brassica Cover Crops

Another study (by Guihua Chen in 2009) showed soybean (*Glycine max*) roots grew through a hardpan by using channels left by canola (*Brassica napus*) cover-crop roots, which alleviated the compaction effect. Chen studied "biodrilling" abilities of three cover crops and the effects of "biodrilling" on corn (*Zea mays*) rotated with soybeans by conducting three experiments.

The first two experiments included three compaction treatments (high, medium and no compaction), four cover crops: forage radish (*Raphanus sativus* var. *longipinnatus*, cultivar 'Daikon'); rape (*Brassica napus*, cultivar 'Essex', a tap-rooted species of the Brassica family); cereal rye (*Secale cereale*, the fibrous-rooted cultivar 'Wheeler'); and a control plot with no cover crops. Chen found that roots of the forage radish were least inhibited by compaction, while rye roots were severely immobilized by it.

The order of biodrilling effectiveness was forage radish, followed by rape and then by the rye roots. Corn and soybean-root penetrations of the compacted soils were enhanced by forage radish and rye treatments. The yield of corn was greater in three cover-crop treatments than in the control crop. In terms of biodrilling, the forage radish and rape were more effective than just the rye cover crop. Chen goes on to say this effect could be useful in no-till farming.

The results of another study suggested that forage radishes benefited maize-root penetration in compacted soils, while rye provided the best availability of surface-soil water; rapeseed tended to provide both benefits. Rye conserved surface-soil water, forage radish enhanced subsoil rooting, while rapeseed tended to provide both benefits.

The radishes in these studies, by the way, are also excellent accumulators of phosphorus (.5%) and potassium (4%), and they accumulate 140 pounds-per-acre of nitrogen—comparable to many leguminous cover crops (see pages 247 and 251).

These radishes don't serve as a host for mycorrhizal fungi, but it was noted that mycorrhizal colonization of corn planted following a radish crop didn't appear to be suppressed.

In another study, "We found that shoots of both forage radish and rye cover crops have the potential to take up significant quantities of phosphorus when these cover crops are managed for maximum dry matter production. And after three years of forage radish cover crops managed with no-till practices, soil test phosphorus of the surface soil moderately increased at the site where forage radish cycled large quantities of phosphorous. In the vicinity of forage radish root holes, however, soil test phosphorus increased dramatically, even when the total Phosphorous cycled by the forage radish cover crops was small. (From: Forage radish cover crops increase soil test phosphorus surrounding radish taproot holes, 2011. Charles White and Ray Weil, *Soil Science Society of America*. Vol. 75, Issue 1, Pages: 121-130.)

One source of biodrilling plant seeds is Sieger's Seed Company. It's expensive, and 10 pounds is the minimum order, so you may want to gang up with friends to biodrill your gardens. Or, better yet, experiment with regular cultivars of the daikon radish. (See below.) Daikon in Japanese means "large root." One type of inexpensive seed comes from Territorial Seed Company; for around eight dollars (or four dollars for four ounces) you can buy a pound of "Groundhog Daikon Radish Cover Crop."

Figure 258. A tool used by Rosa O'Brien to tend her surface-cultivated beds is known in England as a scraper. (Spelling is correct.) In America it's called an onion hoe. The scraper, Dutch hoe, and Hula Hoe® can be used to cut off roots just below the surface—nearly no-till gardening/farming. I use mine all the time for weeding.

The *Raphanus sativus* var. *niger* strain of daikon is referred to as a Tillage Radish®, officially known as Variety CCS-779, and has been awarded U.S. Plant Variety Protection; unauthorized propagation is strictly prohibited. This distinction indicates that Tillage Radish® is a unique variety.

Most of the radish varieties currently sold for cover-cropping (e.g., GroundHog radish™, Nitro radish, Sodbuster, and Bio-till radish) are large-rooted selections of daikon-type radishes. All are similar to the large white daikon radishes traditionally used in Asian cooking. Specialized forage radish seed may cost over $100 per pound for bulk seed. Open-pollinated culinary daikon varieties may have similar forage potential at about five dollars a pound. Standard oilseed radish cultivars (for example, the cultivars Adagio, Colonel, and Defender) tend to have a stubbier, more branched taproot, greater winter hardiness, and lower seed cost than larger-rooted daikons.

Sustainable Food Gardening with *No* Manure

Although most readers won't be growing a market garden, you might be interested to learn that I did discover (in 1976!) a fine example of the efficient use of surface cultivation in a commercial setting, which indicates that it can in fact be adapted to a larger scale of planting. Most of the following techniques can also be adapted to the home garden.

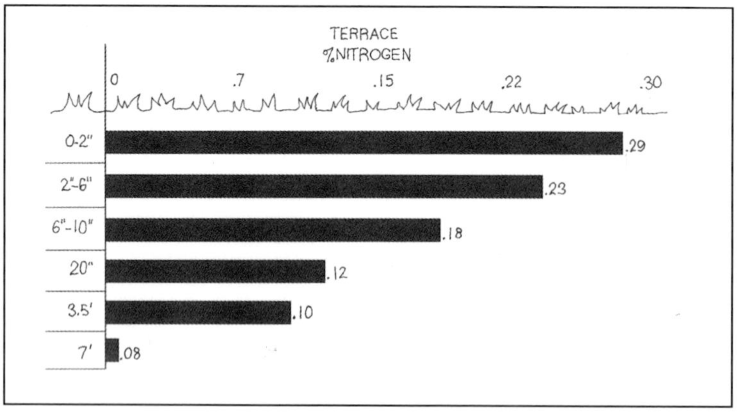

Figure 259. What? Another example of the importance of the upper two inches of the garden's soil? You bet. Here the six-inch-zone has 30 perecent less nitrogen than the upper two inches. Nitrogen is particularly susceptible to tillage as the increase in air helps burn off the nitrogen. Use mulch instead of tillage where possible.

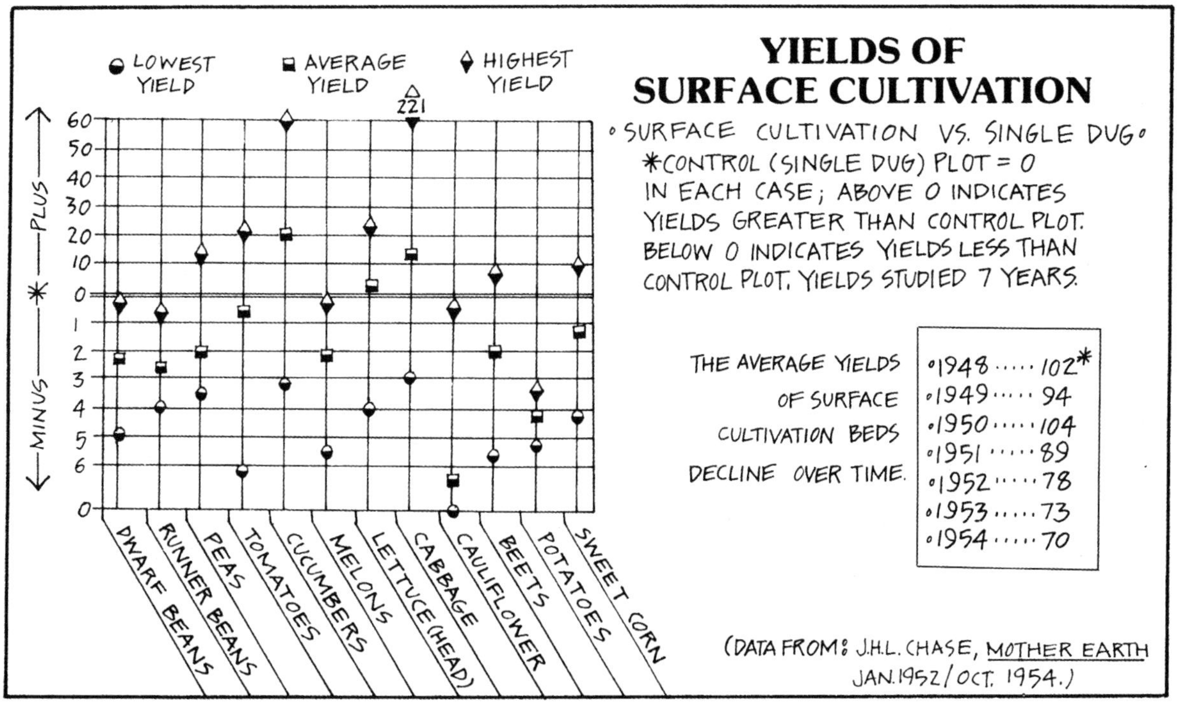

Figure 260. Careful review of this informal study reveals that no-till gardening can out-produce tilled gardens. Even the average yield is high for the first three to four years. After six or seven years I would dig (or double-dig) the soil and start the process again. No Church-of-No-till for me. Be pragmatic.

This anecdotal evidence was collected from an organic market garden in England with 12 years of experience in surface cultivation. The main tool for this approach (in addition to the hand-pulling of weeds or thinning of vegetable plants), is a "scrapper" (not a typo), which is known as an onion hoe in the U.S. The English market gardeners firmly state that "...Preparation for the following crop is one operation with the (scrapper), with the occasional use of a rake, and it should not penetrate the soil more than three-and-one-half inches to four inches for any purpose." (This is not a pure form of no-till gardening, but comes pretty close.) The rationale is that "True aeration of the soil is achieved in undisturbed soil..." The scrapper is always pulled toward the gardener at a 45-degree angle to the path. The tool is used only to pull, not push soil."

This garden was not only successful but "Sales were on a competitive basis." More amazingly, the garden was maintained *without the use of any manures*. (Regardless of your opinion of a vegan diet or lifestyle, the practice of using no manure certainly belongs with the discussion of garden imports and exports in Chapter 4.) The book about this remarkable market garden is entitled – at great length—*Intensive Gardening, using Dutch Lights, Surface Cultivation and Composting for the Commercial Production of Crops, and Introducing a Motion-Study Routine*, by Rosa Dalziel O'Brien, (Faber & Faber Limited, London, UK. 1956).

Here are some highlights from the book, which help to illustrate how this market garden was put together:

Compost, compost, and more compost (keep in mind my earlier caution about too much compost) was one of the key elements in this garden. The composting area consisted of "bins" made of bales of straw—the only significant imported resource—stacked like the letter "E," with two "E"s connected in a row to create four bins. The compost was made with soil layered with grass clippings (from strips planted in access roads and between rows of plots); legumes (grown in the grass pathways and in otherwise fallow vegetable beds); and garden waste. The grass was mowed when long enough to produce clippings of about three-and-

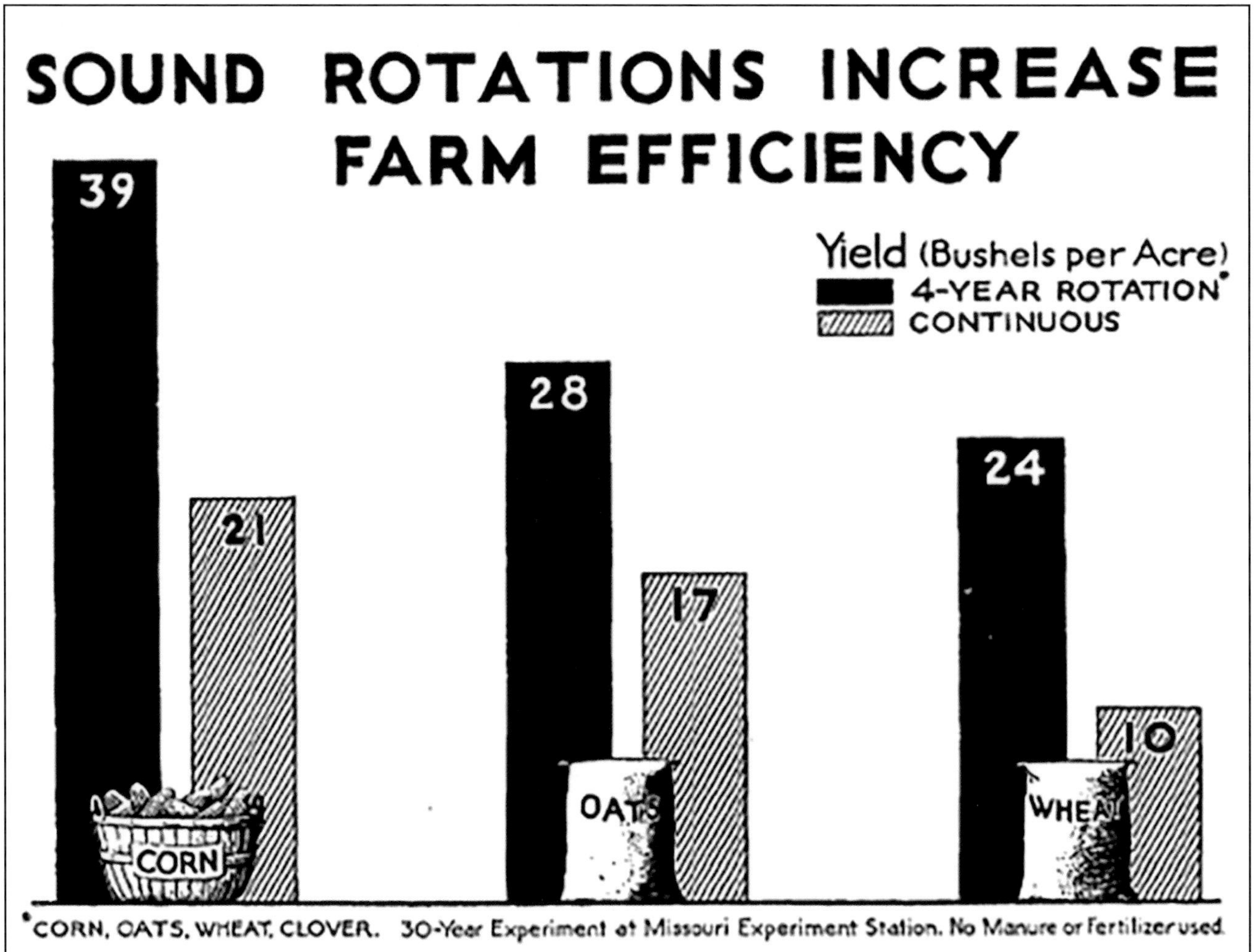

FIGURE 1.—Average yields of corn, oats, and wheat in a 4-year rotation and continuously cropped at the Missouri Agricultural Experiment Station.

Figure 261. This illustration is from the *Soils & Men* book of 1938 from the US Department of Agriculture (USDA.) This book is so important because it was published before a lot of chemical fertilizers were used. It is an organic guide to tillage. There is a good amount of discussion of crop rotations. This is one example. Here clover is used in one year out of four years of rotation. The yeilds are up over 20 percent with clover added to the rotation of wheat compared to growing wheat continuously all four years. Note that this was a 30-year study without any manure or fertilizer! A good way to average out seasonal climate variations. (From my home state of Missouri.)

one half inches. (This is possible in rainy England, the U.S. Midwest, and New England, but not the western United States)

Each compost pile, in its straw-bale "bin," began as a piled-up cubic mass about six feet per side, beginning with a small layer of charcoaled wood (a small import) to "...absorb the impurities from the pile; [it will] serve several successive ones before fresh pieces are needed." (This is an unusual approach, which I'd never encountered before finding this book.) O'Brien describes a very specific technique for layering of grass clippings; old straw from previous "bin walls;" garden waste (thinnings and trimmings, etc.); thin layers of soil; and a very thin dusting of lime—another small import.

The role of the lime in O'Brien's composting recipe was to make the pile slightly alkaline in order to enhance existing populations of the free-living, nitrogen-fixing bacteria *Azotobacter croccoceum*. O'Brien explains: "Our compost heaps include sufficient lime to keep the material alkaline, so that we gain, by bacterial action, about 25% more combined nitrogen than went in with the plant residues, and when we spread the compost on the surface of the soil, the bacteria go on working…[the resulting bacterial action] is quite enough for the crops we grow."

O'Brien also stressed how important it was to use a compost inoculant called "Quick Return"—a special blend of herbs. Each compost pile is pierced from the top by six vertical holes made with a crowbar or broom handle. The diluted inoculant mixture is drained into these holes, after which the pile is capped with two inches of soil. There are five detailed pages in O' Brien's book devoted solely to making the piles, and the text refers frequently to the Quick Return (Q. R.) solution as an essential part of the process. The solution eliminates the need to turn the piles, and can generate finished compost in six to eight weeks in the spring, summer and fall or eight to ten weeks in the winter. Even though the book was published in 1956, the same Q. R. powder is still available in the US (as of 2020, when this book was written) from Biocontrol Network in Tennessee or Harmony Farm and Garden Supply in California.

Other "imports" into this garden included the one-time purchase of so-called "Dutch lights." (Glass coldframes were a traditional feature of Dutch commercial horticulture.) These are panels of glass built to a standardized form and set up like tiny greenhouses. This was a crucial aspect of the market farm, as the Dutch lights allowed for an earlier harvest and extended the harvest time beyond that of crops grown in an open field. They wouldn't be needed in a home garden if the gardener didn't want to push the natural growth cycle of the vegetable.

Also imported, as needed for certain crops; were some seaweed, old domestic soot (from coal-fired chimneys), granite dust, and silver sand (lime-free quartz particles that aren't coated in iron oxides).

O'Brien mentions that it took about three years for the new market garden to make the transition from harboring obnoxious weeds like docks (*Rumex* spp.), thistles, and dandelions to lesser noxious weeds such as nettle, groundsel (a European weed with small yellow flowers, *Senecio vulgaris*) and wild chamomile (probably the low-growing weed *Matricaria discoidea*, not the kind (*Anthemis noblis*) used for tea. It took about the same amount of time for the compost and surface cultivation to enrich the soil, increase its population of beneficial bacteria and enhance its aeration in a natural way. Obviously in this case, patience is a virtue.
O' Brian was a vegan. With its limited number of imported items and its calculated non-use of manures

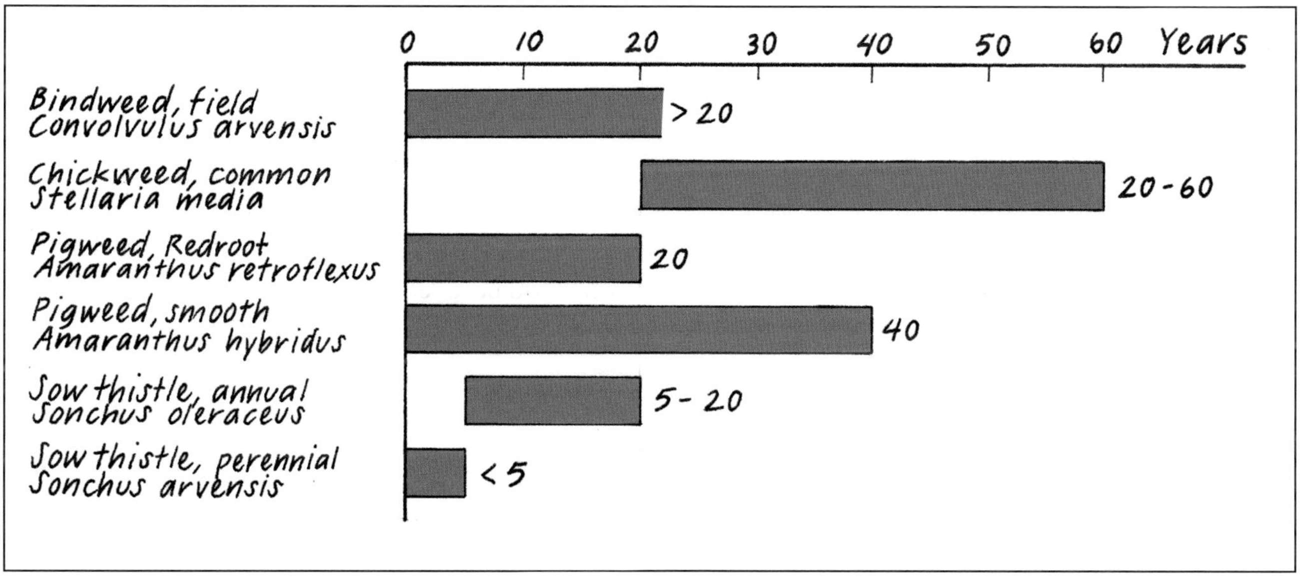

Figure 262. Some weed seeds can last a long time. For example, chickweed seed is viable for 60 years. This means everytime you disturb the soil weed seeds may sprout. Chickweed is edible, but most people can't eat enough to solve their problem. So, they weed. Here is another argument for no-till and plenty of mulch.

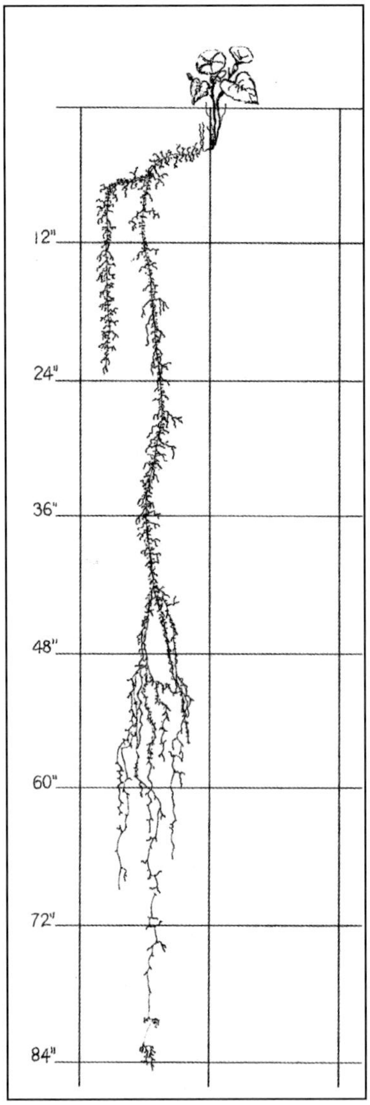

Figure 263. Bindweed (*Convolvulus arvensis*) has amazingly-deep roots. This chart shows why I've never been able to eradicate it in my garden.

("...(no) animal products or by-products, even eggshells were included.")—the fact that this market garden was commercially competitive is truly astonishing.

(As a side note, Jonathan Frei, head gardener of the Rancho La Puerto health spa in Mexico in the '90s, maintained clover pathways in his garden for the guests. (Despite the arid surroundings, he had a decent water supply.) The paths between raised garden beds allowed him to mow a mixture of Dutch white clover and grasses in one sweep of a 21-inch mower about every week or so. The clippings were saved and added to the compost pile.

Frei knew that the mowing helped "shed" some of the nitrogen nodules from the roots of the Dutch white clover and fertilize the vegetable roots growing beneath the surface of the pathways. This very much mimicked O'Brian's methods, even though Frei never had heard of her truck garden. The process did require a bit of attention, with the need to spade the edges of all beds occasionally so the clover didn't spread. Annual cultivation also helped keep the clover in check.

Crop Rotations

There are four reasons to rotate the crops that you grow: disease prevention, crop production relative to soil structure and fertility, control of weeds, and insectary plantings.

Disease Prevention

A number of vegetables shouldn't be grown in the same place every year because of pests and disease. Examples include:

- **Tomatoes** - harbor Fusarium and Verticillium wilt.
- **Eggplants and tomatoes** are all members of the nightshade family and harbor Colorado potato beetles.
- **Brassicas like kale, cauliflower, broccoli, turnip**, etc. can harbor clubfoot disease (*Plasmodiophora brassicae*). Alternaria fungus, Black rot, and Blackleg (a bacterial disease of potato) may also be transmitted to the field by contaminated seeds.
- **Pumpkin** - Fusarium fruit rot is more common in pumpkin when it's planted following a corn crop.
- **Carrots** - The soil-borne fungi *Pythium* spp. and *Rhizoctonia solani* can infect and kill the tip of carrot taproots and result in more misshapen carrots, especially when they're planted after an alfalfa crop.

Crop rotations can, however, be effective in dealing with certain diseases:

- **Lettuce** - Lettuce drop and white mold. Broccoli is a good crop to grow in rotation with lettuce and other crops susceptible to white mold. In studies, the number of sclerotia (hard dark resting body) of certain fungi, of the lettuce-drop fungus, *Sclerotinia sclerotiorum*, decreased after residue from a spring broccoli crop was incorporated during the summer

- **Potatoes -** Verticillium wilt and scab of potato. Both these diseases were reduced when corn or alfalfa was grown the previous year rather than potato. Verticillium wilt severity was also lower when a buckwheat green manure preceded a potato crop than when canola or a fallow preceded potato.
- **Bacterial spot of pepper and tomato** - The bacterium causing spot (*Xanthomonas campestris* pv. *vesicatoria*) can be effectively controlled with a rotation of two years without a host crop.
- **Fusarium wilts of crucifers, cucurbits, pea, spinach and tomato** -These diseases are difficult to manage with rotation because even in the absence of their crop host, the pathogens can persist for many years in soil as dormant chlamydospores and on roots of some non-host plants (symptomless carriers). Rotations of at least five to seven years often prevent the pathogen population from building up

Some plants work both ways: for instance, although a mustard or legume crop can suppress some pathogens, such crops may also promote pathogens that attack plants in those families.

Rotations for Healthy Vegetables

Since pests and diseases spread most readily in monoculture, it's important that you adopt at least a four-course rotation. One way to do this is to split your garden into four plots. Here's a typical example:

1. **Potatoes and curcubits (zucchini, marrow, squash)** - Potatoes are heavy feeders and they are good at breaking and heaving the soil, sort of like gentle good-natured rototillers.

2. **Legumes (peas and beans) and alliums (onions and leeks)** – these replenish the soil with nitrogen.

3. **Brassicas (cabbage, cauliflower, broccoli, Brussels sprouts, turnip, radish, swede, oriental salad leaves)** – all of these love the nitrogen of the legumes. They need at least 0.05 pounds of nitrogen per 100 square feet, and you can use 160 pounds per acre.

4. **Salads, root crops and miscellaneous (lettuce, chicory, beetroot, carrot, parsnip, sweetcorn)** – these are not too hard on the soil, and not as heavy feeders as potatoes and squashes.

Each year the plants you choose from each of these four groupings are rotated into a different plot. For example, in year one potatoes may be grown in one plot, and they can be followed the next year by peas, and so forth.

Another rotation formula to consider is:

1. Legumes
2. Root crops
3. Fruits like tomatoes and squashes
4. Leafy greens

Weed Control (crop rotations)

Cover crops like buckwheat, vetch, soybean, and grain rye can completely smother many annual weeds; be sure to use very high-density sowings, and increase seeding rates by 50 to 100 percent, relative to recommended rates. Planting winter rye or mustard cover crops reduced spring weed cover from 52 percent to nine percent with rye, and to four percent with mustard. But be aware of the flip side of the coin with weed suppressants. If left to go to seed, vetch can become a noxious weed in its own right.

Using tillage as a control gets problematic in that many weeds seeds last in the soil for years and decades.

Alas, if cover crops don't work, you do have to fall back on cultivation. If perennial weeds are cultivated into the soil before seeding takes place, they're weakened. A few such cycles once every two to three years can help suppress those annual weeds whose roots are within shovel-depth, and help keep deeper-rooted perennials in check. (There are, however, limits. Look at the drawing of bindweed (*Convolvulus arvensis*)—Figure 263—with roots as deep as seven feet.) Tillage tends to stimulate the germination of seeds of most weed species, but subsequent cultivation with a Hula Hoe™ may kill the seedlings enough so they will not compete with future crops.

Cover Crops

Cover crops are plants that are grown to improve the soil before planting vegetables or in rotation with them. Grown for one or more seasons and not tilled in, cover crops are often cut off at the ground (or grazed), after which the soil is tilled and planted with vegetables. Unlike a green manure (described below), the cover crop is not turned into the soil. The soil is nonetheless improved by its roots, which can heave the soil and fracture soils at a deeper level that you can comfortably dig—and any roots left behind act like compost as they rot in place. As an example, after five years of using a cover crop, the soil eight to 16 inches deep can be loosened up enough to make it easier for other roots to penetrate into deeper soil.

There are two basic cover crops—grasses and legumes—for nitrogen fixation. For optimum efficiency in N production, the legumes are cut off before they bloom, and the foliage is composted or used directly as a mulch. (See page 251 for why you mustn't let them bloom.) The grasses are important because they add far more fiber to the soil than most legumes do.

Not all beans are equal in N fixation. For example, 75% of the total nitrogen contained in fava bean plants is because of fixation, but in a common bean plant only 39% of the total nitrogen is due to it. (The percentage of nitrogen fixed in cultivars of dry beans is variable.)

In addition to nitrogen, some plants accumulate more phosphorus than others. In a trial in Texas, hairy vetch, crimson clover, red clover, or a crimson/ryegrass mixture served this function. Cutting these cover crops off and adding them to a soil compost releases more of this important nutrient than other plants do.

Phosphorus is especially important for annual crops; it's an essential macro-element, required for metabolic processes such as photosynthesis, energy transfer, and synthesis and breakdown of carbohydrates. The symptoms of phosphorus deficiency are: reduced growth; foliage that's purplish or darker green than normal; thin stems; reduced lateral bud growth; loss of lower leaves; and reduced flowering and fruiting. (Cannabis growers are extremely fixated on phosphorus because of its boosting effect on buds and blossoms.)

The best book on this topic is: *Managing Cover Crops*, edited by Andy Clark (3rd ed., July 1, 2007**;** Sustainable Agriculture Research and Education Program of CSREES; U.S. Department of Agriculture). (Be sure to get the 2007 edition, as the 2nd edition costs over $700 USD!)

Green Manure? How Gross! (Not Really)

Green manuring requires once-a-year digging or rototilling, but it's a great way to improve soil without having to schlep manure into the garden or buy fertilizers.

Is the closest source of animal manure many miles away from your home? Then green manuring might be for you. This is a term for the turning/tilling of fresh green plants into soil to help make it drain better, allow it to hold onto more moisture, and act as a readily available fertilizer. Green manuring is as close to free fertilizer as it gets—discounting the cost of a few seeds and plenty of elbow grease. It's also a valid way to care for the soil for a few years before jump-starting soil improvement on the path to a no-till garden. (Alternatively, you can simply cut the foliage off, take it to a compost heap and let the remaining roots do all the work. This is technically called cover-cropping, as explained above—it takes longer than green manuring to improve the soil, but it's *much* less work.)

Any old plant material, dead or alive, can help improve your soil, but learning just how the natural cycle of decomposition works means that you'll know exactly which part of the cycle to influence, how to speed up the natural processes and how to improve the soil in either the short or long term. "Moderation in all things" is the most beneficial approach with green manures. With grasses, cereals, non-leguminous plants, and legumes all blended together into one seeding, you'll get a broader range of nutrients,

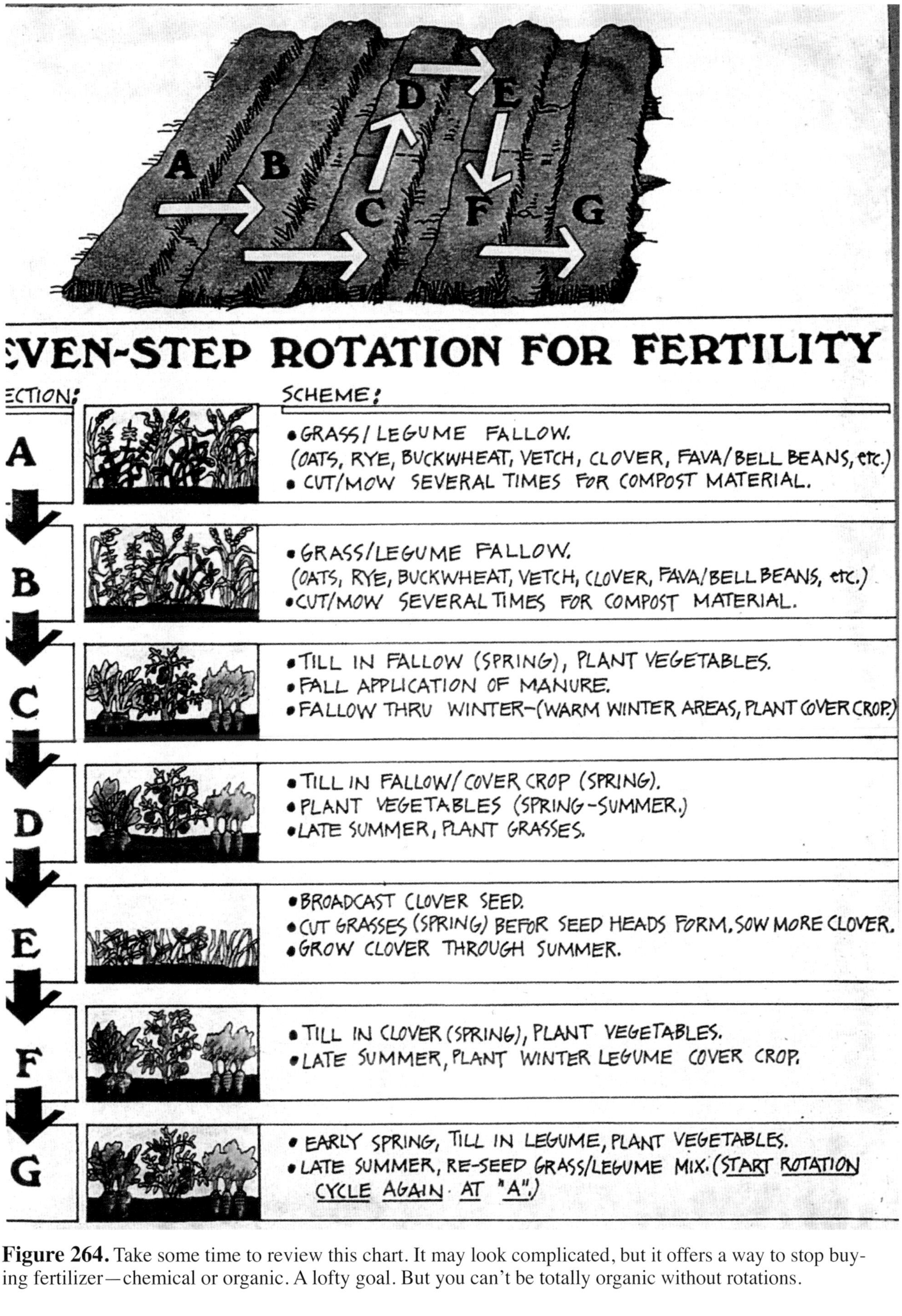

Figure 264. Take some time to review this chart. It may look complicated, but it offers a way to stop buying fertilizer—chemical or organic. A lofty goal. But you can't be totally organic without rotations.

root zones, and secondary benefits such as weed suppression.

If plant refuse is dry and straw-brown when it's turned into the ground, it will still improve soil structure—but over a longer period of time than fresh green material will take. The soil bacteria that digest any kind of plant refuse need moisture, warmth, and nitrogen to increase their populations to levels high enough to decompose the raw materials. If plant matter is dry and crusty, it's low in nitrogen and high in carbon, and will take the bacteria a longer period of time to convert the relatively higher percentage of carbon to a more decomposed, and partially decayed form known simply as "organic matter."

Eventually, the organic matter will be fully decomposed and converted into *humus*—the most wonderful, stable, moisture-holding, nutrient-rich constituent of soil. When chemically reduced fertilizers aren't used, humus plays an essential role in all of the nutrient exchange that takes place in an organic garden. However, because it takes many weeks, or even months, for the high carbon content of dry organic matter to be converted into available nutrients, these nutrients will not be accessible to a recently planted crop. In fact, the incorporated dry material will probably rob nitrogen from the surrounding soil for a bit of time.

The common theory is that soil nitrogen is lost to dry materials because the N is used to help build up the bacterial populations needed to decompose the carbonaceous material (the protein for the exoskeletons of the digesting bacteria and microbes comes from soil nitrogen). (One interesting theory, as discussed in *Mineral Nutrition of Plants* by J. B. Page and G. B. Bodman (University of Wisconsin Press, 1951), postulates "...that the commonly observed depression in plant growth immediately after adding large amounts of green manure...may be due to competition between the higher plants and microorganisms for the limited supply of oxygen than any other single factor.") Either mechanism may cause new crops to appear stunted, produce yellow leaves, and grow poorly. Using dead, dry, and crusty plant matter to improve soil is like putting money into a long-term certificate of deposit; there are few immediate benefits but the future payback is good.

If the plant matter is turned into the soil when it's fresh and green, and, therefore higher in nitrogen, then the soil bacteria can easily and readily decompose its tissues. The premier advantages to green manures are that: 1) when they're applied, their plant matter is decomposed quickly, and 2) you can plant sooner after tilling them in—even the most prudent gardener usually only needs to wait two to four weeks after tilling under a green manure before transplanting or seeding.

Green manure is more like playing the fast-moving stock market; the gains can be immediate and significant, but to put the largest amount of green-manure fertilizer into your "portfolio," you'll need to know how to leverage all aspects of its cycle. With some of the taller and bulkier green manure plants like fava beans and vetches, for instance, the foliage is best cut and hauled away to a compost heap to rot over time—very much less work than tilling.

Green Manure—The High-Fiber Diet

Because of its rapid decomposition, green manure is also a good way to add fiber to your soil. As the fiber decomposes into humus, it will help loosen heavy clay soils and assist sandy soils in retaining moisture. Since the plants are turned under when they're fresh and green, the following crop will also get a healthy boost of nitrogen.

The best green manure plants for sheer bulk of fiber are grasses and cereals. While these plants are growing, their very extensive and massively fibrous root systems help loosen clay deposits by penetrating and growing through the soil.

A Balanced Regimen for Green Manuring

There's really no reason why you can't incorporate all kinds of green-manuring procedures into your gardening regimen. As with any good investment plan, a diversified portfolio is the wisest and safest strategy. Don't forget that roots work 24 hours a day "cultivating" soils far deeper than even the most compulsive and maniacal gardener could till.

Perhaps one of the most classic green manure blends is the combination of oats (or ryegrass), vetch, bell beans and Austrian peas, which has a history that goes back hundreds of years in Europe. Some version of this blend of grasses and legumes has comprised the nutrient backbone for many a farm, and for much longer than chemical fertilizers have been around.

The percentages of each component of these blends vary from supplier to supplier. Peaceful Valley Farm sells a mix composed of 40% bell beans, 20% Austrian field peas, 20% purple vetch, and 20% Oregon annual ryegrass. Gardeners steeped in the history of European gardening methods often add some oats to this mix for more fibrous bulk. Other gardeners figure that their soil already has so much existing wild grass seed (mostly forms of ryegrass) that simply adding a blend of legumes to it produces the same effect as the traditional European mix.

Planting Your Green Manure

Tossing green manure seeds out into rough or untilled soil is like throwing some of your valuable stock certificates into the wind. To get your green manures off to a good start, begin with the same kind of soil preparation you'd use for direct-seeded crops like beans, corn and carrots. As a general rule, seeds should be covered with a layer of soil equal to three times their diameter. Big seeds, such as bell beans, need only a rough seedbed, but fine seed like ryegrass does much better if the soil preparation more closely resembles that needed for a bed of carrots.

Green Manures in the Grand Scheme of Things

Green manures, coupled with fallow periods and cover crops, are an integral part of crop rotations that range from four- to 27-year cycles. (Yes, that's *27 years* with traditional farmers sharing fields in Europe.) With a rotation of sufficient length to allow the soil to "rest" and accumulate nutrients, you can grow all your own fertilizer in place, without having to leave your yard to haul in smelly poop. (A copy of the chart illustrating a "typical" seven-year rotation is found on pg. 279 of my book *Designing and Maintaining Your Edible Landscape - Naturally*.)

My favorite personal story about green manuring harkens back to the drought of 1975 to 1977 in Northern California. Since my clients couldn't afford enough rationed water to keep all their landscape alive, I reasoned that an annual green manure supported by winter rains would help the soil. Each year, therefore, I planted a number of flowerbeds with a fall crop of vetch. The grasses and vetch that sprouted in the fall rains grew just fine throughout the periodic (but skimpy) winter rains.

As the soil began to dry out in the spring, I turned the vetch crop under as a green manure and mulched the beds for the summer. After three winters, the soil had noticeably improved. One former client, whose soil had been a black clay gumbo, even called a year after the drought to thank me for the fantastic improvement in it. I was touched by her gratitude, but I knew that Mother (or Father—equal-opportunity household) Nature had done most of the work.

There's another fascinating dynamic with some grasses that are used for green manures. One study found that a single season of cover-cropping and green-manuring (with barley, oats, rye, Sudan grass, sorghum, and wheat) reduced weeds by 75 percent. (But subsequent yields of lettuce, radish, and tomato crops were also reduced, so they waited a season and fertilized those crops.)

Fertilizing Insects?

The following very succinct review of nitrogen fertilizers and insects was written by Bert Cregg, Ph.D. Professor, Michigan State University Department of Horticulture, Department of Forestry:

"From the *Annals of Applied Biology* entitled 'Fertilisers [English spelling] and Insect Herbivores: a Meta-Analysis' (Butler et al., *Annual of Applied Biology* 161: 223–233, 2012), we learn [that] the authors looked at dozens of studies of the response of insect herbivores to fertilization to answer the question: does fertilization increase insect damage? The answer was absolutely no surprise: It depends.

"What does it depend on? Firstly, what type of insect? Secondly, what kind of fertilizer? For example, fertilizing with nitrogen greatly increases populations of sucking insects. This makes sense when you stop to think that aphids and other sucking insects have to consume a lot of phloem sap –which is essentially sugar water – in order to get sufficient nutrients. Nitrogen fertilization did not significantly increase populations of chewing insects, however. This could be related to off-setting effects of improved nutritional quality of leaves versus increased presence of defense compounds or leaf toughness. For other fertilizer elements Butler et al. found that phosphorus decreased insect populations in two-thirds of the studies (14 out of 21) and that potassium decreased insects in 7 out of 10 cases. As with nitrogen only, complete fertilizers (NPK) tended to increase insect populations, especially for sucking insects.

"I should hasten to point out some limitations of the study as it relates to ornamental tree fertilization. First, the study mainly dealt with fertilization in agronomic crops, not ornamental trees. Secondly, the authors only included studies on insect adults. In many cases insect larvae, not adults, are the most damaging life-stage, especially for insects that affect trees. Nevertheless, the study highlights the difficulty of making generalizations when discussing host stress and insect interactions. In addition to type of insect and type of fertilizer, we could have added nutritional status of the plant before fertilization to the 'It depends' list. A good rule of thumb is that trees shouldn't be fertilized unless a problem is noted by visible symptoms, a soil test, and/or a foliar test – and preferably by more than one of these.

"Bottom line: Before you buy into the notion that fertilizing a tree is going to increase insect problems, make sure you know what type of pest you're dealing with, what type of fertilizer and the current nutrient status of the tree."

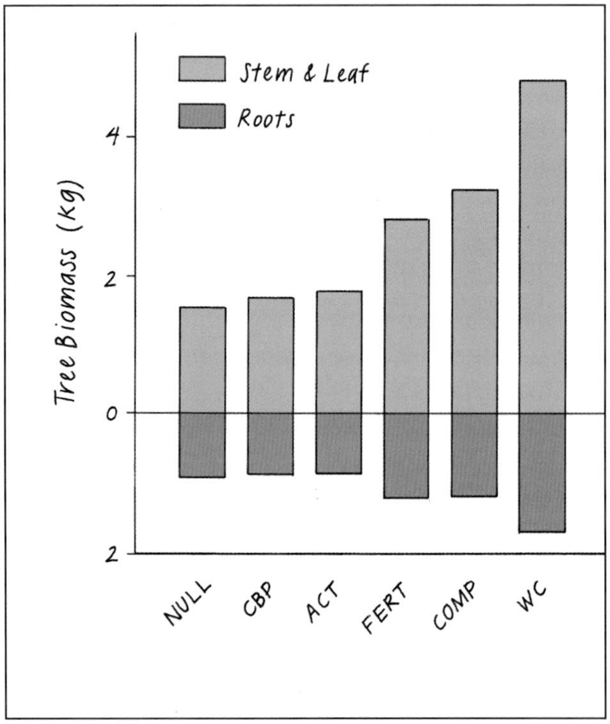

Figure 265. NULL is the control. CBP is a type of sludge. ACT is aerated compost tea. FERT is a chemical fertilizer. COMP is compost and WC is woodchips. It's amazing that the wood chips are superior to the other amendments in promoting new biomass of both the roots and the above-ground growth. And compost is better than ACT by about a third.

Other research study supports this theory. From the chapter "Impacts of Fertilizers on Insect Pests" in *Manage Insects on your Farm, A Guide to Ecological Strategies;* published in 2005 by SARE (Sustainable Agriculture Research and Education of the USDA), we read: "In England, conventional winter wheat fields were plagued with more rose-grain aphids than their organic counterparts. Top-dressed in April with nitrogen, the plants treated with soluble synthetic fertilizers contained higher levels of free protein amino acids in their leaves in June and attracted larger populations of aphids." Researchers concluded that the aphids found the conventionally grown wheat to be more palatable than the organically grown wheat.

In Ohio greenhouse experiments, European cornborer females laid significantly more eggs on sweet corn growing in conventionally fertilized soils than they did on plants growing in organically farmed soils collected nearby. Interestingly, the egg-laying varied significantly among the chemically fertilized treatments but was uniformly low in organically managed soils. The difference appears to be evidence for a form of biological buffering more commonly found under organic conditions.

In California, organically fertilized broccoli consistently developed smaller infestations of flea beetles and cabbage aphids than conventionally fertilized broccoli. Researchers attributed those reduced infestations to lower levels of free nitrogen in plant foliage, further supporting the view that farmers can influence insect pest preferences with the types and amounts of fertilizers they use.

It's a bonus to read that organic methods surpass chemical methods as far as reducing insect problems is concerned.

As mentioned by John Jeavons, too much nitrogen can actually cause insect pest problems.

Do Aphids Really Suck?

The myth that aphids suck sap is a *very* pervasive one. In actuality, their mouth parts can only open and close, so it's the internal pressure of the plant that forces the sap (mostly sugars with some proteins) from the phloem (the part of the plant responsible for transporting the food made in the leaves) into the aphid. This sap pressure can be twice to forty times the atmospheric pressure—the higher the nitrogen content of the plant, the higher the sap pressure (also known as turgid pressure). Over-fertilizing can increase the sap pressure enough to allow the plant to support a lot of aphids. (This information is from "Nitrogen Fertilization of Hemlock Increases Susceptibility to Hemlock Woolly Adelgid" (aphids) by M. S. Mcclure, 1991, *Biology*.) Thus, controlling fertilization rates can help control aphids. This is another reason not to feed fruit trees with compost, manure, or commercial fertilizers.

(And yes, as always there's a flip side to the coin: aphids can actually *improve* soil and *increase* plant growth. Most of the sugar taken up by an aphid is excreted as honeydew, a high proportion of which [over 40% in some species] is a type of sugar that doesn't occur in plants, and that must therefore be synthesized by aphids. It's believed that the sugar provides a carbon source for free-living nitrogen-fixing bacteria, and that the individual plant from which the honeydew came in the first place benefits from this arrangement, particularly if the supply of nitrogen is short. Fun fact.)

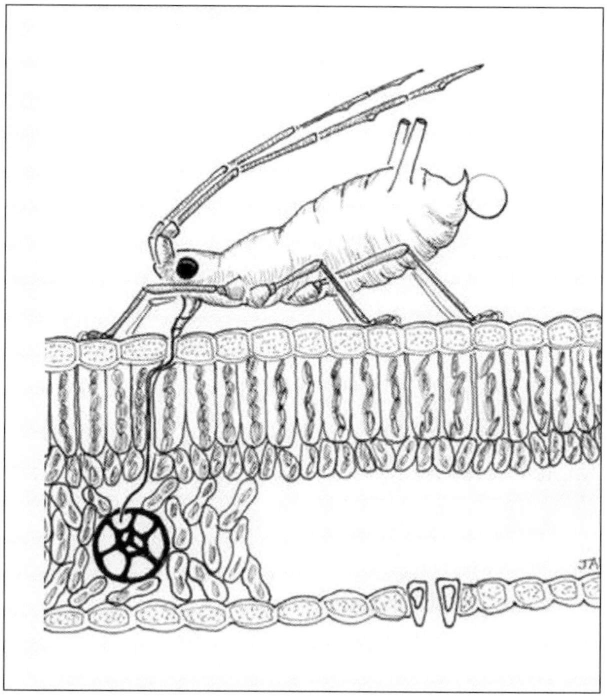

Figure 266. This illustration from the the University of Michigan shows an aphid's long, dark proboscis penetrating cells to get to the sap for nourishment. Notice the drop of honeydew coming out of the rear of the aphid. Sap pressure has a great influence on how much sap the aphid gets. The more nitrogen you fertilize with the greater the sap pressure might be. This allows more aphids to get sap and their populations may go up.

Compost Teas

Compost teas are made by collecting runoff from a compost pile or placing a sack of compost in water and allowing it to soak for several days. Today, we refer to the resulting mixture as non-aerated compost tea or "NCT."

In recent years, *aerated compost tea* (ACT) has come into popular use, although special equipment is required to maintain it in an aerobic (oxygenated) state. While ACTs are marketed as fertilizers, they've also more recently been touted as antimicrobial agents that displace plant-surface pathogens that could cause foliar and fruit disease.

Here's a succinct take on compost teas written in 2011by Jeff Schalau (Agent for Agriculture & Natural Resources, University of Arizona Cooperative Extension, Yavapai County, AZ), and based on the work of Linda Chalker-Scott—Washington Cooperative Extension, Seattle, WA).

What follows is just one example of how a gardening practice can explode onto the gardening scene, fueled by purported "benefits beyond belief"—often in the absence of sound science and research.

"The use of compost as a soil amendment has been applied for centuries to return essential plant nutrients to cultivated soils (i.e. fertilization). Liquid compost extracts have also been used in this manner. Initially, the popular press and trade journals were quick to sing the praises of ACT using statements like 'beautiful blooms,' 'delicious fruits and vegetables,' and 'thick green turf.' The claims also included: 'keep garden plants, turf, and crops free of disease' and 'provide beneficial organisms.' These claims seemed to have some substance, but not all claims are supported by scientific research. In the case of ACT, it has shown positive results in the laboratory, but field studies gave less consistent results.

"In reviewing the available literature, Dr. Linda Chalker-Scott found only four controlled studies using ACT. One study stated that NCT was more effective at preventing disease than ACT—the anaerobic teas worked better than aerobic teas contrary to claims made in non-scientific literature. Another study looked at the fungal disease called apple scab and found that ACT was not effective at preventing it, and in some cases, appeared to enhance apple scab. A third paper reported mixed results in controlling "damping off" in cucumbers. The fourth study reported the occurrence of human pathogens where ACTs were used. This last finding is particularly worrisome as human pathogens in food supplies continue to be a major public-health concern. These findings do little to support claims that ACT prevents plant disease and raise other larger questions.

"To summarize, plain compost, when used as an amendment or mulch, is most effective at disease suppression and NCTs are also somewhat effective. The use of ACT for disease suppression is not supported by published scientific work. In addition, ACT may contribute to environmental contamination and increase risk of exposure to human pathogens. Given the state of the available science, people should not purchase "compost tea brewers" or ready-made

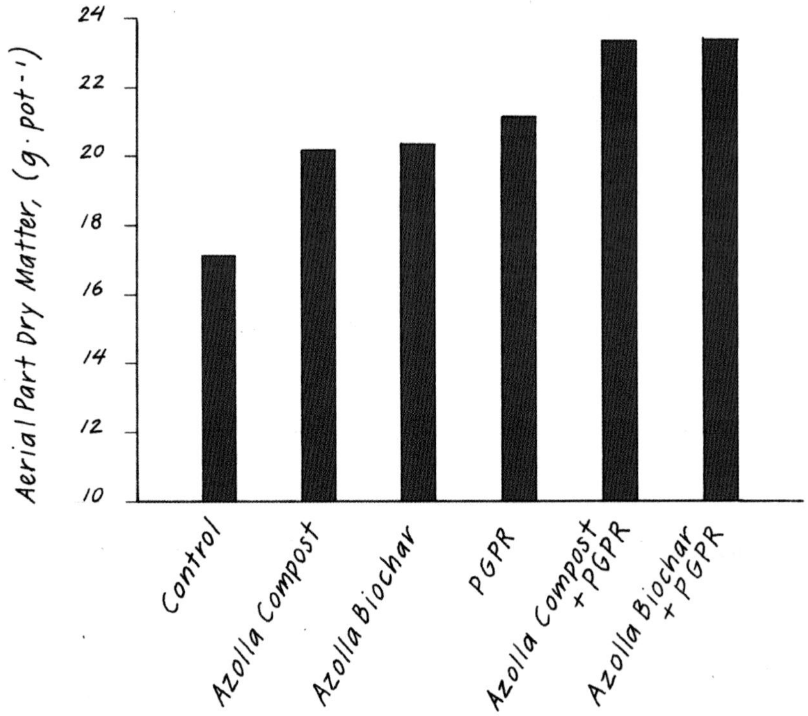

Figure 267. This chart shows the influence in one study of biochar in combination with other fertilizers. The control had no fetilizers added. Azolla (*Azolla* spp.) is a floating plant resembling duckweed or some mosses. It is nitrogen-fixing. PGPR (plant growth-promoting rhizobacteria) is a type of bacteria that can be a local antagonist to soil-borne pathogens or produce a systemic resistance against pathogens throughout the entire plant.

"compost tea." Instead, they should purchase or build a suitable composting system that allows them to recycle kitchen and garden waste into compost.

"Making compost can be an art, but the process is also very forgiving. As with most things, if it sounds too good to be true, it probably is! Products labeled as pesticides (showing active ingredient percentages, EPA registration and establishment numbers, signal words, safety information, etc.) have often been subjected to peer-reviewed science and should perform as noted on the label when used correctly. Other products may make unsubstantiated claims and the buyer must evaluate their merits. Ask product vendors or companies to provide references to peer-reviewed research that support their product's claims."

As I look at it, compost itself is a worthwhile material. Add it to the soil and/or mulch with it and let Nature's rains or sprinkler irrigation soak it in. Simple. Compost teas made at home and used *only* as soil drenches may be helpful, but are extra work. (Also remember that most manure contains potential pathogens, so the teas shouldn't be used on root crops or splashed on the foliage of aboveground plants.) Aerated compost teas—even if homemade—are not worth the cost and may not work as suggested (they're especially suspect as a foliar-disease control).

Biochar

Biochar ("bio" for biomass and "char" for charcoal) is a relatively new kid on the organic-gardening block. It's a sort of glorified charcoal, a carbon-rich solid product resulting from the heating of biomass (such as wood, manure, or leaves) in an oxygen-limited environment.

The long-winded technical definition comes from the International Biochar Initiative, the main biochar advocacy group: "the carbon-rich product [that results] when biomass...is heated with little or no available oxygen. [It is] produced with the intent to be applied to soil as a means to improve soil health, to filter and retain nutrients from percolating soil water, and to provide carbon storage."

The first recent "discovery" of this substance occurred in the tropics. This biochar was actually formed between 450 BC and AD 950, and found in the Amazon Basin. It's also known as "Amazonian dark earth," but is usually called *terra preta*. *Terra preta* is soil that has received large amounts of charred materials and residues from biomass burning, including slash-and-burn agriculture. It may also contain shards of pottery, plant residues, manure, fish, and animal bones.

The modern method of making biochar is called *pyrolysis*. This occurs when biomass (wood chips or other agricultural wastes like cornhusks) is heated up with little or no oxygen supply, causing it to melt into carbon, syngas, and biooil. The carbon is the biochar, which looks like charcoal but is a very different substance.

Biochar, when compared to charcoal, is reputed to have numerous positive effects on soil:

- It acts as a soil conditioner, promoting plant growth by supplying and retaining nutrients
- It sequesters [stabilizes] carbon in the soil, instead of losing it to the atmosphere.
- It provides a place for microorganisms to multiply.
- It increases soil pH.
- It improves soil structure.
- It increases the water-holding capacity of the soil.

While the enhancement of *tropical* soil by biochar is well documented, the benefits of using it in temperate climates for agricultural productivity and financial gain are not yet well established in garden settings.

Biochar can supposedly improve yields by enriching the soil's biomass and furthering soil life (while it should be known that there are many other possible benefits of biochar for temperate-climate crops, I'm considering only these two aspects here).
Alas, the beneficial effect of biochar with respect to mycorrhizal activity in temperate climates remains to be seen. It's thought that the large surface area created by the multitude of pores in biochar particles is attractive to mycorrhizae. However, some studies show biochar actually diminishing the beneficial fungi.

With one study, the VAM root colonization and mycelium lengths *decreased* by 43 to 95 percent with the application of biochar. (This difference depended on the intensity of the original charring temperature, and on how much biochar was being tilled into the soil.) However, in a study of common beans (*Phaseolus vulgaris*) treated with biochar, the biomass left after harvest did increase by 39 percent.

The main question is: does added biochar have as much of a positive effect on soil life and plants as compost alone? Since it was first used in the soils of the tropics, does it provide enough benefits to justify its use in temperate zones?

Some studies find no significant changes. One study concluded: "Biochar on nutrient retention and crop performance in temperate region: a 3-year field trial in a nectarine orchard plant nutrition and soil fertility are hidden or negligible.... biochar was neither harmful to nectarine trees nor detrimental to soil properties... did not result being beneficial to the tree nutritional status, yield, fruit quality or soil properties" (From a 2015 thesis by Giovambattisa Sorrenti, "Biochar In Perennial Crops: Nutritional, Agronomical And Environmental.")

Turns out the answer may be "No." One conclusion was: "Biochar-blended compost had no added value over compost in the field." (This is from an article called "Has Compost With Biochar Applied During the Process Added Value Over Biochar or Compost For Increasing Soil Quality In An Arable Cropping System?;" 2020. T. D. Hose *et al.*, *Applied Soil Ecology*, Volume 156.)

There's much more research to be done on biochar, and the jury is still out. For me, one problem is the cost. This chapter is called "Free Fertilizer." Yikes!—biochar doesn't even enter the picture. In my county (Sonoma, CA) in 2020, a cubic yard of regular organic compost costs $25; with 15 percent-by-volume biochar, it's $45—almost double the cost. If you want to buy only biochar, it'll cost you $180 per cubic yard, more than *seven* times the cost of regular compost. Farmers are told that, in order to conserve 37 percent of soil moisture they should apply 108 tons of biochar —about $360 worth —per 2.5 acres.

And you can't manufacture it on your own. (At least with aerated compost tea—if you "believe" that it works—you can buy enough hardware to make it at home.)

Adding 10- to 50-percent biochar to garden soil has become a common recommendation. One recommendation suggests that a 50/50 mixture of compost and biochar be applied at one-quarter pound per square foot. In my area it could cost almost a $100 per cubic yard for the blend. At a rate of about four inches deep of the 50/50 blend, it would cost you nearly one dollar per square foot. Not cheap.

Biochar also takes a tremendous amount of energy to make. It's "cooked" at 752-932°F for 5–30 min for "slow pyrolysis;" 1562-2282°F for 1–10 seconds for "fast pyrolysis;" and 1652-2192 °F for 0.1–1.0 seconds for "flash pyrolysis." This all has to be done in an anerobic vessel. Try to get those temperatures vat home! Most importantly, the focus of this book is how to create and maintain an organic garden with as few inports from the outside as possible. All things considered, biochar definitely does not fit *any* definition of sustainable gardening.

The Rhizosphere

The rhizosphere—the region of soil influenced by the presence of plant roots—is perhaps the most interesting and important aspect of soil ecology. As mentioned in Chapter 9, recent studies are evaluating plant exudates for use in the process of phytoremediation—the removal or degradation of heavy metals, metalloids, petroleum, hydrocarbons, pesticides, explosives, chlorinated solvents, and industrial byproducts polluting the soil. Plant roots excrete organic compounds (exudates) such as acetic, formic and oxalic acids, as well as organic acids, sugars, aldehydes, ethyl alcohol and other compounds.

Here are some examples of situations, taken from various studies, in which exudates are formed from various plant roots and added to the rhizosphere:

1. Phosphoric acid is excreted by the roots of lupine, peas, buckwheat, mustard and rape (canola). The amount of phosphorus excreted was 14-34 per cent of all the phosphoric acid absorbed by the plant.

2. When detected organic substances were found in the root excretions of lupine, beans, corn, barley, oats and buckwheat, the excretion was observed to reach its maximum amount during the fourth week of growth and at a somewhat earlier period in buckwheat.

3. Nitrogenous and non-nitrogenous organic compounds can be found in the root excretions of corn. Again, the amount of nitrogenous substances in root excretions decreases with the age of the plant.

4. Some researchers have observed phosphatides, amino acids, thiamine, biotin, meso-inositol, paraaminobenzoic acid, carbohydrates, tannins, and alkaloids in the root excretions of plants.

5. Other studies found stuff we can understand—sugars, amino acids, vitamins, and other organic compounds.

6. Small heterotrophic soil bacteria (*Rhizobium* spp.) are capable of forming symbiotic nodules on the roots of leguminous plants (and some non-leguminous shrubs and trees) that fix atmospheric nitrogen on those roots.

Bonus Charts and Photos:

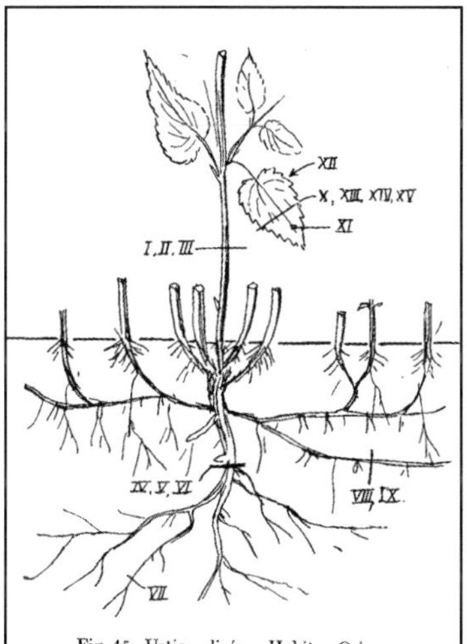

Fig. 45. Urtica dioica. Habit.—Orig.

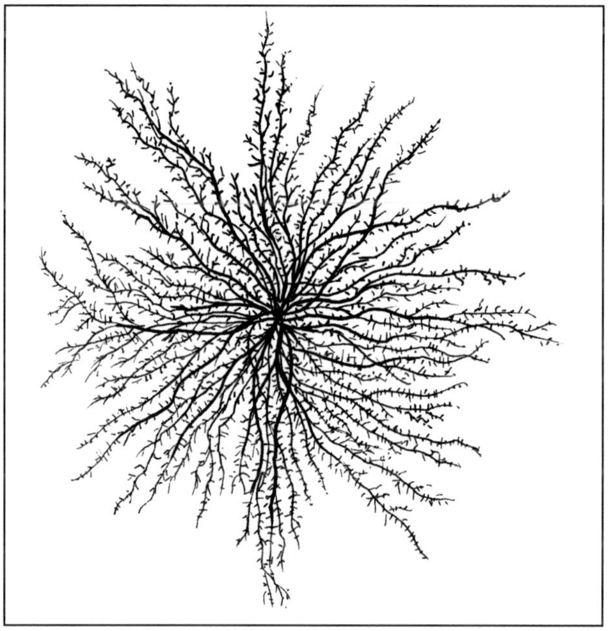

Figure 269. A single stinging nettle plant can have roots five feet in diameter.

Figure 268. Stinging nettle (*Urtica dioica*) is an amazing plant. It is *very* nutritious. Use gloves to harvest and steam or boil the toxin out—like cooking spinach. It is high in calcium, iron, silicon, magnesium, nitrogen, and phosphorus. It is referred to as a dynamic accumulator. Many people think all dynamic accumulators have a taproot. Nettle has an extensive matrix of rhizobia roots that spread wide and up to a foot down, but no taproot. This plant is a good addition to a compost pile or used as a mulch. (The sting disappears when it rots or drys out. The "needles" on dry foliage can irritate.) If left unchecked, it can be invasive. Perhaps gather it only from the wild.

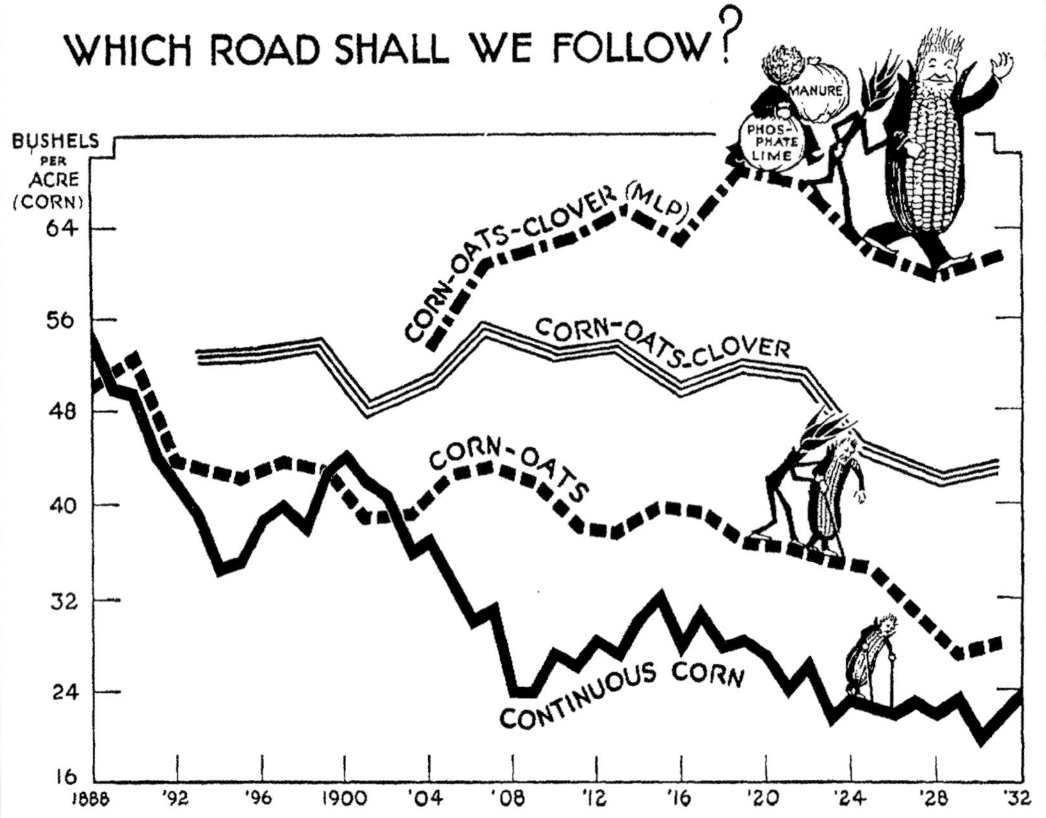

FIGURE 2.—Yields of corn grown continuously; corn grown in a 2-year rotation of corn and oats; and corn grown in a 3-year rotation of corn, oats, and clover, with and without fertilizer, on the Morrow field plots of the University of Illinois.

Figure 270. This is from page 412 of the *Soils and Men Yearbook of Agriculture*, 1938. This book was published before a lot of chemical fertilizers were used, so it applies to organic gardeners and farmers. They did a lot of studies on rotation farmimg with legumes. This study went from 1888 to 1932!! Forty-four years!! I guess the Depression killed it. Or, the guy doing it died. (It's 1938, not many women farming then.)

Anyway, we can learn how much influence the incorporation of a legume in a rotation cycle can be. Of course continous corn with no rotation had the crop drop by about 50 percent. Adding oats was a bit better, but still a serious decline. Adding clover once every three years nearly leveled out the yields of corn. But adding some manure, phosphate, and lime *improved* yields over time. It also had yields higher than the other approaches.

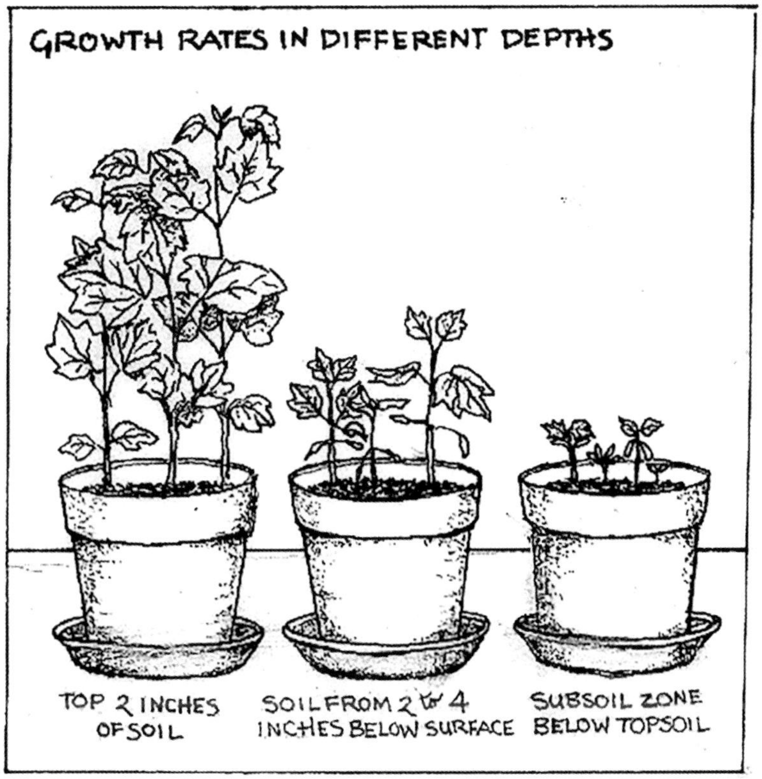

Figure 271. The soil in the pot on the left is gathered only from the top two inches of the forest floor. The middle pot has soil from two to four inches deep. The first pot has foliage almost twice as tall as the second pot. This illustrates how important so-called shallow soils are in the scheme of things. Another aurgument for no-till so the soil's naturally stratified fertility is not disturbed. Yep, more data to support no-till gardening. Start no-till with ornamentals as it is difficult with vegetables. Till for a number of years for vegetables before switching to no-till. In clay soils you may have to till for five or more years to get enough good tilth to allow for switching to no-till. In very sandy soils, you may be able to start the first season.

Chapter 9
Root Dynamics

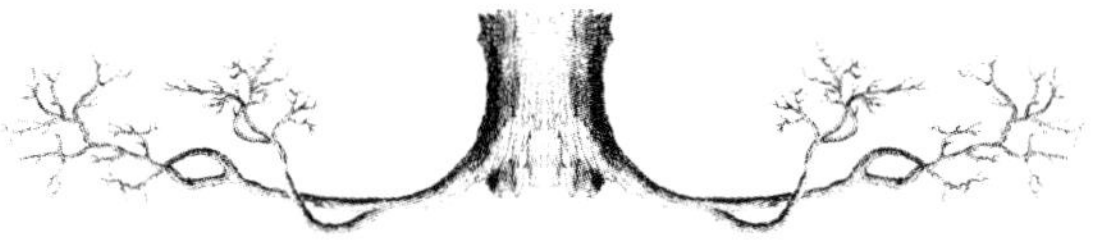

I'm fascinated by roots.

As I cruise down highways, I automatically scan all road-cuts for exposed roots. On Scenic California Highway 1, I'm often not looking at the ocean, but at the beach, snooping for roots creeping out of the soil- and rock-banks. During windstorms, I bundle up in raingear and run around the forest looking for trees about to be buffeted out of their seemingly well-rooted moorings. (As the thigh-thick roots of a falling tree come sifting up through the moist earth, much about true root growth is revealed.)

At one point, I even collected roots. If I spotted an orchard bulldozed to make way for a vineyard or to "plant" some houses, I'd pick through the burn-piles for any unusually revealing root systems. These nakedly exposed twists of wood came home to my house in Occidental, California, to a collection that I suspended like marionettes from the limbs of various forest trees, and enjoyed showing friends and visitors aspects of growth they might never otherwise have seen. (And they also may have discovered my "dirty" secret: I love roots.)

No More Simple Drawings

Many kids' books (and, sadly, some gardening books) show the roots of a tree as a mirror reflection of its foliage canopy. In this dream-world, a pointy pine tree grows roots exactly as wide as the width of its canopy (also known as the dripline), and in a deep triangular shape to match the top. Although these images are deeply ingrained in many people's minds, they have *nothing* to do with reality. The sad truth is that that old stereotypical and misleading mirror-image concept is causing gardeners to waste enormous amounts of water, fertilizer, and mulch, besides harming tree growth, and perhaps killing untold numbers of innocent trees.

Taproot or No Taproot

A *tap-rooted* tree is one with a single, large root growing down from below the trunk to "tap" into deep soil and moisture. From this taproot grow many much smaller horizontal roots called *lateral roots*. Tap-rooted plants often grow very few large-sized lateral roots, thus an accurate illustration of a tap-rooted tree might sort of fit the old mirror-image view of roots, but five things are wrong with this idealized picture:

1. The roots of any tree, even one with a taproot, still grow most of their feeding roots within one to two feet of the soil surface, and these roots roam far beyond the dripline of the tree's canopy to seek food and water. As we'll soon see, this roaming around for food can spread the roots into an area as much as three (or more) times wider than the dripline. *Magnolia grandiflora* trees, for instance, can extend their roots to an area 3.77 times wider than the dripline. In shallow soils with hard-

---

Believe-it-or-not, some trees, as seen above, in the desert have roots that grow back up *toward* the surface to gather intense, but infrequent, rains—flash floods. See page 317.

pans and bedrock, the roots can extend as much as five or more times wider than the old mirror-image concept would have you believe.

2. Very few of the trees most of us grow are actually tap-rooted. Most members of the nut-tree family, such as walnut, pecan, hickory, and butternut, are true tap-rooted trees. (Almond trees belong to the plum family and aren't tap-rooted. Hazelnuts also have no tap-root.) Also, many desert and prairie climate trees, shrubs, and perennials use tap-roots as a survival strategy. Other examples of tap-rooted trees include oak (*Quercus* spp.— primarily when they're young); red-bud (*Cercis* spp.); pines (*Pinus* spp.); persimmon (*Diospyros* spp.); and black walnut (*Juglans niger*). This only applies, however, if the soil is of a good loose texture and deep enough to accommodate a long taproot.

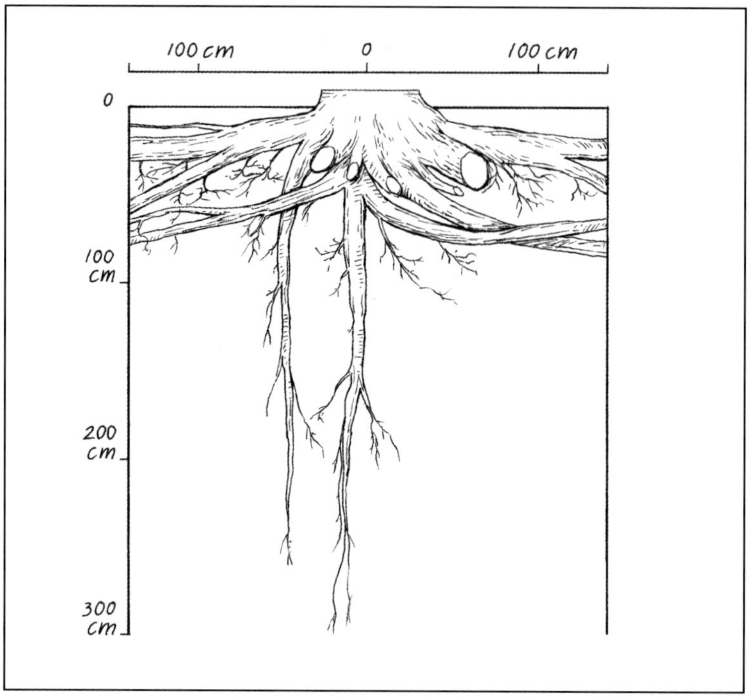

Figure 272. Less than 2 to 5 percent of all trees will grow a taproot. As a rule (where the soil allows), oak, nut, and pine trees—as well as some other trees like persimmons—will form taproots. The taproot of this 59-foot-tall black walnut is 9.8 feet deep.

3. In shallow soils with hardpans, heavy clay, and bedrock, any taproot will be distorted and grow sideways due to the layer of obstruction in the ground.

4. Buying almost any form of containerized (except perhaps tube-grown seedlings) or bareroot tree almost insures that the taproot will be damaged or destroyed— these trees will only grow a natural taproot if planted in deep soils as a seed or nut.

5. Interestingly, some trees start out with a taproot, which then naturally atrophies. Examples include many Californian and western oak trees. Consider the California Coast live oak (*Quercus agrifolia*); when the young seedling is about four to five inches high, the taproot will extend below the acorn for as much as 36 inches or more. After a number of years, the taproot will frequently hit a rock or hard-pan, and the many large horizontal roots (full of short vertical roots) that have been forming all along, take over as 1) stabilizers, 2) as the bulk of root storage of nutrients, and 3) as places from which to launch new feeding roots. Scientists think that the young taproot gives the vulnerable seedling a competitive advantage around where I live, as it can quickly find moisture before the rainless California summer dries out the upper soil. The new root system that forms beneath an oak tree after its taproot has disintegrated represents the most common type of root arrangement—the fibrous root system.

Note: In spite of having a taproot, red oaks gather more nitrogen from the top two inches of soil when compared to soils 10 inches deep. In a 2008 study called "The Vertical Distribution of N and K Uptake in Relation to Root Distribution and Root Uptake," by Hans Göransson, Morton Ingerslev, and Håkan

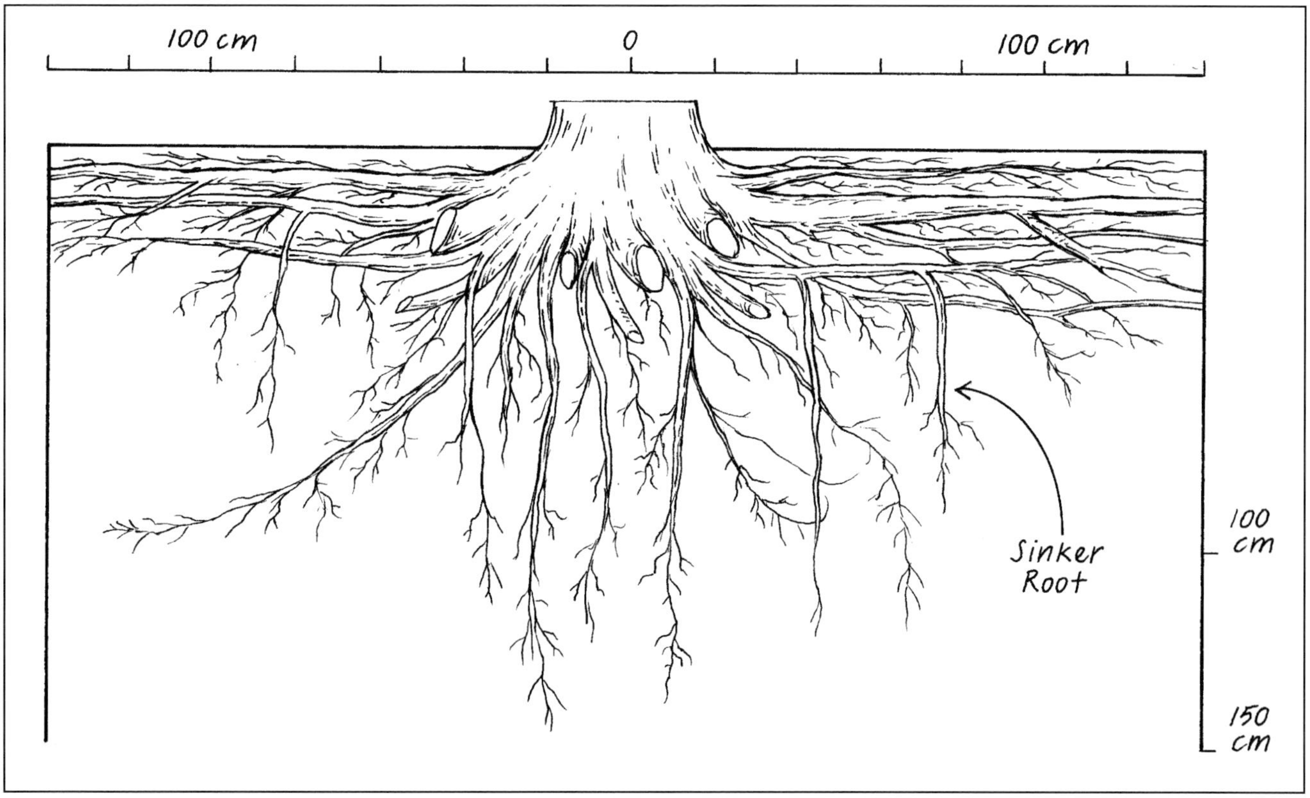

Figure 273. This *Liquidambar styraciflua tree* is 40 feet tall, with roots growing up to 70 inches deep. As with about 95 percent of all trees, it has a fibrous root system and no recognizable taproot. Its vertical roots, growing straight down from some of the lateral roots, are known as "sinker roots," which are also often found on fruit trees. Both Figure 272 and this drawing are based on a book by Noboru Karizumi, a Japanese arborist who never depicted the entire width of the roots, just the crown area. (*Illustrations of Tree Roots*; 1979. This 1121-page book costs $1000 USD. Please buy me one.)

Wallande (*Plant Soil*, 306, P 129-137), it was found that there was no evidence that deep-rooted trees acquired proportionally more nitrogen from deeper soil layers.

Fibrous Roots

A fibrous root system is distinguished by many major horizontal roots that originate at the base of a tree trunk (also known as the "crown" of the root system). Most of these horizontal roots grow within the top one to two feet of the soil, and usually produce numerous vertical roots, called "sinkers," which can arise anywhere along the horizontal roots. In tree parlance, "fibrous" doesn't indicate a very fine, hairy root system such as found on an azalea, heather or fern—it's easier to think of it as meaning "just *not* a taproot."

The natural tendency of a fibrous root system is to grow more sideways than down, and these horizontal roots will *always* grow wider than the width of the aboveground foliage. Our childlike images of perfect mirror reflections above and below the surface of the soil must fade away and be replaced with reality.

Take, for instance, a maturing older oak without a taproot, growing in an ideal soil. The bulk of its roots will be found in the upper one to two feet of earth. If your yard contains heavy clay or a caliche pan (a layer of soil in which the particles are cemented together by calcium carbonate [$CaCO_3$]), say, nine inches down, then the *entire* root system will be found in the upper nine inches.

Wider Than a Spreading Canopy

While all trees are unique, the *very* general guideline for root growth is that in a heavy clay soil, the root system will be half again as wide as the dripline (the edge of the leaf canopy). Conversely, a sandy soil

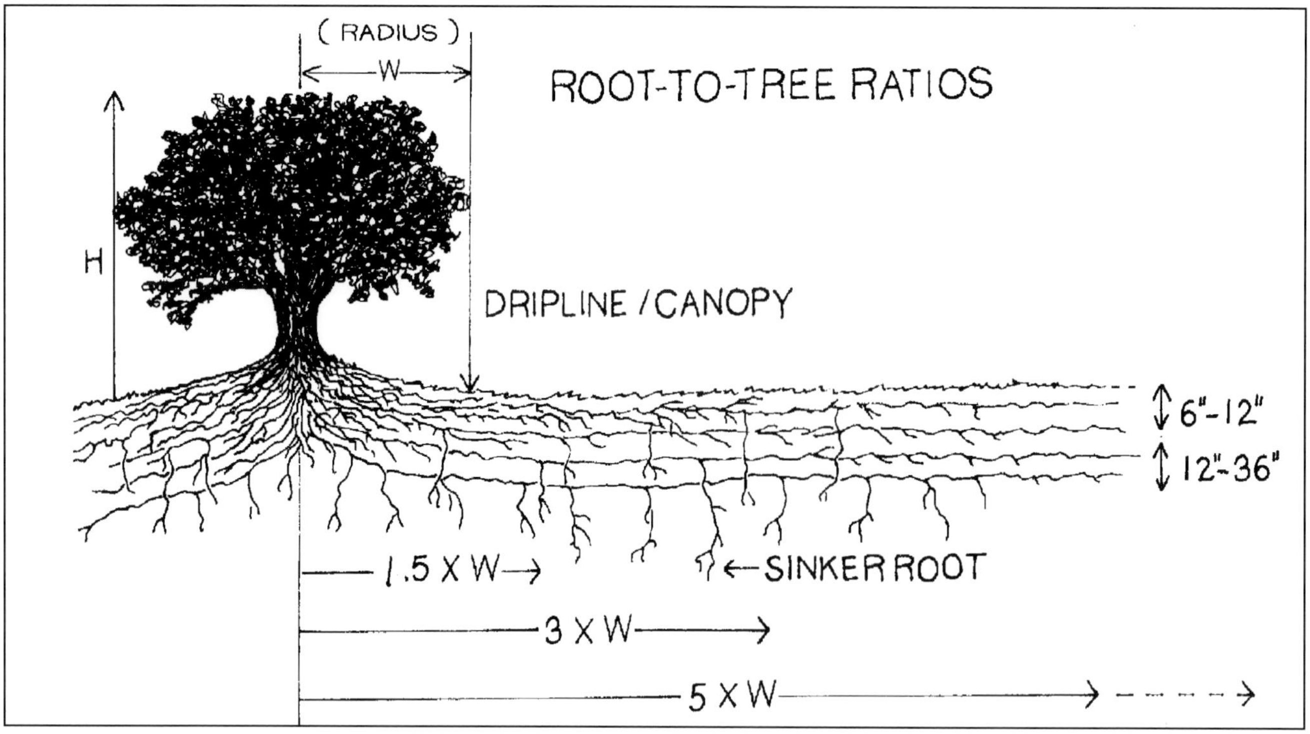

Figure 274. The root growth of a plant in relationship to the soil surface depends primarily on the type of soil in which it's growing. In compacted or heavy soil, roots will grow closer to the surface in order to find an appropriately aerobic zone. In clay soils, 90-95 percent of the roots growing in the top 12 inches of soil (and 50 percent of the plant's absorbing roots) will be found in the top four to six inches. In loamy soils, 90-95 percent of the absorbing roots are found in the top 36 inches of soil, with 50 percent of those in the top 12 inches. As seen here, soil type also affects the spreading width (W) of roots. In very heavy clay soils where the clay is found near the surface, roots can be forced upward and may spread five or more times further than the radius from the trunk to the foliage perimeter (dripline). Roots in loamy soils may spread two to three times wider than the radius of the entire dripline. Heavy loam soils can produce roots that cover an area one-half again as wide as the radius. (Note: not all trees produce vertical sinker roots; they are most common with fruit trees.)

Figure 275. Here's a cross-section of soil at the Pacific coast, with roots growing mostly in the dark soil called the "A horizon." Some roots (but not many) are growing into the tan layer. No roots have grown into the orange clay.

is loose enough to let roots wander with abandon and they can easily grow *three* times wider than the canopy. Again, these simple "rules of root" will vary considerably, depending on the soil.

Let's examine a few amazing examples of aberrant root growth due to specific soil conditions:

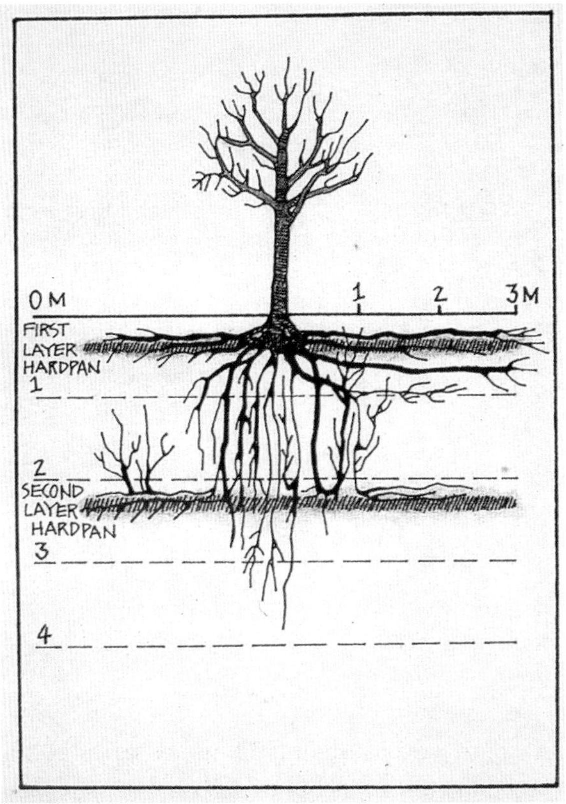

Figure 276. Here, cracks in two hardpan layers allowed some roots through, but the deeper soils were not as hospitable, so some roots started growing back up toward the surface.

Figure 276 shows the roots of a tree growing sideways above a hostile hardpan. One adventuresome root apparently found a crack in the hardpan and grew into the sandy, gravelly soil below. It got as far down as the next hardpan, and started to grow back to where it came from as if to say: "Wrong move, I better go back to those easy, aerobic feeding grounds above."

The walnut tree in Figure 277 didn't find much of a crack in hardpan just eight inches beneath it, so in order to gather enough moisture and nutrients to support this, the roots have traveled an amazing four times wider than the dripline!

The oxygen found in the air we breathe also promotes the health of the many bugs, worms, microbes, bacteria, fungi, and algae living in the soil, and of the plant's slender feeding root-hairs. All of this air-loving (aerobic) soil life liberates soil- and mineral-bound nutrients, directly or indirectly, into a soluble form that is easily absorbed by root hairs. The air flows through tiny passages called *pore spaces*, which

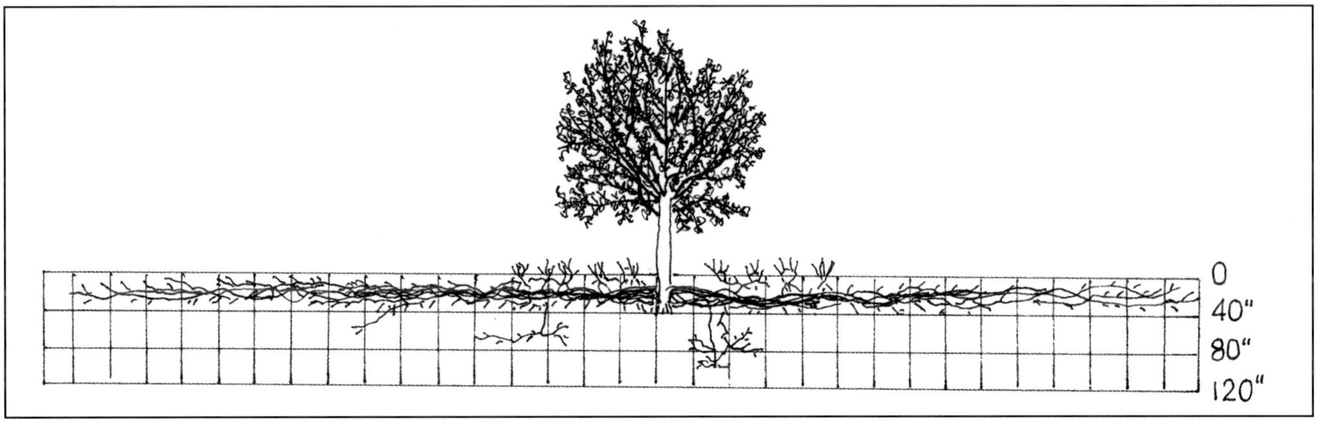

Figure 277. Here's a side view of a walnut tree with root "suckers." The roots are shallow, with 91.75 percent of them found in the top eight inches of soil. This is probably due to the "black clayey silt" located just below the top eight inches, and "iron-spotted silty sand with calcium veins," also unfriendly to roots, below that. As a result, the roots, including a few sinker roots, extend four times wider than the canopy.

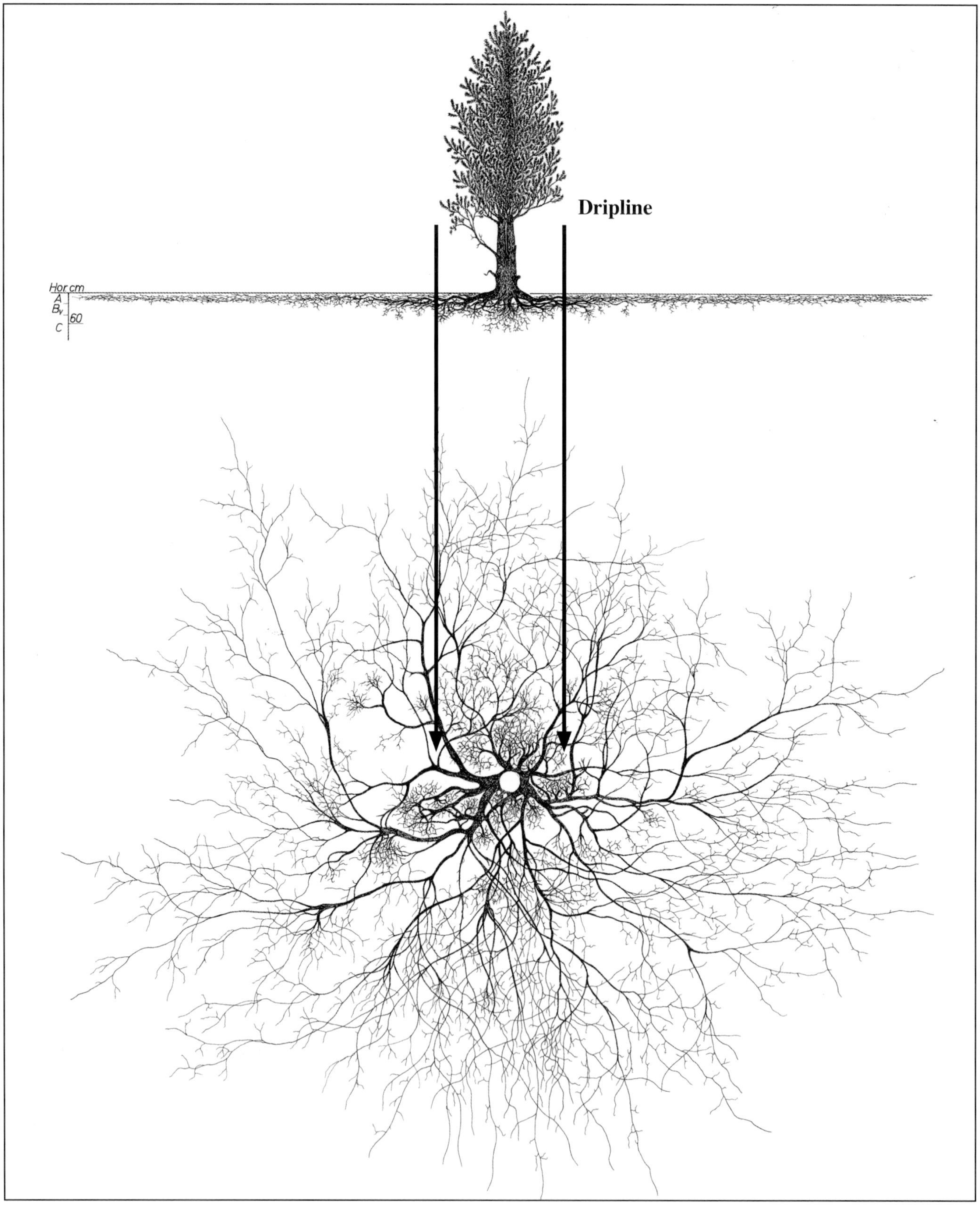

Figure 278. This Swiss stone pine (*Pinus cembra*) is 18 feet tall; its roots (seen as if looking down from the top) extend 3 feet deep and 35 feet wide; the width of the root mass is about twice the height of the tree, and extends far beyond the dripline.

form the natural drainage and air "ducts" of a well-structured soil. Some of the soil's helpful bacteria, called *Rhizobium* bacteria, live in association with roots and have evolved a miraculous process in which those roots directly or indirectly absorb free nitrogen gas. The *Rhizobium* bacteria obtain their nitrogen gas from the air that seeps in through the soil's pore spaces close to the surface. That's why, when you yank a green-bean plant out of the ground, most of the nitrogen-fixing nodules are located near the top of the root system.

Surprisingly, the very soil life that sustains plant growth also produces toxic gases. These gases, released by microorganism and plant growth, must be vented from the soil, their harmful fumes exchanged for life-giving air. Shallow, loose soils quickly perform this exchange, but deep soils take a long time to "inhale" air and "exhale" harmful fumes.

Beneficial bacteria and flora, live near the soil surface because they need plenty of oxygen to fuel their activity, and are responsible for the decomposition of organic matter and the liberation of mineralized (and thus unavailable) nutrients into a soluble form that plants can absorb.

Soil Humus

Soil structure and pore space is greatly dependent upon the soil's organic matter and, more importantly, humus content. As mentioned earlier, humus is a complex compound, much like an organic, spongy plastic. Humus is composed of clumps of soil particles (called colloids) that create good pore space and drainage while simultaneously retaining moisture. Colloidal moisture is greatly responsible for the transport of mineral particles and nutrients from the humus into the root hairs of plants.

The conversion of organic matter into humus is also greatly dependent upon an aerobic environment. The more the soil "breathes," (up to a point), and the more beneficial soil life there is, the greater the production of humus will be, thus increasing the conversion of nutrients into useful forms for healthier plants. If the soil is too hot or too dry, the organic matter can actually start to combust with a form of rapid oxidation that turns most of the organic matter directly into mineralized ash. There is no flame to this combustion, but it does "burn off" valuable fiber and nutrients—especially nitrogen.

Cycles of Tree-Root Growth

A tree can only absorb nutrients that have been dissolved in a water solution—this occurs when the soil is moist, not too wet or too dry. If the soil is too dry, nutrient uptake is inhibited because the soil life can't thrive, and what little water the ground contains is bound tightly to soil particles. Allowing the upper soil to dry out between infrequent irrigations means that nutrient uptake also dries up. Then, when plenty of water is again supplied, the soil is saturated to the point that roots and air-loving soil life may be stressed or killed from *too much* water, and nutrient uptake is inhibited by the lack of air. It also takes some time for the air-loving bacteria to repopulate a soil that's either too wet or too dry, so at either extreme there's a biological lag before the roots get their best meals.

The time of transplanting also has a great impact on root growth. In any climate, fall planting is superior to spring planting, as the roots usually continue growing to some extent during at least part of the winter. In New England, the best time to transplant is October, as it leads to more roots and earlier root growth in the spring. ("Transplant Timing Affects Early Root System Regeneration of Sugar Maple and Northern Red Oak," 2002. Roger Harris, Jody Fanelli and Paul Thrift, *HortScience*, Vol. 7: Issue 6, p 984-987.)

Shallow Suburbia

Many suburban yards come with only a very shallow layer of topsoil, if there's any at all left after house construction. (To make for a good "foundation," many home lots are scraped down to subsoil or rock before building.) If, for example, there's a continuous layer of rich orange clay some 14 inches beneath the surface of your soil, then the 14-inch layer of topsoil is, for all practical purposes, the only place that the roots of your plants will be feeding. Most heavy clay soils, of whatever color—and dark blue or pale white-gray clays are the *worst*—are relatively worthless to deep-feeding roots, as any nutrients are locked up in their tight anaerobic structure. (There are places where glacially deposited topsoil extends down for dozens of feet, but these are more the exception than the rule. Moisture and nutrient absorption still happen mostly in the top one to two feet of the soil.) Again note that the Rutgers Cooperative Extension found that "Withholding water does not send peach roots deeper into the soil, nor do light, frequent irrigations encourage shallow rooting." (*Best Management Practices for Irrigating Peach Trees*, Jerome Frecon, 2002. Rutgers Cooperative Extension: Rutgers; The State University of New Jersey.)

This, by the way, is a *very* important publication as it dispels 14 myths about tree growth and irrigation. Be sure to Google the title and author. Another important point is: "Irrigating one side of a peach tree will not confine the root system to that side of the tree." A must-read.

Role Reversal in the Botanical World

We imagine, in our simplistic way, that big trees have deeper roots than little old vegetables or short, lowly grasses. This concept fails to hold water in the dishpan of reality. Here we see that the spicy but small perennial horseradish plant extends its roots to 12 or more feet, while a cherry tree is rooted to only eight feet. In reality, no consistent relationship exists between the aboveground size of a plant and the depth of its roots, thus we must treat plants as individuals and make no assumptions.

In the tropics, root systems can be even more remarkable. Robert Sanford of Stanford University found 12 species of trees with roots growing *up* their trunks in pursuit of nutrients that trickle down the surface of the bark as accumulated moisture. (*New Scientist*, April 1967, Vol. 9.)

Barking Down the Wrong Root

Sometimes our actions really defy logic. Consider the tendency of many gardeners and lawn-keepers to water a 10- or 50-year-old tree by putting a fully turned-on hose right at the base of it. Although this seems to be everyone's instinct, it's nearly useless to the tree, as anyone who has dug up an older tree's roots can demonstrate. The older roots near the base of the tree are covered with bark, which doesn't allow the formation of moisture-and nutrient-absorbing root hairs. Old roots don't take up any water—period.

To confirm this, I turned to the scientific literature. This phenomenon hasn't been well studied, but finally, after five years of casual searching, I found a study in England that employed radioactive phosphorus isotopes to trace the uptake of moisture; it found that the roots extending in a 4.5-foot radius from the tree trunk accounted for less than 10 percent of the absorbed water and nutrients. This is why putting a sprinkler right up against the trunk of a mature tree is a futile act. You may feel good, but the tree doesn't. (Again, there are exceptions. Trees such as *Eucalyptus* spp., the leaves of which gather a fair amount of rain and fog-drip that flows toward and down their trunks, may well have feeder roots near their bases. I haven't seen any "maps" of their roots.)

Castigating Oscillating Sprinklers

Another approach to watering trees is the use of sprinklers. This is far superior to digging a moat around the tree because the full root-zone has a greater chance of being watered—if the spray isn't localized directly around the trunk. The difficulty with sprinklers is that most of them, and all of the oscillating types, throw out too much water per minute. In addition to the problem of runoff, the fact is that by applying water faster than the soil can absorb it, you guarantee that much of the pore space near the surface of the soil will be flooded. Flooded soil pores means dead root hairs; roots stressed from drowning; little or no nutrient uptake while the pores are flooded; the destruction or temporary disabling of much of the soil's biota; and an anaerobic soil condition.

All of this is compensated for after the soil has drained enough to revive an aerobic condition in its pore spaces, but in the meantime, the plants are under some degree of stress, which makes for less than ideal growth. Because most sprinklers put out water faster than loamy or clayey soils can absorb, people tend to turn them off when the runoff becomes intolerable. Often, this means that the water hasn't percolated very deeply, and that only the upper very few inches of soil are actually getting watered.

This is a common situation with trees planted in lawns. The ideal solution is to "pulse" your sprinkler irrigation. Pulsed watering means turning the sprinkler on for a short period (say, 15 minutes), then off (maybe for 30 minutes), and then on again for a brief time. These intervals of irrigation-drainage-irrigation-drainage continue for as long as is required to moisten the soil to the depth of a foot or more. (The specific cycle lengths and intervals can be determined by careful observation and probe-testing in each individual location.) Obviously, this deep-soaking pulsing of sprinklers is best used with a drip-irrigation controller that can time the on-and-off cycles automatically. Attempting to water this way by manually controlling the sprinkler(s) would take a *lot* of patience.

Recycling Root Feeders

Can I just say that I hate those metal-probe root feeders that seem to be sold everywhere—even by the "gourmet" mail-order companies that claim to only sell high-quality tools? My advice on these items is: if you have one lying around the garage or gardening shed, recycle it or mail it back to the manufacturer. These devices are a waste of money, because they force the water too deeply into the ground for the actual feeding roots to use, and at such a high volume that the soil's pore space rapidly gets flooded.

Fibrous Roots

To re-capitulate: a fibrous root system is simply not one that "just doesn't have a taproot." Fibrous root systems are distinguished by the many major horizontal roots that originate at the base of the tree-trunk (the "crown" of the roots). Most of these horizontal roots are found within the top two to three feet of the soil, and, since aerobic conditions favor root growth, most of the feeding and water-absorbing root hairs are usually found close to the soil surface.

Fibrous root systems may also grow vertical roots, called "sinkers," which can descend from many places along the horizontal roots. Sinker roots, when combined with the total width of the horizontal roots, help to anchor a tree against strong winds, support a heavy fruit or nut crop or expansive limb growth, and counterbalance the weight of snow and ice. Sinkers are often found on the roots of fruit trees.

As mentioned above, many suburban yards have been reduced, by the house-building process, to a very shallow layer of topsoil or none at all. If the soil in your yard includes heavy clay; solid rock; a "plowpan" (formed from rototilling to the same depth every year in a clayey soil); a permanent saturated wet zone ("standing water" or "waterlogging")' or a caliche pan 6 or 12-inches down, then the *entire* root system of your plants will most likely be confined to the upper six or 12 inches of soil.

As a general rule, if your good-soil layer is five inches or less deep, tree and shrub growth will be poor. Ten inches allows for fair growth, 16 inches

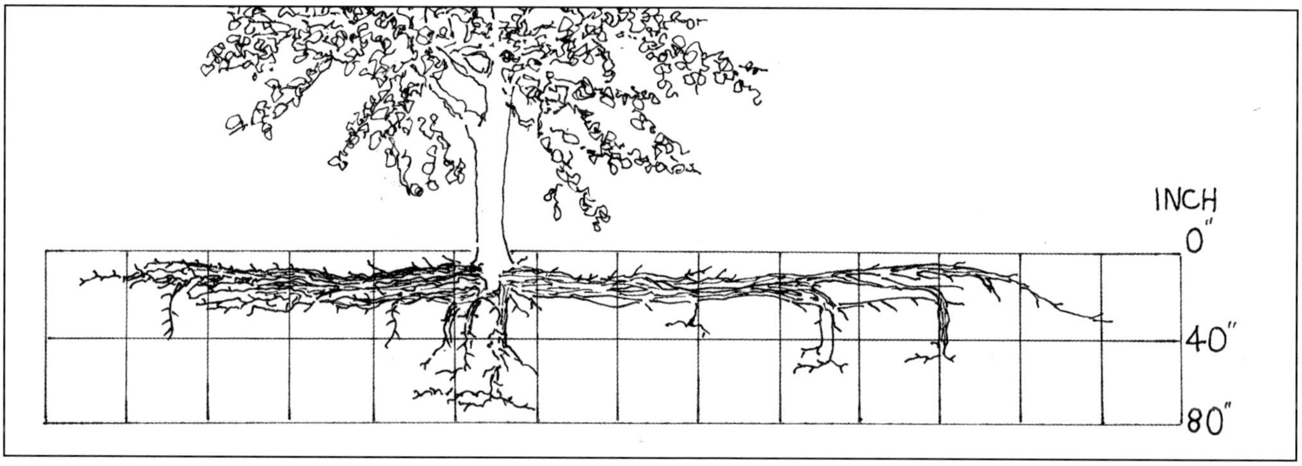

Figure 279. This 25-year-old apricot tree has produced some sinker roots and a few roots that appear to be taproots, but are not true taproots. The lateral roots are found mostly between the top eight inches and the top 40 inches of the soil. (Each square = about one square meter.) The shallow eight-inch zone with no roots is probably the result of cultivation to keep the orchard weed-free. The roots are about double the width of the canopy's drip-line. (This drawing and that in Figure 277, were done by Janos Tamasi. (From: *Root Location of Fruit Trees and its Agrotechnical Consequences*; Akadémiai Kiadó, Budapest, 1986.) This is the best book in the world about fruit tree roots. It's based on 25 years of research in Hungary, and is *very* rare.)

Figure 280. Most trees (this oak tree is an excellent example) show a noticeable root flare where the crown of the root system meets the trunk. If you don't see a flare, the tree was planted too deep—not the healthiest situation.

for good growth, and 20 to 30 inches provides excellent growth. Oddly enough, soils deeper than 30 inches often *decrease* the vigor of plants due to the tendency of such soils to have already been leached of valuable nutrients. (*The Landscape Below Ground*, 1994, American Society of Aboriculturists, P 9.)

The Whole Truth

All of the preceeding points are meant to explain why the *depth* of a planting hole is unimportant. The goal is to allow your tree, shrub, etc., to extend lots of wide lateral roots with abundant sinker roots.

The role of fibrous roots was clearly demonstrated in a survey of peach trees conducted by the Rutgers Cooperative Extension in New Jersey. They found that 80-90% of the fine feeder roots grow in the top 12 inches of soil, and that the top six inches of the soil held 70% of the moisture absorbed. (Again note: "Withholding water does not send peach roots deeper into the soil, nor do light, frequent irrigations encourage shallow rooting." (*Best Management Practices for Irrigating Peach Trees*, Jerome Frecon, 2002. Rutgers Cooperative Extension. Rutgers, The State University of New Jersey.)

Not Your Typical Planting Hole

Planting a tree seems easy enough; just dig an enormous hole, fill it up with lots of amendments, compost, and fertilizers, bind a stake to the trunk, water at the base of the trunk like there's no tomorrow, and sit back. According to many recent studies, all this is literally overkill or just plain wrong.

When I was running a landscape business, I was amazed to discover over the years that the trees planted with the most amendments were most likely to blow over, and that perfectly good-looking small perennial shrubs could be pulled out of the ground years after planting, their roots still conforming to the shape of the original planting hole.

Many of the outdated planting guidelines about amendments continue to be praised with great fervor on radio and television gardening shows throughout the country. (It's little surprise, since soil-amendment companies are often some of the main sponsors of such radio programs.) They're perpetuating profitable myths.

Thanks to research, we now know: that proper planting begins by choosing the best rootstock or plant for your soil; that we needn't use *any* amendments, compost, or fertilizers; that an actual planting hole may not be worthwhile (planting on a mound can be healthier); and that mulch shouldn't be piled up against the trunk or stem of the plant.

The Right Roots for Your Soil

You can't possibly dig a hole wide enough for the natural root-to-crown ratio that I mentioned earlier. What's essential is the compatibility of the roots to the soil in which they're planted. In the case of nearly all trees, most gardening books will advise: "prefers a rich, deep soil," yet few suburban yards are blessed with such ideal soil. I recommend taking a soil sample to your local nursery or Cooperative Extension and asking them which trees are best suited to your situation.

The most difficult situation, as mentioned, is heavy clay soil; its anaerobic condition often overrides any fertility it may contain, and the roots of many trees and shrubs can't thrive in such conditions. Plant

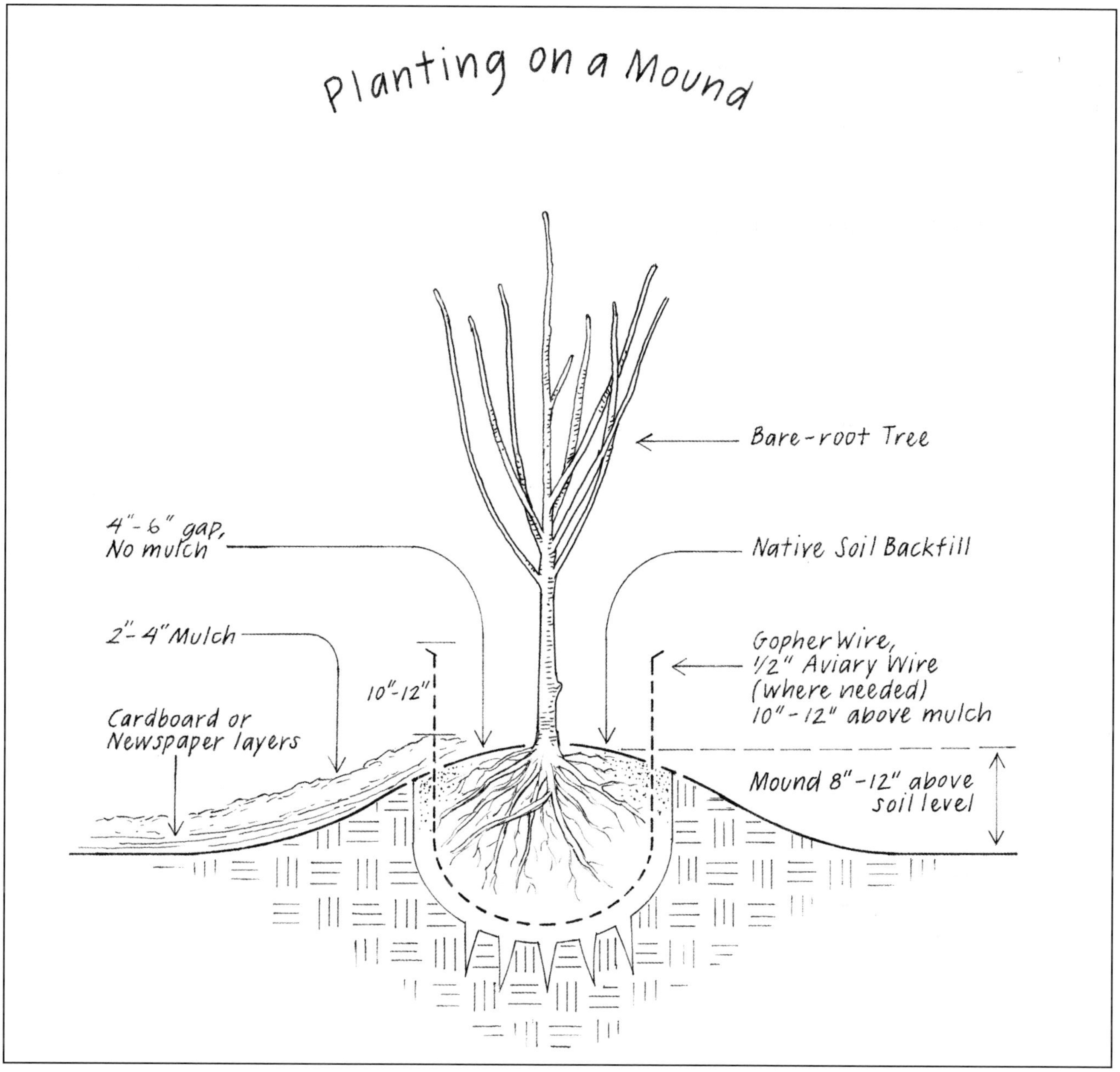

Figure 281. I like to plant trees and shrubs on a mound for three important reasons:
1 - The water drains away from the crown of the plant's root system, preventing crown rot.
2 - The roots are encouraged to grow away from the planting area in search of water and nutrients, thus forming a wider root system with more feeding root hairs, making for a healthier tree.
3 - The wider the root system gets, the better the tree can stand up to wind, snow, and ice.
(If you have a gopher problem, you'll have to dig a hole, put the soil back in the basket, as shown here, and then form your mound and plant.)

them on a raised mound (see Figure 281) for extra protection.

The Scoop on Planting

Digging a hole seems like a simple task, but this is misleading. A poorly dug hole will have a detrimental effect for decades on any tree planted in it, or may even kill the tree.

Remember that the roots of most trees and shrubs will, if unrestricted, extend outward horizontally to cover an area 1.5 times wider than the canopy dripline in clayey soil, and up to 3 times wider in sandy soil. For a wonderful shade tree like a pin oak (*Quercus palustris*) with a 40-foot spread, this means roots extending 60 feet across in clay soil and 120 feet across in sandy soil! The goal of planting is to encourage the tree's or shrub's roots to grow beyond the planting hole into the native soil as quickly as possible; thus the size of even a well-dug hole has little influence on the growth of the tree.

Dig a planting hole only when the soil is slightly moist, but not wet. Turning over a shovel full of wet clay soil causes the plate-like structure of the clay to compress, excluding air and further destroying what little loose structure the soil may have had. The result is a stickier and nearly anaerobic clay soil—a condition that's tough on young root hairs. Sandier soils can be worked when they're moister.

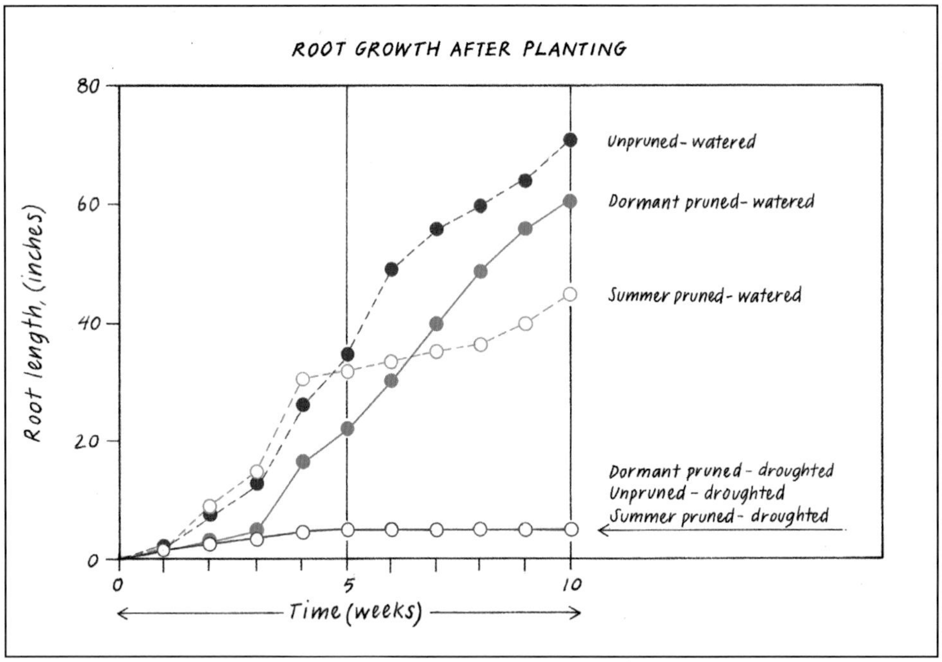

Figure 282. Research showed that the unpruned trees in this study grew 10 inches longer than pruned trees. The wider the root system, the more feeding root hairs and a healthier tree with greater growth. (Based on "One Root, Two Root, Good Root, Dead Root: Myths and Mythology of Tree Root Systems"), a 2011 PowerPoint presentation by Richard J. Hauer, Ph.D. Associate Professor of Urban Forestry, College of Natural Resources, University of Wisconsin at Stevens Point.

Dig when a shovelful of soil will crumble if tapped slightly; to actually dig the hole, switch to a long-handled spading fork with flat forged metal tines.

Make a hole only as deep and wide as you'll need to fit the spread of the root system of your tree or shrub. (Some researchers have found that a hole much wider than it is deep is also helpful, but if your rootstock is well chosen for the soil, a wider hole isn't worth the effort.) In the case of balled-and-burlapped (B&B) plants, the hole must be wide enough for you to pack the soil around the ball. (Be sure to cut off any wire used to hold the B&B together before planting.) If you have gophers, see Figure 281.

It's important to remember to finish off the digging of the hole by using the spading fork to fracture the sides and bottom. This provides nooks and crannies that the young root hairs can use to creep out of the hole and into native soil.

Tree roots can grow a considerable distance if left alone.

In Figure 282 we see that roots grow faster and further than many people expect. At four weeks, the roots of both pruned and unpruned (and irrigated) trees have grown to about 75 cm (two feet, five inches). This puts them well beyond the boundaries of a typical planting hole or mound. By 10 weeks; the roots of the tree that's been irrigated, but not pruned, have grown about five feet, 6 inches (175 cm). Placing your drip irrigation or watering moat less than five feet from the tree will confine the roots (I call this water bondage), so that they can't live up to their potential to explore the largest volume of soil and absorb the greatest amount of nutrients. This is also a way to produce a slight dwarfing effect on the tree, and may leave it more dependent upon you for good fertility.

Once again, remember that 60 percent of the absorbing roots generally extend beyond the dripline—well beyond the planting hole.

No Amendments are Good Amendments

Amendments are defined here for our purposes as materials like sand, peat moss, compost, and rice hulls, which are added to a planting hole, supposedly to improve drainage and keep the soil loose and friable. *Fertilizers*, such as blood meal, cottonseed meal, greensand, and wood ashes, are ingredients that are added to provide mostly nutrients. Some amendments, such as compost, also act as mild fertilizers.

One of the best studies of the effect of amendments on fertility was done at Oklahoma State University at Stillwater by Joseph Schulte and Carl Whitcomb. They planted 108 silver maple trees with 11 different soil treatments and a control (untreated planting hole). One conclusion was: "No benefit was derived from the use of soil amendments either with a good clay loam soil or a very poor silt loam subsoil." They found that the control plantings with no additional amendments generally outperformed the plantings with amendments for drainage and fertility.

In a hard rain, the loose soil of the amendments will turn your planting hole into something like an underground swimming pool full of water, drowning soil life and important root hairs. Adding a lot of amendments and fertilizers only leaves the roots unprepared for the shock of what lies beyond the amended area. Often the roots fail to make it out of the well-amended hole and merely circle around in the loose planting medium, rendering the trees likely to blow over during a storm. The trees most tolerant to wind are those with the widest root systems.

Sandy soils can tolerate some amendments because their structure before and after the addition isn't so different, but since the fiber of the amendments absorbs plenty of water, adding a lot of amendments to drought-prone sandy soils will concentrate the roots in the more moist, amended area, and again leave the tree vulnerable to wind damage.

No Fertilizer Is Good Fertilization

Roots are relatively lazy; they feed where it's easiest. Fertilizers in the planting hole encourage roots to stay in the planting hole. Compost, especially in quantity, turns out to be one of the worst additions when planting trees because it's both an amendment and a fertilizer. If you feel you must fertilize, add the material *beyond* the planting hole to encourage the roots to spread into the native, unamended soils.

Wired for Pests

If your garden is full of gophers, you'll need a bigger hole to allow for a protective wire basket. The bigger the basket, the more roots you'll protect from those hungry little "ground rats." Use 1/2-inch aviary wire or hardware cloth that's four feet wide and comes in rolls. (Be forewarned: I once dug up a six-year-old plum tree that was planted in a wire basket and found that many of the older roots had been damaged by the wire.

Also, in the case of two four-year-old semi-dwarf apple trees in wire baskets that I dug up, most of the roots were growing only *inside* the basket, as the gophers simply ate the roots as they grew outside of it. (See Figure 283) Be sure that the upper edge of the basket rises 12 or more inches above the soil and mulch to keep nocturnally wandering gophers from climbing over the wire. To install a basket, simply dig a hole two to three feet deep and wide, put the basket in, place the tree or shrub, and fill the hole again.

The basket won't, however, protect your plants from other varmints like ground squirrels, mice, rabbits, badgers, chipmunks, beavers, or groundhogs. A cylinder of one-quarter-inch wire hardware cloth extending from several inches below the ground to 18 or more inches above the ground (depending on the average snow depth) *may* prevent rabbits and mice from chewing on the bark.

Planting Success by Heaps and Mounds

As noted in Chapter 4, many ornamental and fruiting trees die of crown rot (*Phytophthora* spp.), a fungal disease of the upper portion of the roots, near the soil surface—without the gardener even suspecting the culprit. The fungus damages the sapwood of the tree, either killing limbs or the entire plant, and once the symptoms appear (pale-yellow and wilted leaves), it's too late to do anything. Observant horticulturists have noticed that the trees that survive this disease were either in well-drained soil or planted on mounds. Because the crown-rot fungus thrives in damp *and* warm soil, the mounds act as preventative measures in climates with rainy summers.

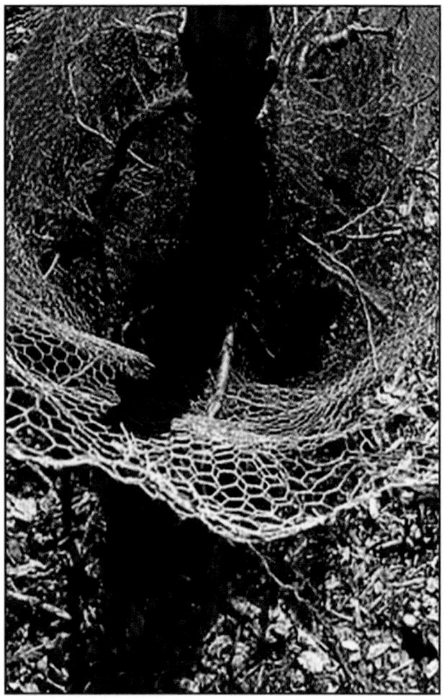

Figure 283. I traveled to Beltsville, MD six times over 10 years to use the USDA Library for research. With each visit I spent some time looking for studies on the subject of foiling gophers with the use of wire baskets. I finally found a study in which the researchers were attempting to use live traps to survey gopher activity. A V-shaped fence was meant to funnel the ground rats into a live trap. Hah! The gophers merely climbed over the 10-inch-tall wire fence. On the right we see that almost no roots have survived beyond the wire basket, the top edge of which was about six inches above the mulch. Make your own baskets from 1/2-inch aviary wire.

Step-by-Step Mound Planting

When planting a bareroot tree on an 18 inch-high mound, a planting hole usually isn't required. While you're making the mound, soak the bare roots in water; this rehydrates the tissue and washes off any foreign matter clinging to the roots.

Next, remove all the grass and weeds from the three- to four-foot diameter of the mound-to-be. Then, with a spading fork, heave and crack the native soil within the entire circle. Scrape up good soil from the surrounding area (after the weeds have been skimmed off) and use it to make a small cone of dirt, 10 to 18 inches wide and tall, in the center of the circle. Spread the roots of the bareroot tree over the top of this cone of soil.

If you're planting a container-grown or tube-grown plant, make sure to tear apart any well-knitted roots, knock off most of the planting mix, and spread any circling roots out on top of the soil cone.

Either way, make sure some of the exposed roots are tucked into the native soil that you fractured open with the spading fork. After the roots of the trees are spread evenly out on the cone of soil, gather plenty of topsoil to construct the rest of the mound. Cover the trunk of the tree to the same depth as it was in the growing grounds; this is noted by a marked change in color on the trunk. Tamp the soil with your shoes to eliminate air pockets that can desiccate young roots. Rake the finished mound into the shape of a gentle knoll.

After planting, soak the entire mound once or twice. Then mulch the mound and beyond about four inches deep, making sure that the mulch stays at least six inches *away* from the trunk.

Many people have disrespected the mounding method, even though I've used it with success with ornamentals and edibles for nearly 20 years, so I decided to call my two favorite authorities on trees—Carl Whitcomb, formerly of the Oklahoma State University at Stillwater, who pioneered the work on using less amendments; and Alex Shigo, formerly with the USDA Forest Service, the man who retaught the world how to prune properly.

When I asked Whitcomb if he'd had any experience with planting on mounds, he replied: "Yes! We [at Oklahoma State University, Stillwater, OK] planted

Figure 284. An attractive way to use a planting mound. This tree is actually planted on a six-foot-wide mound that measures six inches high where the trunk meets the ground.

Figure 285. These trees are in an orchard. The higher mound really provides healthy drainage, but may be less attractive in a home landscape.

out some sycamores seven to eight years ago right on the ground with no planting hole or depression. We just poured a large cone of topsoil over the roots, with a steep angle. They needed more water the first several months during dry spells, but they rooted out well and one of them remains [the others were removed] and has a five-inch caliper. It hasn't received any supplemental watering for seven to eight years."

Shigo replied: "I'm all for it. The best botanical gardens in England and Australia have *all* their trees planted on small mounds. It's easier to regulate watering from a position of dryness than from wetness." When I asked him why this technique hasn't caught on in America, he answered: "Our pioneering spirit still prevails. It's easier to treat trees as expendable than to show them the respect they deserve. Now that we've run out of land and we're trying to plant in tougher settings, like the prairies (where trees don't naturally occur in great abundance) we're finding out how important techniques like this one are."

Proper Staking

You can avoid a lot of problems by purchasing only trees with sturdy trunks that don't require staking. However, if the choice is limited and you're forced to buy a spindly tree, remember that the more a tree flexes in the wind, the stronger it gets.

The stakes, if required, should be placed nowhere near the trunk. The best support comes from two pressure-treated pine, cedar, or redwood stakes, each two to three inches in diameter, placed 8 to 10 inches away from the trunk. They should be placed so that the tree doesn't hit the stake when it blows, and so that the stakes don't hit the roots or root ball when they're pounded into place. (Although if you're planting a large B&B tree, you may have no choice but to drive the stakes through the root ball.) In windy areas, place the stakes so that a line drawn between them is *perpendicular* to the direction of the strongest winds.

Tie the tree as low down as possible on its trunk to allow for the most flexibility. To determine the flex point, grab the trunk firmly with your right hand near its top, and bend the top over gently with your left hand; the tree will probably be able to return to an upright position. Move your right hand down the trunk, grasp it again, and again bend and release the top with your left hand. At a certain point, the top will remain flopped over and not regain its upright position. Tie the tree to the stakes at a point six inches higher than the last position of your right hand.

Finally, use small-gauge (#22) 6- or 10-strand wire to run from each stake to make a figure eight loop around the trunk, passing it through a piece of old garden hose or plastic aquarium hose to keep from damaging the tree. Make sure the combination of wires allows the trunk to sway at least six to eight inches. Then let 'er blow!

Figure 287. This tree-in-bondage may never be free of a stake; its trunk is tied too rigidly to the same stake that was used in the plant container, and the black straps are cinched too tightly. Bamboo stakes like this are used to hold up trees with stems too weak and spindly to stand up straight, but all trees need to be able to move in the wind to develop strong trunks.

Figure 286. It's very hard to find a proper staking in any public setting; here's one of the few I've ever seen. The black-hose support ties are loose enough to allow the tree to blow around a bit, which will help build up its trunk.

Figure 288. This tree is certainly well staked, but the black rubber ties are too stiff to allow a healthy jostling of the trunk. (The green skirt is for protection from deer.)

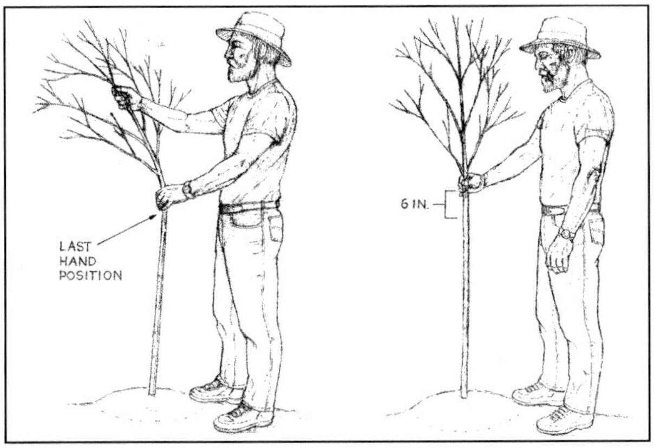

Figure 299. Use this method to find where to use restraining straps on a young tree: hold the trunk in one hand and move your hand up gradually while wiggling the tree back and forth, until the trunk begins to bend more easily and flop a bit. Place the rubber restraints six inches below this point. If the tree still won't stand up straight, buy another tree.

An Idea That Should Be Demoated

One very popular tree-watering method is the construction of a shallow moat around the trunk of a tree and six to twelve inches away from it. As a result, in rainy weather or as a result of improper watering, the trunk may be struggling to survive in a basin of water.

Also, roots are lazy: if there's plenty of water in one spot, they won't waste much energy trying to find another source of moisture. If a tree has been watered this way since it was transplanted, then the roots will be localized to a very small percentage of their natural soil volume—near the moat. If the tree survives the first few months, it can easily become a water and fertilizer "junkie." Also, because the roots don't spread far and wide for proper anchoring, trees surrounded by moats will be the first ones to blow over in wind or fall over in ice storms. Such trees haven't formed a root system extensive enough to forage through a large soil volume for their own nutrients, nor to absorb enough moisture without irrigation.

Let's consider a young moat-watered tree with four cubic feet of soil supporting its root system—a surface area of four square feet and a watering depth of one foot. If its canopy is four feet in diameter and the soil is sandy, the roots would naturally extend to a six-foot radius. But even keeping the moat-irrigation depth to only one foot, the volume of soil this tree would explore is (Pi R-squared, 36 X 3.14) 113 cubic feet—28 times more soil volume than just four square feet from which to gather moisture and fertility.

If the tree is watered *properly* down to two feet, than the root-explored soil volume doubles to 226 cubic feet—56 times the moat-stunted version. The moated tree can't possibly grow enough roots in its lousy four-square-foot area to compensate for what would be the natural extension of its root growth. But the time, water and fertilizer needed will be much greater than that of a properly planted and watered tree.

In Figure 282 we learn that irrigation is *very* important to new root growth. This chart records that the un-watered roots grew about four inches (10 cm) in 10 weeks. The irrigated (also unpruned) roots grew an amazing five feet, six inches (169 cm).

Buying Bareroot Trees

In the early spring you can buy bareroot trees and shrubs, vines, cane fruits, and roses while they're dormant. The plants can cost 25 to 40 percent less than container plants, and with planning on your part, this can be the only time of the year that you buy trees—especially fruit trees.

When sizing up a prospective tree, most people look at the branches first; then (maybe) they check out the roots. Personally, I start with the graft union. You can correct a lot of aspects of the branches and roots with pruning and shaping, but the callous of the graft can't be changed, it's a part of the tree forever. So, let's take a look at what makes a healthy graft union.

On the tree in Figure 300, the cut above the rootstock was made in such a way that it's not forming any callus, so this is a tree I would never consider buying. I don't even need to look at the roots or branches.

Now, here's a graft union that I like in Figure 301. You can see that the callus goes all the way around the perimeter of the graft, and will seal over the rootstock in a short period of time. The newly grafted

Figure 300. This is a very poor grafting union, with the scion growing far to the side of the rootstock. A bad graft can't be changed; the tree is deformed for life.

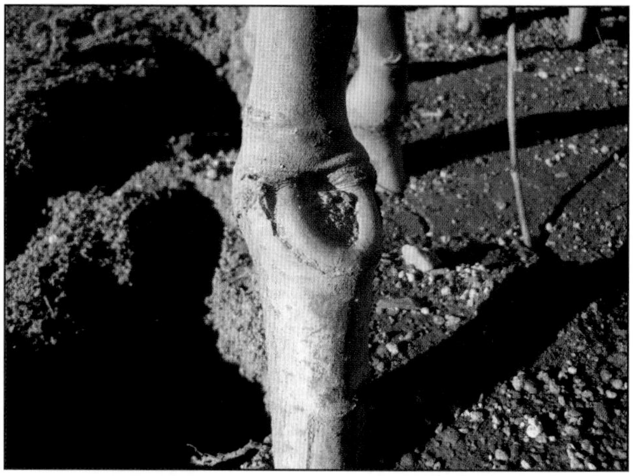

Figure 301. Here's an ideal graft union, with the healing callus beginning to form a circle. The callus will continue to grow to cover the exposed rootstock.

piece has also formed a nice straight shoot of the part of the wood that gives you the variety of fruit you want to eat. Now that I've determined that this tree has formed a good callus, I'll definitely want to look at the roots.

Look at the roots in Figure 302: very sturdy, well formed, and with a nice open area to spread and place over a planting mound. (After placing the roots on top of the mound and spreading them, gather the good soil you've collected from preparing the mound area and use it to cover all of the root system.)

Here's a walnut tree with a good graft and a nice callus, so let's look at its root system. Walnut trees, like some other nut trees, are an exception to the general rule in that they grow a taproot. (Less than five percent of all trees have a taproot; the exceptions are nut trees, oaks, and pines.) In order to harvest this rootstock, the nursery had to cut the taproot off, and it won't grow back. You generally won't get a taprooted tree from either a container-grown or a bare-root tree, even if it originally had one.

So once we've found a tree with a good graft and a good rootstock, we can check out its top. If a tree has almost no branches of any consequence up to almost two-and-one-half feet off the ground it's a problem. If you bought this tree, you wouldn't get any lower

Figure 302. On the left is a bare-root tree with good roots that have wide angles to spread over a planting mound. The roots in the middle are too stiff to be easily spread and should be avoided. The walnut tree on the right has a noticeable taproot, which won't re-sprout and requires a very deep planting hole.

branches; all of them would form above the two-and-one-half-foot level.

A lot of horticulture books recommend leaving a space of at least three feet between the graft and the first branch above it, but that's based on agricultural situations where eventually a tractor must be able to pass under the first row of limbs. Since savvy homeowners are going to mulch—maybe with newspaper—right up to within a few inches from the trunk, they don't need a tree with limbs that start branching a yard above the ground.

Since it's very hard to find a tree with perfect limb placement from the get-go, let's look at what's called a "maiden" or "whip"—a young tree that consists of a single shoot and no branches. If a maiden tree has no branches below four feet, it should be cut off one foot above its graft. When you do this, leaving four or five buds above the calloused union, many shoots will "wake up" and sprout. Having stimulated a lot of shoots to grow along the potential trunk in its first foot above the soil, you can then train them to form sideways branches. Watch the early growth on a maiden tree to make sure that the first and highest bud grows vertically. If it starts to grow out to the side, you can use a stick to tie the shoot up until it's vertical, and it will harden up in that position within a year to maintain the plant's vertical growth.

This is called "pruning the maiden." You may think it's ironic that you've just paid $30 for the tree and now have to cut off $20 worth of its wood, but you really have to do this to force limbs to grow where you want them. I've done this, and wound up with some of the limbs coming out just 6-8 inches above the graft. The fruit was so heavy that one apple that I picked that year was actually sitting on the mulch! That means you may have to bend over to harvest the first few crops, but as the tree grows, the fruiting branches rise higher and higher up the tree. Once this happens, if you don't like the lower branches, you can just cut them off during the summer.

What happens when you order bare-root trees by mail? Well, basically, you get no choice. You wind up with odd shapes, malformed and raggedy limbs, no shoots down below, and branches crossing over. You have no choice as to the health of the graft union. They just give you whatever they pull out of the ground. I avoid mail-order suppliers whenever possible. Go to a local nursery that has a good choice of bare-root stock and find a clean graft, a healthy root system, and a top that's either already well-formed or one that you can cut back.

More and more fruit trees are now sold with the roots all bundled up so you can't review their health and shape. In this case, you look at the graft union first, then the branches. Not ideal, but your only choices.

Planting from Tube Containers

Trees and shrubs grown in various shaped tubes or deep containers are especially useful for planting hedges, windbreaks, trees needed to last a long time, or large woodland-like plantings. The cost of container-grown plants is low, and the roots are less

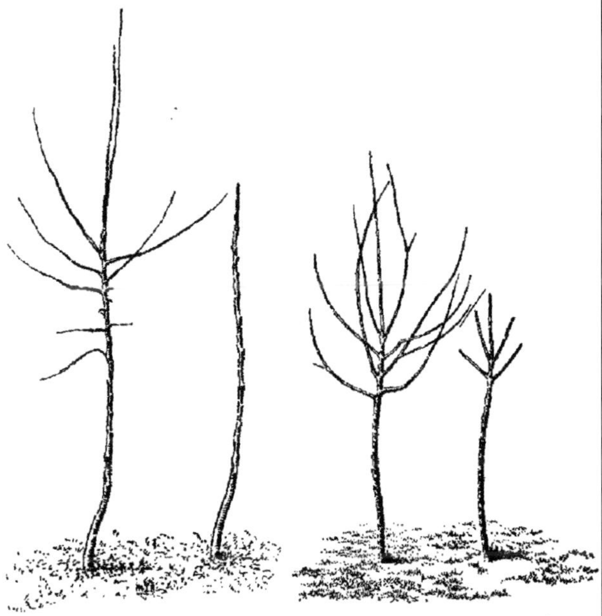

3200. Pruning peach tree at time of transplanting.

3201. Pruning of a young tree on transplanting.

Figure 303. The tree on the right of each illustration is a "maiden" tree as it looks when you buy it. Cutting it back stimulates side shoots (here four of them). The harder you prune back the maiden, the more likely it is to stimulate lower branches and more branches in general. The other tree was also pruned back some and lower branches have formed. (From: L. H. Bailey's *Standard Cyclopedia of Horticulture,* 1914.)

likely to be damaged.

The tubes are usually six to 18 inches deep, with one or more holes or slots at the bottom of each tube. When the taproots reach the hole, they get "air-pruned," which encourages the growth of many lateral roots. When buying tube-grown seedlings, look for just a few air-pruned roots and plenty of lateral ones. The top may be just four to eight inches tall. The roots easily slip out of the tube to be viewed because the holes cause air-pruning that prevents them from growing out of the tubing. The roots usually hold together and are also easy to slip back into the tube.

To plant, turn the tube upside-down and jostle or tug the seedling out. Shake off as much of the potting soil as possible, and spread or fan out the roots. With a spading fork, insert the fork and heave open a crack in the soil (leave the fork in place), and insert the plant, placing the trunk of the seedling in the soil at the same level it was in the tube. To close the planting hole, lift the fork out, place it 4-6 inches away from the seedling and parallel to the previous insertion, insert it again, and press the handle toward the seedling to compress the soil around the roots. Water thoroughly and add a ring of mulch, leaving the trunk uncovered.

The biggest problem with tube-grown plants is finding them; I seldom see them in regular nurseries. In southern California they're occasionally used for tropical food trees like citrus, avocados, and cherimoyas. Many varieties of tube-grown trees can be found in nurseries that are growing native plants for revegetation.

Figure 304. Buy tube-grown trees whenever possible. After two to four years, these trees will have grown as tall, or taller, than more expensive one-gallon plants. When planting perennials, try to find those in four-inch pots; they'll be bigger after a few years than plants from larger containers, especially if the larger plants are root-bound.

Trees & Hardscape Pavement, Trees with Shallow Roots

Whether in the city or the suburbs, tree roots can heave hardscape features such as patios, decks, and walkways. Here are some general guidelines to allow adequate space for root growth (there are many exceptions) when planning your hardscape:

1. When planting trees, allow at least three feet of soil between paved areas so that the roots have room to grow and expand.
2. When you're planting in a space with three to four feet between paved areas, as between a sidewalk and the street, or next to a patio or foundation, select trees that mature at less than 30 feet tall.
3. If you're lucky enough to have a space of five to six feet between paved areas you can plant trees with mature heights of up to 50 feet tall.
4. For trees higher than 50 feet you'll need at least eight feet between paved areas.

(Prepared by Sherry Rindels, Department of Horticulture, Iowa State University, Ames, Iowa. March 31, 1995 issue, P 36.)

Also, consider installing "expansion joints" in sidewalks or in the patio near trees; this will limit possible sidewalk replacements to just a few sections rather than large areas. You might be able to reduce heaving by building curving and narrow sidewalk sections near potentially hazardous trees.

Trees Incompatible With Hardscape:
Here I list the **Common name,** botanical name, and cultivar, followed by comments, Zones and Height x Width**.** This information is provided for placement in woodland plantings or in large lawns.

Norway Maple - *Acer platanoides* 'Superform'. Tolerates dry summer soils. Yellow fall color. Considered to be an invasive tree in Virginia. Ask your local Cooperative Extension about how it performs in your climate. Zones 3-7, 50'x 40'.

Red Maple - *Acer rubrum* 'Franksred'. Vigorous growing tree; very invasive. Great fall color—intense shades of orange-red and red. Prefers moist areas. Some rate it as satisfactory for roots in confined space; others say it needs six feet of planter width. Zones 3-9, 45'x 35'.

Silver Maple - *Acer saccharinum*, The fastest-growing of all the maples, but easily damaged in rain- and ice-storms Well known for its ability to buckle concrete. The roots are also invasive and clog drainpipes. Yellow to rich red in the fall. Zones 3-9, 60'x 40'.

Sugar Maple - *Acer saccharum*. Fall color reveals brilliant shades of yellows, oranges and reds. Leaves are medium to dark green and follow the greenish-yellow flowers of spring. Needs *lots* of room to grow. Zones 3-8, 70'x 45'.

European Beech - *Fagus sylvatica*. A large tree with large roots, often seen planted a shade tree in a lawn. Attractive smooth, gray bark revealed after the golden-brown leaves of fall. Zones (4)5-7, 55' x 40'.

White Ash - *Fraxinus americana*. Grows best in deep, moist soils. Needs a lot of room for its roots. Fall color changes from reddish purple at the outer and upper portion of the canopy to almost yellow down low toward the middle. Zones 3-9, 60'x 60'.

Tulip Tree - *Liriodendron tulipifera* is a handsome specimen tree with stature and good yellow fall color. Susceptible to drought. Due to its size, not a practical tree near walkways, patios or house foundations. Zones 4-9, 80'x40'.

Sweetgum - *Liquidambar styraciflua*. Nice fall color—the color depends upon the cultivar. (The cultivar 'Rotundiloba' is fruitless, thus avoiding the obnoxious prickly "sweetgum balls.") Well known as having invasive roots. Prefers moist soils. Zones 5-9, 70'x 45'.

Southern Magnolia - *Magnolia grandiflora* and *Magnolia* 'Little Gem'. Evergreen. The species Magnolia is a native tree with gorgeous waxy-white flowers up to 15" wide. Blooms in late spring and sometimes off and on during the summer. ('Little Gem' is a compact tree, 20'x8' with three- to four-inch flowers and shallow roots. Grows well in narrow places or by patios.) Both Zones 6-9, *M. grandiflora* 60'x 50'.

London Planetree - *Platanus* x *acerifolia*. Fascinating mottled brown, tan and pale-green bark. Has been listed as an approved street tree in some communities. If planted in a narrow space between two paved areas, it will often heave concrete walkways and curbs. Zones 4-8(9), 80'x 70'.

American Sycamore - *Platanus occidentalis* is a native tree throughout the Eastern and Midwestern parts of this country. Evolved as a riparian tree. Needs at least eight feet between hardscape surfaces, or away from a single hardscape like a patio. (Warning: it drops a ton of large leaves, which is a drag with a patio below.) Best planted in deep, moist soil or in a well-watered lawn with deep soil. Zones 4-9, 80'x 80'.

Lombardy Poplar - *Populus nigra*. Used to a riparian habitat, the roots of this tree grow near the surface. Less of a problem with heaving hardscape than with lawnmowers scalping its surface roots, but still should be kept away from patios or other paved surfaces. Especially in Europe, often seen in *allées,* with the trees lining both sides of a road or driveway—with plenty of room left for the roots. Zones 3-9, 80'x 12'.

Weeping Willow - *Salix babylonica*. Not used as a street tree, as its limbs cascade down to the ground. Prefers to be—and looks more in its element—next to a water feature. In a lawn, the shallow roots will "surface" and be easily scalped by the lawnmower, but probably won't crack pavement. Zones 6-8, 30'x 40'.

Hardscape-Friendly Trees

The following with the notation *, are recommendations for street trees with minimum planting width as established in Chico, CA. Other tree lists for acceptable street trees come from St. Helena, CA; St. Louis, MO; Iowa; and Seattle, WA. Although generally hardscape-friendly, some of these trees will cause problems if they are planted too close to paved surfaces. In some cases I've cited the specific minimum distance from the hardscape.

The varieties of trees allowed in street-side plantings may be very different from those in this list—check with your local planning department. Be certain that the mature height, spread, and cold- and heat-tolerance, are appropriate for your garden or yard before planting. Check to make sure the mature tree won't interfere with phone and power lines. Ask your neighbors what *hasn't* worked. Talk to Cooperative Extension people. Visit old cemeteries to see how big some trees will get.

Trees Near Hardscape, taken from various street-tree approval lists.

Common name, botanical name, cultivar (if any) followed by comments, Zones (using the USDA hardiness zone map) and H x W.
(All trees are deciduous unless noted.)

European Hornbeam - *Carpinus betulus.* 'Fastigiata'. 'Fastigiata' usually indicates a columnar shape, but this tree matures into an oval-vase shape with deep-green leaves and smooth silver-gray bark. This sturdy tree with strong branch attachment prefers a seven-foot area between hardscapes.* Can get scales (sap-sucking insects that resemble small bumps on twigs and branches and prey on most kinds of trees and plants). Zones 4-7(8), 50'x 40'.

Cornelian Cherry - *Cornus mas.* Oval rounded canopy of dark-green leaves that turn a slight purple-red in the fall. This is one of the first trees or shrubs to bloom in early spring—as early as March. Its red fruit is usually eaten by birds in mid-summer. No serious pests. Tolerates alkaline soil. Zones 4-8, 25'x 20'.

Green Ash - *Fraxinus pennyslvanica.* Tolerates high pH, wet soils, dry soils and hot, arid winds. Often used as a windbreak, and very commonly as a large street tree. Yellow fall color. Needs a minimum of seven feet for a healthy root-system.* Don't plant both male and female trees, as the seed pods will be abundant and cause quite a bit of litter. Zones 3-9, 55'x 40'.

Ginkgo - *Ginkgo biloba* 'Fastigiata' and 'Autumn Gold'. An ancient tree and "living fossil," with "roots" that can be traced back some 150-200 million years. Each fall there's a glorious display of hot yellow fan-shaped leaves. Be sure to buy a male tree such as 'Fastigiata' to avoid the smelly fruit of the female tree. 'Autumn Gold' is a male tree with spreading canopy. Ginkgos are pest- and disease-resistant. They're also a good city tree as they tolerate ozone and sulfur dioxide pollutants. Prefers at least seven feet for its roots.* Zones 4-9, 60' x 40'.

Thornless Common Honey Locust - *Gleditsia triacanthos* 'Shademaster'. A popular street tree with bright-green foliage. Prefers at least seven feet of root space.* After leaf-fall, its many long, twisted fruit pods can be messy. Good yellow fall color. Medium-sensitive to salt from winter de-icing. Bark on new trees is tender and will probably get sunburn on the southwest side unless protected with paint. Zones 4-9, 25'x 16'.

Rocky Mountain Juniper - *Juniperus scopulorum* 'Skyrocket'. Evergreen with blue-gray foliage. The species is pyramidal but rather wide. 'Skyrocket' is the narrowest of all the columnar junipers, and can be used in a classic landscape design with parallel rows along a driveway, walkway, (called an allée) or around a pool. Zone 3-7(9), 15'x 2'.

Eastern Red Cedar - *Juniperus virginiana.* Evergreen. Grows well under many conditions where other trees fade. Can be pruned to form a great hedge. Female trees produce berries the birds love to eat. Zones 2-9, 50'x 20'.

Goldenrain Tree - *Koelreuteria paniculata.* A wonderful tree that blooms with bright-yellow flowers from June through late July. Makes interesting pale-green seed pods that turn brown and look a bit frayed. Blooms three to four years after planting. Makes a great tree for the patio or street (low-to-medium salt tolerance). Zones 5-9, 35'x 35'.

Crabapple - *Malus* spp. Many sizes, colors and shapes to choose from, usually with yellow-red fall color. Birds eat the fruit. Some are disease resistant.

Figure 305. The Chinese pistache (*Pistacia chinensis*, not the tree the nuts come from) is a popular ornamental street tree in the west for its fantastic fall colors—a mix of hot oranges and reds.

Medium to high tolerance to salt. Many need five feet of space for its planting area*. Zones 4-8, Many choices; consult with your local nursery.

American Hornbeam - *Ostrya virginiana*. When first planted, the shape of the canopy is pyramidal, as it grows older it becomes more rounded. Good tree for smaller spaces. Needs only four feet as a minimum planting area.* The bark has an interesting shaggy look. Sensitive to salt. Zones 4-9, 35'x 35'.

Persian Parrotia - *Parrotia persica*. One of the top ten trees for fall color, with yellows, orange and scarlet reds. The new leaves bud in shades of red-purple and then turn to a glossy green. It also has fascinating brown shaggy bark. With well-drained soil and a bit of moisture, this tree is virtually free of pests and diseases. Prefers slightly acidic soil, but will tolerate some alkalinity. Zones 4-8, 30'x 20'.

Chinese Pistache - *Pistacia chinensis*. Not the tree the nuts come from. A popular ornamental street tree in the west for its fantastic fall colors, a mix of hot oranges and reds. Be sure to buy a male tree if you don't want red berries to drop onto your patio. Endures drought and is relatively free of pests. It's resistant to oak-root fungus, but if overwatered in a lawn area, it may get verticillium wilt. Zones, 9-10, 40'x 40'.

Scarlet Oak - *Quercus coccinea*. *Q. rubra* and *Q. palustris* are often confused with this tree, but this species flaunts a more glorious red fall color that may last up to a month. Most large oaks need at least a 7' wide distance from hardscape.* It's hard to transplant; transplant large-sized trees (B&B or 24" to 36" box) and give it plenty of room. Zones 4-8(9), 70'x 45'.

Pin Oak/Swamp Oak - *Quercus palustris*. This is a very popular street tree. Has red-gold fall leaves that, in milder climates, turn brown and hang on the tree for most of the winter. One of the better oaks for growing in a lawn. Zones 4-7, 50'x 30'.

Willow Oak - *Quercus phellos*. Needs at least six feet between two hardscapes to grow well.* Considered to have the most graceful foliage of all the oaks. Leaves may remain on the tree though the winter. Established trees are drought-resistant. 50'x 35'.

English Oak - *Quercus robur* 'Fastigiata'. It's said by many that this is one of the best-looking oaks if you want a more columnar tree. The species form is a grand sight—very noble looking—up to 60' tall and 35' wide. Has a high resistance to salt. Most large oaks need at least a 7' wide distance from hardscape.* This cultivar and the species are prone to mildew, but the cultivars 'Rose Hill' and 'Skyocket' are more resistant to it. 'Fastigiata' is 60'x15'. Both Zones 3-7.

Japanese Pagoda Tree - *Styphnolobium japonicum*, formerly *Sophora japonica*. This is another outstandingly good-looking tree, with fragrant cream-colored flowers. In fall, after the July-August bloom, it produces pods that turn from pea-green to yellow and then brown. Flowers and pods can stain hard surfaces. You need a really big space for this tree. Resistant to oak-root fungus. From the Seattle Urban Forest website: (www.ci.seattle.wa.us/transportation/treeswithreservations.htm) "In Seattle, [this tree] is highly susceptible to canker attack." Zones 4-7, 70'x70'.

American Arborvitae - *Thuja occidentalis*. This evergreen is a common tree in home landscapes; due to its "well-behaved" roots, it's often planted near houses. It's a very sturdy tree that's often sheared to make a low hedge or windbreak. Crisp, rich green in the spring and summer, fading to a brownish-green color in the winter. Hardy to -40 F. Takes all kinds of soils. Zones 2-7, 50'x15'.

Figure 306. Root barriers are commonly used in urban street plantings in an effort to prevent hardscape heaving, but roots will always try to grow into the most aerobic soils—something this manufacturer's drawing fails to show. See Figure 307.

Figure 307. Here we see that the root barrier was useless; the roots merely grew right over the top of the collar. This happens often when the barrier is covered by mulch.

Littleleaf Linden -*Tilia cordata* can't take heat, and prefers a cool climate. In areas where it thrives, it's a common street tree. Is prone to pests and diseases like Japanese beetles, and aphids, with their subsequent sooty mold. Zones 3-7, 60'x45'.

Again, be certain of the mature height, spread, and cold- and heat-tolerance of a tree before planting it in your yard or garden. Ask your neighbors what *hasn't* worked. Talk to Cooperative Extension people. Again, visit old cemeteries to see how big some trees get.

Controlling Roots

In cities, another way to protect hardscape from tree roots is a device called a Sidewalk Shield, also known as a Tree Root Barrier. (See Figure 306.) The concept behind these is that the cylinder forces the roots down and they exit it down low, well under the sidewalk, and stay there. Unfortunately this isn't realistic. All roots need aerobic soils, and will grow up from most depths to reach a soil level with enough air to keep them happy. In Figure 276, page 287, we saw that roots that grew down to a hardpan, and even to a second hardpan, often *grew back up* to reach better soil and more air.

In a study by the tree-root guru Laurence Costello, my prejudices were confirmed: "Root depth 30 cm (12 inches) outside barriers was greater for trees with barriers, but at 90cm (25 inches) and 150 cm (five feet) away, depth was equivalent to controls [no root barriers]. Roots tended to grow toward the soil surface after growing under the barriers... suggesting that the subsurface barrier did not have a direct effect on root depth." ("Tree Root Response to Circling Root Barriers," November 1997. Laurence R. Costello, Clyde L. Elmore, and Scott Steinmaus, *Journal of Arboriculture* 23(6) p 211-218.)

A 2008 summary containing the results of seven studies (Barker 1997; Gilman 1995, 1996; Costello et al. 1997; Peper 1998; Peper and Mori 1999; Gilman 2006) in "A Review of Root Barrier Research" by Justin Morgenrot, (*Arboriculture & Urban Forestry.* 34(2):p 84–88), found that "...roots tend toward upward growth once they have passed under (the) barriers.

Other studies found that the roots stayed nearly as deep as the barrier but grew up toward the surface about four feet away from it. This may be enough to protect a narrow sidewalk, but not a patio. These root barriers, which have not been very available in typical retail nurseries or hardware stores, are now showing up there. Don't waste your money.

The Suburbs vs. the City

The soils in urban areas often tend to be very constricted, with a lot of concrete and asphalt. If you're working with shallow soil, this important graph shows how much soil volume you'll need in order to promote good tree growth. The vertical axis shows the projected diameter (DBH=diameter at breast height) of the mature tree. The horizontal axis shows the potential cubic-foot volume of soil needed. Trace from the appropriate spot on the vertical axis to the diagonal line and then down to the bottom axis to reveal how much cubic footage the roots of a healthy mature tree will need. In this example, a tree with a mature diameter of 16 inches requires 1,000 cubic feet of soil. If there's a hardpan, rock, or clay layer in the soil one foot down, for example, a 16-inch-diameter tree would need to extend its roots sideways through 1,000 cubic feet of soil to allow it to grow to a healthy maturity.

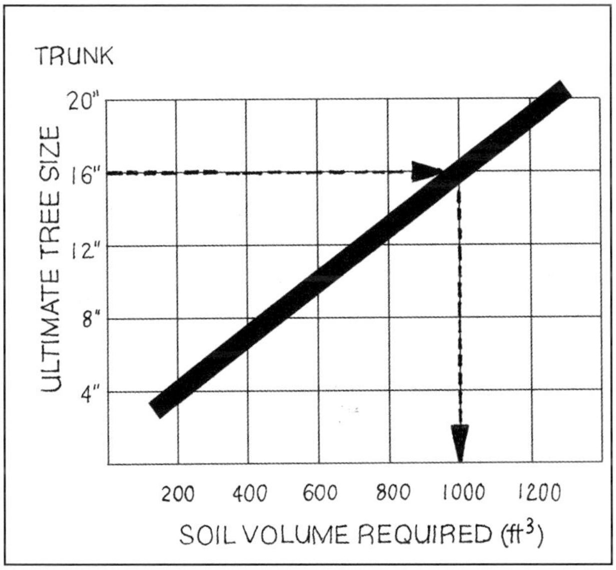

Figure 308. This chart is used to calculate the minimum volume of soil needed to keep a tree's roots healthy. It is used a lot in urban plantings. Here a tree with a mature, 16-inch diameter trunk needs 1000 cubic feet of soil. (Or, 37 cubic yards.)

In the suburbs, patios or walkways near your home may also constrict root mass; this chart can help you determine how far away from patios, paths, and sidewalks to plant in order to provide a healthy cubic-foot volume of soil. From *Diagnosing Root and Soil Disorders on Landscape Trees*, by David Whiting, CMG #103, Colorado State University Extension Service, 2014. This is a great publication on roots, and is available from: www.cmg.colostate.edu.

Exudates

Root exudates create a microbial "soup" in the *rhizosphere* (the first few millimeters outside the "skin" of the roots) that helps to liberate all manner of nutrients. These special chemicals also help roots absorb these, nutrients, develop a healthy environment for microbes; and feed a variety of bacterial and microbial life in the soil. Here are some of the dynamics, as discovered in various scientific studies:

- Plant roots excrete organic compounds (also called exudates) such as acetic, formic, and oxalic acids, as well as organic acids, sugars, aldehydes, ethyl alcohol and other compounds.

- Phosphoric acid is found in the root excretions of lupine, peas, buckwheat, mustard, and rape (canola). The amount of phosphorus excreted was found to equal 14-34 percent of all the phosphoric acid absorbed by the plant.

- When organic substances were detected in the root excretions of lupine, beans, corn, barley, and oats, the excretion reached its maximum during the fourth week of growth; in buckwheat, this maximum was reached somewhat earlier. Upon the ripening and aging of the plants, the amount of root excretions decreases, and stops altogether toward the end of the growth period.

- Roots of germinating seeds of rice excrete sugars, aldehydes, ethyl alcohol, and other compounds.

- Nitrogenous and non-nitrogenous organic compounds can be found in the root excretions of corn. As mentioned, the amount of nitrogenous substances in root excretions decreases with the age of the plant.

- Some researchers observed phosphatides, amino acids, thiamine, biotin, meso-inositol, paraaminobenzoic acid, carbohydrates, tannins and alkaloids in the root excretions of plants. (Don't ask me what most of the items on this list mean; it's just fascinating how many there are.)

- Intercropping fava beans (*Vicia faba*) and maize (*Zea mays*) results in enhanced nitrogen nodulation by the fava beans. Root–root interactions between fava beans and maize significantly increase both nodulation and symbiotic N2 fixation in intercropped fava beans. Furthermore, root exudates from maize promote fava-bean nodulation, whereas root exudates from wheat and barley don't.
- Even pest-resistance comes into play: *Elytrigia repens* (couch-grass) produces several *phytotoxic* compounds in its root exudates. *Hordeum vulgare* (barley) treated with *E. repens* root exudates was significantly less likely to be chosen as a host by aphids than control *H. vulgare* plants.

- Some root exudates in the rhizosphere can increase the availability of soil micronutrients, including iron, manganese, copper, and zinc. Effects of root exudates provide a competitive advantage over neighboring plants that lack the same abilities.

- During the early growth stages of legumes, the majority of belowground nitrogen transfer appears to occur via root exudates. Roots can release organic forms of nitrogen, primarily through root nodules and root tips, and neighboring plants are able to take up these forms of nitrogen.

- "Plant-neighbor detection is usually mediated by direct contact or chemical release. In belowground plant-plant signaling interactions, the detection of neighbors may occur through contact of root tips, soil microbes, especially common mycorrhizal network, or root-secreted signaling chemicals (exudates.)" ("Plant-Neighbor Detection and Allelochemical Response are Driven by Root-Secreted Signaling Chemicals;" 2018. Chui-Hua Kong et. al.; *Nature Communications*, 9:3867.)

- Collectively, blends of defensive root exudate build a diverse and flexible protective layer of five chemical compounds in the rhizosphere that helps fight against disease.

- The root exudates of marigold (*Tagetes* spp.) are known to contain toxic chemicals having *nematicidal* (nematode-killing), insecticidal, fungicidal, and antiviral properties. Thus, as a method of biocontrol of nematodes, growing of marigolds is not only attractive but also highly economical, and helps in environmental amelioration.

- "Nitrogen fertilization did significantly increase nitrogen exudation rates. These results suggest that exudation of dissolved nitrogen from roots or nodules of nitrogen-fixing trees could be a significant, but minor, pathway of transferring N to neighboring plants in a much more rapid and direct way than cycling through death, decomposition and mineralization of plant residues." ("A Test of a Potential Short Cut in the Nitrogen Cycle: the Role of Exudation of Symbiotically Fixed Nitrogen from the Roots of a N-Fixing Tree, and The Effects of Increased Atmospheric CO2 and Temperature," 1999. Shauna M. Uselman, Robert G. Qualls; and Richard B. Thomas, *Plant and Soil*, 210: 21=32.)

- "Some native species grown from the seed of individuals that survived *Centaurea maculosa* (spotted knapweed) invasion were more resistant to the general competitive effects of *Centaurea*; to the root exudates from *Centaurea*; and to a chemical specific to the root exudates of *Centaurea*." ("Natural Selection for Resistance to the Allelopathic Effects of Invasive Plants;" 2005: Ragan M. Callaway, et.al, *Journal of Ecology*, Vol. 93, No. 3 (Jun., 2005), p 576-583.)

- Corn roots emit special chemical signals that attract certain growth-promoting microbes. Some of the bacteria improve the absorption of iron and phosphorus and actually make it difficult for harmful bacteria to colonize the roots. The corn also exudes chemicals called *benzoxazinoids* that help suppress soil-borne diseases.

- Malic acid is a plant exudate that attracts *Bacillus subtilis,* an organism that improves the defense response of roots.

- Another chemical exudate, a peptide, actually *suppresses* immune response, but in some plants, this makes nitrogen fixation more efficient.
- Legumes can release phenolics that suppress soil pathogens and even deter herbivores from feeding on them.

Recent studies are looking at plant exudates as agents of *phytoremediation*—the use of roots to absorb nasty chemicals and compounds—such as heavy metals, metalloids, petroleum, hydrocarbons, pesticides, explosives, chlorinated solvents and industrial by-products—and tie up the poisons in the tissue of the plant. Removal or degradation of these compounds is thought to occur in the rhizosphere.

Allelopathy

Allelopathy is the chemical inhibition of one plant (or other organism) by another, due to the release into the environment of substances acting as germination or growth inhibitors. In other words, it's the science of using root exudates (those special chemicals given off by roots) to limit the growth of plants growing nearby. It's a bit of a stretch from the topic of fertilizers, but it does involve the interaction of chemicals with roots. So, here it goes.

Allelopathy includes chemicals washed from the leaves to the soil; products of decomposition of the dead plant; and volatile chemicals which are released into the atmosphere. The chemicals can include scopoletin, ferulic acid, *p*-hydroxybenzoic acid, catechin, amygdalin, patulin (a "relative" of aflatoxin), phenols, camphor, cineole, flavonoids, alkaloids, coumarins, steroids, cyanogenic glycosides, carbohydrates, amino acids, or terpenes found anywhere in the plant. (I won't even try to define some of those!) Allelopathy can even affect vegetables directly; for instance, broccoli residue interferes with growth of other cruciferous crops that follow. The science of allelopathy is exhaustive and is well founded in research.

One of the most published examples of allelopathy is the black walnut (*Juglans nigra*). (*Not* the English walnut.) Its roots exude a chemical (known as *juglone* (5-hydroxy-alphanapthaquinone!). that's also found in its wood and leaches from the leaves. However, not all plants are suppressed; tomato plants are perhaps the best known for succumbing to juglone. As an example, when tomatoes are grown near black walnuts, and their roots contact the walnut roots, the tomato seedlings will die. Black walnut leaves, bark, or wood chips shouldn't be used to mulch plants or pathways. Even after composting, black walnut refuse may release small amounts of juglone.

Here's a good list of juglone-susceptible and -tolerant plants (from John and Lewis Jett of the West Virginia Extension Service):

PLANTS SENSITIVE TO JUGLONE:

Vegetables: cabbage, and the Solanum Family (eggplant, pepper, potato, tomato).

Fruits: apple, blackberry, blueberry.

Landscape plants: black alder (*Alnus* spp.), azalea, basswood, white birches, Hopa crabapple, hackberry, Amur honeysuckle, Japanese larch, lespedeza, lilac, saucer magnolia, silver maple, mountain laurel, loblolly pine, red pine, scotch pine, white pine, potentilla, privet, rhododendron, Norway spruce, sassafras.

Flowers & herbaceous plants: autumn crocus (*Colchichum* spp.), peony.

Field crops: alfalfa, crimson clover, tobacco.

PLANTS TOLERANT OF JUGLONE

Vegetables: lima beans, snap beans, beets, corn, onions, parsnips.

Fruits: cherry, black raspberry.

Landscape plants: red cedar, crabapple, elm, winged euonymus, forsythia, hawthorn, hemlock, hickories, black locust, most maples, oaks, autumn olive, pachysandra, pawpaw, persimmon, wild rose, sycamore, most viburnums, Virginia creeper.

Flowers and herbaceous plants: bluebells, Kentucky bluegrass, daffodil, daylily, ferns, fescue, iris, Jack-in-the-pulpit, liriope, narcissus, phlox, poison ivy, Shasta daisy, trillium.

Field crops: white clover, red-top grass, orchard grass, soybean, timothy, wheat.

Butternut, English walnut, pecan, shagbark hickory and bitternut hickory trees are all related to black walnut trees, but they make such limited quantities of juglone that toxicity to other plants is hardly ever observed.

Another possibly allelopathic tree is the black elderberry (*Sambucas nigra*). Its allelochemicals, aka *cyanogenic glycosides* can stunt some plants, such as grasses, that grow around it, but stimulate some other plants. ("Potential Allelochemicals from *Sambucus nigra;*" Ana M. Simonet and Armando Zarrelli- http://www.sciencedirect.com/science)

Allelopathic annual plants can even offer weed suppression. Consider barley (*Hordeum vulgare*), rye (*Secale cereale*), sorghum (*Sorgum bicolor*), pea (*Pisum sativum*), buckwheat (*Fagopyrum sagittatum*) Sudangrass, and sorghum-Sudangrass hybrids; all of these plants suppress weeds by releasing natural exudates that inhibit the growth of neighboring plants. Allelopathic substances from all of them have been shown to inhibit the seed growth and early growth of many weeds.

Other allelopathic plants include *Brassica* spp. (such as rapeseed, mustard, broccoli, and cauliflower), castor beans, chrysanthemums, marigolds (dwarf French), and sesame. Because of their stunting effect, such plants should never be planted among other plants as a "traditional" companion plant

Sunflowers are another good example of allelopathic plants. According to a 2008. paper called "Allelopathic Effects of Sunflower (*Helianthus annuus*) on Germination and Growth of Wild Barley (*Hordeum spontaneum*)": "The inhibitory substances present in sunflower plants causing this allellopathicity could be used as a potential natural herbicide resource." (Ashrafi, Y.Z. , Sadeghi, S., Mashhadi, R.H. and Hassan, A.M. Faculty of Agronomy and Animal Science, University College of Agricultural and Natural Resources, University of Tehran, Karaj, Iran. *Journal of Agricultural Technology* 4(1): p219-229.

"Allelopathic Potential of Sunflower (*Helianthus annuus*)," a 2006 report from Colorado State University professor Amanda Broz, states: "Sunflowers release a variety of allelochemicals that could be used to control weeds, however, caution must be used when applying sunflower residues or allelochemicals to the field because they often have inhibitory effects on desired crop species. There are some promising results for certain chemicals, such as the norsesquiterpenes that inhibit dicot weed germination but have little effect on barley"

Some plants, according to the UC Cooperative Extension in Fresno County, CA, are actually toxic to the pest known as root-knot nematodes. These include both Sudan grass and sorghum, which contain a substance called *dhurrin* that degrades into hydrogen cyanide

Phytoremediation & Hyperaccumulation

Phytoremediation (also called hyperaccumulation) is the science of using living plants to accumulate heavy metals and other toxins from the soil. This wasn't even being researched when I wrote my first book in 1986, but great strides have been made since then.

It's a very hot topic because of the heavy-metal contaminated soils in urban areas, especially in those areas where there's a significant amount of lead pollution in the soil. The EPA reports that 400 parts per million is the threshold limit of toxicity in areas where kids play, and 1200 parts per million /kg of dry weight or higher in non-play areas. (EPA *Testing Your Home For Lead in Paint, Dust, And Soil* http://www.epa.gov/oppt/lead/).

The EPA also states that 18% of all our homes rest on soil with lead levels above the recommended limit, from the remains of lead paint, old leaded-gasoline pollution, and nearby factories. If your house was built earlier than than 1978, (when legislation made it illegal to use lead in white paint), or if you live in an urban area, you may want to have your soil tested for lead (cost: about $50) If your garden soil has a high lead content, you can't eat the vegetables listed below. You also can't compost them; they need to be disposed of by special incineration or hazardous-waste facilities.

The cost of mechanical or chemical lead-removal and

Plants and the Elements They Clean from the Soil

Alfalfa	Chromium
Barley	Aluminum
Common Wheat	Lead
Crucifer Broccoli	Lead, Cadmium, Zinc, Cesium or Caesium, Strontium-90, Uranium
Crucifer Cabbage	Lead, Cadmium, Zinc, Cesium or Caesium, Strontium-90, Uranium
Crucifer Cabbage Family	Lead, Cadmium, Zinc, Cesium or Caesium, Strontium-90, Uranium
Crucifer Collards	Lead, Cadmium, Zinc, Cesium or Caesium, Strontium-90, Uranium
Crucifer Kale	Lead, Cadmium, Zinc, Cesium or Caesium, Strontium-90, Uranium
Crucifer mustard flowers	Lead, Cadmium, Zinc, Cesium or Caesium, Strontium-90, Uranium
Crucifer Mustard Greens	Lead, Cadmium, Zinc, Cesium or Caesium, Strontium-90, Uranium
Crucifer or Cabbage family	Lead, Cadmium, Zinc, Cesium or Caesium, Strontium-90, Uranium
Crucifer Rapeseed	Chromium, Mercury, Lead
Crucifer Mustards	Cadmium, Zinc, Cesium or Caesium, Strontium-90, Uranium
Highland Bent Grass	Aluminum, Arsenic, Lead, Zinc
Hydrangea	Aluminum
Indian Mustard	Chromium, Copper, Lead, Zinc, Cesium or Caesium, Strontium-90
Pennygrass, Pennycress	Chromium, Lead
Ragweed	Lead
Red Clover	Zinc
Seapink Thrift	Lead
Smooth Water Hyssop	Copper, Mercury
Sunflower	Arsenic, Chromium, Copper, Lead, Cadmium, Zinc, Cesium or Caesium Strontium-90, Uranium, Methyl Bromide
Violet Crown-Cup	Arsenic
Violet Star Cup	Arsenic

Figure 309. Many plants, acting as so-called "dynamic accumulators," gather and store more of certain metals/minerals than other plants do. The next page features a chart of edible plants and flowers than can accumulate toxic minerals and metals—but only if there is a high level of the contaminant present. In many areas, contaminated foliage can't be composted or dumped, but must be incinerated properly or hauled to a toxic-disposal facility. My county isn't worried about small amounts of extra minerals/metals, and says that I can put the plants in the green-waste bin— but someone down the line will wind up with them in compost from the recycling company.

Edible Hyperaccumulation Examples Include:

Botanical name **Common Name**	**Metal**	**Process**
Allium fistulosum Onion	Lead	Hyperaccumulation
Allium schoernoprasum Chives	Cadmium	Hyperaccumulation
Brassica juncea Indian mustard	Metals Including: lead, cadmium, & uranium	Hyperaccumulation
Brassica oleracea Kale, collards, broccoli, cabbage, Brussel sprouts, cauliflower, *B.* 'Romanesco'	Lead	Hyperaccumulation
Brassica rapa Field mustard	Cadmium & Zinc	Hyperaccumulation
Claytonia perfoliata Miner's lettuce (edible "weed")	Cadmium	Uptake/Accumulation
Helianthus annus Sunflower	See Indian mustard	Hyperaccumulation
Spinacia oleracea Spinach	lead	Uptake/Accumulation
Viola spp. Violets (edible flowers)	metals & cadmium	Hyperaccumulation

Allium fistulosum Onion	Lead	Hyperaccumulation
Allium schoernoprasum Chives	Cadmium	Hyperaccumulation
Brassica juncea Indian mustard	Metals Including: lead, cadmium, & uranium	Hyperaccumulation
Brassica oleracea Kale, collards, broccoli, cabbage, Brussel sprouts, cauliflower, *B.* 'Romanesco'	Lead	Hyperaccumulation
Brassica rapa Field mustard	Cadmium & Zinc	Hyperaccumulation
Claytonia perfoliata Miner's lettuce (edible "weed")	Cadmium	Uptake/Accumulation
Spinacia oleracea Spinach	lead	Uptake/Accumulation
Viola spp. Violets (edible flowers)	metals & cadmium	Hyperaccumulation

restoration of soil is six to ten times more expensive than using plants. Lead levels were, in one study, reduced in 75% of the sites by mechanical/chemical means. Based on this test, costs of between $60,000 to $100,000 per acre were projected to clean an acre of urban soil to a depth of twenty inches. More significantly, planting and harvesting crops of Indian mustard over three growing seasons in lead-laced soil would result in 500 tons of plant biomass to be disposed of by incineration. This figure is only 0.25 percent of the total 20,000 tons of soil that would require excavation and haulage to the nearest hazardous landfill, at a cost of $600,000. ("Cleaning Toxic Sites from the Roots Up;" Niall Kirkwood. *Harvard Design Magazine*, Fall 2002/Winter 2003 Number 17.)

Important Note:

If you eat any of the above plants grown where metals (or PCBs) exist, you're eating a dose of poison. Do *not* compost such plants; composting *cannot* make these elements less hazardous. Technically they must be sent to a hazardous-materials dump (landfill), which may be hard to find and will be costly. Sadly, using plants to clean up soil is not for homeowners.

As another example, consider *Amaranthus retroflexus*, often called "pigweed," a popular nutritious leafy vegetable crop in Africa, Asia, and South America. It's also used by contemporary gardeners in the US (where it's an exotic invasive plant) as a source of edible leaves and grain. *A. retrofllexus* is a C4 plant (an especially efficient photosynthetic plant), and is also an excellent indicator of heavy metal pollution; problems may arise if contaminated plants are cooked and eaten by humans. *Amaranthus retroflexus* is used in phytoremediation for the uptake of cadmium, mercury, and zinc.

For a more interesting revelation: Leon Kochian and Mark Fuhrmann, a DOE scientist at Brookhaven National Laboratory in Upton, New York, found significant variation in the effectiveness of plant species for cleaning up contaminated sites:

"One species, a pigweed called *Amaranthus retroflexus,* was up to 40 times more effective than others tested in removing radiocesium from soil. We were able to remove 3 percent of the total amount in just one three-month growing season," said Kochian, "With two or three yearly crops, the plant could clean up the contaminated site in less than 15 years."

Tobacco and rye grass (I doubt you'll eat these!), along with corn, sunflower, and spinach, have been shown in studies to be capable of removing lead from water, with sunflower having the greatest ability. In one 2000 study, after only one hour of treatment, sunflowers reduced lead concentrations significantly. Indian mustard removed a wide concentration range of lead (4 mg/ Liter-500 mg/Liter). (Raskin, I. And B. D. Ensley: *Phytoremediation of Toxic Metals: Using Plants to Clean Up the Environment*. John Wiley & Sons, Inc., New York.)

In a study of hyperaccumulation in India, it was found that vegetable-crop plants like guar (*Cyamopsis psoralioides*), cauliflower (*Brassica oleracea* var. *botrytis*), chili (*Capsicum annuum*), dill (*Anethum graveolens*), and mint (*Mentha piperata*), were hyper-accumulative. Ivy gourd (*Coccinia indica*), (*Abelmoschus esculentus*), and luffa (*Luffa acutangula*) were hypo-accumulative [slightly accumulative]. It was suggested that guar (*Cyamopsis soralioides*), could potentially be used to decontaminate soils polluted by heavy metals. The authors say: "The data on accumulation and mobility of heavy metals such as Cd [Cadmium], Co [Cobalt], Cu [Copper], Fe [Iron], Ni [Nickel], Pb [Lead] and Zn [Zinc] from soil to leaves through roots and stems, suggested that all the metals were highly mobile."

A report on "Comparative Phytoremediation of Chromium-Contaminated Soils by Fenugreek, Spinach, and Raya" *(Communications in Soil Science and Plant Analysis, Volume 38, Issue 11-12, 2007*) states: "The overall mean uptake of Cr in shoots was almost four times, and in roots was about two times higher in raya [Indian name for mustard plant] compared to fenugreek. The findings indicated that family *Cruciferae* (raya) was most tolerant to Cr toxicity, followed by Chenopodiaceae (spinach) and Fabaceae (fenugreek). Because raya removed the highest amount of Cr from soil, it could be used for pytoremediation of mildly Cr-contaminated soils."

(Hyperaccumulation and Mobility of Heavy Metals in Vegetable Crops in India, June, 2009. Nirmal Kumar J.I., Hiren Soni, Rita N. Kumar and Ira Bhatt, *The Journal of Agriculture and Environment* Vol:10, Technical Paper 29.)

One of the most highly studied hyperaccumulating plants goes by the charming name of "alpine pennycress" (*Thlaspi caerulescens*). In work done by Leon V. Kochian, an international expert on plant responses to environmental stress, over a 13-year period, he found that "…*Thlaspi* accumulates these metals [zinc and cadmium] in its shoots at astoundingly high levels. A typical plant may accumulate about 100 parts per million (ppm) zinc and 1ppm cadmium. *Thlaspi* can accumulate up to 30,000 ppm zinc and 1,500 ppm cadmium in its shoots, while exhibiting few or no toxicity symptoms….a normal plant can be poisoned with as little as 1,000 ppm of zinc or 20 to 50 ppm of cadmium in its shoots." *Thlaspi* is a small, weedy member of the broccoli and cabbage family, so this plant is a hyperaccumulation indicator for all of the Brassicaceae (Cruciferae) family of edible plants.

A short summary:

Copper: Smooth Water Hyssop, Thyme-Leafed Gratiola, Water Hyssop, Indian Mustard, Tape Grass, Water Hyacinth, Copper Flower, Sunflower, Creosote Bush, Duckweed, Water Lettuce, Alpine Pennycress.

Mercury: Water Hyssop, Smooth Water Hyssop, Thyme-leafed Gratiola, Rapeseed, Water Hyacinth, Hydrilla, Water Lettuce, Osier.

Lead: Highland Bentgrass, Ragweed, Seapink Thrift, Pacific Mosquitofern, Kale, Mustard Greens, Collards, Broccoli, Cabbage, Water Hyssop, Indian Mustard, Rapeseed, Tape Grass, Water Hyacinth, Blue Sheep Fescue, Sunflower, Hydrilla, Duckweed, Osier, Kariba Weeds or Water Ferns, Pennycress, Common Wheat.

Root Communication

It gets better. In one study, scientists tried infecting the leaves of a plant with a pathogenic bacterium, *Pseudomonas syringae*. The leaves of the infected plants began to yellow and show other symptoms. However, infected plants that were inoculated with the beneficial microbe *Bacillus subtilis* were unharmed by the disease, because the microbe formed a shielding biofilm with antimicrobial features around the roots. The researchers were actually able to detect the transmission of a signal from the leaves to the roots with *B. subtilis* in the soil. The roots reacted to this by secreting a beneficial compound called malic acid (an exudate) that attracted *B. subtilis*. As a result of this new knowledge, farmers can now add the microbe *B. subtilis* to their soil to increase plant immunity.

I'm also fascinated to find out that corn crops even emit chemical signals that attract growth-promoting beneficial microbes to live among their roots. On the minus side, some of the microbes can influence which plants survive in certain environments; this may assist invasive species to get a foothold. There are always two sides to any coin.

Bonus Charts and Photos:

Figure 310. Photographs from Paris, France, in 2000. It seems that the understanding of roots hasn't improved enough since 1785 to warrant changing the sign, which says: "*L'Arbre Prend Racines*" (The Tree Takes Root). The erosion of the slope of the "labyrinth" (apparently a feature in the garden where this sign is found) has exposed the roots of this common plane tree, planted in 1785.

Usually invisible, the root systems of trees can be immense—but are often thought to have a volume only equal to that of the visible part of the tree. The roots anchor a tree in the soil and are responsible for the absorption of water and minerals necessary to its development."

Figure 311. And now, a pop quiz: Q. What's wrong with this photograph? A. The tree trunk doesn't flare at the bottom. Here at the Missouri Botanic Garden, someone either planted the tree too deep or mulched too high—both posing a potential root rot problem.

Figure 312. These photos show two ways of using in-line drip-irrigation tubing to support good root growth, and to irrigate most, if not all, of the roots of the plantings shown. The spacing of parallel lines on the left is presumably appropriate for the soil type—the better the drainage, the closer the lengths of tubing should be. On the right (before mulch was applied), you can see that the tubing snakes from plant to plant throughout the planting. The wet spots not near the transplants will water extending roots as the plants grow; in a short while, the entire root system will be fully irrigated. I prefer the parallel lines on the left, as they're much easier to install and probably take less tubing.

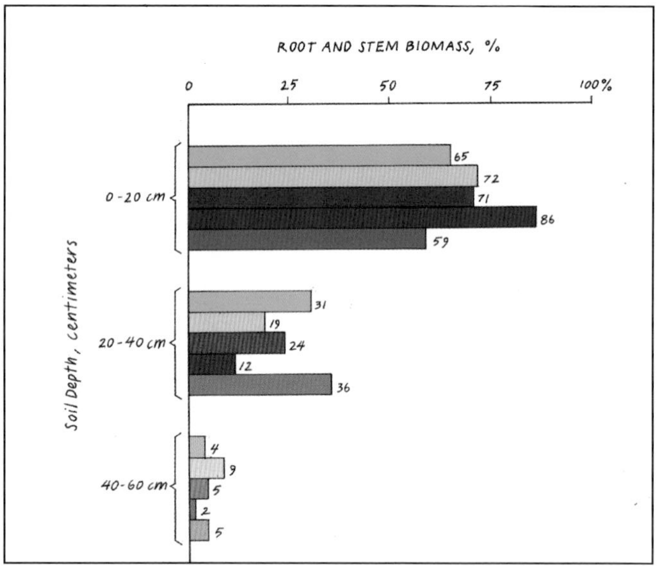

Figure 313. This chart shows the percentage of the root-and-stem mass of five California chaparral plants. Even though these are very drought-resistant plants, most of their roots are growing in the top 20 centimeters (eight inches) of the soil. (From top to bottom of each zone.)
1 - *Arctostaphylos pugens* (Green)
2 - *Ceanothus greggii* (Orange)
3 - *Eriogonum fasciculatum* (Red)
4 - *Adenostoma fasciculatum* (Purple)
5 - *Haplopappus pinifolius* (Blue)

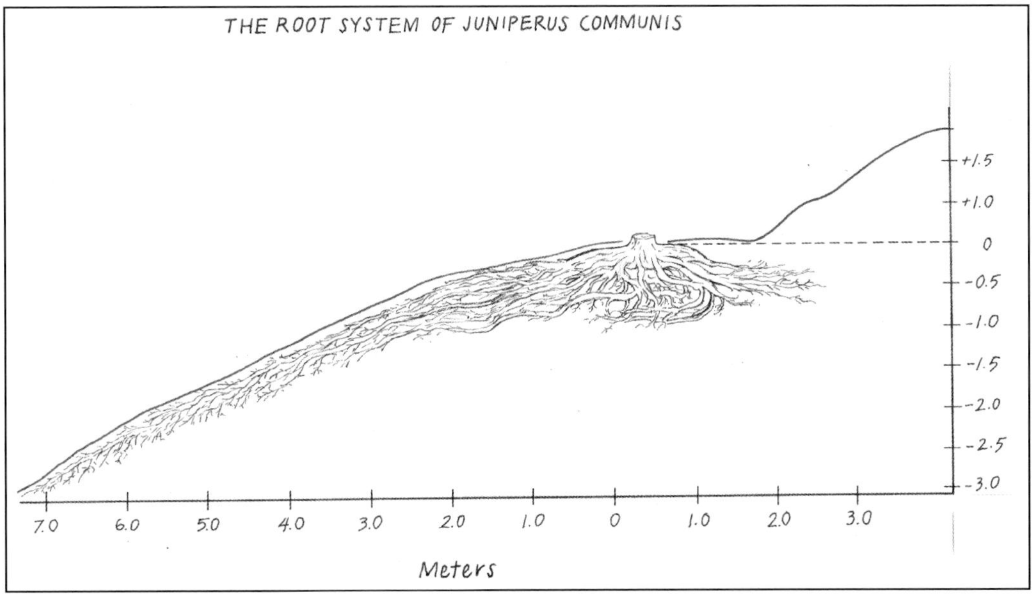

Figure 314. Although the root system of the juniper (*Juniperus communis*) shown here extends as far as 23 feet downhill, the longer-downhill-than-uphill scenario is not necessarily the rule for trees on slopes. I have accurate "from life" drawings in my files that also show roots growing uphill. I can't see a "rule of thumb" about roots and slopes.

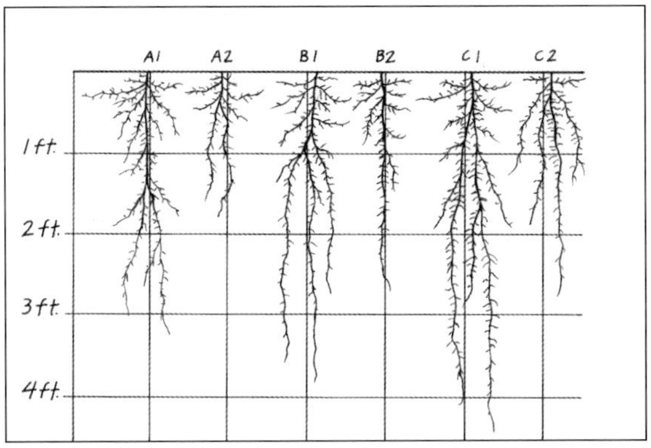

Figure 315. The roots designated here as A1, B1, and C1 are those of plants growing in full sun; the others grew in half-shade conditions. When light is limited, roots don't grow longer and wider in search of nutrition and moisture, and trees become stunted.

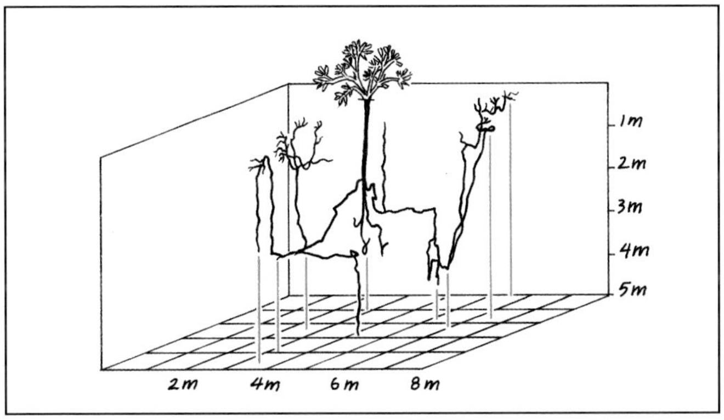

Figure 316. Some of the roots of this desert plant (*Koberlinia spinosa*, the crucifixion thorn) may grow over 15 feet deep to gather moisture from a slow rain after it has soaked into the deeper soil. Here we see roots growing up to within four inches of the soil surface in order to absorb rains from flash floods that don't percolate deeply.

(Root systems of some Chihuahuan Desert plants, 2001. Robert Gibbins and James Lenz. R., *Journal of Arid Environments*, 49: P. 221-263.)

Figure 317. There are two flare root zones on this single tree. One flare is down by the wheelbarrow. The other is in my Dad's left hand. The tree was planted too deep, but managed to resprout some sizeable lateral roots near the soil's surface. (At the Missouri Botanic Gardens, no less.)

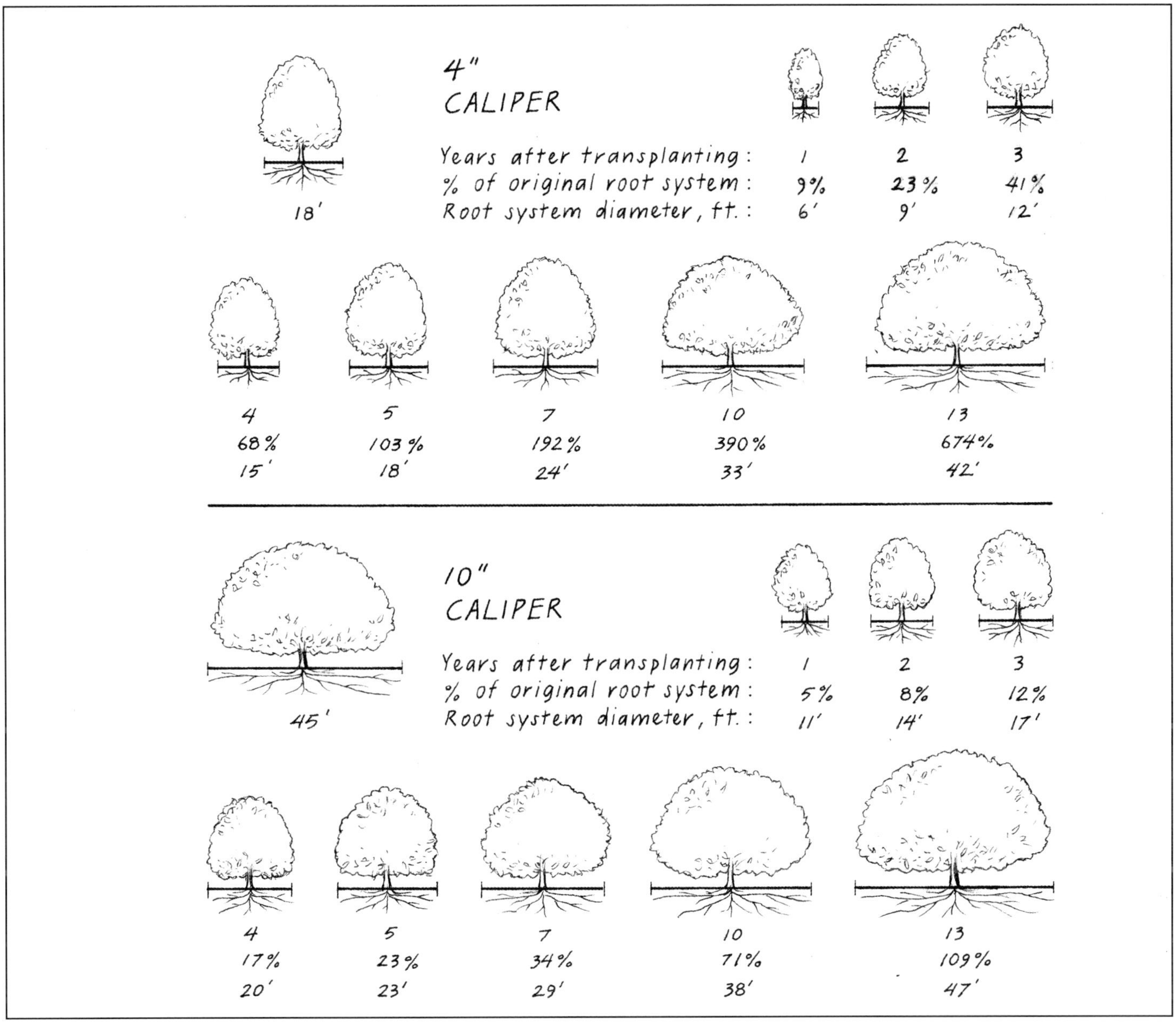

Figure 318. There is a myth that, when buying trees, bigger/taller is better. Not so. For many years earlier my observation was that smaller plants outgrew larger plants within one to five years depending on the plant and the root system in the container or the B&B rootball. I finally found this study published in 1985. Sure wish I had found it earlier to corelate my observations with science. This chart needs a careful review. The top two rows represent a four-inch caliper dug-up tree 18 feet in diameter. The roots were cut back to six feet in diameter. After five years the root system was just a bit bigger than the size of the original diameter at 103%. The lower two rows show a much bigger tree with a ten-inch diameter trunk. The roots were cut back to five percent at 11 feet. It took *13* years for the roots to grow 109%. Save money and buy the smaller plant. Most gardeners I work with now prefer perennial plants in four-inch containers. They outgrow one-gallon container plants in one to two years.

(Tree size affects root regeneration and top growth after transplanting, 1985. Gary Watson, *The Journal of Aboriculture*, Vol. 11, No. 2, P. 37-40.)

Chapter 10
Companion Planting

Here's what Wikipedia says about companion planting: "Companion planting in gardening and agriculture is the planting of different crops in proximity for any of a number of different reasons, including pest control, pollination, providing habitat for beneficial insects, maximizing use of space, and to otherwise increase crop productivity."

In the 1970s, I really wanted to believe in the version of companion planting extolled in Rodale Press publications and by various gardening gurus. And I did, until I looked for the science to back up what turned out too often to be anecdotal information. Luckily, there's been a lot of important research since then.

There are actually many definitions for companion planting. The one I like is: "a specific type of polyculture when two plants are grown together because they are thought to have a beneficial, synergistic improvement on the growth of each other. This not to be confused with intercropping, which is the good use of space by planting vegetables close together or in close sequence to each other."

Figure 319. A lithograph from 1772 shows a marigold before being bred for larger flower petals.

Another definition comes from *Companion Planting and Insect Pest Control* (by Joyce E. Parker, William E. Snyder, George C. Hamilton and Cesar Rodriguez-Saona, 2013.) "We define companion plants as interplantings of one crop (the companion) within another (the protection target), where the companion directly benefits the target through a specific known (or suspected) mechanism."

To this day my favorite example of the complexity of companion plants is a study (published in 1980 by Rodale Press!) of intercropping bush beans with marigolds—one of the most common "guidelines" of companion-planting books.

The concept behind the study was that both French marigolds (*Tagetes patula*) and African marigolds (*T. erecta*) help keep Mexican bean beetles (*Epilachna varivestis*) away from green beans. A control plot, not planted with marigolds, sustained more damage than the beans with

Above: a myth - "Roses are healthier when planted with fragrant hardy salvias to repel cabbage moths and carrot rust flies." Ha! An old, old fable. Read on for the facts based on science.

marigold rows, *but* the control plot produced more beans.

It's now thought that this was due to the exudates that marigolds produce (*thiopene* and *alpha-terthienyl*) in order to stunt the growth of plants nearby—a form of evolutionary competitiveness. ("Effects of Companionate Planting on Snap Bean Insects, *Epilachna varivestis* and *Heliothis", April 1980; M.* A. Latheef, R. D. Irwin: Volume 9, Issue 2, *Entomological Society of America*, Dept. of Agriculture, Virginia State University, Petersburg, VA 23803.)

Companion Planting Mechanisms

There actually are five very different ways in which a companion plant (or plants) might work:

- As a "trap crop"
- To repel the pest
- To mask the "target" plant
- By physically blocking the pest
- To attract, harbor, and feed beneficial insects with pollen and nectar

Figure 320. Cabbage was used as a trap crop to manage pollen beetles. (1772)

Trap Crops

Trap plants are those that attract pests away from the vegetable plant you're trying to protect. Having a lot of pests in one place can, with timing, allow you to kill a good number of them, using your hands, a water spray, or an organic pesticide. The pests tend to stay on the trap plant and don't frequent those growing nearby. I've seen this every spring with fava beans. The first few inches of a fava bean stem can be covered in a dark mass of the earliest aphids, which seem to attract beneficial wasps and ladybugs to the plant. In this scenario, I usually don't see aphids on other plants, and the fava beans always recover enough to make a good crop.

Here are some examples where this dynamic has been shown to be effective:

- Soybeans can be protected from Mexican bean beetles (*Epilachna varivestis*) with a trap planting of snap beans.
- Even the same plant can help; an early-planted potato crop was used to protect later plantings of potato.
- Collards have been used to draw diamond-backed moths (*Plutella xylostella*) away from cabbage.
- Four examples from the Connecticut Cooperative Extension: Collards around the perimeter of a field reduced diamondback moth by 89%. To stop pepper maggots on bell peppers, cherry peppers were planted as a trap crop—producing 98% pest-free fruit. Blue Hubbard squash reduced the cucumber beetles on yellow summer squash by 95%. The same Blue Hubbard squash planting reduced squash-vine borers by 88% when the gardeners were able to kill the borer populations in the perimeter trap crop.
- Diverse trap crops consisting of all three trap-crop species (Pacific gold mustard, pac choi, and rape) proved the most effective trap-crop mixture.
- Plant collards near cabbage to draw the diamondback moth away from the cabbage crop. ("Management of Diamondback Moth (Lepidoptera: Plutellidae) in Cabbage Using Collard as a Trap Crop;" 2000. Mitchell, E.R., G. Hu, and D. Joanowicz. 2000. *HortScience*. Vol. 35 p. 875-879.)
- The munching effect of Mexican bean

beetles on soybeans can be controlled by using a trap crop of snap beans, ("Evaluation of Trap Crop Procedures for Control of Mexican Bean Beetle in Soybeans and Lima Beans", (Rust, R W, *Journal of Economic Entomology*, 70: 630-632.)

- For over 50 years in Belorussia, early-planted potato trap crops have protected later plantings of potatoes from Colorado potato beetles. ("Combined Protection of Potatoes" 1975. Dorozhkin N. A. Bel'skaya, S. I. and Meleshkevich. A. A. *Zasch. Rast*; 7: 6-8.)

- In Finland, mixtures of Chinese cabbage, marigolds, rape (*Brassica napus*, variety *napus*, a type of mustard green) and sunflower were used successfully as a diverse trap crop to manage the pollen beetle (*Brassicogethes aeneus*) in cauliflower. ("Biological and Agrotechnical Control of the Rape Blossom Beetle *Meligethes aeneus* (*Coleoptera, Nitidulidae*)": 1989. Hokkanen, H. M. T. *Acta Entomologica Fennica*; 53:25-29.)

- Three trap-crop species (Pacific gold mustard, pac choi, and rape) provided the most effective trap-crop mixture to control the crucifer flea beetle (*Phyllotreta cruciferae* Goeze) in broccoli (*Brassica oleracea var. italica*). ("Diversity by Design. Exploring Trap Crops and Companion Plants to Control *Phyllotreta cruciferae*, the Crucifer Flea Beetle, in Broccoli"; 2012. Arker JE., Washington State University, (Dissertation).

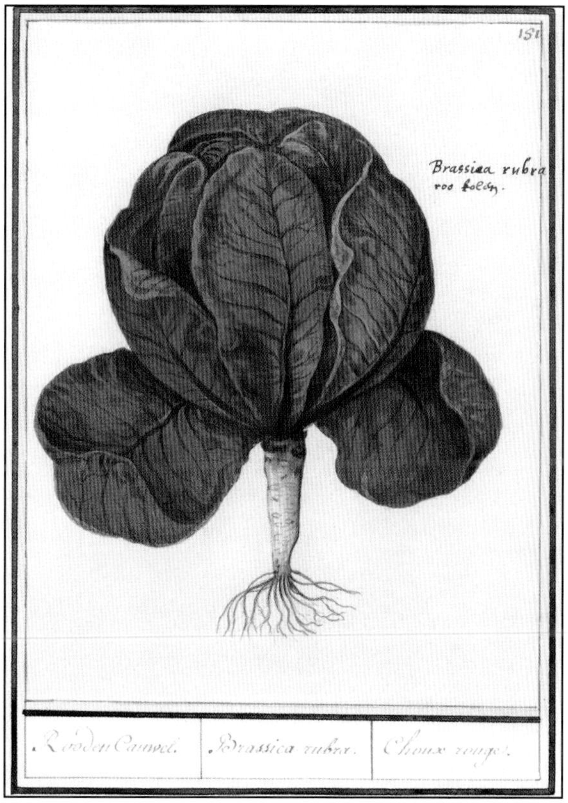

Figure 321. Cabbage planted before cabbage can be a trap plant.

In the above examples, where the trap is the same as the main crop (or a crop in the same genus) it's essential to plant the trap crop *before* you put in the main crop. This allows you to spray the trap-crop pests to keep them from migrating to the intended crop. Be sure to control the pests in the trap crop so that they don't build up to become what is referred to as a "nursery" crop—growing unchecked and spreading the pest to the crop to be harvested. The area dedicated to the trap plant should be no more than 20% (some say 10%) of your garden.

Timing is everything. Take the example of strawberries, wheat, and berry-killing nematodes: if you plant strawberries and no wheat, 43 percent of the berries will be attacked by nematodes, but only five percent will be killed if wheat is planted just two weeks ahead of the strawberries. [Vernon, R.S., T. Kabaluk, and A. Behringer. 2000. Movement of *Agriotes obscurus* (*Coleoptera: Elateridae*) in Strawberry (*Rosaceae*) Plantings with Wheat (*Gramineae*) as a Trap Crop", *The Canadian Entomologist*, Vol. 132. p. 231–241].

Using Plants to Repel Pests

OK, sometimes using plants to repel pests actually does work:

In the abovementioned tests of Brassica crops, the companion plants that helped repel insects included sage (*Salvia officinalis*); rosemary (*Rosmarinus officinalis*); hyssop (*Hyssopus officinalis*); thyme (*Thymus vulgaris*); dill (*Anethum graveolens*); southernwood (*Artemisia abrotanum*); mint (*Mentha* spp.); tansy (*Tanacetum vulgare*); chamomile (several genera); and orange nasturtium (*Tropaeolum majus*). (7)

- Intercropping of spring onions (companion plant) was found to protect broccoli (target crop) from pest attack. Spring onions were used as a repellent to "push" insect pests away from the broccoli. (8)

- Basil (*Ocimum basilicum*), when planted with tomatoes, has been recorded to repel thrips and tomato hornworms. (Anon. "Companion Planting." 2004a [Online]. Available: http://www.ghorganics.com/page2.html.

- The genus *Allium* (onion, garlic, leeks etc.) has been observed to exhibit repellent properties against a variety of insects and arthropods, including moths, as detailed in "Plant Essential Oils as Arrestants and Repellents for Larvae of the Codling Moth (*Lepidoptera: tortricidae*);" (1999. Landolt, P. J., Hofstetter, R. W., and Biddick, L. L.; *Environmental Entomology*; 28: 954-960).

- Alliums also work against cockroaches (Behavioural Responses of *Blattella Germanica* L. (*Orthopt., Blattelidae*) Induced by Repellent Plant Extracts", Scheffler, I., and Dombrowski M.. *Journal of Applied Entomology* 115: 499,-505).
- Not to mention mites ("Characterisation of the Two-Spotted Spider Mite (*Tetranychus urticae* Koch, *Acari: Tetranychidae*) Response to Aqueous Extracts from Selected Plant Species, 2007. Dabrowski, Z. T. and Seredynska U., *Journal of Plant Protection Research*, 47: 113-124.)
-

And, oh, yes, aphids ("Olfactory Response of *Myzus persicae* [*Homoptera: Aphididae*] to Volatiles from Leek And Chives—Potential for Intercropping With Sweet Pepper", 2007; Amarawardana, L., Bandara, P., Kumar V., Pettersson, J., Ninkovic, V., and Glinwood, R..; *Acta Agriculturæ Scandinavica*, B 57: 87-91.)

Figure 322. Rosemary (*Rosmarinus officinalis*) has been used to protect celery and tomatoes.

- The companion plants sage (*Salvia officinalis* L.); rosemary (*Rosmarinus officinalis* L.); hyssop (*Hyssopus officinalis* L.); thyme (*Thymus vulgaris* L.); dill (*Anethum graveolens* L.); southernwood (*Artemisia abrotanum* L.); mint (*Mentha* L. spp.); tansy (*Tanacetum vulgare* L.); chamomile (several genera); and orange nasturtium (*Tropaeolum Majus*.) protected celery and tomatoes. (*Companion planting in Australia*. Little B.; Reed Books Pty Ltd, New South Wales; 1989.)

- Intercropping tomatoes with cabbages has been suggested as a strategy to repel the diamondback moth. ("Effects of Cabbage-Tomato Intercropping on the Incidence and Oviposition of the Diamondback Moth, *Plutella xylostella*" (L.); 1975. Buranday, R. P., and Raros, R. S., *Phillippine Entomology*;2: p 369-375.) [Note: Oviposition means egg-laying]

- Ragweed (*Ambrosia artemisifolia* L.) has been used to repel the crucifer flea beetle (*Phyllotreta cruciferae*) from collards (*Brassica oleracea* L. var. *acephala*) ("The Influence of Vegetational Diversity on Population Ecology of a Specialized Herbivore, *Phyllotreta cruciferae*",

(Coleoptera: 1972. *Chrysomelidae Oecologia*;10: p 321-346.)

And, face it, sometimes companion planting just doesn't work as a pest repellant:

- No significant differences were found in the number of eggs, larvae, pupae, or damage by cabbage pests between the following companion plants; French marigold (*Tagetes patula* L.); garden nasturtium *(Tropaeolum majus);* pennyroyal (*Mentha pulegium* L.); peppermint (*Mentha piperita* L.); garden sage; thyme; and control treatments.("The Effect of Companionate Planting on Lepidopteran Pests of Cabbage", 1979. Latheef, M. A., and Irwin, R. D.; *The Canadian Entomologist*;111: p 863-864.)

- French marigolds (*Tagetes patula* L.) intercropped with carrots did not repel the carrot fly (*Psila rosae.)* (Effect of Mixed Cropping on Some Insect Pests of Carrots And Onions;" 1884. Uvah, I. I., and Coaker, T. H.; *Entomologia Experimentalis et Applicata*; 36: p 159-167.)

- Sage and thyme represent two common companion plants noted for their pungent odors. There was no difference in diamondback moth oviposition between Brussels sprouts (*B. oleracea*) intercropped with sage (*S. officinalis*) and thyme (*T. vulgaris*.) ("The Effects of Labiate Herbs and White Clover on *Plutella xylostella* Oviposition", 1986. *Entomologia Experimentalis Applicata*; 42: p 243-247.)

- Marigolds (*Tagetes erecta* L.) and mint (*Mentha piperita* L.) did not repel the onion fly (*Delia antiqua* Meigen) or the cabbage-root fly (*D. radicum* L.) (Laboruntersuchungen zu Massenanzucht, Lebensweise, Eiablage and Eiablageverhaltender Kohlfliege, Chortophila brassicae (Bouché) (Diptera, Anthomyiidae), 1968. Zohren, E., *Zeitschrift furAangewandte Entomologie*; 62: p 139-188.)

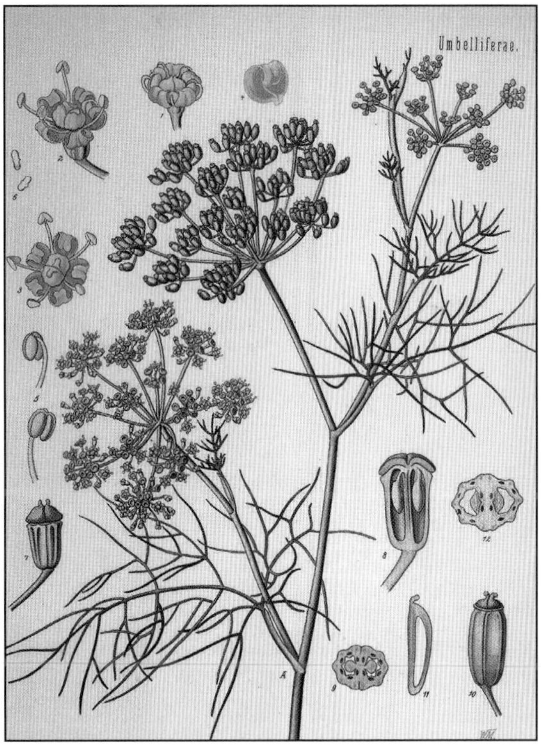

Figure 323. Dill (*Anethum graveolens*) is a great umbel used to mask crops from bean beetles.

Masking or Camouflaging Your Crop

- A study of beans, frequently damaged by the lygus bug (*Lygus* spp.) was done with summer savory (*Satureja hortensis*); marjoram (*Origanum majorana*); sage (*Salvia officinalis*); big marigold (*Tagetes erecta*); red beet (*Beta vulgaris*); onion (*Allium cepa*); and dill (*Anethum graveolens*). The lowest percentage of damaged beans came from samples obtained from plots cultivated in the close proximity of dill and marigold. [Note: *not* the French marigold used in home gardens. Ed.]

- Garlic and other alliums have been found to deter the green-peach aphid (Amarawardana et al., 2007). If an insect searching for its host encounters a field with vegetables companion-planted with alliums, the pest primarily smells the overpowering volatile allium *semiochemicals*, and will continue its search because the scent of the susceptible host plant is 'masked.'"(Olfactory Response of *Myzus persicae* (*Homoptera*: *aphididae*) to Volatiles from Leek and Chive: Potential for Intercropping with Sweet Pepper", 2007. Amarawardana, L., P. Bandara, V. Kumar, J. Pettersson, V. Ninkovic, and R. Glinwood; 2007. *Acta Agriculturae Scandinavica, Soil and Plant Science*; Vol. 57. p 87–91.)

Blocking Pests

Five examples of studies in which physical blocking has worked:

- Tall plants can impede pest movement within a cropping system. For example, maize has been used to protect bean plants from pest attack, and dill has been used as a vegetative barrier to inhibit pest movement in organic farms (author's personal observation). (11)

- Research showed that, in a cowpea-sorghum companion planting, cowpeas were protected from the striped bean weevil by the physical barrier of the tillering [setting seed] sorghum. ("Observations on the Pest Status of the Striped Bean Weevil *Alcidodes leucogrammus erichs*. On Cowpea Under Intercropping Systems in Kenya;" International, Amoako-Atta, B. 1983. *Journal of Tropical Insect Science*, Vol. 11, No. 4. p 351-356.)

- The cabbage-root fly (*D. radicum*) was disrupted when host plants were surrounded by a wide variety of plants, including weeds and marketable crops such as spurry (*Spergula arvensis*); peas (*Pisum sativum*); rye-grass (*Lolium perenne*); or clover (*Trifolium* spp.). ("Influence of Background on Host Plant Selection And Subsequent Oviposition by The Cabbage Root Fly (*Delia radicum*)", 1994. Kostal, V., and Finch, S. *Entomologia Experimentalis et Applicata*; 70: p 153-163.)

Figure 324. Clover flowers attract bees and can help block cabbage root fly (*D. radicum*).

Dill (companion plant) is also used as a physical barrier to protect broccoli (target crop) from pest attack. The height of the dill can impede pest movement. (Personal observation, Joyce E. Parker Washington State University.)

- Corn has also been used to protect bean plants from pest attack. ("Some Limitations of Weed Biocontrol in Tropical Ecosystems in Colombia", Altieri, M. A. and Doll, J. D. , T.E. Freeman [ed.], *Proceedings, IV International symposium on Biological Control of Weeds*. University of Florida, Gainesville p 74-82; 1978.)

Companion-Planting Realities

Many people think of companion plants as those that repel pests when planted next to a plant that attracts them. The recommendations in most books are not always accurate, often implying that the pest-repelling effect is achieved by planting a certain plant next to a pest target. This effect is actually a relatively rare phenomenon, but here are three examples of it that I've found in the scientific literature:

- In tests of Brassica crops, the companion plants that helped repel pests included sage (*Salvia officinalis*); rosemary (*Rosmarinus officinalis*); hyssop (*Hyssopus officinalis*); thyme (*Thymus vulgaris*); dill (*Anethum graveolens*); southernwood (*Artemisia abrotanum*); mint (*Mentha* spp.); tansy (*Tanacetum vulgare*); chamomile (several genera) and orange nasturtium (*Tropaeolum majus*). (Isman, B., "Botanical Insecticides, Deterrents, and Repellents in Modern Agriculture and an Increasingly Regulated World", *Annual Review of Entomology* 2006; #51: 45-66.)

- Basil (*Ocimum basilicum*) planted with tomatoes has been recorded to repel thrips and tomato hornworms. (Anon. *Organic Vegetable IPM Guide*, 2004.)

- Marigolds are a proven control of nematodes. Some of the cultivars that work the best for controlling nematodes are 'Nemagold'; 'Petite Blanc'; 'QueenSophia'; 'Tangerine'; 'Golden Guardian'; 'Single Gold'; and 'Nema Gone'. These are best planted as a green manure one or more seasons before planting the crop. Intercropping has little effect. Once the soil is planted to a crop the nematodes begin to reappear.

Figure 325. The garden nasturtium (*Tropaeolum majus*) can be an useful companion *and* has edible leaves, flowers, and seeds.

And, well, sometimes it just doesn't work:

- When three companion plants, rue (*Ruta graveolens*), zonal geranium (*Pelargonium hortorum*), and garlic chives (*Allium schoenoprasum*) were interplanted with roses, the roses were not protected from the Japanese beetle (*Popillia japonica.*) In fact, the geraniums *increased* the number of beetles. In the same study, surrounding roses with sachets of fennel seeds, cedar shavings, crushed red pepper, or osage orange fruits significantly increased the number of beetles over those that infested the control plants. ("Evaluating Companion Planting and Non-host Masking Odors for Protecting Roses from the Japanese Beetle (*Coleoptera: Scarabaeidae*"); D. W. Held, P. Gonsiska, and D. A. Potter; *Journal of. Economic Entomology*: 96(1): 81-87, 2003)

- Several putative companion plants were studied for their ability to discourage Japanese beetles (*Popillia japonica*) from damaging roses, and it was concluded that companion plants were unlikely to help. (Held, D. W., Gonsiska, P., and Potter, D. A. "Evaluating Companion Planting and Non-Host Masking Odors for Protecting Roses from the Japanese Beetle *(Coleoptera: scarabaeidae)*", *Journal of Economic Entomology*: 2003; 96: 81-87.)

- Another study, testing French marigold (*Tagetes patula L.*); garden nasturtium (*Tropaeolum majus*); pennyroyal (*Mentha pulegium L.*); peppermint (*Mentha piperita L.*); garden sage; thyme; and control treatments "...found no significant differences in the number of eggs, larvae, pupae, or the amount of damage by cabbage pests between companion plants." (Latheef, M. A. and Irwin, R. D. "The Effect of Companionate Planting on Lepidopteron Pests Of Cabbage", *The Canadian Entomologist* 1979; #111: 863-864.)

- French marigolds (*Tagetes patula*) intercropped with carrots did not repel the carrot fly (*Psila rosae*). (Uvah, I. I. I., and Coaker, T. H.; "Effect of Mixed Cropping on Some Insect Pests of Carrots And Onions", *Entomologia Experimentalist et Applicata* 1984; #36: 159-167.)

- Even though sage (*S. officinalis*) and thyme (*T. vulgaris*). represent two common companion plants noted for their pungent odors, there were no differences in diamondback moth (*B. oleracea*) oviposition between control plantings and Brussels sprouts intercropped with sage and thyme (Dover, J. W.; "The Effects of Labiate Herbs and White Clover on *Plutella Xylostella* Oviposition;" *Entomologia Experimentalis Applicata* 1986;42: 243-247.)

- Coriander/cilantro (*Coriandrum sativum*)
- Cowbane (genus *Oxypolis*)
- Cow parsnip (genus *Heracleum*)
- Hogweed (*H. mantegazzianum* and *H. sphondylium*)
- Cumin (*Cuminum cyminum*)
- Dill (*Anethum graveolens*)
- Pignut (*Bunium bulbocastanum*)
- Fennel (*Foeniculum vulgare*)
- Lovage (*Levisticum officinale*)
- Parsley (*Petroselinum crispum*)
- Parsnip (*Pastinaca sativa*)
- Poison hemlock (*Conium maculatum*)
- Queen Anne's lace (*Daucus carota*)
- Carrot (*Daucus carota*, ssp. *sativus*)
- Water hemlock (genus *Cicuta*)
- Water parsnip (genus *Sium*)

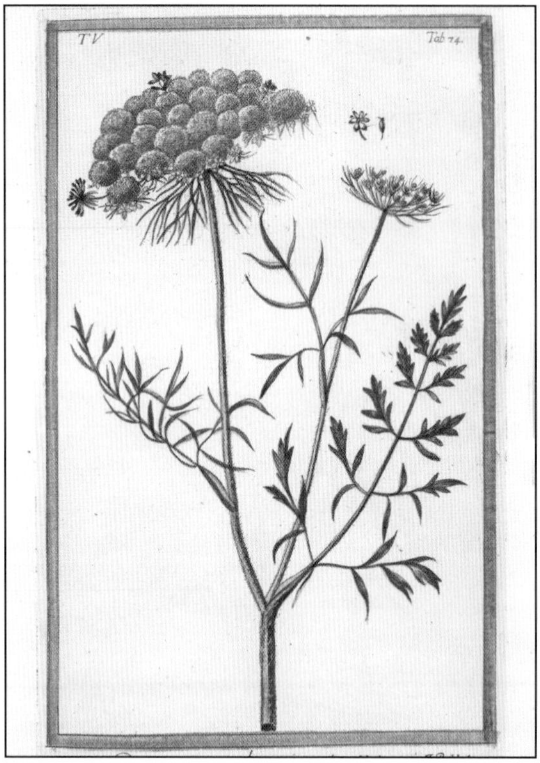

Figure 326. Another lithograph from 1772 shows a Queen Anne's lace flower. (*Daucus carota*)

Plant Companions for Attracting Beneficial Insects

Attracting beneficial insects is actually the most important aspect of companion planting, with the flowers of companions providing either pollen or nectar to sustain the beneficial bugs. Easy access to nectar as fuel for flying in pursuit of pests is important for bugs that prey on them. There are two families of plants that provide abundant nectar in small flowers close to the open air. These are the parsley or carrot family (*Apiaceae,* formerly *Umbelliferae*), and the sunflower family (*Asteraceae* family, formerly *Compositae*).

Examples of the *Apiaceae* family are:

- Angelica (genus *Angelica*)
- Burnt saxifrage (*Pimpinella anisum*)
- Asafoetida (*Ferula assa-foetida*)
- Caraway (*Carum carvi*)
- Celery (*Apium graveolens*)
- Celeriac (*A. graveolens* var. *rapaceum*)
- Sweet cicely (*Osmorhiza berteri*)
- Chervil (*Anthriscus cerefolium*)

Some examples of composites are:

- *Aster* (Michaelmas daisy)
- *Bellis* (English daisy)
- *Callistephus* (China aster)
- *Chrysanthemum* (chrysanthemum, daisy)
- *Cosmos* (cosmos)
- *Dahlia* (dahlia)
- *Echinacea* (coneflower)
- *Helianthus* (sunflower)
- *Rudbeckia* (black-eyed Susan)
- *Tagetes* (marigold)
- *Zinnia* (zinnia)
- *Achillea* (yarrow)
- *Ageratum* (ageratum)

- *Anthemis* (golden marguerite)
- *Artemisia* (wormwood)
- *Calendula* (pot marigold)
- *Centaurea* (bachelor's button, or cornflower)
- *Echinops* (globe thistle)
- *Erigeron* (fleabane)
- *Eupatorium* (joe-pye weed, boneset white snakeroot)
- *Gaillardia* blanketflower)
- *Helichrysum* (strawflower, everlasting)
- *Liatris* (blazing star, button snakeroot)
- *Ratibida* (prarie coneflower)
- *Santolina* (lavender cotton)

Sometimes some of these don't seem to work to attract beneficial insects:

- "The use of lacy phacelia (*Phacelia tanacetifolia*), buckwheat (*Fagopyrum esculentum*), and mixed grasses (*Poaceae* spp.) had no impact on aphid populations. ("Apple Aphid [*Aphis* spp.], and Predator Populations in an Apple Orchard at the Non-Bearing Stage: The Impact of Ground Cover and Cultivar;" 2008. *European Journal of Entomology*, 105: 521-529.)

Sometimes they do:

- "Sunflowers (*Helianthus* spp.) can attract a significant number and type (30) of beneficials when intercropped with five different crops as early as when the sunflowers are six inches tall." ("Intercropping with Sunflowers to Attract Beneficial Insects in Organic Agriculture;" March 2005. Gregory Jones; Jennifer Gillet, *The Florida Entomologist*, Vol. 88, No. 1 p 91-96.)

Figure 327. There are hundreds of varieties of sunflowers, all make good composite insectiaries.

- Organic farms with clovers (*Trifolium* spp.) had more bees than conventional farms. ("Organic Dairy Farming: Impacts on Insect-Flower Interaction Networks and Pollination", 2011. Eileen Power and Jane Stout, *Journal of Applied Ecology*, 48: 561-569.)

- Earth ridges planted with orchard grass harbored two predators of cereal aphids. These "beetle banks" were so effective that the savings on pesticides outweighed the labor and seed costs required to establish them. ("Managing Vegetation Surrounding the Field"; *Agroecology and the Search for a Truly Sustainable Agriculture*, p 253-261.)

- In New Mexico, an increase in predators and parasitic insects was attracted by a mixture of: California bluebell (*Phacelia campanularia*); buckwheat (*Fagopyrum esculentum*); dill (*Anethum graveolens*); plains coreopsis (*Coreopsis tinctoria*); garden cosmos (*Cosmos bipinnatus*); and sweet alyssum (*Lobularia maritima*). ("Using Insectary Plants to Attract and Sustain Beneficial Insects for Biological Pest Control"; Tessa Grasswitz, Guide H-169.

- "...Floral resource plants can be used to increase the abundance of natural enemies in ornamental urban landscapes. Clover increased natural enemies in May, while clover and goldenrod did so in June and July." ("Manipulating the Abundance of Natural Enemies in Ornamental Landscapes with Floral Resource Plants", 2005. *Biological Control*, 33, p.203-216.)

- *Fagopyrum esculentum* (buckwheat), *Centaurea cyanus* (cornflower), *Vicia sativa* (common vetch), and *Pastinca sativa* (parsnip) increased parasitoids populations in cabbages, while *Daucus carota* (Queen Ann's Lace), *Carveum carvi* (caraway), *Pastinaca sativa*, and *Vicia sepium* (bush vetch) were rated as "promising plants" in apple orchards. ("Functional Biodiversity to Improve Pest Control in Organic Cropping Systems"; Lukas Pfiffner, Hans-Jacob Schärer and Henryk Luka.)

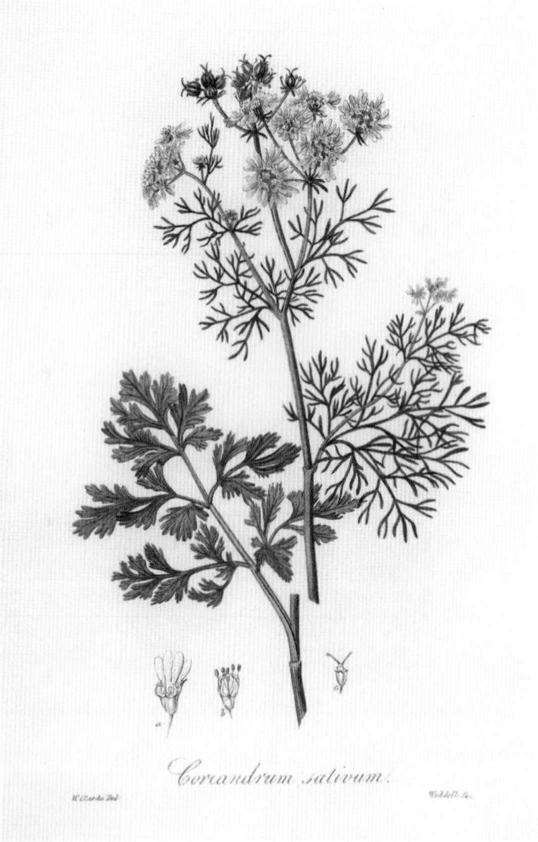

Figure 328. Let some of your cilantro go to seed and you'll have coriander (*Coriandrum sativum*) seed. Bloom is a great insectary flower.

- Aphids in sugar beets were reduced by a mixture of *Sinapis alba* (white mustard), *Phacelia tanacetifolia* (lacy phacelia) and *Coriandrum sativum* (coriander). ("The Effect of Strip Management on Reduction of *Aphis Fabae* Populations by Predators on Sugar Beet Crops"; 2005. Jacek Twardowski, Michal Hurej, Zdzislaw Klukowski, *Journal of Plant Protection Research,* Vol. 4, No. 3.)

- Sweet alyssum attracted the most (beneficial) syrphid flies, stayed in bloom the longest, and attracted the highest numbers of predators. ("Attractiveness of Common Insectary and Harvestable Floral Resources to Beneficial Insects"; 2011. Brian Hogg, Robert Bugg, Kent Daane; *Biological Control*, 56 p 76-84.)

    - Syrphid flies visited coriander (*Coriandrum sativum*) more than buckwheat (*Fagopyrum esculentum*) and phacelia (*Phacelia tanacetifolia*). ("Frequencies of Visits to Selected Insectary Plants by Predatory Hoverflies, Other Beneficial Insects, Herbivores"; April 2006; M. D. Ambrosino, J.M. Luna, P. C. Jepson, S. D. Wratten; *Environmental Entomology*, Vol. 35, Issue 2, p 394-400.)

    - In a season-by-season study by Oregon State University, early-season flowering coriander (*Coriandrum sativum*) attracted the most syrphid flies. In the late season it was: yarrow (*Achillea millefolium*), fennel (*Foeniculum vulgare*), and Korean licorice mint (*Agastache rugosa*). ("Relative Attractiveness of Potential Beneficial Insectary Plants to Aphidophagous Hoverflies"; October 2000. M. R. Colley, J. M. Luna; *Environmental Entomology*, Vol. 29, Issue 5, p 1054-1059.)

- *Centaurea cyanus*, commonly known as cornflower or bachelor's button, increased parasitism and predation of the cabbage moth eggs in one study. The plant helped increase the cabbage biomass in one year. ("Wildflower Companion Plants Increase Pest Parasitation and yield in Cabbage

Fields;" Experimental demonstration and call for caution, September 2014. Oliver Balmer, et al., *Biological Control*, Volume 76, p 19-27.)

- In England, plantings of coriander, corn marigold, fennel, and phacelia attracted beneficials in a study based on when they bloomed. Phacelia was the only flower to attract beneficials at the end of summer. (Plants for predators – a participatory experiment, 2006. HC Sievwright, GL Sutton and A Rosenfield; *Aspects of Applied Biology*; #78.)

- Using sweet alyssum (*Lobularia maritima*), phacelia (*Phacelia tanacetifolia*), and buckwheat, leafroller damage in apple orchards was reduced 29%, and there were twice as many wasps in the fields with alyssum and buckwheat than in the control field with no flowers. ("The Effects of Floral Understories on Parasitism of Leafrollers on Apples in New Zealand"; 2006. Irvin, et al., *Agricultural Entomology*, #8 p 25-34.)

Figure 329. A heavy-hitter in the beneficial insectiary world. In the top three, at least by my judgement. Alyssum (*Lobularia maritima*).

- Seven of twelve species planted next to fields of feed-corn and rye attracted predatory tachinid flies better than the edges of the field without the flowers. (*Tachinidea* Associated with Flowering Plants: Estimating Floral Attractiveness"; 2012. Shoki Al-Dobai et al., *Biological Control*, 61, p 230-239.)

- Buckwheat (*Fagopyrum esculentum*) increased nectar and pollen available for hoverfly longevity and egg production. ("The Contributions of Floral Resources and Honeydew to the Performance of Predatory Hoverflies"; October 2013. Paul van Rijn, et al., *Biological Control*, Volume 67, Issue 1, p 32-38.)

- Sweet alyssum (*Lobularia maritima*) plants attracted natural enemies and indirectly suppressed densities of woolly-apple aphids. ("Flowers Promote Aphid Suppression in Apple Orchards", 2013. Lessando Gontijo, et al., *Biological Control*, 66, p 8-15.)

- Among early-season flowering species, cilantro was the most attractive to hoverflies, followed by alyssum and buckwheat. Mustard, buckwheat, and Korean licorice mint were most attractive to parasitic hymenoptera [wasps]. Among late-season flowers, fennel and phacelia were most attractive to hoverflies, but attractiveness to parasitic wasps did not differ among evaluated plants. (*Enhancing Biological Control with Beneficial Insectary Plants*; J. Luna, M. Colley, and M. Staben, Department of Horticulture, Oregon State University.

- Plots of the following plants were seeded and irrigated on September 23, 1997 at North Willamette Experiment Station (NWREC): *Achillea* sp. (yarrow); *Hesperis matronalis* L. (dame's rocket);

Cheiranthus allionii (wallflower); *Phacelia tanaecetifolia*; *Carum carvi* (caraway); and *Avena sativa* ('Amity' oat). In this trial, the Amity oat appeared particularly attractive to leafhoppers, and phacelia appeared most attractive to plant bugs. *Evaluation of Potential Fall-Seeded Cover Crops for Weed Suppression and Beneficial Insect Habitat* D. Kaufman, G. Koskela, L. Royce, North Willamette Experiment Station, Oregon State University.

- The objective of one study was to evaluate the benefit of adding plant resources to a lettuce field in order to ensure the biological control of aphids (*Nasonovia ribisnigri*) and thrips (*Frankliniella occidentalis*). Providing plant resources [*Lobularia maritima* (alyssum), and a plant mixture of *L. maritima, Centaurea cyanus*, *Vicia sativa*, and *Lupinus hispanicus*] in the field ensured the presence of key predators: adult hoverflies were attracted and aggregated on flower patches, and predatory bugs did establish on the insectary plants (those grown to attract, feed, and shelter insect parasites and predators to enhance biological pest control). Being highly mobile, both predators did disperse to plots without flower patches and established on lettuce plants unless limited by pesticide treatments. As a result of predator establishment, prey populations were reduced below the economic threshold. ("Insectary Plants to Enhance the Biological Control of *Nasonovia Ribisnigri* and *Frankliniella Occidentalis* In Lettuce"; 2008. Oscar Alomar, Judit Arnó, Rosa Gabarra, *Landscape Management for Functional Biodiversity IOBC*. Bulletin Vol. 34, 2008 p 9-12.)

Figure 330. *Phacelia tanecetifolia (*lacy phacelia) is a heavy-hitter in the beneficial insectiary world. In the top five, at least by my judgement.

- Planted among low-growing plants, apple and prune trees, hazel, nettle, and mint were good host plants for predatory mites. A relatively higher density of predators in border rows of the plantation was a good indication of the direct effect of surrounding vegetation on *phytoseiid* population on blackcurrants (Jaworski, 2000). Regular collection of the mites on various herbaceous plants showed that the following species harbor large populations of predatory mites: *Urtica dioica* (common nettle); *Geum urbanum* (wood avens); *Fragaria vesca* (wild strawberry); *Ballota nigra* (black horehound); and *Lamium album* (white or dead nettle) in Poland (Kropczynska & Miazga, in press). ("Vegetation on Field Margins as a Source of Predatory Mites (*Phytoseiidae*) for Strawberry Plantations in Poland: Preliminary Observations"; 2008. Zbigniew T. Dabrowski, Danuta Kropczynska, Jakub Garnis; *Landscape Management for Functional Biodiversity IOBC* Bulletin Vol. 34, p 25-28.)

Sometimes the process of attracting beneficial insects depends on a mixture of mechanisms. For instance, to control spotted stem-borer (*Chilo partellus*) in maize (*Zea mays*), the repellent plants in a case study included a variety of non-host plants such as molasses grass (*Melinis minutiflora* P. Beauv.); silverleaf desmodium (*Desmodium sandwicense*); or green-leaf desmodium (*Desmodium intortum*), and the trap-crop plantings included Napier grass (*Pennisetum purpureum* Schumach) or Sudan grass (*Sorghumbicolor ssp. drummondii*) (Khan Z R and Pickett J A. ("The 'Push-Pull' Strategy for Stem-borer Management: a Case Study in Exploiting Biodiversity and Chemical Ecology"; *Ecological Engineering For Pest Management-Advances In Habitat Manipulation For Arthropods* (ed. by GM Gurr, SD Wratten and MA Altieri), CABI Publishing, Wallingford, UK; 2004; p 155-164.)

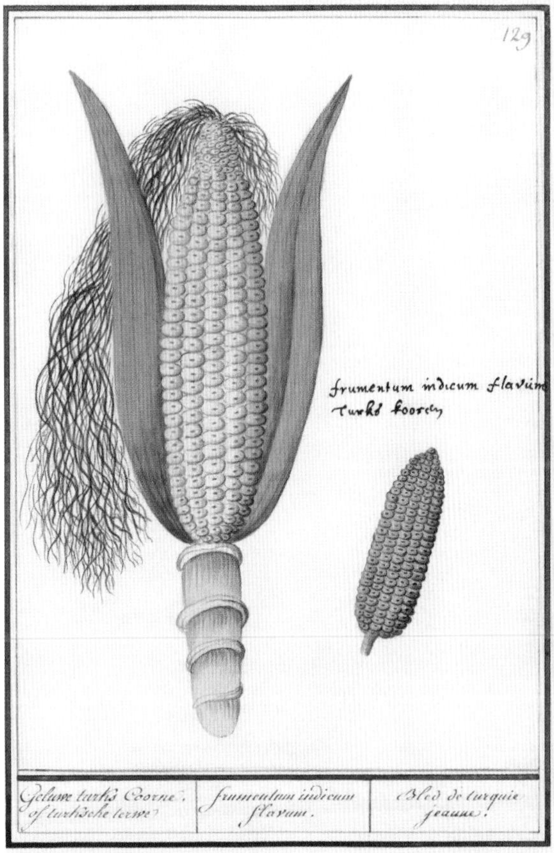

Figure 331. Three sisters gardens grew mostly grain for flour or flint corn for the animals. Here in a lithograph from 1892.

Note: The sources of some of the above information about plant-to-plant protection include:

- *Companion Planting and Insect Control*, 2013. Joyce E. Parker, William E. Snyder, George C. Hamilton, and Cesar Rodriguez-Saona, Chapter
- *Weed and Pest Control – Conventional and New Challenges*. Edited by Sonia Soloneski, IntechOpen. (This review states: "A repellent plant that can be effective for one pest might not provide effective control for another.")

The Three Sisters—Corn, Beans, and Squash

The so-called "Three Sisters" plantings of corn, beans, and squash consist of all annual vegetables. This combination is often pointed to as a good model for polyculture, and as *the* model for companion planting. "Throughout Central America, a common intercrop of corn, beans, and squash is traditionally grown. Grown together, these three crops optimize available resources. The corn towers high over the other two crops while the beans climb up the cornstalks. The squash plants sprawl along the ground, capturing light that filters down through the canopy and shades the ground. The shading discourages weeds from growing." ("Intercropping Principles and Production Practices Agronomy Systems Guide," 1998. Preston Sullivan, Technical Specialist, ATTRA, Fayetteville, AR.)

Does this really work? Consider the cultivation of corn in the arid southwest. The Hopi Indians, from centuries of experience, know that corn roots are extensive and greedy. Thus they plant hills of corn with multiple seeds, but widely spaced to prevent depletion of soil nutrients and to conserve moisture. This is less of a problem in moist temperate areas, where nitrogen recycles faster, due to greater rainfall and faster decomposition of plant matter retained in the soil. The trade-off of intensive (vertical) with imported fertilizers versus wider (horizontal) spacing still applies.

"A perennial or forage-legume crop only adds significant nitrogen for the following crop if the entire biomass [stems, leaves and roots] is incorporated into the soil. If a forage plant is cut and removed from the field, most of the nitrogen fixed by the forage is also removed. Roots and crowns add little

Figure 332. One view of a classsic three-sisters planting. The corn provides a "pole" for the beans to climb up. Thc bcans add some nitrogen in the second year. And the squash may supress weeds. (One study found the squash helped keep racoons off the corn.) Illustration copyrighted by Charlotte A Ricker, used with permission.

soil nitrogen compared with the above ground biomass." *Nitrogen Fixation by Legumes Guide* A-129 W.C. Lindemann, Soil Microbiologist C.R. Glover, Extension Agronomist, New Mexico State University.) Unbeknownst to most gardeners, the residue from the corn plants contains more nitrogen than the residue from a bean crop, simply because the corn crop has more biomass.

When the grain from a grain/legume crop (in this case beans) is harvested, little nitrogen is returned for the following crop, as most of the nitrogen fixed by the grain during the season is removed from the field. The stalks, leaves, and roots of grain-companion legumes such as soybeans and shell beans contain about the same concentration of nitrogen as found in non-legume crop residue. (See my explanation of the nitrogen cycle on pages 249 - 255). Thus, the myth that the beans supply nitrogen to the corn in the current season crumbles. As a positive aside, beans intercropped with corn have been shown to decrease outbreaks of leafhoppers and fall armyworm.

Along with the weed-suppressing properties of squash plants, there's another important interaction between squash and beans. To produce large nodules, roots prefer an environment of 54° to 75°F—the formation of nodules decreases in temperatures above 80°F and up to 86°F. A soil that's too dry will also reduce nodule formation, so the shade from the squash leaves is important. Squash also reduces losses of corn to spider mites and aphids. Over-concentrating on corn should be avoided, as this would be too attractive to pests.

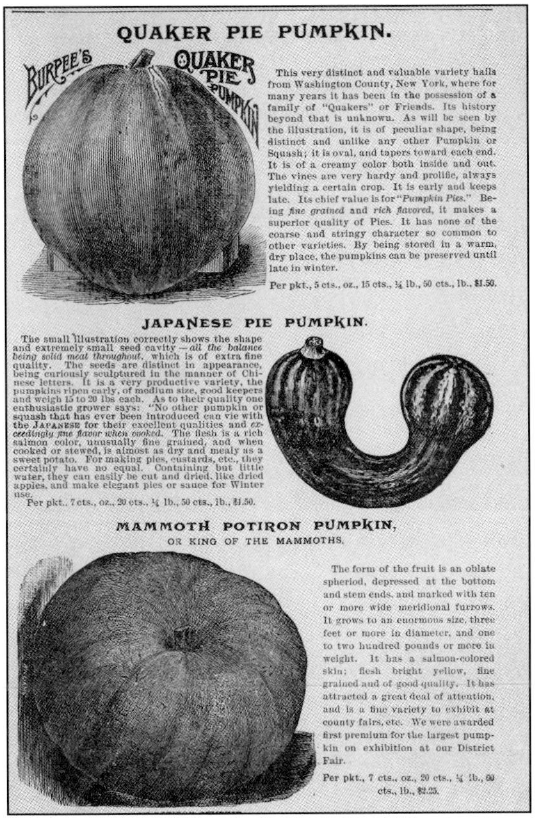
QUAKER PIE PUMPKIN.

BURPEE'S QUAKER PIE PUMPKIN

This very distinct and valuable variety hails from Washington County, New York, where for many years it has been in the possession of a family of "Quakers" or Friends. Its history beyond that is unknown. As will be seen by the illustration, it is of peculiar shape, being distinct and unlike any other Pumpkin or Squash; it is oval, and tapers toward each end. It is of a creamy color both inside and out. The vines are very hardy and prolific, always yielding a certain crop. It is early and keeps late. Its chief value is for "*Pumpkin Pies*." Being *fine grained* and *rich flavored*, it makes a superior quality of Pies. It has none of the coarse and stringy character so common to other varieties. By being stored in a warm, dry place, the pumpkins can be preserved until late in winter.

Per pkt., 5 cts., oz., 15 cts., ¼ lb., 50 cts., lb., $1.50.

JAPANESE PIE PUMPKIN.

The small illustration correctly shows the shape and extremely small seed cavity—*all the balance being solid meat throughout*, which is of extra fine quality. The seeds are distinct in appearance, being curiously sculptured in the manner of Chinese letters. It is a very productive variety, the pumpkins ripen early, of medium size, good keepers and weigh 15 to 20 lbs each. As to their quality one enthusiastic grower says: "No other pumpkin or squash that has ever been introduced can vie with the JAPANESE for their excellent qualities and *exceedingly fine flavor when cooked*. The flesh is a rich salmon color, unusually fine grained, and when cooked or stewed, is almost as dry and mealy as a sweet potato. For making pies, custards, etc., they certainly have no equal. Containing but little water, they can easily be cut and dried, like dried apples, and make elegant pies or sauce for Winter use.

Per pkt., 7 cts., oz., 20 cts., ¼ lb., 50 cts., lb., $1.50.

MAMMOTH POTIRON PUMPKIN,
OR KING OF THE MAMMOTHS.

The form of the fruit is an oblate spheriod, depressed at the bottom and stem ends, and marked with ten or more wide meridional furrows. It grows to an enormous size, three feet or more in diameter, and one to two hundred pounds or more in weight. It has a salmon-colored skin; flesh bright yellow, fine grained and of good quality. It has attracted a great deal of attention, and is a fine variety to exhibit at county fairs, etc. We were awarded first premium for the largest pumpkin on exhibition at our District Fair.

Per pkt., 7 cts., oz., 20 cts., ¼ lb., 60 cts., lb., $2.25.

Figure 333. In the three-sisters plantings, the squash were winter squash. Here are some choices in the 1887 W. Atlee Burpee seed catalog.

Rhea Trotman Martinez, a researcher at the University of Wisconsin-Stevens Point, studied the effect of three-sisters plantings in a climate far removed from Tabasco, Mexico, and in an area with more frequent summer rains (in fact, there was a severe "rain event" during the study). She reports: "…conditions in monoculture versus Three Sisters plots (were) not in the original plans for this project; however, following a storm with heavy rains and high winds, it was noted that several of the corn plants had been knocked over, and it appeared that a greater number of plants had been affected in monoculture plots than in Three Sisters plots. Although the difference was not significant, a higher percentage of corn plants were affected in monoculture plots than in Three Sisters plots."

The summary of this research was less encouraging. Martinez found that "No significant difference in the number of cucumber beetles on squash plants, and damage caused by corn smut and corn earworm on corn plants was noted between monoculture and Three Sisters treatments. Yield per plant was not significantly greater in Three Sisters treatments than in monoculture treatments. There was also no significant difference in changes in total soil nitrogen between treatments from 2006-2007." Was this due to the fact that the study was not done in a dry desert climate? It's hard to say (but it's easy to say the results of these studies are all over the map).

You can also try intercropping with fava (faba) beans. "The grain (seeds) yield of the faba bean intercropped with maize was greater than that of faba bean monoculture due to increases of the stems per plant and the pods per stem of faba bean." ("Nitrogen Fixation of Faba Bean (*Vicia faba* L.) Interacting with a Non-legume in Two Contrasting Intercropping Systems"; Fenliang Fan; Fusuo Zhang; Yana Song; Jianhao Sun; Xingguo Bao; Tianwen Guo; Long Li, *Plant & Soil*; May 2006, Vol. 283, Issue 1/2, p 275.)

An added benefit of the squash groundcover is, of course, the food it yields. Dr. Jane Mt. Pleasant of Cornell University has run trials of three-sisters plots against conventionally grown monocultural corn plots and found that the calories produced in the three-sisters system were 17% higher per unit area.

The best book on this subject is: *Native American Gardening, Stories, Projects, and Recipes for Families*, by Michael Caduto and Joseph Bruchac (Fulcrum Publishing, 1996). This is a fantastic book and covers diagrams for eight different planting schemes used by the Wampanoag and Hidatsa tribes of the Southwest. The recommended distance between mounds is usually three feet.

Intercropping with Good Fungi

In a study done in Nigeria, it was found that the beneficial fungi called mycorrhizae can improve the yields of a intercrop of maize (corn) and pigeon peas (*Cajanus cajan*, family *Fabaceae*): "*Glomus clarum* [a type or species of mycorrhizae] significantly improved their growth, shoot biomass, grain [maize] yield and nutrient uptake of maize and pigeon pea. Intercrop maize with or without mycorrhiza had higher yield in the residual experiment compared to monoculture. The grain yield was 48% higher in mycorrhizal pigeon pea compared non-mycorrhizal pigeon pea for both the intercrop and the sole pigeon pea." ("Response of Maize-Pigeon Pea Intercrop to Arbuscular Mycorrhizal Fungi in Nutrient Depleted Soil"; Dania, S. O.; Fagbola, O.; Dania, Margaret I., *Annual Review & Research in Biology*; Vol. 3 Issue 3, p 232, 2013.)

In another study of the beneficial fungi, "Results showed that all treatments that included legumes, whether grown as a sole crop, in legume-cereal [maize] or in legume-legume cropping systems in the previous year, had significantly higher ...fungal colonization of the rotational maize crop roots by a range 39% to 50% and 19% to 47% than those in maize supplied and not supplied with Nitrogen fertilizer, respectively, in a maize-maize short rotation...("Pigeon Pea and Cowpea-Based [*Vigna unguiculata*] Cropping Systems Improve Vesicular Arbuscular Mycorrhizal Fungal Colonisation of Subsequent Maize on the Alfisols in Central Malawa"; Keston. O. W. Njira, et al. Department of Soil and Geological Sciences, Sokoine University of Agriculture, Morogoro, Tanzania and the Depart. of Crop and Soil Sciences, Lilongwe University, Lilongwe, Malawi.)

Figure 334. This pole lima bean variety has been in the family of Dr. Bradshaw for 150 years. Here it is in the 1887 W. Atlee Burpee seed catalog. It produced over 1700 pods then. It's still available as a short-pole bean.

Bonus Photos and Graphs:

Figure 335. The minimum area needed to ensure good corn pollination in a Three Sisters garden is at least a 10 by 10 foot square of space. If you're just planting a dense corn patch, the minimum size is four by four feet. (Photo courtesy of UConn Extension Master Gardener Program, Bethel, CT.)

Figure 336. Many of the native corn cobbs have multiple colored kernels. The Zuni tribe use the colors to represent the Four Directions. The colors are: yellow -north, blue - west, red - south, and white - east. The Arikara people plant four seeds in each mound, each corn kernal is a differenet color: red, white, yellow, and blue.

Figure 337. Probably not three sisters. But a good picture of how much of the corn from three-sister farms was used.

In one study, the number of calories per acre for a monoculture of corn equaled 21,400. for monocrop of beans, 47,000. And a monocrop of squash produced 107,000 calories.

By comparison, the three-sisters plot equaled 198,000 calories per acre. The total for all the calories per acre for the three monocrops combined equaled 175,400. This equals a nine percent greater yield with Three Sisters plantings.

(*An evaluation of the productivity of the Native American 'Three Sisters' Agricultural System in Northeren Wisconsin*, 2007. Rhea Trotman Martinez, Thesis. The College of Natural Resources, the Univerity of Wisconsin-Stevens Point.)

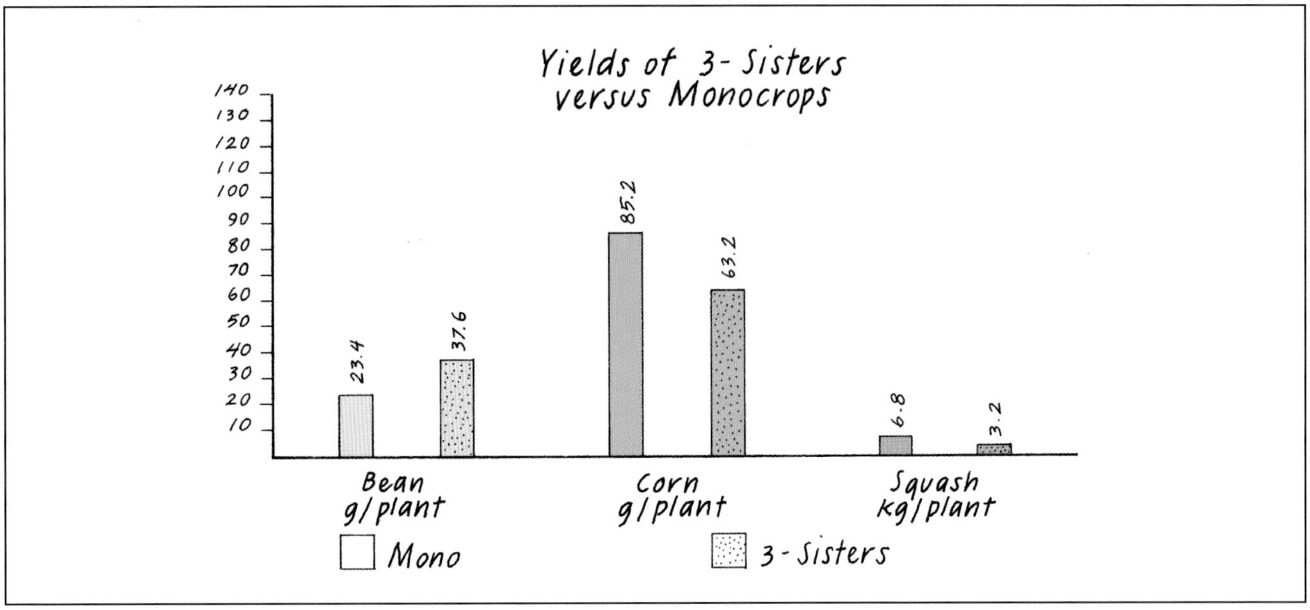

Figure 338. This chart shows that only beans had greater yields in grams per plant in a Three Sisters planting compared to a monocrop of beans. Three Sister planting had about one-half of the squash grams per plant compared to the monocrop planting.

(*An evaluation of the productivity of the Native American 'Three Sisters' Agricultural System in Northeren Wisconsin*, 2007. Rhea Trotman Martinez, Thesis. The College of Natural Resources, the Univerity of Wisconsin-Stevens Point.)

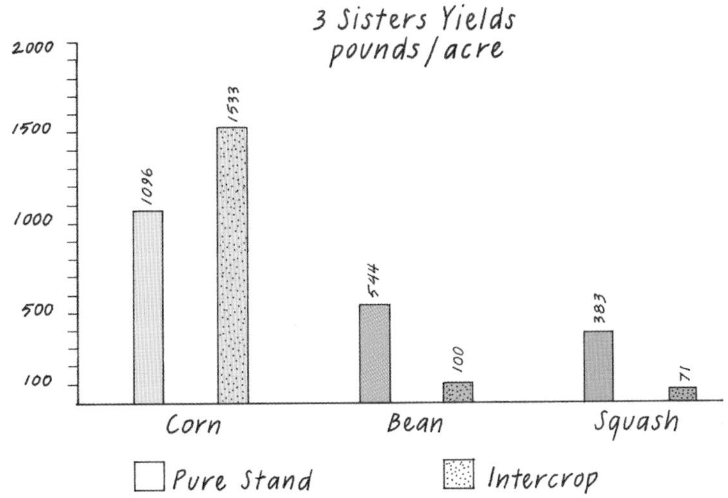

Figure 339. Here we see in a Three Sisters planting the harvest of beans was one-fifth that of a pure stand (monocrop.) The corn harvest, in pounds per acre, was *greater* in the Three Sisters planting compared to a monocrop. If corn was the most important crop, it was shown to be beneficial to grow it in mixture with squash and beans, considering the fruits of the other two simply as a bonus.
(*An evaluation of the productivity of the Native American 'Three Sisters' Agricultural System in Northeren Wisconsin*, 2007. Rhea Trotman Martinez, Thesis. The College of Natural Resources, the Univerity of Wisconsin-Stevens Point.)

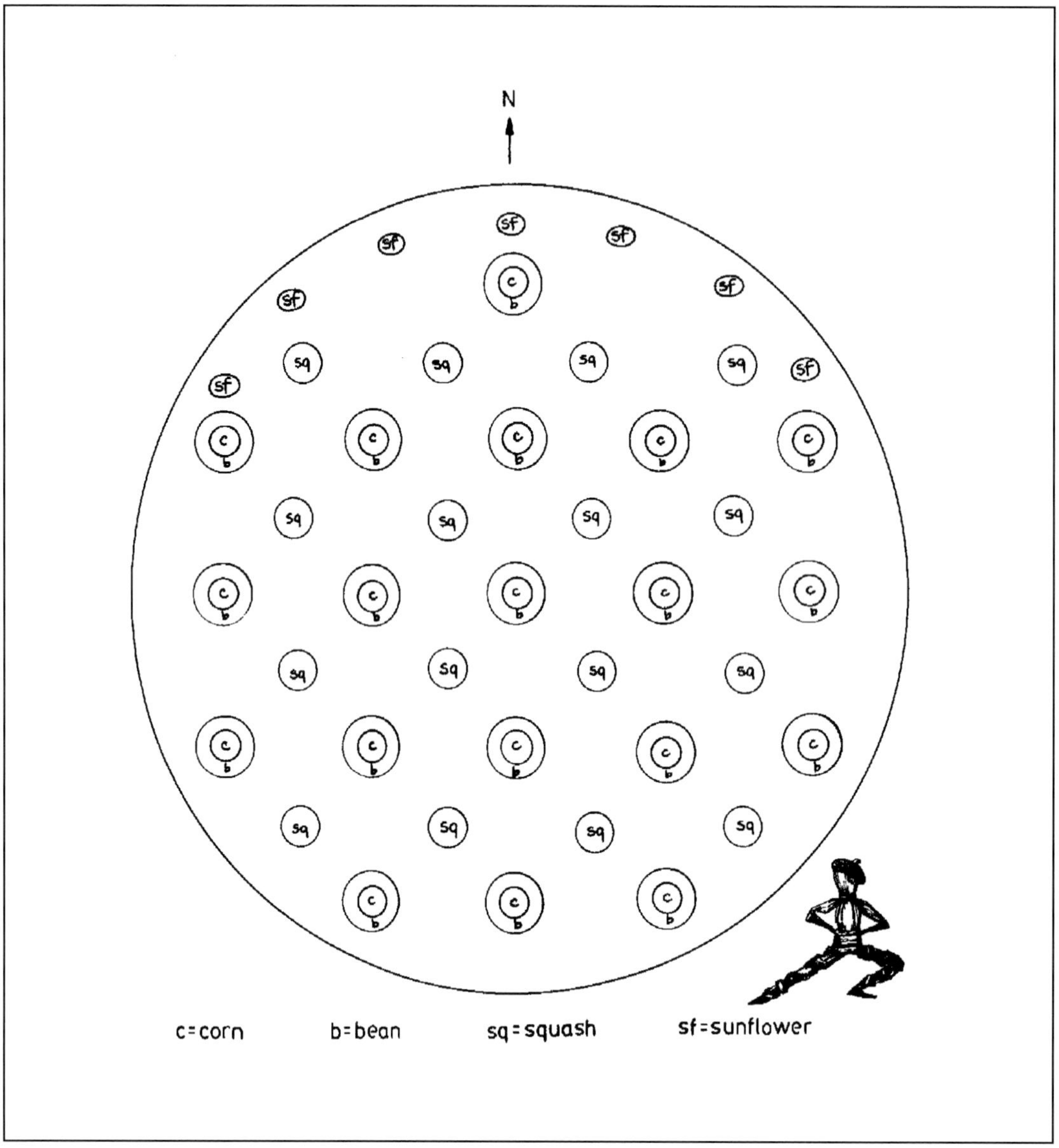

1" = 5'

Figure 340. This is a scale drawing of the mound planting approach to a three-sisters garden by the Wamapanoag Native American tribe. The scale is exactly one inch equals five feet. This approach includes sunflowers—an edible composite plant that also attracts benefical insects. The tall sunflowers are planted only on the North side so they don't shade the garden. Hopi gardeners made holes on mounds to plant the corn about four paces apart.

(Used with the permission of Fulcrum Publishing, from: *Native American Gardening, Stories, Projects, and Recipes for Families*, by Michael Caduto and Joseph Bruchac, Fulcrum Publishing, 1996.)

Chapter 11
Attract Beneficial Insects

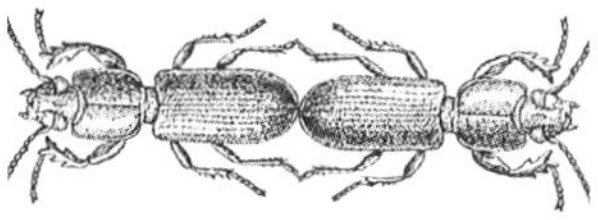

Insects are sometimes called bugs. But that's not very scientific; so I'll stick with insects, mostly.

Flower diversity equals insect diversity. A range of flowering plants is important for attracting beneficial insects. But, like a two-sided coin, flower diversity attracts what we call pests. For example, in early spring aphids colonize my blooming fava beans. Their predators—ladybugs, syrphid flies, and parasitic wasps—follow shortly. As spring progresses, the aphids move on to other plants, as do the good guys.

Ladybugs

Of all the bugs in the world, *Hippodamia convergens*, the ladybug (also known as the ladybird or lady beetle) is practically *the* poster insect for beneficial organic-garden insects. It's become well known for its aphid-eating proclivities—both the adult beetle and the larvae eat only aphids. Unfortunately, according to entomologist Dr. Ken Hagen of UC Berkeley: "Purchasing ladybugs for aphid control means that you're simply throwing [your] money away."

This is why: the ladybugs sold by mail order and through nurseries are captured from large winter congregations in the Sierra mountains of California. Entomologists have discovered that when these insects gather each fall, they're packed with enough stored fat for the winter and part of the spring. Many sources keep ladybugs in refrigeration so they stay less active and store longer for sales. Unless the ladybugs are defatted before they're sold, they're not even hungry when they wake up. According to Hagen: "*Hippodamia convergens* gathered from aggregations in the Sierras (are) ineffectual predators; green lacewings are better." The ladybug is more inclined to leave your yard than to start snacking on aphids, although those that do hang around will eat up to 5000 aphids in their lifetime. Each young larva will eat up to 400 aphids before it forms a pupae.

Another thing to know when considering a ladybug purchase is that up to 10% of the overwintering ladybugs are infested with parasites, and selling them only speeds up the spread of these parasites. While parasitized beetles do eat, they can't reproduce.
You can lure ladybugs to your garden with the same beneficial plants that attract most good bugs—mem-

Figure 341. These orange eggs are those of a ladybug, not of a pest.

Ground beetles, like those at the top, are predators of slugs, snails, aphids, insect larvae, insects, caterpillars, and insect eggs. One of the few predators of snails and slugs, they hide under rocks.

bers of the sunflower and parsley families. Also, you must stop spraying with *anything*. Yes, skip even the "organic" sprays as they're often too broad-based to protect the ladybugs and their larvae. The soft-bodied larvae, which look like long, creepy, orange-and-black bugs and are often mistaken for a pest, are very susceptible to all manner of pesticide sprays, organic or not.

Figure 342. Here, two ants have "herded" some aphids, which they'll "farm," eating the honeydew that the aphids produce. They may even move aphids from stem to stem or plant to plant.

If you find ladybugs that are black with one orange spot on each wing, treasure them, as *Chilocorus stigma* (meaning "twice-stabbed ladybug) won't fly away to congregate in the mountains each winter, but will hibernate in ground litter in your garden to revive in the spring. They're found only east of the Rocky Mountains. Another type *Chilocorus orbus* (also black, with a circular red spot on each wing), has a range from California to Washington. Both beetles are mainly predators on scale bugs and prefer arboreal habitats.

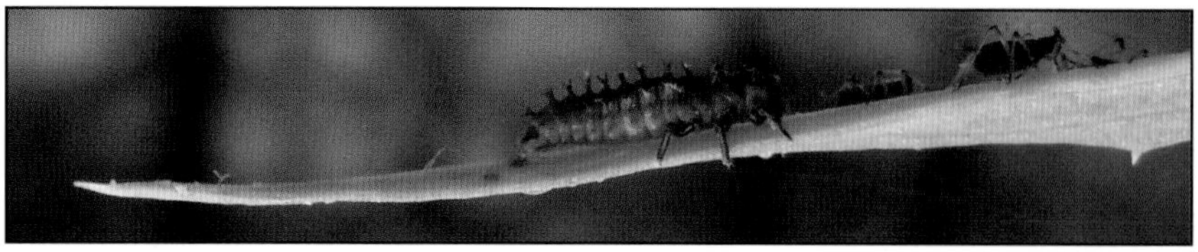

Figure 343. This is the larval stage of a ladybug; some people think it looks like a pest and will destroy it. Even organic sprays will kill it, and dead larvae can't eat the 400 aphids they would otherwise each consume before reaching their adult stage.

Figure 344. These "twice-stabbed" ladybugs don't migrate, but hang around in mulch during the winter. Here the larvae and eggs look similar to those of a convergent ladybug. That aphid (left) better run.

Figure 345. Here we see two insects hunting aphids for very different reasons. The ant wants to harvest the honeydew from aphids it has rounded up. The adult ladybug just wants to gobble up the aphids—it may eat up to 4000 aphids before laying its eggs and dying.

Figure 346. This fava-bean plant reveals a lot about ladybugs and aphids. #1 shows the brown bloated body of an aphid with a parasitic-wasp larva inside; spraying the aphids will kill off the beneficial parasites as well. (A swollen brown aphid body with a hole that looks like an open hatch door means that the wasp has already emerged.) #2 points to a winged adult aphid, while #3 indicates a cluster of young wingless aphids. The ladybug larva is in pursuit of aphids. When checking for pests, leave some aphids alone so that the beneficial wasps can multiply.

Syrphid Flies

The larvae of syrphid flies (of the family *Syrphidae*, also called hoverflies) are some of the most voracious predators of aphids. Compared to ladybugs, however, they're little known, and their larvae, easily mistaken for slugs, are often mistakenly killed. Fortunately there are plenty of them—one common genus of hoverfly includes over 300 species on the American west coast alone.

The syrphid fly lives up to its alternate name—it can hover in the air in front of flowers like a helicopter, something that bees can't do. Syrphid flies are often mistaken for bees or wasps because of their various types of horizontal yellow-and-black striping and thus mistakenly killed. If they survive, their larvae eat *lots* of aphids. A single larva may eat much more than its body weight in aphids in its life, making them much better predators of aphids than ladybugs. The adults fly from flower to flower to gather the nectar and pollen they need to survive, and should be treasured as pollinators as well

The careful observer will note, however, when comparing bees and hoverflies, that the latter has shorter antennae, a stout waist, and only one pair of wings (the bee has two pairs). Some common species lack the bee-like markings and/or shape. *Sphaerophoria cylindrica* is slender, and, as the name implies, cylindrical; except for its markings, and really doesn't resemble a bee at all. A large hoverfly generally has a wide, somewhat flattened abdomen, not at all like the plump tummies of bees. Most hoverflies have glossy bodies, but a few species are fuzzy and do resemble bumblebees. A defining characteristic of the Syrphidae is a "spurious" wing vein—a fold in the middle of the wing membrane that's best seen with a hand lens.

Figure 347. The larvae of a syrphid fly, here with an aphid in its mouth, looks like a slug. Don't kill it.

Because the larvae look like tiny slugs, they're often mistaken for pests and squished or sprayed—the poor little guys get it on all fronts. All sprays, organic or chemical, will not only kill syrphid larvae but also their favorite food—aphids, not to mention all other insects, whether beneficial or not. Check to see if the plant has such a low level of aphids that it's not a problem, and leave the plant alone as nature maintains its natural balance. The good guys can't survive without some of the bad guys to eat

In an article by the respected entomologist Frederique Laviopierre, (Originally published in *Pacific Horticulture Magazine*, 2007, Vol. 68, No. 2.) we read:

"Adult hoverflies feed on nectar and pollen and, occasionally, on the honeydew exuded by aphids. They need nectar to fuel their high-energy flight, and females need pollen to produce their eggs. Most hoverflies, and the majority of the common garden

Figure 348. How do you tell the difference between a syrphid fly and a bee? The syrphid has much shorter antennae, and only one pair of wings—the honeybee has two pairs, a fore and hind wing. The eyes on the syrphid fly are larger and cover most of its head, while the honeybee's eyes are located at the side of the head and are proportionally smaller.

If you've ever paused near a patch of pollen-laden flowers on a sunny summer afternoon, there's no mistaking the unique flight of the hoverfly as it zigzags about the blossoms, occasionally stopping in mid-air, its shimmering wings barely visible. Unlike most other winged insects, flies in the order *Diptera* (meaning two-winged) have only one pair of flying wings. The second pair of wings, known as "halters," consists of two little knobs that function as miniscule gyroscopes, allowing the flies to quickly change direction—a useful ability when being chased by predators or by a biped armed with a swatter. (Not that any conscientious gardener should want to swat syrphids! With few exceptions, hoverflies are among our most abundant and helpful garden allies.)

species, have aphid-feeding larvae, while a few prey on leaf-beetle larvae and eggs, or on other insects. Resembling small slugs or caterpillars, the larvae are easily mistaken for pests. Lacking eyes, hoverfly larvae were long thought to blindly detect their prey by raising the front half of their body and swinging it from side to side until they bumped into a tasty morsel. Evidence now shows that at least some species detect their prey through chemical odors.

Hoverflies are one of the most useful natural enemies of plant pests; some species have been estimated to eat up to 1,200 aphids during the larval stage. In one study, the diminutive chevroned hoverfly was found to exert from seventy to one hundred percent control of

Figure 349. This European honeybee is gathering pollen or nectar from an English lavender blossom.

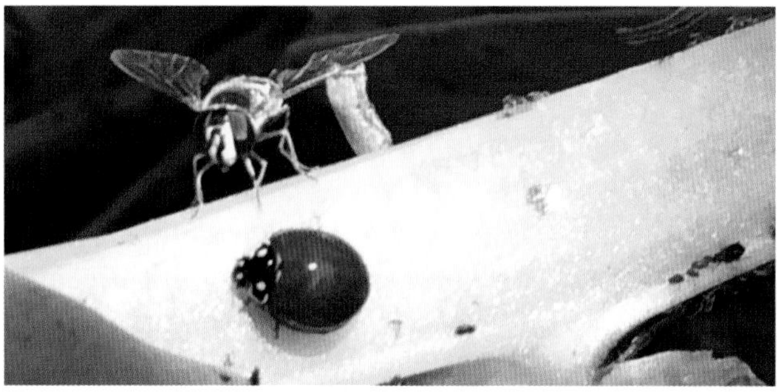

Figure 350. Two predators of aphids, a syrphid fly and a ladybug. Watch out aphids!

aphids. After capturing its victim, the hoverfly larva holds it in the air, sucking it dry of juices; nearby aphids or beetle larvae appear to pay no attention to the demise of their unfortunate neighbor. Not all hoverflies prey on pests. The larvae of some species dwell in the nests of social insects, such as bees, wasps, and ants, where they scavenge refuse. Some are pests of roots, stems, and bulbs (narcissus fly larvae) but, for the most part, these are of no concern to the gardener, who is unlikely to suffer any significant damage. Those that frequent decaying organic matter aid in decomposition.

A complex habitat and diversity of flowering plants found in most gardens create an ideal environment for attracting hoverflies. Because many species pupate (and overwinter) in leaf litter, a permanent mulch will encourage persistent populations. Hoverflies are also sensitive to wind; if your garden occupies a windy site, provide some shelter near flowerbeds. Predatory hoverflies seek out pest colonies in which to lay their eggs; females of these species are bound to find a multitude of egg-laying sites in gardens.

In yet another study, phacelia (*Phacelia tanacetifolia*, a flowering plant in the borage family, also known as

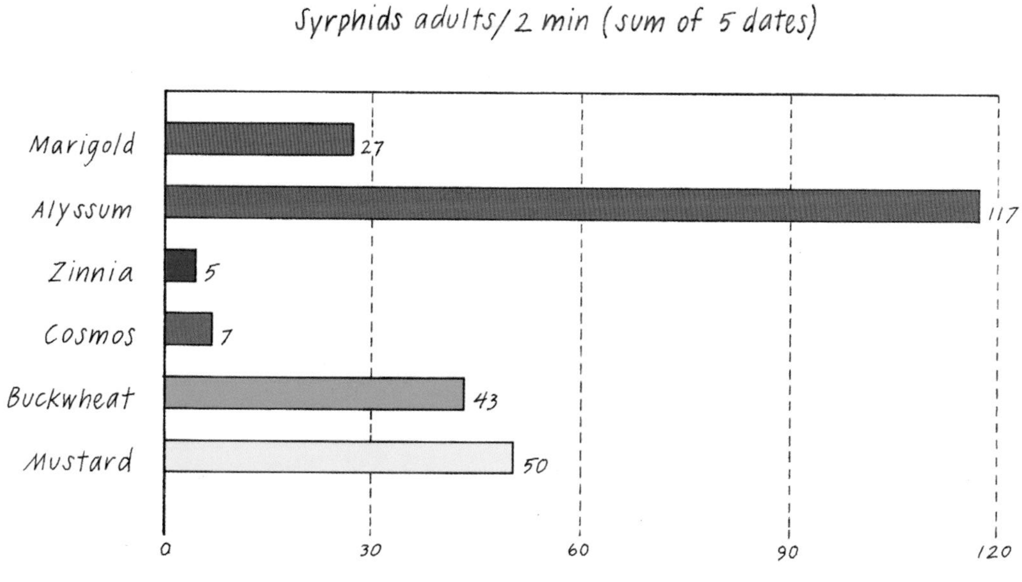

Figure 351. In this study, alyssum attracted two times more syrphid flies than mustard. At my house, 60 miles north of San Francisco, CA, alyssum blooms scantily through the winter—elsewhere it's at least a very early spring flower. It can attract beneficial wasps to terrorize aphids on early flowering fava beans as they mature (see Figure 346). Soon after the alyssum blooms fully in February, mustard plants will flower throughout many California vineyards.

lacy phacelia, blue tansy or purple tansy) attracted both bees and hoverflies. The favorite flowers of syrphid flies are the daisy (*Asteraceae*) and umbel (*Apiaceae*) families. They also feed on nectar and pollen in a wide range of other plant families, including roses (*Rosaceae*), buckthorns (*Rhamnaceae*), borages (*Boraginaceae*), and willows (*Salicaceae*).

Figure 352. Borage (*Borago officinalis*) can attract some beneficial insects, but is best known for attracting bees, especially in early spring. It can bloom all winter in very mild climates.

Figure 353. This is a collection of *Phacelia tanacetifolia* (lacy phacelia or purple tansy) blossoms. Many studies in my files conclude that this flower is one of the best for attracting beneficial insects.

In a thesis by Elsa Laubertie, it was found that, in order of attractiveness, phacelia (*Phacelia tanacetifolia*), followed by buckwheat (*Fagopyrum esculentum*), and then coriander (*Coriandrum sativum*) had the best potential to attract feeding female syrphids. The hoverflies that had fed on phacelia and buckwheat pollen were found up to 17.5 meters (57 feet) from the actual floral areas. (For more on this see pages 352 - 356.)

Figure 354. The flowers produced by buckwheat harbor many beneficial insects.

Beneficial wasps

By planting aromatic plants near your ornamental or crop plants, you'll find that the incidence of pests and diseases may fall. Plants such as borage, limanthes (poached egg plant), *Echium* spp. (can be invasive in coastal California, but easy to manage in a garden), and members of the umbel family (as an example, a few parsnips allowed to go to seed) attract predators such as hoverflies, parasitic wasps, and other beneficial insects.

In another study, the fields under apple trees were planted with alyssum (*Lobularia maritima*), phacelia (*Phacelia tanacetifolia*); and buckwheat (*Fagopyrum esculentum*) for their effect on the larvae of the light-brown apple moth, *Epiphyas postvittana*. The numbers of larvae were much lower in both phacelia and control treatments. In the alyssum and buckwheat treatments, the damage was up to 29% lower and there were more than twice as many braconid (a beneficial wasp) cocoons than in control plantings. ("The Effects of Floral Understoreys on Parasitism of Leafrollers (*Lepidoptera tortricidae*) on Apples in New Zealand"; N. A. Irvin, S. L. Scarratt, S. D. Wratten, C. M. Frampton, R. B. Chapman, J. M. Tylianakis. From: *California Agriculture,* September-October 1998, 52(5):23-26.)

Also:

Marking studies demonstrated that lady beetles, lacewings, syrphid flies and parasitic wasps fed on nectar or pollen provided by borders of flowering plants around farms; many insects moved *250 feet* [italics mine] into adjacent field crops. Studies also showed that syrphid flies, parasitic wasps and

Figure 355. This tomato hornworm is covered in the white cocoons of parasitic wasp larvae, which eat the caterpillar nearly to death and then makes cocoons on the outside of its body. Spraying or killing the hornworm will also kill and reduce the number of wasps.

Figure 356. Beneficial wasp, known as a "potter" or "mason" wasp, is shown here inserting an egg into a caterpillar. It's a member of the *Eumeninae* family.

lacewings fed on flowering cover crops in orchards and that some moved as much as six feet high in the tree canopy and 100 feet away. "... The use of nectar or pollen by beneficial insects helps them survive and reproduce. Therefore, planting flowering plants and perennial grasses around farms [also around your garden—ed.] may lead to better biological control of pests in nearby crops [vegetables]."

Spiders

Many people are scared of spiders (more properly called arachnids), and kill them and knock down all spider webs on sight. Gardeners should certainly reconsider this behavior, as spiders are very good predators of beetles, caterpillars, aphids, bees, wasps, butterflies and moths. (Some of these are "good" insects, but you can't have it all.) There are 3,500 species of spiders, so you have plenty of options. In gardens, mulch and flowering plants can encourage occupation by spiders, especially if you leave some grass stems to support their webs. Flower borders and orchards with ground covers can also harbor the busy arachnids. I once, as an experiment, left all of a lavender plant's stems unpruned, and was blessed one morning with dozens of spiderwebs. See Figure 358. (But, alas, it never happened again.) It's nice to know spiders often catch more insects than they eat. Some build lairs below the ground surface and race out to catch insects that walk nearby, and because of this they tend to avoid areas under tillage. Some are "jumping" spiders.

Figure 357. The paper wasp terrifies many people, but it does take bits of caterpillar to its mud-and-paper nest to feed the larvae growing there.

Figure 358. As an experiment, I left the stems of lavender in my garden uncut, and woke up one day to this miracle of spiderwebs! I was blessed this day, but never again.

Figure 359. This orb-weaver spider (*Argiope aurantia*) is fairly common. Since it's so large, people tend to leave it alone.

Ground Beetles

Many people never consider the great importance of beetles in the horticultural scheme of things. Many predaceous beetles spend all their time on the ground, and thus are known as (duh) "ground beetles" (genus *Diaelus* spp.). Some of these are predators of slugs, snails, aphids, insect larvae, small insects, caterpillars, and insect eggs. Beetles are one of the few predators of snails and slugs; tiger beetles, for instance, can eat their body weight in prey every day, burrowing into the ground and "hunting" whatever gets close to their lair. Take good care of these helpful predators, as they can live up to four years of slug patrol. Since they need bare soil to do this, however, 100% mulch is a problem for them (so is tillage). Leave some open soil for these predators that prefer it.

To encourage these helpful predators, you can consider building a "beetle bank."

"In England, in an attempt to provide suitable overwintering habitat within fields for predators of cereal aphids, researchers created "beetle banks" sown with perennial grasses such as *Dactylis glomerata* [cat grass] and *Holcus lanatus* [Yorkshire fog]. When these banks run parallel with the crop rows, great enhancement of predators (up to 1500 beetles per square meter) can be achieved in only two years. ("Designing Species-Rich, Pest-Suppressive Agroecosystems through Habitat Management"; Clara I. Nicholls and Miguel A. Altieri, 2004. American Society of Agronomy, Monograph no. 43.)

To construct your beetle bank, use small mounds up to six feet wide and one foot high. Perennial native grasses, such as switchgrass (*Panicum virgatum*) or California oatgrass (*Danthonia californica*), are good candidates for planting on top of the bank, but be sure to keep woody perennials from colonizing it. Replacing the grasses with perennial shrubs reduces the habitat for the beetles. Another approach to beetle banking is to construct a simple pile of fieldstones to provide shelter for beetles and other predators like lizards. Both need to be sited near the garden, but out of major "viewsheds"—the view of an area from a specific vantage point. (Out of sight, but not out of mind.)

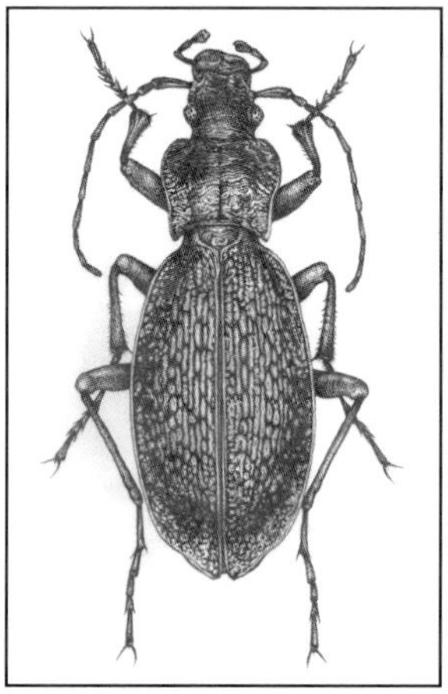

Figure 360. This is an old lithograph of a Carabid ground beetle (*Carabus nemoralis*). They live under duff, sticks, and stones, and don't travel far.

Figure 361. Ground beetles aren't often noticed, or thought of as predators, but they're one of the few insects that attack slugs and snails.

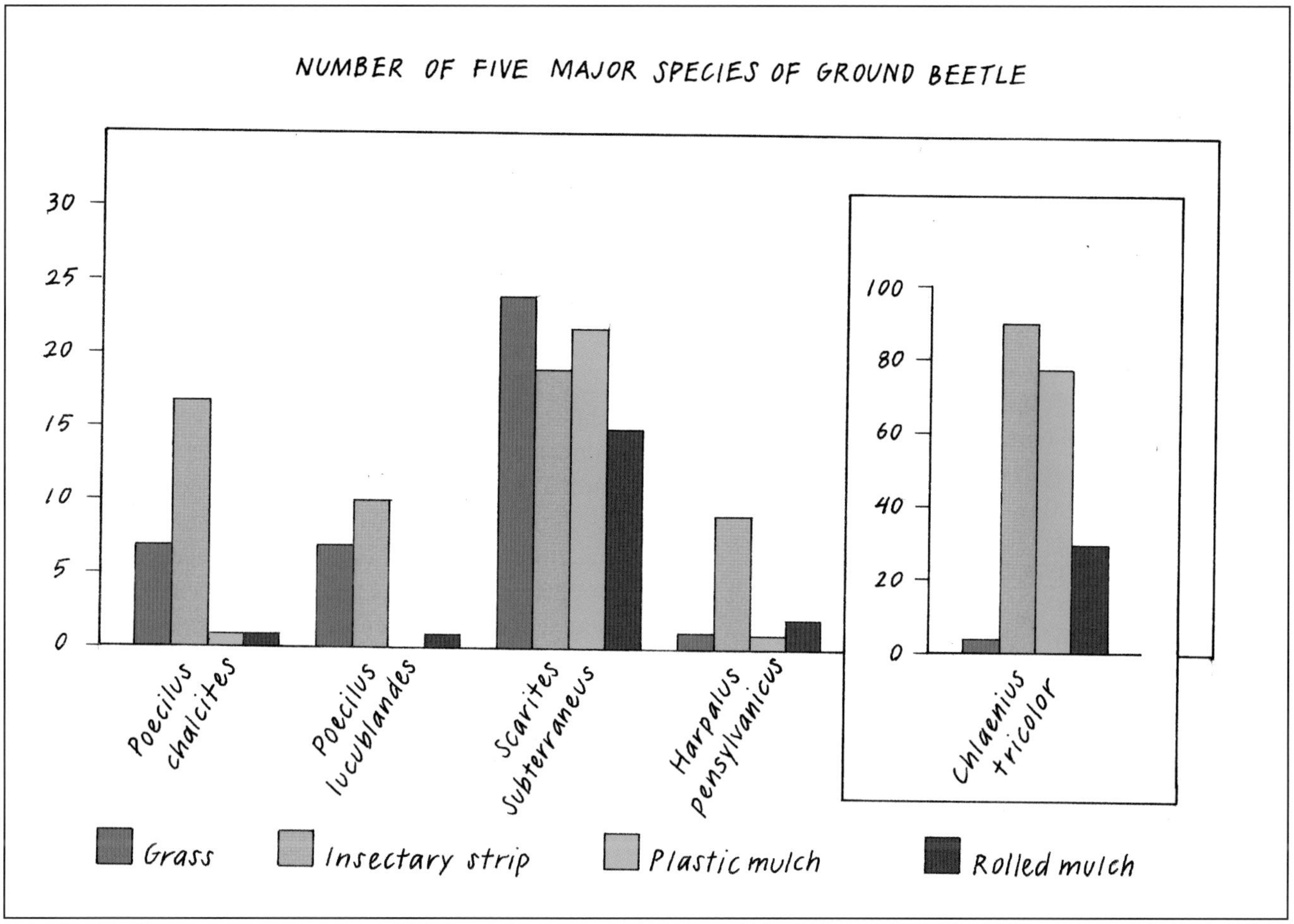

Figure 362. This study shows the importance of an insectary strip in cucumber fields. The mix of plants in the strip: Rocky Mountain penstemon (*Penstemon strictus*); goldenrod (*Solidago* spp.); marigold (*Tagetes tenuifolia*); Queen Anne's lace (*Daucus carota*); and bachelor's button or cornflower (*Centaurea cyanus*). (From: "Invite Insect Allies to Your Farm Using Insectary Strips," by Dr. Gladis Zinati, Director of Vegetable Systems Trial, Associate Research Scientist, Rodale Institute. 2018, Kutztown, PA.)

Hedges and Flower Borders for Companion Pest Control

What are the potential benefits of cultivating hedgerows (a process also called "farmscaping"), formal hedges, and floral plantings in the home garden?

Hedgerows enrich wildlife, providing habitat for mammals like foxes and squirrels, for birds and reptiles that eat insects, and for insects that act as pollinators and beneficial predators of plant pests.

Here's an example of mammal interaction: "A relationship exists among squirrels, fungi, and trees. The squirrels feed on the fungi, and then assist in its reproduction by dropping fecal pellets containing viable fungal spores onto the forest floor. There, new fungal colonies establish. Tree feeder roots search out the fungi and form a symbiotic association that enables the tree roots to increase their nutrient uptake. The fungi, in turn, derive food from the tree roots. Each benefits from the other's presence or actions. [Intercropping Principles and Production Agronomy Systems Guide by Preston Sullivan, Appropriate Technology Transfer for Rural Areas.]

And here's how birds get into the act: In a study by Rachael Freeman Long of the Yolo County, California Cooperative Extension, it was found that bird populations increased 35 percent with protective hedgerows, and some preyed on the codling moth larvae that ravage apple cores. All the way back in 1911, the U.S. Department of Agriculture published a report about birds and codling moth: "This little bird [the downy woodpecker] finds the concealed larvae under the bark, not from any noise the insect makes; it is not a grub of a beetle having a boring habit and liable to make a sound that might betray its retreat, in seasons of the year when not torpid [inactive]. A caterpillar makes scarcely an appreciable noise, even when spinning its cocoon, and when that is finished it rests as quietly within as an Egyptian mummy in its sarcophagus." [Pretty lyrical for a government publication.]

"There is no evidence that the downy woodpecker ever makes a mistake; it has some way of judging. Other woodpeckers known to feed on the codling-moth larvae are the hairy woodpecker (*Dryobates villosus*), Texan woodpecker (*Dryobates scalaris hairdi*), red-headed woodpecker (*Melanerpes erythracephalus*), red-shafted flicker (*Colapteso cafers couaris*), and the pileated woodpecker (*Phloeotomus pileatus*). We find that in the United States there are 36 species of birds, belonging to 13 families, which feed on the codling moth. The three important families seem to be woodpeckers, titmice, and sparrows.")

- Hedgrows **reduce soil erosion and condition soil** – hedgerows slow water and prevent erosion. Some plants improve the water-holding capacity of the soil. Others are nitrogen-fixing plants.

- They **conserve water** – hedgerows retain water and reduce drying winds in summer, thereby reducing irrigation needs.

- They decrease wind damage – wind disturbs pollination and damages fruit and flowers. Properly designed hedgerows dramatically reduce wind speed, thereby improving crop yields.

- They create borders and privacy screens – hedgerows provide privacy. As they mature and become denser, they can reduce noise and serve as fencing.

- They **help create biodiversity** in gardens – it's the overall diversity that can encourage a large cross-section of beneficial insects. But remember: it must be carefully chosen diversity—willy-nilly diversity can work against you.

Hedges for Companion Plantings

When it comes to enhancing the habitat for beneficial insects in the home garden, the use of hedgerows is one of the most important crossovers from agricultural research, and one that can truly be scaled down from an agricultural setting to a suburban garden or yard. They don't have to be long hedges, just clusters or rows with a mixture of plants confirmed as species that attract beneficial insects. Plants can be can be grouped and clustered to form short borders instead of long hedges; the beneficial insects attracted to these clusters can easily fly to find pests elsewhere in the garden, and bees can collect pollen.

In 2016, the Xerces Society for Invertebrate Conservation produced a fantastic publication on hedgerows called Habitat Planning for Beneficial Insects, Guidelines for Conservation Biological Control, by Jennifer Hopwood et. al. (Digital copies are free from www.xerces.org.) It contains great lists of flowering plants to use, as well as descriptions of beneficial insects, and it's fairly easy to figure out from this work how to downsize the concept to a home landscape.

There's been a lot of research done by the University of California at Davis on using borders as shelters for beneficial insects. This has mostly taken place in a suburban setting—the backyards of the nearby Village Homes planned community. All of their findings indicate that a hedgerow provides a great deal of habitat for beneficial insects to pursue pests throughout the garden. For instance:

• Significantly more hoverfly eggs were found in fields surrounded with *Phacelia tanacetifolia* (*lacy phacelia)* than in control fields. It was clear that biological control of aphids by syrphid larvae can reduce the use of insecticides. (Use of *Phacelia tanacetifolia* Strips To Enhance Biological Control of Aphids by Hoverfly Larvae in Cereal Fields, by Janice M Hickman, Stephen D Wratten; 1996.)

• When predators of Colorado potato beetle (*Leptinotarsa decemlineata*) in eggplant fields were studied, it was found that dill (*Anethum graveolens*) and coriander (*Coriandrum sativum*) produced flowers that were complementary to Colorado potato-beetle predators. The number of predators was significantly higher in the fields interplanted with dill and coriander than in the control field without flowers. (Impact of Strip-Insectary Intercropping with Flowers on Conservation Biological Control of the Colorado Potato Beetle"; By Joseph M. Patt, George C. Hamilton, James H. Lashomb, 1997. *Advances in Horticultural Science,* Vol. 11, No. 4, P 175-181.

• When of the flowers of sweet alyssum (Lobularia maritima) were studied for their effects on the suppression of aphids in California lettuce fields, the presence of alyssum resulted in more hoverfly larvae and fewer aphids (*Hemiptera*: *Aphididae*). The results of this study showed that increasing the numbers of flowers can enhance aphid suppression and crop quality due to the elevated levels of natural enemies. ("*Increasing Syrphid Fly Diversity and Density in Sown Flower Strips within Simple vs. Complex Landscape";* Sebastian Haenke, Barbara Scheid, Matthias Schaefer, Teja Tscharntke and Carsten, 2009. *Journal of Applied Ecology,* Vol. 46, No. 5, P 1106-1114.

Figure 363. Lacewing eggs dangle from thin threads amid a mass of aphids. (Lacewing larvae eat aphids, but are also cannibals.)

- Researchers J.M. Holland, S.R. Thomas, & S. Courts found that the plant type that attracted the most species of syrphid flies was *Rubus* spp. (brambles such as blackberry, raspberry, etc.). The number of syrphids visiting *Convolvulus arvensis* (field bindweed) was unexpected, as it doesn't seem to have the usual characteristics of a beneficial attractant. They also recommended the use of *Phacelia tanacetifolia* (lacy phacelia) strips as a part of an integrated farm.

- "...Vetch planted in the understory of pecan orchards increased convergent lady beetle populations in the canopy, resulting in control of black pecan aphid (*Melanocallis caryaefoliae*) populations." (Manipulating Plant Resources to Enhance Beneficial Arthropods in Agricultural Landscapes, Dec 2005. Douglas A. Landis, Fabián D. Menalled, Alejandro C. Costamagna and Tammy K.Wilkinson, *Weed Science*, Vol. 53, No. 6, p 902-908.)

Distance

I was delighted to get more specific observations of exactly how far beneficial insects can fly in search of nourishment:

Miguel Altieri (Professor of Agroecology at the University of California, Berkeley in the Department of Environmental Science, Policy and Management)

- found that many beneficial insects migrated as much as 25 meters (**82 feet**) from a hedgerow into a neighboring field of corn. (Personal communication 10-29-10) Obviously a range of 164 feet (two times 82, to take in both directions) covers most gardens if not all landscapes.

- Here's the conclusion of another important study done on the dynamics of locating a hedgerow next to a farmer's field (but applicable to home gardens): "...Many (beneficial) insects moved ***250 feet into*** adjacent field crops. Studies... showed that syrphid flies (70% of the introduced flies), parasitic wasps and lacewings fed on flowering cover crops in orchards and that some moved ***6 feet high*** into the tree canopy [good for dwarf fruit trees] and ***100 feet away*** from the treated area (ed. where the beneficial insects where released.). The use of nectar or pollen by beneficial insects helps them survive and reproduce. Therefore, planting flowering plants and perennial grasses *around* [my italics] farms may lead to better biological control of pests in nearby crops." ("Beneficial Insects Move from Flowering Plants to Nearby Crops;" Sept/Oct. 1998. Rachael Freeman Long, Andrew Corbett, Celia Lamb, Chris Reberg-Horton, Jeff Chandler, Michael Stimmann; *California Agriculture* 52(5): p 23-26.)
- In a study entitled "Hedgerow Restoration Promotes Pollinator Populations and Exports Native Bees to Adjacent Fields" (June

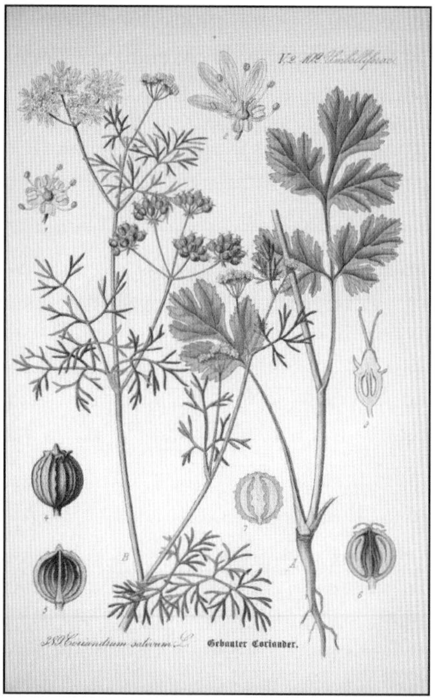

Figure 364. Let some of your cilantro bloom; the umbel-shaped flowerheads attract a lot of beneficial insects, and the subsequent seeds are coriander.

2013. Lora A. Morandin and Claire Kremen *Ecological Applications*, Vol. 23, No. 4,, p 829-839), researchers found a greater number of native bees up to at least ***100 meters*** **(328 feet)** into the field from the hedgerow. There were seven times more native bees in the hedgerow than in the weedy, grassy edges.

- In another study, centered on eggplants, the insectary plants dill and coriander, planted ***108 feet apart*** caused a significant reduction in Colorado potato beetle. The report stated: "... [This] has the potential to become a very important component of Integrated Pest Management."("Impact of Strip-Insectary Intercropping with Flowers on Conservation Biological Control of the Colorado Potato Beetle"; 1997. J. M. Patt, G. C. Hamilton and J. H. Lashomb, *Advance Horticultural Science*, 11, p 175-181.)

- Entomologist Rachael Freeman Long and fellow observers found that beneficial insects

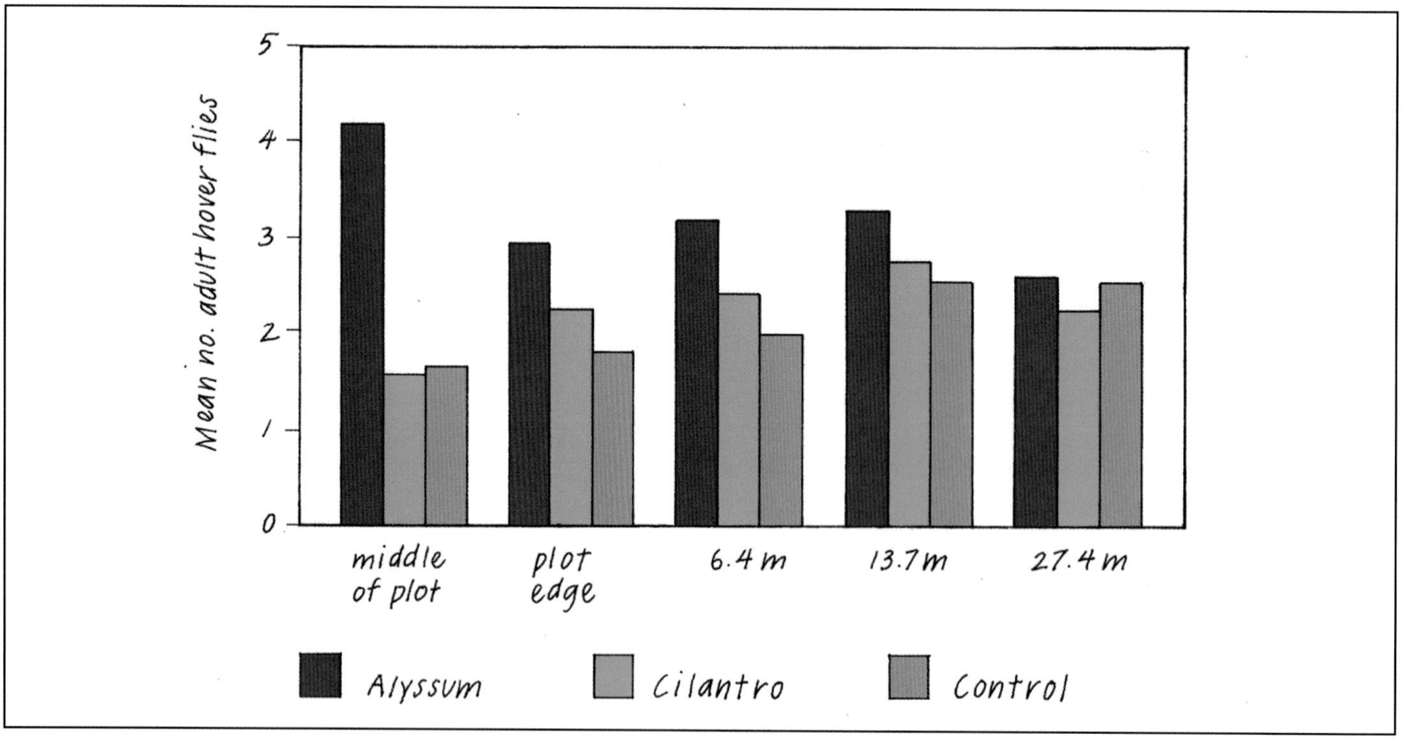

Figure 365. Here's another study showing how important alyssum can be in protecting plants at a distance from insects. The alyssum attracted more syrphid flies than either cilantro or the control plot of mixed grasses, all the way out to 27.4 meters (90 feet.) It was most effective in the crop area, but more so at 13.7 meters than at the edge of the plot.

moved up to ***250 feet*** into almond trees from the insectary plantings. (*California Agriculture*, Volume 52, Number 5, p 26.)

- A 1998 thesis by Micaela Ruth Colley reviewed the impact of distance, noting that hoverflies moved ***13.7 m* (41 feet)** into a field

Figure 366. In many studies *Phacelia tanacetifolia* (lacy phacelia or blue tansy) is shown to be an effective insectary plant. (Photo from The Watershed Nursery, Richmond, CA.)

of broccoli adjacent to alyssum plots:

- "Biological control of *Aphis glycines* (soybean aphid) is reduced in simple landscapes dominated by corn and soybean production. Members of the *Coccinellidae* (ladybird beetles or ladybugs) an important family of natural enemies of *A. glycines,* were also found to decline in soybean fields located in agriculturally dominated landscapes. Therefore, maintaining overall landscape diversity, and especially perennial grassland and forested habitats, in the landscape will favor suppression of soybean aphid. Moreover, these habitats may also enhance biocontrol services in other crops because the generalist natural enemies found in soybean fields are also important antagonists of many other herbivores. Our results suggest that such non-crop habitats do not need to directly border agricultural fields to promote biocontrol services; we found that occurrence of these habitats within **(about) *1.5 km* (.9 miles)** explained the highest percentage of the variation." ("Landscape Diversity Enhances Biological Control of an Introduced Crop Pest in the North-Central USA"; 2009. M. M. Gardiner, et. el., *Ecological Applications*, 19(1), p 143-154.)

- Phacelia and buckwheat were studied as an insectary resource in the field in New Zealand. "The effect of incorporating phacelia or buckwheat in the margins of 5 m x 5 m broccoli plots was tested for hoverfly activity and floral 'preferences.' Hoverflies that had fed on phacelia and buckwheat pollen were found up to ***17.5 m* (58 ft.)** from the floral strips and females of [syrphid flies] *M. fasciatum* and *M. novaezelandiae* consumed more phacelia pollen than that of buckwheat in the field." ("*The Role Of Resource Subsidies In Enhancing Biological Control Of Aphids By Hoverflies (Diptera syrphidae)*"; Thesis, Elsa Laubertie, Lincoln University.)

- More beneficial rove beetles (*Staphylinidae*) and arthropods were found in raised boundaries covered with tussock grass (*Demeirtrias atricapillus*). (Structural Features of Field Boundaries Which Influence The Overwintering Densities of Beneficial Arthropod Predators"; 1994. P. Dennis, M. B. Thomas, and N.W. Southerton, *Journal of Applied Ecology*, 31. p 361-370.)

- Syrphid flies were found ***50m* (164 ft.)** from strips of alyssum in fields of lettuce. ("Manipulating Floral Resources Dispersion for Hoverflies in a California Lettuce Agro-Ecosystem"; 2011. Mark Gillispe, et al, *Biological Control*, 59 – p 215-220.)

I have dozens more instructive citations, but you get the idea. All this is to say, most beneficial insectary plants can be planted *anywhere* in the average yard and still attract enough predators of pests. There's *no* reason most companion plants need to be planted close together; this makes for easier gardening.

Occasionally, however, hedgerow proximity doesn't make much difference:

- In another study of *Phacelia tanacetifolia*, syrphid flies moved up to 100m into the field, and it was found that beneficial *ichneumonid* wasps (parasitoids) were more abundant in the *Phacelia* strip than in the nearby wheat crop. Aphid populations and the percentage of those parasitized were unaffected by differences in the wasps' distribution. Syrphids are stimulated to move into the crop by the presence of flowering plants like buckwheat and cornflower. However, during summer the differences in syrphid populations were not reflected in aphid populations perhaps because of other natural enemies that are less dependent on flower resources. ("*The Impact of Floral Resources on Syrphid Performance and Cabbage Aphid Biological Control;"* Paul C.J. van Rijn1, Jurgen Kooijman1, Felix L. Wäckers. 2006. *IOBC Bulletin*, Vol. 29 (6).

- "Although hoverflies have the potential to slow aphid population growth, in most of the cases observed, in the field but also in laboratory experiments, their activity has not been sufficient to reduce efficiently aphid populations." ("*The Role of Resource Subsidies in Enhancing Biological Control of Aphids by Hoverflies (Diptera: Syrphidae)"*, Thesis, Elsa Laubertie, Lincoln University.)

- With the pollination of cucumbers by bees, the research did not fulfill expectations. The bees tended to stay in the strips of yellow mustard (*Brassica hirta*), sweet alyssum (*Lobularia maritima*), and buckwheat (*Fagopyrum esculentum)* instead of flying into the cucumber crop. ("Floral Strips in Cucumber Fields;" Bill and Mary Weaver.)

- "...extra plant diversity and greater area of semi-natural habitats on organic farms offer no advantage for increasing cereal aphid parasitoid diversity..." ("Parasitoid Control of Aphids in Organic And Conventional Farming Systems;" Sept. 2009. *Agricultural Ecosystems and Environment*, Vol. 133, Issue 1-2, p 14-18.)

- "...the number of hoverfly larvae did not differ with distance...over ***64 m*** **(210 ft.)** away." ("Manipulating Floral Resources Dispersion for Hover Flies in a California Lettuce Agro-Ecosystem;" 2011. Mark Gillispe, et al, *Biological Control*, 59, p 215-220.)

And then there were mixed results:

- Three flower treatments were tested: 1) buckwheat (*Fagopyrum esculentum*); 2) yellow mustard (*Brassica hirta*); 3) sweet alyssum (*Lobularia maritima*); and cucumbers as a control. Flowers were planted within a commercial cucumber field in *20-m-long strips* (66 ft.). Cucumber yield was unaffected by flowers with one exception, in one year cucumber yield in the sweet-alyssum plots were greater than those in plots with no flowers. Our research indicates that adding flowers to cucumber fields to increase services from beneficial insects needs to be further investigated to better understand the effect of factors such as relative flowering strip size. ("Floral Strips Attract Beneficial Insects but Do Not Enhance Yield in Cucumber Fields;" April 2017. N. F. Quinn, D. C. Brainard, Z. Szendrei; Journal of Economic Entomology, Volume 110, Issue 2, p 517–524.)

In summary, in all the research of beneficial insect attracting plants four annual plants stand out as major attractors of beneficials: Phacelia (*Phacelia tanacetifolia*), sweet alyssum (*Lobularia maritima*), coriander (*Coriadrum sativa*) and the buckwheat grain (*Fagopyrum esculentum*).

Alyssum Insectary Plantings

Alyssum is one of the most studied plants for improving attraction of beneficial insects. From one such study: "Alyssum is one of the most frequently studied species in habitat management for conservation

Figure 367. Sweet alyssum (*Lobularia maritima*) is perhaps the most studied insectary plant. In mild-winter climates it blooms most of the year.

biological control. Alyssum as a promising insectary plant identified to intercrop with lettuce because alyssum flowered quickly after planting, was not overly aggressive or likely to become a weed, and attracted several beneficial species but few pest species. For more than 10 years, alyssum has been planted as an insectary plant in organic-lettuce fields in California to attract adult hoverflies (*Diptera*: *Syrphidae*) that feed on pollen and nectar."

"In summary: this study provides the first information on agronomic aspects of strip intercropping transplanted romaine lettuce with alyssum for biological control of aphids in organic systems...alyssum dry matter was highly correlated with open inflorescences of alyssum, and alyssum flower production increased with alyssum transplant density. Additive intercropping [of alyssum and lettuce] appears to be a more efficient intercropping approach to produce alyssum floral resources for beneficial insects." ("Agronomic Aspects of Strip Intercropping Lettuce with Alyssum for Biological Control of Aphids," April 2013. Eric B. Brennan, United States Department of Agriculture/Agricultural Research Service, U.S. Agricultural Research Station, *Biological Control*; 65, p 302–311.)

As another alyssum example: "The results [of this study] show that the alyssum flowers' attractiveness contributed to increasing the abundance of generalist predators during both experiments, which in turn translated into a significant reduction of collards pests, especially aphids." ("Alyssum Flowers Promote Biological Control of Collard Pests;" August 2017; André L. Ribeiro, Lessando M. Gontijo, *BioControl*; Vol. 62, p 185–196.)

According to biodiversity consultant Jude Bloom of Cascadia Permaculture in Cottage Grove, Oregon,

Figure 368. Here's a planting of buckwheat (*Fagopyrum esculentum*), shown before its blooming cycle. It also makes a great green manure.

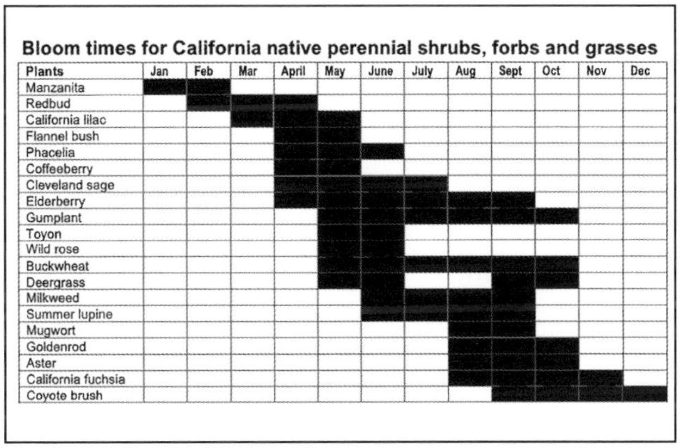

Figure 369. This chart shows the bloom pattern of native plants (in Yolo County, CA) that attract beneficial insects and provide nectar and pollen during the entire year. All are very drought-resistant.

Bloom Period of Insectary Plants

Common Name	*Botanical Name*	Jan.	Feb.	Mar.	Apr.	May	June	July	Aug.	Sept.	Oct.	Nov.	Dec.
Agastache[24, 28]	*Agastache foeniculum*						■	■	■				
Alfalfa[1,7,9]	*Medicago sativa*						■	■	■	■	■	■	
Alpine Cinquefoil[1]	*Potentilla concinna*			■	■	■							
Alyssum, Sweet[1, 3, 7, 9, 13, 14, 16, 21a, 2]	*Lobularia maritima*	■	■	■	■	■	■	■	■	■	■	■	■
Angelica[1, 3, 5, 9, 22, 26, 30]	*Angelica* spp.				■	■	■	■	■	■			
Anise Hyssop[1, 9, 10, 22, 30]	*Agastache foeniculum*						■	■	■				
Columbine, Aquelegia[21a]	*Aquelegia* x *hybrida* 'Blue Bird'					■	■						
Arugula[23, 25]	*Eruca sativa*				■	■	■	■	■	■	■		
Asters[3, 6, 22, 30]	*Aster lateriflorus*							■	■	■	■		
Aster, Alpine[21a]	*Aster alpinus*						■	■	■	■	■		
Baby Blue Eyes[3, 4, 7, 9]	*Nemophila menziesii*			■	■	■	■						
Baby's Perfoliate Pink[21b, 23]	*Gypsophila pacifica*			■	■	■	■	■					
Basket-of-Gold[1, 21a, 22, 25, 30]	*Aurinia saxatilis*			■	■								
Bee Balm, Lemon[3]	*Melissa officinalis*						■	■	■	■			
Bee Balm, Scarlet[21b]	*Monarda didyma*							■	■				
Bidens, Smooth Beggartick[3]	*Bidens laevis*								■	■			
False Queen-Anne's Lace[22, 23, 25]	*Ammi majus*						■	■	■				
Basil, Sweet[22, 23, 30]	*Ocimum basilicum*						■	■	■	■	■		
Bittercress, Woodland[5]	*Cardamine flexuosa*		■	■	■								
Borage, Common[3, 7, 22, 23, 30]	*Borago officinalis*				■	■	■	■	■	■			

Figure 370. I anticipated needing to know when insectary plants bloom and produced an insectary "map" showing the bloom cycle for 191 ornamental and California native plants, and showing which insects are attracted by each. Here's a sample spreadsheet page; it's easy to scan down, for example, from March or October and determine all the plants you should be growing then if you want to attract beneficial insects during the early spring and late fall months.

The spreadsheet also lists (in the accompanying handbook, from formal peer-reviewed-studies) which flowers attract beneficial insects, and what insects are attracted by each. (To purchase the Insectary Plant List, go to: https://robertkourik.com/drip_irrigation_garden_books.html and scroll down.)

who often works with with farmers in that area, you can take five percent of a field out of production and plant alyssum and still get the same yields while reducing the pest-spray programs. She often recommends the biodiversity of alyssum strips to her clients who are commercial growers.

Seasonal Bloom

Pests infiltrate our gardens all season long. To thwart their attacks, gardeners need beneficial insects all season long as well. To attract same, you need a handy supply of pollen and nectar to feed the beneficials while they pursue the pests. To achieve this, you need to have the right kinds of beneficial flowers blooming over the duration of the growing season in your area. It's easy to find members of the parsley and sunflower family plants blooming in June through August, but more difficult in the very early and very late seasons of the year. This is why you need to carefully select plants so you have blooms appearing over the entire growing season. Figure 369 is an example of seasonal bloom planning near Yolo County, California. (These plants are all California native plants, chosen because they're drought-resistant.)

Notice that Ceanothus griseus (California lilac) blooms as early as March, and Baccharis pilularis (coyote brush) blooms into mid-winter. The other plants have overlapping bloom cycles so there is always some pollen and nectar available.

Hedgerow plants are also good candidates to replace weedy "pioneer" plants, (which can be vectors of pest and disease) in disturbed habitats—the hedgerow shrubs can overgrow many weeds and shade them

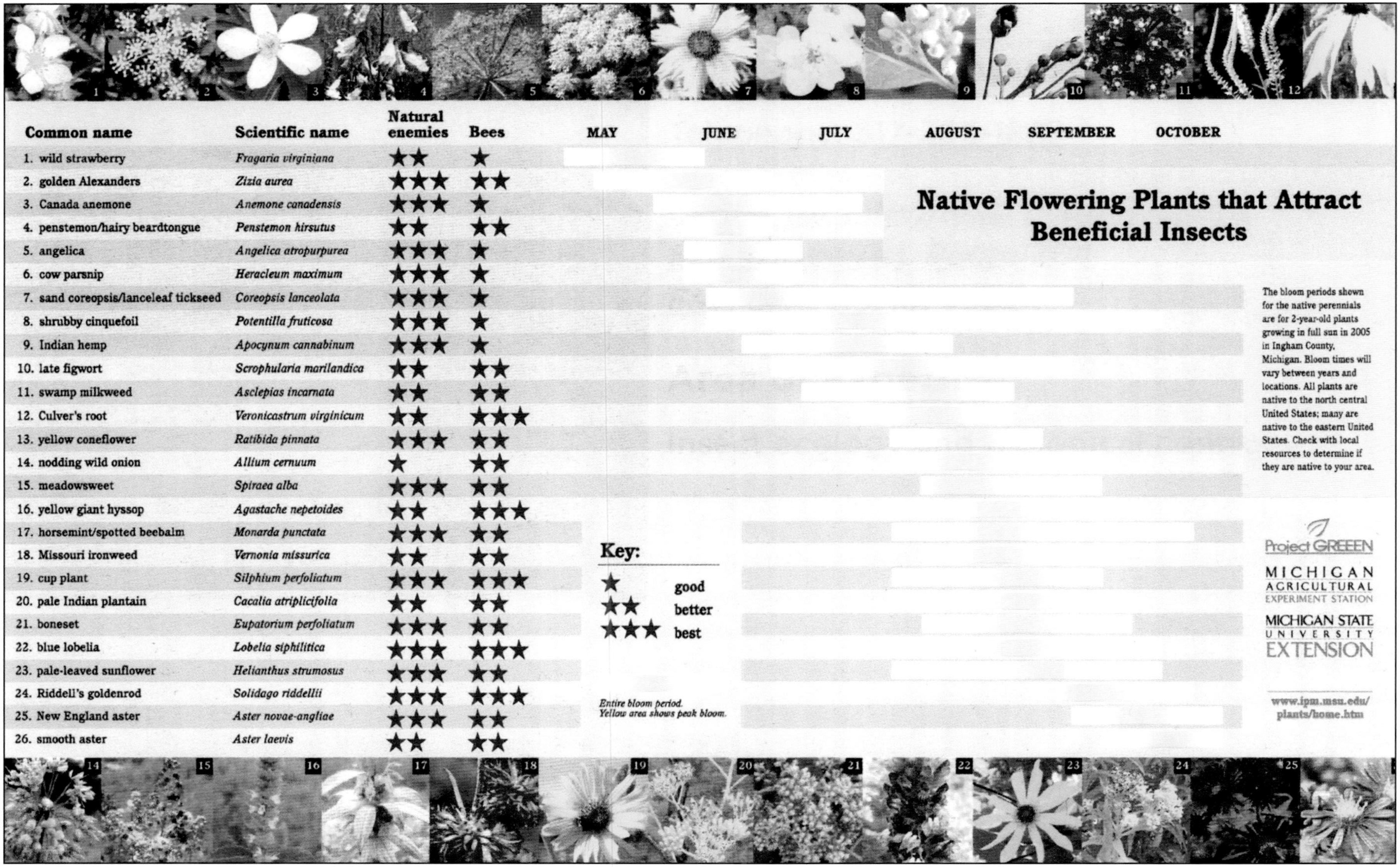

Native Flowering Plants that Attract Beneficial Insects

Common name	Scientific name	Natural enemies	Bees
1. wild strawberry	*Fragaria virginiana*	★★	★
2. golden Alexanders	*Zizia aurea*	★★★	★★
3. Canada anemone	*Anemone canadensis*	★★★	★
4. penstemon/hairy beardtongue	*Penstemon hirsutus*	★★	★★
5. angelica	*Angelica atropurpurea*	★★★	★
6. cow parsnip	*Heracleum maximum*	★★★	★
7. sand coreopsis/lanceleaf tickseed	*Coreopsis lanceolata*	★★★	★
8. shrubby cinquefoil	*Potentilla fruticosa*	★★★	★
9. Indian hemp	*Apocynum cannabinum*	★★★	★
10. late figwort	*Scrophularia marilandica*	★★	★★
11. swamp milkweed	*Asclepias incarnata*	★★	★★
12. Culver's root	*Veronicastrum virginicum*	★★	★★★
13. yellow coneflower	*Ratibida pinnata*	★★★	★★
14. nodding wild onion	*Allium cernuum*	★	★★
15. meadowsweet	*Spiraea alba*	★★★	★★
16. yellow giant hyssop	*Agastache nepetoides*	★★	★★★
17. horsemint/spotted beebalm	*Monarda punctata*	★★★	★★
18. Missouri ironweed	*Vernonia missurica*	★★	★★
19. cup plant	*Silphium perfoliatum*	★★★	★★★
20. pale Indian plantain	*Cacalia atriplicifolia*	★★	★★
21. boneset	*Eupatorium perfoliatum*	★★★	★★
22. blue lobelia	*Lobelia siphilitica*	★★★	★★★
23. pale-leaved sunflower	*Helianthus strumosus*	★★★	★★
24. Riddell's goldenrod	*Solidago riddellii*	★★★	★★★
25. New England aster	*Aster novae-angliae*	★★★	★★
26. smooth aster	*Aster laevis*	★★	★★

Figure 371. *Attracting Beneficial Insects with Native Flowering Plants,* Anna Fiedler, Julianna Tuell, Rufus Isaacs, and Doug Landis Department of Entomology, Michigan State University. Extension Bulletin E-2973 Revised January 2008.

out. The best strategy may be to replace the pioneer shrubbery with plants native to your area. In California, using natives that are planted and established at the correct time can reduce or eliminate the need for summer irrigation. (See page __ for techniques on establishing plants so no summer irrigation is needed.).

In The Permaculture Activist Spring issue 2015, Rico Zook, a permaculture consultant, included the concept of beauty as one "harvest" of a well-designed hedgerow. You can attract bees and other beneficial insects with native and exotic shrubbery that also serves as a pleasing part of your landscape. Such perennial plants in California (marked *), at least, include:

- Blue elderberry (*Sambucus nigra*) *
- California coffeeberry (*Rhamnus californica*) *
- California fuchsias (*Epilobium canum*) *
- California lilac (*Ceanothus* spp.) *
- California wild rose (*Rosa californica*) *
- Cleveland sage (*Salvia clevandii*) *
- Flannel bush (*Fremontodendron californicum*) *
- Hollyleaf cherry (*Prunus ilicifolia*) *
- Huckleberry (*Vaccinium* spp.) *
- Lavender (*Lavandula* spp.)
- Manzanitas (*Acrtostaphylos* spp.) *
- Mock orange (*Philadelphus lewisii*)
- Oregon crabapple (*Malus fusca*)
- Oregon grape (*Mahonia aquifolium*) *
- Red current (*Ribes sanguineum*) *
- Rosemary (*Rosmarinus officinalis*)
- Salmon berry (*Rubus spectabilis*)
- Serviceberry (*Amelanchier alnnifolia*) *
- Soapbark tree (*Quilaja saponaria*)
- Toyon (*Heteromeles arbutifolia*) *
- Vine maple (*Acer circinatum*) *
- Western redbud (*Cercis occidentalis*.) *
- Wild buckwheat (*Eriogonum* spp. - no relative of grain buckwheat) *
- Yarrows (*Achillea* spp.)

Seeding Hedges, Hedgerows, and Borders

We can now buy many mixtures of seed specifically meant to attract beneficial insects. The three mentioned here are taken from the only study that I know of that was done by someone unconnected with seed companies. ("Insect Communities Associated with Beneficial Insect Habitat Plants in North Carolina;" 2006. L. M. Forehand, D. B. Orr, and H. M. Linker, Environmental Entomology, 35 (6), p 1541-1549.)

The plants in the **Good Bug Blend** attracted the greatest number and variety of beneficial parasitoids, predators, and ground beetles. However, it also attracted a great number and variety of crop-feeding pests, as did the **Border Patrol** blend. In addition to attracting beneficial insects, the **Border Patrol** plantings attracted the highest variety and numbers of herbivore pests and the highest number of pest moths in the early evening; perhaps that's why it's no longer for sale.

Peaceful Valley Farm & Garden Supply, **Good Bug Blend**:

- Baby's Breath
- Calendula
- California Buckwheat
- Caraway
- Carrot
- Celery
- Coriander
- Crimson Clover
- Daikon Radish
- Dill, Red Radish
- Fennel
- Hykon Rose Clover
- Nungarin Subclover
- Parsley.
- Red Clover
- Semi & Non-dormant Alfalfa
- Sweet Alyssum
- White Clover
- White Yarrow
- Yellow Sweet Clover

Territorial Seeds: **Beneficial Insect Mix**

- Alyssum, Tall White Sweet
- Aster, New England
- Baby Blue-Eyes

- Bishop's Flower
- Black-Eyed Susan
- Candytuft, Annual
- Cilantro
- Coreopsis, Lance-Leaved
- Daisy, Shasta
- Dill
- Forget-Me-Not
- Gaillardia, Annual
- Gayfeather
- Gilia, Globe
- Poppy, California
- Prairie Clover, Purple
- Rockcress, White
- Wallflower, Siberian

Arbico **Good Bug Power Blend –** Contains yellow, red, crimson, rose, and white clovers; grizzly alfalfa; white alyssum; nasturtium; white yarrow; carrot; dill; daikon radish; and celery seeds

Clyde Robin's **Border Patrol –** No longer for sale, but here's what was in it:
Achillea millefolium (yarrow); *Ammi* majus, Angelica *atropurpurea* (angelica), *Fagopyrum esculentum (buckwheat)*, Moench, *Helichrysum bracteatum (everlasting or strawflower)* (Ventenant), *Iberis* spp. (candytuft), *Nasturtium* spp., *Nemophila menziesii (baby blue eyes)*, *Oenothera biennis* (evening primrose), and *Rudbeckia hirta (black-eyed Susan)*.

Insects and Flowers

There are also various other lists detailing which flowers attract which insects. The following are from Fred 'Farmer Fred' Hoffman, Mary Louise Flint, Peaceful Valley Farm Supply, Cornflower Farms, and Steve Zien.

Plants That Attract Lacewings:

Achillea filipendulina (Fern-leaf yarrow)
Anethum graveolens (Dill)
Angelica gigas Angelica
Anthemis tinctoria Golden marguerite
Atriplex canescens Four-wing saltbush
Callirhoe involucrata Purple poppy mallow
Carum carvi Caraway
Coriandrum sativum Coriander
Cosmos bipinnatus Cosmos white sensation
Daucus carota Queen Anne's lace
Foeniculum vulgare Fennel
Helianthus maximilianii Prairie sunflower
Tanacetum vulgare Tansy
Taraxacum officinale Dandelion

Plants That Attract Hoverflies:

Achillea filipendulina Fern-leaf yarrow
Achillea millefolium Common yarrow
Ajuga reptans Carpet bugleweed
Allium tanguticum Lavender globe lily
Alyssum saxatilis Basket of Gold
Anethum graveolens Dill
Anthemis tinctoria Golden marguerite
Aster alpinus Dwarf alpine aster
Astrantia major Masterwort
Atriplex canescens Four-wing saltbush
Callirhoe involucrata Purple poppy mallow
Carum carvi Caraway
Chrysanthemum parthenium Feverfew
Coriandrum sativum Coriander
Cosmos bipinnatus Cosmos white sensation
Daucus carota Queen Anne's lace
Eriogonum fasciculatum CA Buckwheat
Foeniculum vulgare Fennel
Lavandula angustifolia English lavender
Limnanthes douglasii Poached egg plant
Limonium latifolium Statice
Linaria vulgaris Butter and eggs
Lobelia erinus Edging lobelia
Lobularia maritima Sweet alyssum white
Melissa officinalis Lemon balm
Mentha pulegium Pennyroyal
Mentha spicata Spearmint
Monarda fistulosa Wild bergamot
Penstemon strictus Rocky Mt. penstemon
Petroselinum crispum Parsley
Potentilla recta 'warrenii' Sulfur cinquefoil
Potentilla villosa Alpine cinquefoil
Rudbeckia fulgida Gloriosa daisy
Sedum kamtschaticum Orange stonecrop
Sedum spurium Stonecrops
Solidago virgaurea Peter Pan goldenrod
Stachys officinalis Wood betony
Tagetes tenuifolia Marigold "lemon gem"
Thymus serpylum coccineus Crimson thyme
Veronica spicata Spike speedwell
Zinnia elegans Zinnia "liliput"

Plants That Attract Ladybugs:

Achillea filipendulina Fern-leaf yarrow
Achillea millefolium Common yarrow

Ajuga reptans Carpet bugleweed
Alyssum saxatilis Basket of Gold
Anethum graveolens Dill
Anthemis tinctoria Golden marguerite
Asclepias tuberosa Butterfly weed
Atriplex canescens Four-wing saltbush
Coriandrum sativum Coriander
Daucus carota Queen Anne's lace
Eriogonum fasciculatum CA Buckwheat
Foeniculum vulgare Fennel
Helianthus maximilianii Prairie sunflower
Penstemon strictus Rocky Mt. penstemon
Potentilla recta 'warrenii' Sulfur cinquefoil
Potentilla villosa Alpine cinquefoil
Tagetes tenuifolia Marigold "lemon gem"
Tanacetum vulgare Tansy
Taraxacum officinale Dandelion
Veronica spicata Spike speedwell
Vicia villosa Hairy vetch

Plants That Attract Parasitic Mini-Wasps:

Achillea filipendulina Fern-leaf yarrow
Achillea millefolium Common yarrow
Allium tanguticum Lavender globe lily
Anethum graveolens Dill
Anthemis tinctoria Golden marguerite
Astrantia major Masterwort
Callirhoe involucrata Purple poppy mallow
Carum carvi Caraway
Coriandrum sativum Coriander
Cosmos bipinnatus Cosmos white sensation
Daucus carota Queen Anne's lace
Foeniculum vulgare Fennel
Limonium latifolium Statice
Linaria vulgaris Butter and eggs
Lobelia erinus Edging lobelia

Lobularia maritima Sweet alyssum white

Melissa officinalis Lemon balm
*Mentha pule*gium Pennyroyal
Petroselinum crispum Parsley
Potentilla recta 'warrenii' Sulfur cinquefoil
Potentilla villosa Alpine cinquefoil
Sedum kamtschaticum Orange stonecrop
Tagetes tenuifolia Marigold - lemon gem
Tanacetum vulgare Tansy
Thymus serpylum coccineus Crimson thyme
Zinnia elegans Zinnia

Plants That Attract Tachinid Flies:

Anthemis tinctoria Golden marguerite
Eriogonum fasciculatum California Buckwheat
Melissa officinalis Lemon balm
Mentha pulegium Pennyroyal
Petroselinum crispum Parsley
Phacelia tanacetifolia Phacelia
Tanacetum vulgare Tansy
Thymus serpyllum coccineus Crimson thyme

Plants That Attract Minute Pirate Bugs, Damsel Bugs and Big-Eyed Bugs:

Carum carvi Caraway
Cosmos bipinnatus Cosmos "white sensation"
Foeniculum vulgare Fennel
Medicago sativa Alfalfa
Mentha spicata Spearmint
Solidago virgaurea Peter Pan goldenrod
Tagetes tenuifolia Marigold "lemon gem"

For more information about beneficial predatory insects, read: *Natural Enemies Handbook: The Illustrated Guide to Biological Pest Control*, from the University of California Division of Agriculture and Natural Resources.

Native and Exotic Blooms for Pollinators and Beneficial Predators

A big debate on the topic of beneficial pollen and nectar plants is whether to use only native plants, only exotic beneficial plants, or a mixture of the two.

The native-plant enthusiasts say that the native plants are best suited to the insects found locally. Others say, European honeybees (*Apis mellifera*) are not native bees, so it doesn't matter if all the plants are native to the area. (Although native bees and bumblebees play a big role in pollination, they don't offer a "crop," i. e. honey.) Proponents of mixing the two categories say that in urban areas there's no control over the mix of plants; in fact, you can count on a lot of exotics. They also say that a lot of commercially named honeys are made from the nectar of exotics. Examples include: clover, orange blossom, star thistle (an invasive), herbal, blackberry, blueberry, acacia (another invasive), and eucalyptus (another invasive). Whether you use just one plant source or a mix, you'll get honey. Just figure out the best bee-attracting and

pollen-forming flowers for each category.
"Researchers know that non-native plants can interact with pollinator populations, and mixing native and non-native plants in designed landscapes benefits pollinators in both positive and negative ways; however, field studies in public and private gardens have shown no consequential evidence that the introduction of some non-native plants has any significant negative impact on resources available for local pollinators. In fact, creating near-native designed landscapes may have a mutualistic benefit in the maintenance of pollinator populations." ("Pollinators May Benefit from Mixing Native and Non-Native Plants;" by Scott Stewart.) Of course, it's best to choose your exotics carefully so as to avoid invasive plants.

An Example for Michigan Bees

A study in Michigan found that: "...the best midsummer native plants that were popular to native bees and honeybees included wild bergamot or bee balm (*Monarda fistulosa*); mountain mint (*Pycnanthemum incanum*); gray goldenrod (*Solidago nemoralis*); early goldenrod (*Solidago juncea*); whorled milkweed (*Asclepias verticillata*), and spotted knapweed (*Centaurea stoebe*)." ("Michigan State University Pollination Study Reveals Best Attractants for Bees that You May Want to Plant Near Your [Apple] Trees;" March 20, 2017.)

The same debate applies to beneficial insectary plants. From Rodale Institute's publication: "Invite Insect Allies to Your Farm Using Insectary Strips," (Dr. Gladis Zinati, Ph.D., Director—Vegetable Systems Trial and Associate Research Scientist.)
To make an insectary planting work, follow these guidelines.
You should know:

- "The pest you're targeting.
- The beneficial insects you want to populate.
- The duration of the flowering time per plant species.
- The structure of the plants and type of flowers they produce.
- The soil conditions where the insectary plants will grow.
- The management practices that best support populations of beneficial organisms."

This study didn't use native plants.

This was the summary of the insectary-strip study for the control of cucumber beetles: "Results deduced from the two-year research showed that the integration of insectary flowering strips aided in inviting a diverse array of beneficial insect species, including ground beetle, ladybeetle, tachinid fly and minute pirate bug. The presence of these beneficial insects such as ground beetles, ladybeetles, and tachinid flies reduced survival of striped cucumber beetle. By enhancing the environmental habitat for beneficial insects that control striped cucumber beetle, the culprit for transmitting the bacterial wilt disease, the latter can potentially be controlled by using natural biological systems without the use of pesticides."

Biodiversity

Don't forget Linda Chalker-Scott's paper in Chapter 3, (page 71) reviewing the literature about whether or not non-native, woody shrubs and trees were a problem for, or a benefit to, biodiversity. ("Nonnative, Noninvasive Woody Species Can Enhance Urban Landscape Biodiversity;" July 2015. *Arboriculture & Urban Forestry*, 41(4), p 173-186.)

Chalker-Scott's insight is revealing and thought-provoking, "Natives-only...reduces the planting palette and ultimately community biodiversity of *urban* [my emphasis] areas." She discusses the biodiversity for insects and pollinators, and more, and notes that in some places it's possible that biodiversity in general may increase while native-plant diversity is reduced.

Dr. Gladis Zinat's study at the Rodale Institute didn't use native plants. Here's a list of the plants used and the beneficial insects they attracted.

Insectary Plants and the Beneficial Insects They Attract:

Alfalfa	Ground beetles, wolf spiders, lacewings, ladybeetles, minute pirate bugs, damsel bugs
Alyssum	Hoverflies
Amaranthus	Ground beetles
Calendula	Ladybeetles, hoverflies, butterflies, bees
Dill	Ladybeetles, lacewings, Ichneumon wasp
Fava bean	Ladybeetles, bees, pirate bugs
Goldenrod	Leatherwing soldier beetle, hoverflies, and butterflies
Lemon balm (**very invasive**, I'd leave it out)	Parasitic wasps, tachinid flies
Marigold	Ladybeetles and lacewings
Parsley	Parasitic wasps, hoverflies, tachinid flies
Queen Anne's Lace (invasive)	Lacewings, ladybeetles, hoverflies
Shasta Daisy	Minute pirate bugs, beneficial mites
Sunflower	Pirate bugs

Beneficial Insects and Pests They Attack:

Hoverflies	Aphids, small caterpillars, thrips
Ground beetles	Moth and beetle larvae, slugs
Ladybeetles, lacewings, Ichneumon wasps	Aphids, mites and whiteflies. The wasp lays eggs and larvae feed on host such as beetle.
Lacewings, ladybeetles, hoverflies	Aphids, small caterpillars, soft bodied insects
Parasitic wasps, tachinid flies	Beetle and fly larvae, caterpillars

Parasitic wasps, hoverflies, tachinid flies	Moth, beetle and fly larvae, whiteflies
Minute pirate bugs, beneficial mites	Thrips, aphids, mites, scales, whiteflies
Minute pirate bugs	Moth, beetle and fly larvae, whiteflies
Lady beetles and lacewings	Aphids, mites and scales

Bonus Charts and Photos:

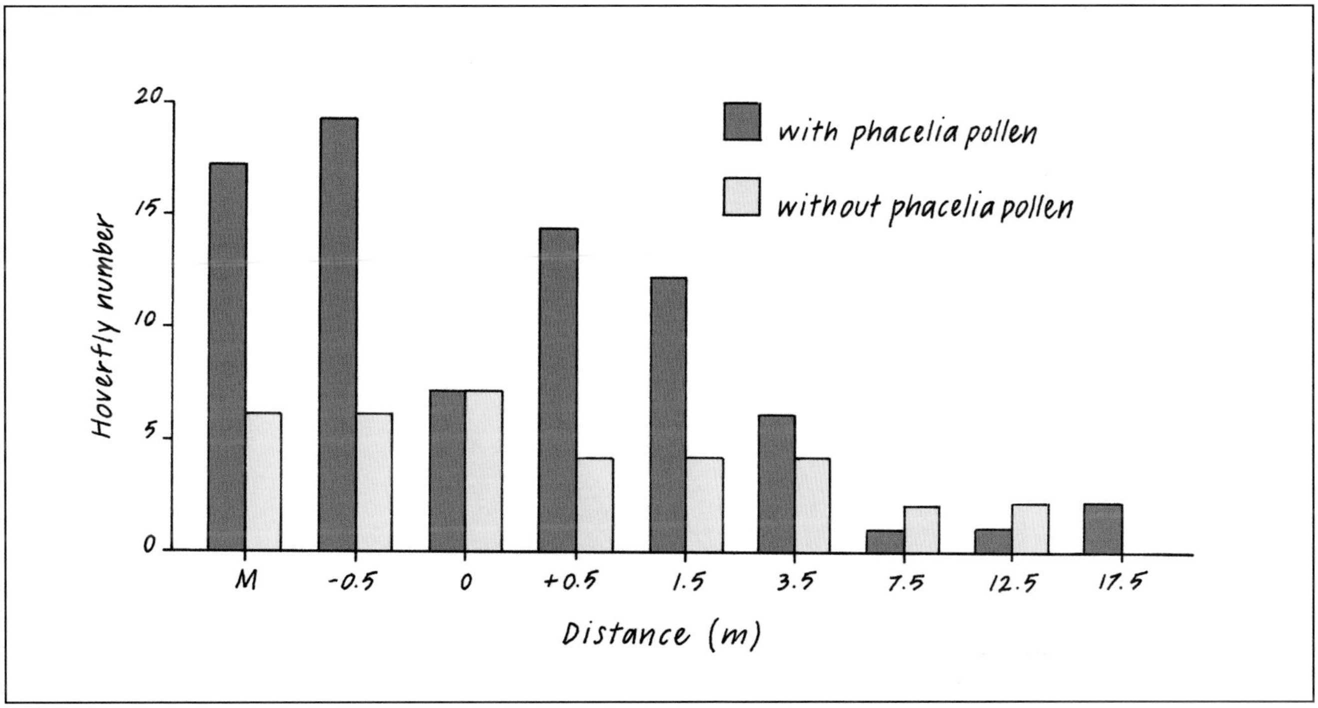

Figure 372. This study shows how far syrphid flies will fly from the center of a planting. (Zero stands for the middle of the crop planting; the distance is in meters; and "M" stands for the margin of the field with the insectary planting. Even at three meters (9.8 feet) there are almost as many syrphids in the margins as there are in the middle of the planted area.

Figure 373. This ant is about to gather honeydew from the back end of an aphid. Ants usually tend to herd aphids together for ease of "harvesting."

Figure 374. This bumblebee is shown entering the throat of a foxglove. These insects are beneficial in that they are good pollinators, although they don't eat or parasitize any insects. Later in their life-cycle the bumblebees will swarm lavender flowers.

Figure 375. The tomato hornworm moth (known as a Sphinx moth) flits around at dusk and is often mistaken for a hummingbird. Its pupa stage is spent in the ground (d.). None of these hornworms have been parasitized.

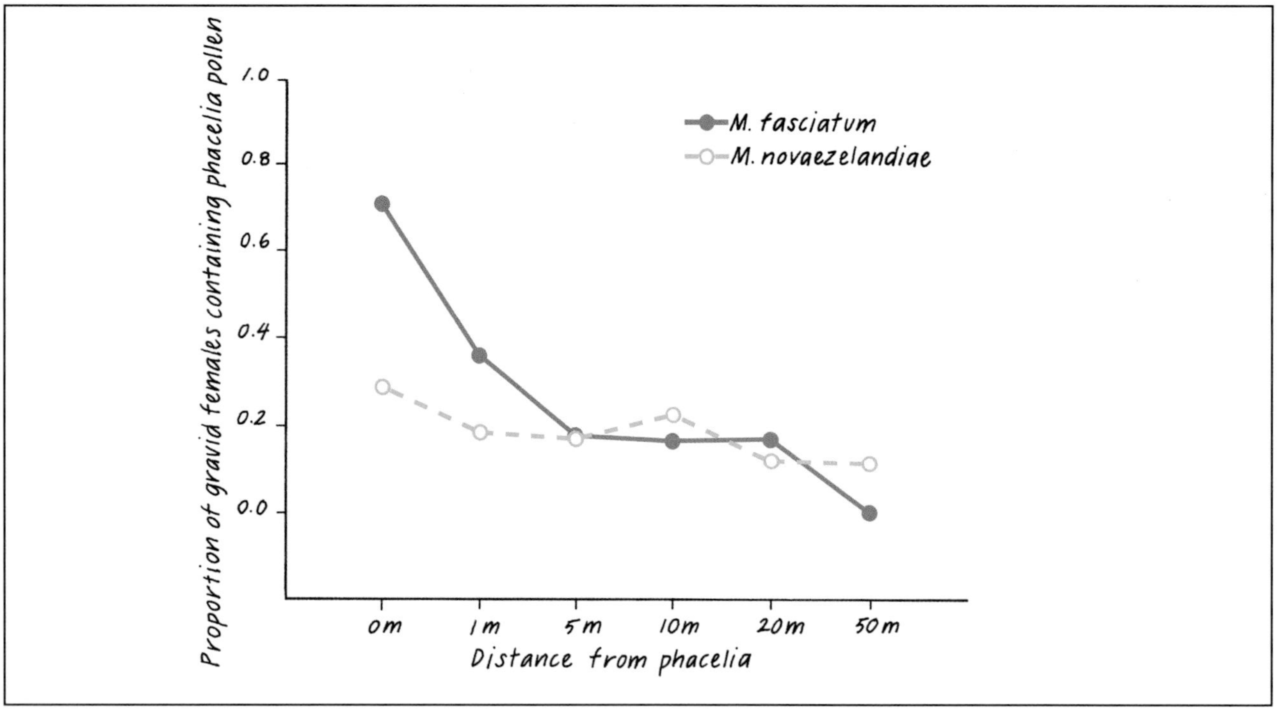

Figure 376. The "M" on this chart stands for *Malanostum*—a genus of syrphid flies. A gravid female is one carrying eggs. The chart shows how the quantity of *M. fasciatum* drops by half in the first meter away from a planting of phacelia. The number of gravid female *M. novaezelandiae* syrphids, even 50 meters beyond from the insectary planting, is about only one-third less.

Figure 377. *Ammi visage,* a tap-rooted member of the carrot family that resembles Queen Anne's lace, is known by many common names, including toothpick-plant, toothpickweed, bisnaga, khella, or Bishop's weed. It's native to Europe, Asia, and North Africa, but can be found throughout the world as an invasive species.

Figure 378. This spider is eating a bumblebee on a French lavender blossom. Life is a two-edged sword. (Can't have everything.)

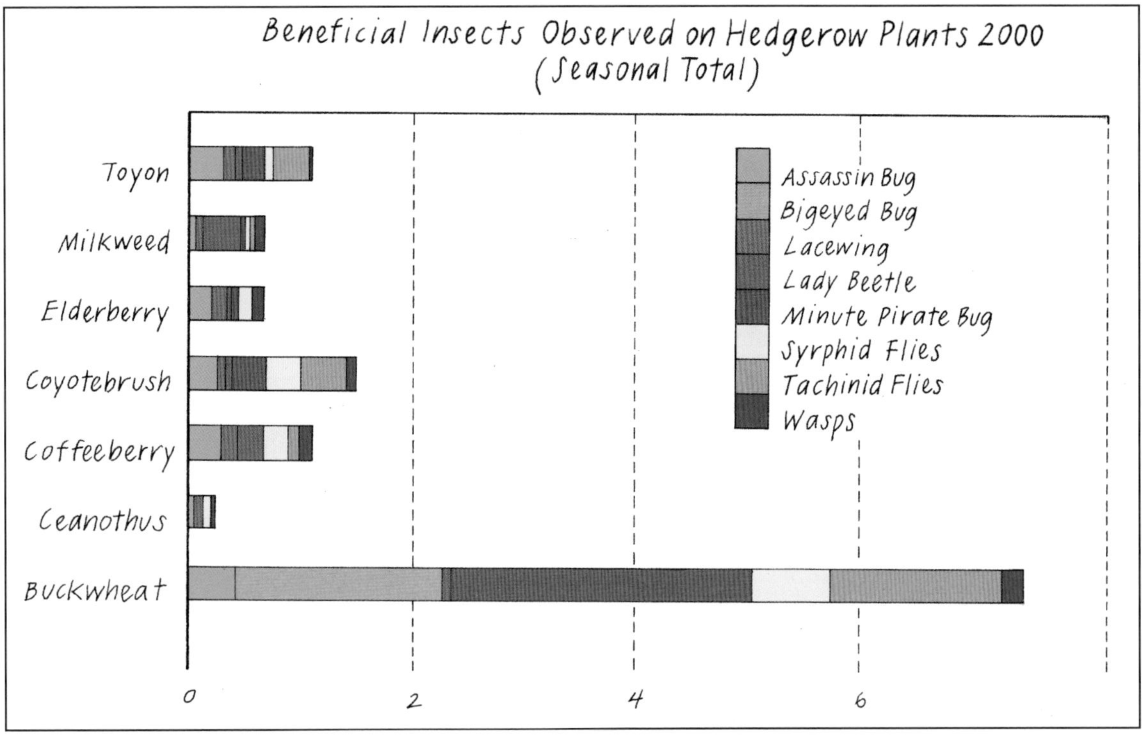

Figure 379. The buckwheat in this study was not the edible pseudocereal buckwheat (*Fagopyrum esculentum*). Like all the other plants on the list, it's a California native plant and is part of the *Eriogonum* genus.

Figure 380. There are many species of *Eriogonums*, all very drought-resistant. Although the flowers of some Eriogonums are drab, this *Eriogonum umbellatum* is very showy, and I use it a lot in garden designs. (The client loves the color and I love the insectary influence.) Common names include: Sulfur Buckwheat, Sulfur Eriogonum, and Sulphur-flower Buckwheat.

Chapter 12
Designing Sustainable Gardens

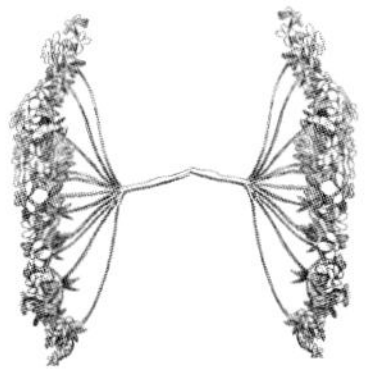

Natural Landscape Design

First, a bit of a sidestep: here's a look at the main elements of how to begin designing *any* kind of natural-looking garden or landscape, followed by a section on how to integrate perennial edible plants and vegetable beds into it.

There's really no such thing as a natural gardener-made landscape. *All* gardens are constructs made by people, not by nature, to ultimately serve our own needs and our own sense of aesthetics. I think this means that, in a sense, and perhaps unconsciously, all gardeners are really servicing their own egos. Our gardens might look lovely, but they usually don't look "natural." A natural garden (oxymoron) would look as if nobody planted or tended it, so producing a truly natural-looking landscape would not satisfy most gardeners. (If nobody recognizes it as the product of another person's planning and work, then nobody gets any compliments.) A pure and unpretentious landscape would look so naturalized that most peo-

Figure 381. Josh Williams designed and planted a mix of California native grasses, sedges, Douglas iris, and some carex at his home in Sebastopol, CA. His goal was to make the scene appear as if the plants had already been there and the bridge was added. I think he succeeded. (Photo © Philip Van Soelen.)

Above: Umbel flowers like the ones above are good landing platforms for small insects. They attract many types of beneficial insects, especially small parasitic wasps.

ple would just walk by without noticing. I've never (knowingly) seen such a garden except for Josh's.

Unnatural, but Nice

Our gardens will always be a constant dialogue about the following terms: nature-like, naturalized, earth-friendly, and environmentally sound. Unnatural doesn't mean malevolent. Reasonably nature-like gardens can meet every definition of beautiful, and be both nice *and* unnatural. Given that we're growing artificial constructs, exactly how nature-like can we get? With care and attention, our gardens can be quite environmentally responsive, but the key to achieving this is in the details. Here are some pointers:

A Process of Elimination

Gardeners usually try to do too much. Some properties have plenty of natural landscape to begin with, but the typical approach is to remove almost everything that's there, buy a bunch of topsoil and plants and put completely new stuff back. This is not only far from natural; it's wasteful and costly.

So here's my advice: if you've got a backyard or property with lots of natural features, don't even think of planting until you've fully realized the potential of what nature has already provided. All that's often required is to selectively thin, prune, or remove existing trees, shrubs and plantings. The idea is to sculpt your existing plants to reveal their hidden form and texture. (Think of Michelangelo carving away bits of stone to reveal the magnificence of the *David*.)

Here are some examples of backyard sculpting: Carefully trim limbs from trees to reveal serpentine or wonderfully textured trunk(s), or shape existing shrubs to reflect the line of a nearby tree or a distant hillock. Pull out weedy exotic or invasive shrubs to display the rugged shape of a boulder. Remove (or relocate) ferns or shrubs to open up a pattern beneath a spreading tree. Weed out exotic grasses to leave behind a scattered pattern of native bunch-grasses. Carefully open up an oddly shaped hole in the canopy of a tree to provide a slice of the distant horizon, or (if you're lucky) a nearby lake or the silvery ocean. Thin some of a tree's foliage to reveal a few hollow dead limbs that etch a pattern in the sky and enable you to watch the nightly flight of bats. Plant inside any remaining hollow stumps.

Figure 382. The Stewarts selectively pruned their coast live oak (*Quercus agrifolia*) and California lilac (*Ceanothus* spp.) to reveal this breathtaking view of the Big Sur, CA coastline.

Figure 383. Michelangelo's 1525-1530 sculpture the 'Atlas Slave'. There is a line of four of these magnificent, unfinished statues leading up to the 'David' in the Galleria dell'Accademia in Florence, Italy.
Photo - Dr. Jörg Bittner Unna, Germany, Wikipedia Commons.

Figure 384. Another view created by pruning in the Stewarts' garden. Some *Phormium* spp. have been added for a striking vertical accent.

Mix it Up

For some reason, people have this compulsion to plant everything equidistant and far enough apart to match the plant's theoretical mature width—usually taken from a dog-eared favorite gardening book. But nature seldom "plants" anything on equidistant centers, so to do so with each transplant is unnatural. Some plants belong together. Planting at odd intervals, without any regard to the mature width, is perhaps the first "rule" of natural planting. Let plants mingle with abandon. In the wild, few plants get to grow to their ideal, mature width. Trees may grow just a few feet apart, sometimes in clusters of three or more.

So don't be cheap. Buy a bunch of trees or shrubs, and consider planting odd-numbered clusters of them, like small groves, in random spacings. To be fully random, throw out a fistful of golf or tennis balls, one for each plant you bought, and let them fall where they may. Some balls may be very close and others far apart. The hard part is to not move any of the balls. If one seems "too close" to another, that's random. Let it lie. This same technique works for perennials and bulbs (just toss the bulbs, although if you're dealing with a busy landscape, you might want to attach a cloth "streamer" to each so you don't lose any). Keep it random.

Little or No Pruning

The forest doesn't prune itself, and all too often in our landscapes we cut away more than we need to. Although I shear my lavender heavily and religiously each year, prune my fruit trees every summer (sometimes in the winter), and shape (not shear) my Jerusalem sage (*Phlomis fruticosa*) once every year or so, I've never pruned the strawberry tree (*Arbutus unedo)* I planted 25 years ago. It's lopsided, but so are some trees in the forest just 50 feet away. It produces edible fruit that I have to gather from a ladder due to not pruning it, but that's a compromise I'm happy to make.

In a natural forest, if a limb grows with a weakness, it will eventually just fall from the tree naturally; this is why we should always prune a young tree that needs help to get off to a sturdy start. One example of when to prune is shown by the "included" or "invaginated" bark on the trunk of this tree. (Figure 385.) When two potential trunks share the same spot, the two shoots compete for tissue. The trunk can't share the same circumference without having one of the shoots

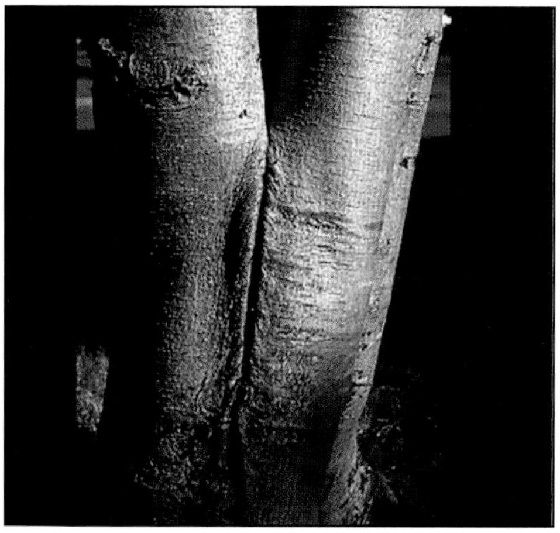

Figure 385. "Invaginated" and "included" are two words used for a tree configuration in which two limbs growing too closely together form a weak union that can easily split.

weaken or fall off, so pruning off one of the parallel shoots when the tree is very young provides a strong start to a tree that may never have to be pruned again.

(Since limbs do fall in the forest, consider not picking up, chain sawing, or composting some of them. Leaving them makes habitat for lizards, beneficial beetles—a beetle "bank"—and other insects that aid in the process of decomposition

Keep those hand pruners and saws sharp, oiled, and ready to go, but maybe think twice before shearing in a "natural" landscape.

Random Seedlings

Photographs of my spring garden over the years show foxgloves (*Digitalis purpurea*) growing in very different places. I've never planted digitalis since I first transplanted it over 30 years ago; rather, I let it go to seed in my well-mulched no-till organic garden, and find that the seed really "walks around" every year. I find new plants five, 10, 20 feet away from where they grew the previous year, and how they got there I have no idea. I don't have to use golf balls, as they do this random sprouting all on their own.

In order to perpetuate the unplanned beauty of these walking foxgloves, I have to tolerate the amount of time the brown post-flowering stalks need to be left standing while they're making seed, but it's a good tradeoff for the delight of surprise blooms every

Figure 386. Jerusalem sage (*Phlomis fruticosa*) flourishes in front of a strawberry tree (*Arbutus unedo*).

year. I do sometimes shake a seed-head over a new area, but I never thin them. (Foxgloves can escape from the garden but I don't think of them as truly invasive. I just pull out the few wayward plants before they bloom.) I've "planted" mullein the same way.

Watered to Death

I live in the Mediterranean zone of California, where in some years, many species of oaks get no rain from May through September or October. To native oaks, summer water is a poison that can kill. These trees have spent a millennium or more evolving to thrive without summer rain. In the process, they were able to tolerate a number of hitchhiking fungal hosts. Predatory *mycelia*—examples include crown, armillaria and heart rots—need two conditions in order to thrive: moisture and warm soil. As evolution arranged it, the soil is plenty moist during the wet California winters, but too cool to promote much fungal growth. During the dry summer, the soil is warm enough to stimulate fungal growth, but the upper layer of the soil has dried out enough to retard most fungi. These seasonal alternations insure that there's very little time when both necessary ingredients are present together. Thus, whatever limited amount of fungus survives is not enough to threaten the 300-year life expectancy of some oaks.

It's totally unnatural to water beneath an oak tree during the California summer, and no better way to promote the rapid growth of the fatal fungi. Such trees can die from rot within 25 to 50 years—a quick demise for a tree that could otherwise live another 100 to 300 years. In this move-around society, most homes will have had several owners in the quarter-century it takes the tree to slowly die, thus, the actual culprit who installed the irrigation system usually isn't around when the tree meets its sad and unnecessary end.

One of the debates about our native Californian oak trees revolves around how far away summer irrigation should be kept from the trunks of oaks. The conservative answer is: a zone should be free from summer irrigation for a radius of at least eight feet from the trunk. The more prudent gardener would decree that an unirrigated area equal to the width of the canopy (at the minimum) should be maintained.

However, as discussed earlier, the greater portion of any tree's roots, especially the majority of the tiny absorbing root hairs, extends well beyond the dripline of the canopy. In a heavy clay soil, roots, as a general rule, will extend half again as wide as the dripline. A sandy soil, because of its lower resistance to root penetration, allows tree roots to grow as much as three times wider than the canopy.

Since nature selected oaks to require no summer water, than the truly natural landscaper should avoid all summer water within the entire root system—up to three times the canopy's width, or more. Alas, if there are two or more oak trees in the typical suburban yard, this means the entire yard should be irrigation-free—period. This brings up the delicate art of denial. Most homeowners rationalize, if they even give it any consideration, that some lawn or a few irrigated shrubs under part of the oak's foliage is all right, but this may shorten the oak's natural life. To prevent or minimize this potential problem, install drip irrigation *at least* eight feet away from the trunk of any oak.

Figure 387. The roots of this oak tree in a public complex are protected from root rot by a permanent mulch covering a plant-free zone that extends all the way to its dripline. The driveway at right avoids the area under the dripline.

Anti-Deadhead

Any act of gardening that differs from what nature would do has a negative impact, however slight. Most gardening techniques are *de*-evolutionary—they set things back, and the wrong technique can set the garden itself back considerably. This doesn't mean we can't compensate, such as using more fertilizer to compensate for compacted clay soil.

As another example, I grow a lavender called *Lavandula* x *intermedia* 'Provence,' which has tall-stemmed blossoms. After their fragrant bloom is spent, the flower parts fall from the stalks fairly quickly. Most gardeners would cut back the stems (a practice called "deadheading") as soon as the color begins to fade. Years ago, as mentioned in an earlier chapter, I left the slender woody stems through the winter as an experiment in low-maintenance via avoidance. To my pleasant surprise, one of the first places to be garlanded with spider webs on a dewy spring morning were the remaining naked, dead 'Provence' stems. Now, I cut all of the spent 'Provence' blooms back to within three inches

Figure 388. I left a small area in my garden unpruned to see what it would look like. It looked ugly, but I woke up to this glorious sight one morning. (I know I've used this photo before, but I love it.)

of the bare stem so as to insure plenty of tidy bloom, and the spiders have no doubt found other places to sling their webs.

Wander Around

Deer often travel in a pathway that they wear down with repeated use. This "main highway" is used to travel to areas where they wander to feed. One way to make a natural pathway through trees and vegetation is to use these ungulate corridors as a basis. (Of course, you'll probably annoy the deer)

If you're making a natural path, consider simply using the dirt that's already there. (However this seems to require a "mud room" in one's house.) I consider the next best material to be aged chips of random shapes and lengths, obtainable from a tree service. Green shredded leaves and chips from an arbor service add another natural pattern and texture, will compost down as the path is used, and can be renewed as needed.

You can design your path with random widths. If you want two people to be able to walk side by side, make sure the narrowest portion of the path is at least five feet wide. You can make the path even wider and then allow some plants to extend their foliage in a random pattern over the edge of the path. I did this in my garden, but planted very close to the edge of the path; as a result, over a ten-year period, the path went from five feet wide to less than one foot in width. *But,* I have lavenders along the edge and walking through them releases their wonderful scent, so I don't mind. The foliage hides much of the path, allowing it to slip into the background. (See Figure __ for guidelines on other spacings.)

Getting Stoned

Natural stones, installed in a random pattern, add a great touch to many landscapes. I like to place stones to form an unobtrusive "spine" that curves through my garden. Placing stones on both sides of a path makes it appear that the stone was there before the garden was put in. An even more natural look can be obtained by allowing a stone to protrude slightly *into* the edge of the pathway. This is even more natural looking than other kinds of stone placements

Many people like to think they're getting their money's worth by just plunking stones on top of the soil. Instead, look for the natural change in color that shows how deeply the stone was previously embedded in the ground. Sometimes using this as a guideline will place more than half of a stone below ground level, but it's *so* much more natural looking. Hint: don't waste your money on plants that will grow up to obscure these wonderful features. To set off stones in my garden, I plant very prostrate ground covers like thymes (*Thymus* spp.), a low-growing specimen of manzanita (*Arctostaphylos uva-ursi*), sea pinks (*Armeria maritima*), ajuga (*Ajuga reptans*), or, in a confined place, dichondra (*Dichondra argentea*).

Meandering Mounds

A few decades ago, creating oddly shaped mounds throughout the landscape was all the rage. (Most looked like what I call the turd-from-the-sky effect—just lumps.) Mounds can add a sense of flow and interest, especially where the adjacent land is flat—watch out, Kansas—but to make them look settled in and more natural, construct them them with very wide gentle slopes, and not too high.

Natural Space

Whatever the climate in which you're planting,

Figure 389. These carefully placed stones are sunk one-quarter to three-quarters deep in the ground for the most natural look. The horizontal stones allow for walking over the "creek" formed by runoff during winter rains.

Figure 390. An attractive planting of exclusively native plants at the Santa Barbara Botanic Garden. Front to back: *Erigeron* 'Bountiful', *Diplacus* 'Jelly Bean Lemon', *Diplacus* 'Jelly Bean Gold', and *Diplacus* 'Trish.' (*Diplacus* used to be *Mimulus* spp.) Unfortunately they all bloom at once. (Photo © Frederique Lavoipierre.)

don't cram all plants together willy-nilly without considering the appropriate use of, and need for, space. Some ecosystems naturally require space around some or all of the resident plants. The chaparral, for instance, is a mosaic of space and shrubbery. In the desert, the pattern includes plenty of space.

Natural, Momentary Color

The seasonal flow of color is not consistent. In every climate, color highlights pass through the garden at different times of the year. Nature's colors are often fleeting, so it takes some real planning to have a beautiful moving pattern of continuous highlights.
I often rely on a great handbook that shows the different colors of each plant for Northern California, along with a chart that shows the weeks and months of bloom. I've never seen this done for other regions (maybe you can make money drafting such a handbook for yours). If you don't have access to such a handbook, you can talk to gardeners, neighbors, garden-nursery attendants, and garden-club members.

Appropriate Lawns (Not Always the Devil in Disguise)

I've never actually planted a lawn (I took care of some in 1975-1980—with organic fertilizers), but I have specified lawns for clients who had kids, back before iPads or cell phones existed! These lawns were for the outdoor play that kids need and (used to) enjoy. For the adults, I sometimes designed lawns that were big enough to play croquet and badminton—20 by 44 feet, *plus* at least eight feet all around for players to stumble off the court and rescue wayward shuttlecocks.

It *is* possible to have and enjoy a lawn these days without resorting to the use of broad-leaf herbicides, pre-emergents, harsh chemical fertilizers, or as much water as we used decades ago. For example, subsurface drip irrigation prevents *all* evaporative losses because the tubing is buried up to 18 inches deep. You can even sunbathe while the lawn is being irrigated!

If Dutch white clover shows up in your lawn, no need to get obsessive about spraying it away. Instead, let its nitrogen-fixing nodules feed the grasses when you mow. According to "Effects of White Clover Inclusion on Turf Characteristics, Nitrogen Fixation, and Nitrogen Transfer from White Clover to Grass

Species in Turf Mixtures" (in *Communications in Soil Science and Plant Analysis*, Volume 38. Issue 13-14, 2007), "Nitrogen transfer from white clover to the associated turfgrass varied from 4.2 to 13.7% of the total N that the white clover fixed annually." Don't forget that lawns take less time to maintain than vegetable gardens (see pages 6-7). Get a copy of *The Organic Lawn Care Manual* by Paul Tukey (Storey Publishing, January 30, 2007).

Fertilizing Lawns Under Shade Trees

Not all places available for planting in a yard are sunny. Shade from a tree can make it very difficult to grow a good-looking lawn. Fertilizing shaded areas in a lawn, especially with compost or bloodmeal, takes a bit more attention to detail. To deal with these problems:

- First, choose a grass type or blend that has been selected to grow in shade. Examples include fine fescues—rough and tall—and Kentucky bluegrasses.
- Mow the grass with an electric-powered mower set about one inch higher than the setting for sun-loving turf.
- Consider opening the canopy of some of your shade trees with selective summer pruning—prune off the lower branches for better morning and afternoon sun.
- Fertilize at a rate slightly higher than you do the grass in sunny areas. Rough and Kentucky bluegrasses need more nitrogen than other grasses such as fine or tall fescues.
- When mowing, leave the shredded clippings as mulch; they act as a very slow-releasing and low-NPK fertilizer. You can also use compost with an NPK of .5-1.5% nitrogen, 1-3.5% phosphorus, and 1-2% potash.

Organic and environmentally sound lawns are often fertilized with slow-release compost. Top-dress turf up to 2,000 square feet with about a quarter to three-eighths of an inch of well-aged compost by raking it in. (Larger areas may require a spreader.)

So, to sum up, lawns have their place in a natural home landscape. They need not require a lot of work, water or amendments, and they're a great place to watch shooting stars or neck—or both!

Now on to introducing edibles into your landscape.

Figure 391. Joy Larkcom's garden features huge fava bean leaves, roses (for edible flowers), and a woven seat made from willow twigs.

Figure 392. This is part of my design for the Preston Farm and Winery in Healdsburg, CA. Each bed is watered with drip irrigation, and herbs are planted in the corners of the beds to soften the angular lines of the boxes.

Designing Sustainable, Productive, and Attractive Edible Landscapes

Overall, one of my most important gardening tenets is that I don't find it necessary to have a forest-farm/garden, and I don't layer everything on top of each other. The foremost idea behind *all* my designs is that each component or zone of the garden is often separated slightly from the others for easier maintenance, adequate fertilization, reasonable access to sunlight, and proper irrigation.

Each element of each zone is within the same landscape, which works as an ecological whole. This means, for instance, that my fruit and nut trees are planted as far away from the vegetables as is practical, to prevent root competition, conflicts in water use, and shading. The strawberries and cane fruits are also in separate areas so as to not conflict with pruning and harvesting the fruit and nut trees. Vegetables and herbs are in yet another zone, as close to the kitchen as possible—remember my guideline: never plant the veggies further from the house than you can throw the kitchen sink—and often in raised beds with a wire-mesh bottom; these make the plants more accessible, keep the area tidier, and exclude marauding gophers. As you might assume from all this, I create nothing that resembles a "guild."

An Edible History

The rest of this chapter is about how I design a sustainable garden to be edible, productive and attractive.

First, a bit of history. In the mid-1970s, along with Rosalind Creasy, I developed the concept of "edible landscaping" (which Ros called "foodscaping" at first). I had been designing ornamental landscapes for clients for some time, when a friend connected me with a person who was quite ill and wanted only the freshest vegetables for his diet. Prior to that time (1978), most food gardens consisted of veggies planted in rows wide enough to till with a rototiller (every other ad in the 1960s and 1970s *Organic Gardening* magazine was for Troy-Bilt™ rototillers).

My new client's yard was small and demanded intensive raised beds. The sunniest site was right below his dining room, so I asked myself, why not make it beautiful as well as useful by combining the best of my ornamental design ideas with edible crops?

Figure 393. Rosalind Creasy loves to use strong, dramatic colors and structures. A red geranium and ceramic sun brighten the entry to the front yard; the second arbor frames a view of the patio entrance. The boxwood hedges give the garden a permanent shape, no matter what's happening in the vegetable beds.

After an article about me and about this unusual approach to gardening was published in the 1978 winter issue of *CoEvolution Quarterly*, the phrase took off. All this is to say that I've been doing this for a while. And both Ros and I have always used organic, sustainable methods of growing and tending our edible landscapes.

Edible Landscaping

Edible landscaping is easily defined: a productive garden that tastes great *and* looks fantastic, as well as a source of relaxation and fun. There's an emphasis on using edible plants and as many perennial edibles as possible, (such as lemon trees in my San Francisco Bay Area climate or maybe blueberries and rhubarb elsewhere) in place of ornamentals, using the same guidelines as you'd apply to an ornamental garden to make sure it's attractive. At first the edible-landscaping approach focused solely on perennials, but soon it expanded to include veggies—but not long ugly rototilled rows of corn!

Figure 394. Here a red bench in Ros's front yard camouflages the chicken yard and picks up the color theme. (Check with the powers that be to see if you can legally keep chickens in your front yard.)

Rosalind Creasy is the *grande dame* of edible landscaping and the best designer I know—yes, me included. (She lives in Silicon Valley and has designed edible gardens and consulted with a number of the companies and their executives for decades.) I talked with Ros about her approach to designing edible landscapes.

Ros' interest in gardening began early, when her dad gave her a ten-foot-square garden plot when she was five years old. She began designing the layout of the vegetables (alas, by moving them to different locations at different times over the same season), and harvested next to nothing, but she attributes that little plot to giving her a lifelong interest in food gardening. A statement she read years later declared that: "Gardening should be caught, not taught."

Fast-forward to 1978, when she presented her final paper in the local Foothill College landscape-design

Figure 395. Ros's back yard has a yellow theme; another ceramic sun plaque provides a focal point, no matter how the vegetables and annual flowers look. The riot of color masks eggplant foliage that's often damaged by insects and tends to look ratty.

Figure 396. Ros planted onions and lettuce in a diamond pattern in this raised bed. After the lettuce is harvested, the onions will fill in the bed.

class. She had been frustrated for three semesters because out of the 450 plants they had studied, only two—pineapple guava and citrus—were edible. Her designs in those classes contained the first elements of an edible landscape. (At the beginning of the semester when she suggested designing with edibles, her professor called it "salad-bowl gardening," but after reviewing her designs over time, he was convinced, and in the end gave her an A grade on her final design and for the course.)

In the '80s, Ros tried to include edibles in traditional landscape designs for her clients. At first there was a lot of resistance, but as she brought samples of luscious fresh fruits from her garden to each meeting, the tide slowly began to turn. It also helped that business trips to Europe with her husband Robert (not me) opened her eyes to many more design possibilities. In Germany and England she saw artful plantings of cabbages combined with flowers; in Spain she saw Seville oranges lining city plazas; and discovered persimmon trees gracing temple gardens in Taiwan. She found the most validation in France when she visited Chateau Villandry, with its famous ornamental and edible formal gardens that feature tightly clipped boxwood hedges that provide a theme and structure regardless of the season. All were the early inspiration for her subsequent calling.

By 1979 she was at work on her first book on edible landscaping—*The Complete Book of Edible Landscaping: Home Landscaping with Food-Bearing Plants and Resource-Saving Techniques*. Since then she has written numerous edible-landscaping books, many now available in revised editions. As more of her clients integrated edibles into their designs, she realized that the joy she felt in harvesting her sun-warmed tomatoes and blackberries was shared by others.

Due to an injured back, it wasn't until 1982, with the help of Wendy Krupnick, that she took out her front lawn and replaced it with an edible landscape; its original design embraced many of the elements she uses to this day in designing gardens.

Ros begins not with the plants but with the hardscape, often the pathways. She prefers curved lines to guide the eyes, and begins by laying out hoses to delineate where the paths should go. Hardscape elements like arbors, benches, chairs, and tables soon follow. These elements are meant, along with tightly clipped boxwood hedges, to add structure to the garden so it can "hold together" during off seasons and the winter.

Hardscape elements, hedges, and espalier fruit trees unify the garden and "make sure it doesn't look like a hippie garden patch." For example, a layer of

Figure 397. Proper soil management means that Ros gets really good-sized vegetables like this cabbage, an heirloom variety called 'Early Jersey Wakefield'.

protective cloth for early freeze protection looks less noticeable and intimidating behind a low hedge than in an open plot of just veggies. Ros loves strong, bright colors—red, deep blue, and bright orange for starters. She loves to place foliage, edible or not, in places where the early morning or late-day sun can radiate clear, bright greens. Part of the color also comes from the many flowers incorporated within the garden, flowers that she knows will attract important beneficial insects. She grows cosmos, zinnias, dwarf marigold, alyssum, salvias, and many herbs, especially fennel, thyme, and rosemary, for the kitchen and for the beneficial insects.

For the last 40 years, she has redesigned portions of her garden every season as theme gardens to photograph for books and articles and to give gardeners and homeowners inspiration, as well as a wide palette of edible plants to inspire their landscapes. Examples include her *Rainbow Garden* book, with colorful vegetables and herbs like ruby chard, yellow beets, purple cabbage and basil, and orange peppers; and the *Edible Mexican Garden,* exuberantly filled with chilis of course, but also with roselle, chayote, blue grinding corn, tomatillos and Mexican herbs. The point she's making is that edibles, both common and exotics, belong in all of our ornamental landscapes.

For Ros, it's been a four-decade relationship with her front yard that continues to be her learning and photography studio. She feels "I was placed here to do this one thing." And she's done it with flare, style, chutzpah, and delight.

Figure 398. Ros's rich blue patio furniture provides a counterpoint to the garden's many red accents. The pots are planted with edible flowers and a dwarf lime tree.

Designing Style

With every new client, I begin with a scale-drawing base map. I try to use a scale of one inch = four or five feet (four feet is better). You can begin your own design by sketching in every permanent part of your existing landscape: walkways, patios, hose-bibs, compost piles, all shrubbery that you plan on keeping, etc. Be sure to note where the windows and doors of your house are so you don't block them with tall shrubs or trees, and so you can plan for extra color outside the windows. Don't worry about being too precise, this is only for a general plan—the details will be determined as the garden takes shape.
One way to plot out these details is to place a sheet of tracing paper over the base map. I'll often scribble "bubble drawings," showing where additional major components—vegetable boxes, arbors, fruit trees, etc.— might be placed.

Then I take the base map with the sketch out into the garden, and use hoses to outline the paths, areas for the vegetable boxes, and any patios or fire-pits I envision. As I do this, I'll start to refine the plan on another piece of tracing paper.

Paths come first and are placed where existing trails are already worn into the lawn or dirt. Don't lay out any sharp turns, as people will just cut across the corners. To invite the walker, make each path wider where it enters the garden and also where it leaves. Remember, even though paths are designated as hardscape, they can take the form of soft curving walkways covered with tree chips.

Next, I draw an accurate-to-scale map of every component of the hardscape: raised beds, paths, patios, compost areas, fire-pits, valve-boxes, hose-bibs (if you're starting from scratch, put plenty of those in the garden so that you can enjoy the convenience of short hoses), planting troughs or pottery, gates, fences, and the like. Then it's back into the garden. You can use a tool that landscape contractors use—an upside down spray-paint canister (called inverted spray paint) or hoses to define all the hardscape elements on the ground, and adjust them as required.

Note: ***Avoid circles and mandalas***! They look great on paper, but a circle on a plan doesn't look like a circle when you're standing in the garden—you'd have to construct a complex ellipse to get that effect. Mandalas and circular paths are also very hard to navigate; usually add many more steps to getting around the garden than is practical; and make it hard to drag hoses around, necessitating the added cost of drip irrigation. Keep it simple.

Raised Beds

Next I locate the veggie beds. I only use raised beds with hardware-cloth wire bottoms because of our "ground rats" (gophers). If possible, I try to place the beds running north and south to receive the best sunlight. As with Lou and Susan Preston's garden at the Preston Farm and Winery in Healdsburg, CA, I often try to fan the beds out in a graceful arch facing south, as shown in Figure 402.

I prefer to make the sides of raised beds two feet high, as it makes it easier for this baby-boomer (and anybody else for that matter!) to garden. This also

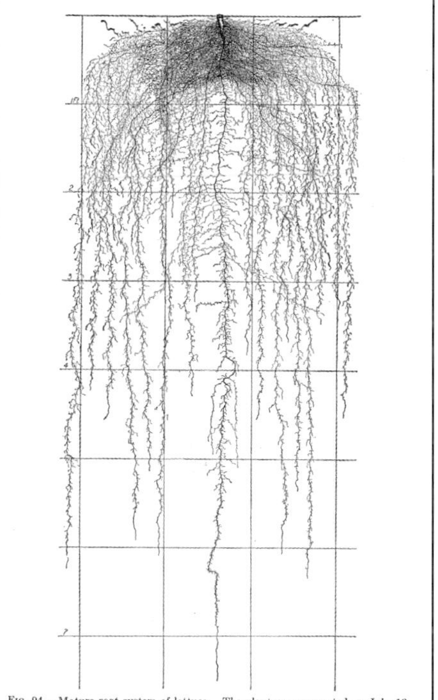

FIG. 94.—Mature root system of lettuce. The plant was excavated on July 13 and had flower stalks nearly 3 feet tall.

Figure 399. This image shows lettuce roots over seven feet deep (each square equals one foot). Notice that many roots are growing in the top two feet of soil, a good reason to make raised beds at least two feet deep.

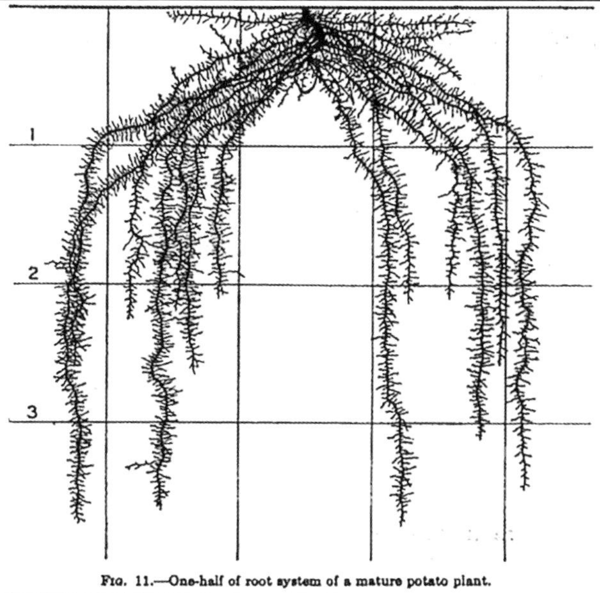

FIG. 11.—One-half of root system of a mature potato plant.

Figure 400. Much of a potato's roots grow in the top two feet of soil. These two drawings of root excavations are by John Weaver in Nebraska, who painstakingly dug trenches to follow the roots and sketch their depth.

provides soil-room enough to contain 80 to 90 percent of most vegetables' roots. (See Figures 399 and 400.)

In the Prestons' garden, I stuck with one-foot-high sides, which was a better scale for this small garden. I prefer the same six-foot by three-foot box dimensions as the Prestons', but I'll sometimes make beds as long

Figure 401. With the use of in-line emitter tubing, your cat can relax in the vegetable bed and stay dry while the drip system is on! (Taken at the Preston Farm and Winery, Healdsburg, CA.)

as eight feet (any longer makes it tougher to walk around in the garden).

I like to avoid, as nature often does, rigid, straight lines when possible, so I try to find ways to hide the rigid geometry of the boxes. Like Ros Creasey, I find curves more soothing. If I can, I'll often use an ornamental hedge or flower border on one side. At the Prestons', for example, I used an English lavender (*Lavandula angustifolia*) hedge to obscure the rigid architecture of the southern side of the boxes. (The tubing for the irrigation of each bed comes up the south end of each box, and the lavender hedge blocks the view without being too high to shade the vegetables.) To hide or soften the geometric corners, I plant perennial herbs like thyme and chives to cascade over them. After filling the beds, but before planting, I eat any edible weeds—it amuses me to digest these tasty aliens before planting. (This step is optional.)

Figure 402. There are five beds to the left of the door at the Preston Farm and Winery, and three to the right; this helps maintain the subtle asymmetry that the clients requested. The lavender is young, but will soon grow to soften the front of the boxes.

To add grace to the look of the garden, I plotted the raised beds on two curves, one on each side of the landing. Raised beds were required because the area is only two to three feet above the floodplain of the nearby creek, and a wire bottom was necessary to keep gophers out.

The Prestons wanted an informal, semi-symmetrical garden design, so I placed three beds on one side of the landing and five on the other side, as I think that odd numbers are more natural-looking. The focal point (radius) for each curve was different, resulting in another very subtle hint of asymmetry.

Locating Perennial Edibles

Back to planning: I return to my tracing-paper "bubbles" to decide on the area for the orchard of deciduous fruit trees, for which I'll select the best rootstocks for the varieties I want to eat. I try to multiple-graft as many trees as possible to insure a varied selection of fruits and nuts throughout the year. (I was inspired by a woman I met in 1976 who had grafted a single plum tree with a veritable fruit salad of of plums, peaches, and almonds.)

Next, I choose locations for edible shrubs; for tasty ground covers like herbs and strawberries; for cane fruits; and for evergreen fruit trees like oranges, avocadoes, lemons, and pineapple guavas. (Apologies to my home state of Missouri and all the other areas where these don't flourish.) These often take the place of "foundation plantings," which are just what they sound like—plants grown close to the house to disguise its foundation.

The Orchard

In my former orchard, I had an informal, woodland-like planting of only 22 fruit and nut trees, but they produced 31 varieties of fruit, because I grafted nine varieties of apples onto one tree. One of my favorite fruit trees is a rare and original Luther Burbank plumcot (a cross between a plum and an apricot) with the sumptuous fragrance of a warm apricot and the mellow flavor of an even-tempered Japanese plum. Alas, it blooms so early that I only got a reasonable crop every three to four years. Actually, the plumcot is the second-favorite fruit in my planting; the winner is the beet-red fleshed Elephant Heart plum, which produces a great crop every year. (Asian pears, especially the 'Hosui' cultivar, are another major

Figure 403. My favorite Asian pear is the 'Hosui.'

U.S. Department of Agriculture Pomologicatl Watercolor Collection. Rare and Special Collections, National Agricultural Library, Beltsville, MD 20705.

contender for first place.)

These apple and plum trees were grafted onto standard-sized rootstock so they would grow tall enough to put the fruit out of reach of the many deer roaming in my area. In its youth, each tree was surrounded by a five-foot-high and 10-foot-diameter heavy-gauge cylinder of the wire mesh normally used to reinforce concrete pads for driveways. Since a full-grown deer will stand on its hind feet to gobble leaves and fruit, each tree is enclosed until a good portion of the fruit-bearing canopy has grown to above seven feet. Limbs at the five- to six-foot level might look high enough, but just wait until the weight of the fruit lowers them to the four-foot level—the deer's favorite height for grazing. Dwarf trees would require the wire basket for the life of the tree—a total drag for pruning and harvesting.

No deer? Plant some dwarf fruit trees or summer-prune standard-sized trees to keep them below six feet in height. Dave Wilson Nursery has some great YouTube videos on this topic. It also shows you how to plant more than one tree in a hole. With summer pruning you can keep the trees at six feet or so for easy harvesting. With dwarf trees it's hard to justify multiple grafting. as the resulting fruits can get very small.

Each of my orchard trees was watered with graywater from my shower for one to two years, and then they were on their own. They never needed fertilizing or compost. The yield may have been reduced somewhat, but I still had plenty of fruit.

No Gridlock

Here's an important feature of my little "orchard": the trees were not planted on a grid as in a commercial orchard. There's no reason a homeowner needs to plant in a grid, which is done mostly to encourage a full, globe-shaped canopy in order to to maximize fruit production and make it easier to mow or till between the trees. I lived in a forest. The fruit trees were planted in a small open meadow. I figured a grid would be too unnatural as an aesthetic, so I planted the trees randomly—some closer together than others—to create a grove.

Multiple Grafting

I figure it also might be a good idea to *reduce* the size of the crop. Since I was then a bachelor, a full crop from a standard-sized tree was more fruit than I wanted to process. I now recommend, for nearly all of my clients, a single standard-sized fruit-tree rootstock with multiple varieties grafted onto the limbs. I figure that most people don't want to come home from work and process 100 pounds of fruit, or feel guilty when they can't use it all. With the smaller amount of fruit ripening over a longer period of time, they can graze the different fruits over many months. In my orchard, I had ripe plums from late May (the plumcots) through late summer with the 'Elephant Heart' plum. For apples, I grafted 'Akane' for a July crop, and through December I harvested the wonderful 'Arkansas Black' apple. (It could even have been through January with a 'Virginia Beauty' apple.)

Figure 404. One of my apple trees had nine varieties of apples grafted on to it. Here I grafted a red-flowering 'Pink Pearl', which has marbled pink flesh, to a white-blooming 'Sierra Beauty' apple tree.

Figure 405. Here, I used what's called a "cleft graft" to graft four different varieties of apple onto an older branch of this apple tree. This is how it looks the second year after winter grafting.

My orchard was mulched with used carpet (topped with a covering of wood chips to hide the backing) in order to suppress the four-foot-tall wild blackberry vines and poison oak that used to thrive there. My organic-purist friends freaked out. I have no idea if any chemicals leached out —they were *very* used carpets from dumpsters behind rug stores—and were absorbed by the tree and transferred to the fruit. I do know that most other nasty elements concentrate mostly in the seeds, so I wasn't too worried. Corrugated cardboard from large boxes is an option, but needs to be replaced every two to four years. I've since switched to cardboard, after digging out *all* the blackberry roots, as I now have far fewer trees.

As my orchard was across a dirt driveway from my other edible plants, this formed a barrier or border that mostly kept the roots from overlapping. Because the edibles require more irrigation, it would have been a conflict if the trees were closer. This social distancing for trees and vegetables is not often possible in many suburban yards, so whatever you do to fertilize and water the vegetables will affect the trees. This means your trees would never need fertilization because their roots will probably spread way beyond their foliage into the veggie garden. And, as explained in the first chapter, I've hardly ever seen a fruit tree that needs extra nitrogen. Perhaps you can plant the fruit trees on one side of a flower border (to be described below) that faces away from the vegetables. I've created flower borders with carefully chosen perennials that never had to be irrigated.

Flower Borders

As described in great detail in the previous chapter, insectary plants can help reduce the need for pest sprays, whether chemical or organic. Again, these plants do *not* have to be planted among the veggies to have an effect, as beneficial insects can fly far distances to attack pests. Clustering all the insectary plants into one grove or border means they can be easily cared for and don't interfere with transplanting and seeding crops during the season. To integrate insectary plants, I usually use the flower-border element. In one area or more, I plant a wide (up to three feet) and long strip of flowers. Most of my clients think it's a display for creating flower arrangements in the house. If the floral display is primarily what they desire, I don't even need to tell them about the insectary plants.

Figure 406. The foliage, flower, and root of Queen Anne's Lace (*Daucus carota*). This can be an invasive plant.

Flowers that serve both ornamental and insectary purposes include:
From the composite or daisy family (*Asteraceae*, formerly called *Compositae*):

- Artichoke flowers (*Cynara scolymus*)
- Asters (*Aster* spp.)
- Chrysanthemums (*Chrysanthemum* and other-spp.)
- Cosmos (*Cosmos* spp.)
- Dahlia (*Dahlia* spp.)
- Dandelion (*Taraxacum* spp. Before they make seed they're floral displays.)
- Feverfew (*Tanacetum parthenium*)
- Marigolds (*Calendula* spp., may be too stinky for some)
- Oxeye daisy (*Leucanthemum vulgare*)
- Pot marigolds, (*Tagetes* spp.)
- Shasta Daisy (*Leucanthemum* x *superbum*)
- Sunflower (*Helianthus* spp.)
- Yarrow (*Acheilia* spp.))
- Yarrow (*Achillea* spp.)
- Zinnia (*Zinnia* spp.)

From the umbel or parsley family (*Apiaceae*, formerly called *Umbelliferae*):

- Anise (*Pimpinella anisum*) "smelly"
- Caraway (*Carum* spp.)
- Dill (*Anethum graveolens*) "smelly"
- Lovage (*Levisticum officinale*)
- Queen Anne's Lace (*Daucus carota*)

Figure 407. Dill (*Anethum graveolens*) can spread a bit but is easy to weed out. Use for cooking or salads before it blooms.blooms.

It's best if the borders are no wider than three feet so it's easy to clip flowers from both sides. If there's room, I run the border north/south for the most even sunlight. Since this is also ideal for veggies and greenhouses, if there's a conflict for space in my plan, then the veggies win.

Deterring Deer

To avoid the horrible concentration-camp-like look of most deer fences in my area, I built a double four-foot deer fence for one client. The two four-foot-high fences are placed five feet apart with no plants in between, and the highest part of each fence is four feet above the ground. This really works. I can't interview deer to find out their reasoning, but the theory is that they can't figure out how to land between the fences and rebound over the inside fence into the garden.

I built that double-four-foot deer fence for Myra Portwood of Sonoma County in 1998; and it still works. Myra wanted to face the most visible portion of the fence with split-wood grape-stakes. Staggering the tops of alternate grape stakes and leaving a two-inch gap between each created a fence with a light, airy, and rustic feeling. The inner fence, masked by honeysuckle (*Lonicera japonica*), is my tried-and-true four-foot tall, 2"x4" wire-mesh barrier. (The rest of the 600-foot fence runs through the forest and is a traditional eight-foot wire fence because it's easier and cheaper to build and is hidden by the trees.)

The entrance to the garden (for humans) is an arbor built with two laminated wooden arches, and a gate made of two four-foot grape stakes cut to mirror the arches. The arbor is six feet wide to allow a wheelbarrow through. A deer-proof hedge of English lavender (*Lavandula angustifolia*) runs along the outside fence to mask the bottom.

It would seem that this structure could be adapted into any well-designed landscape anywhere that has no winter snow.

Figure 408. This double fence I built for Myra Portwood avoids the concentration-camp look of most deer fences. The upper right photo shows both the outer four-foot-tall wooden fence that faces the driveway and the four-foot inner metal barrier spaced five feet from the wooden section. The spacing of the slats and the staggered top of the outer fence give it a light and airy look. The bottom photos show the fences three years later. The deer-resistant lavender (*Lavandula angustifolia*) is masking the bottom third of the wooden fence, lessening its mass, and the inside metal fence is covered with honeysuckle (*Lonicera japonica*.)

Drip Irrigation

In 1983, when I first moved into my house in Occidental, CA, I installed drip irrigation for my flowers and vegetables in the form of in-line emitter tubing, the least well known of drip-irrigation technologies. The tubing is about 1/2 inch in diameter and comes with an emitter pre-installed inside the tubing at regular intervals. These internal emitters utilize what is known as a "tortuous path," which forms a continuous vortex, a kind of horizontal tornado that keeps any sediment, sand, or silt in suspension so it won't clog the emitter.

In-line emitter tubing moistens the soil along the entire length of the line, but slightly below the surface, where bulbous-shaped wet spots come together to form one nearly continuous moist zone. The pre-installed emitters come in pre-spaced, 12-inch intervals. The emitters inside the hose are rated to dispense either ½ gallon or one gallon per hour (gph), and the hose is available in both non-compensating and pressure-compensating versions.

The benefits of pressure-compensated in-line emitters are: it's easy to install; simple to snake around your existing plantings; suffers less clogging than porous tubing and some punched-in emitters; works at the greatest range of pressures (10-25 psi); provides consistent rates of irrigation without regard to slope or length; and most importantly, has no external parts to snap off.

Drip irrigation can be as simple or as complex as you want it to be. If you're intimidated by any device more complicated than a hand-operated can opener, rest assured that drip irrigation is nowhere near as complicated as it looks. Think of all the drip parts as toys for grown-ups, an adult version of Lincoln Logs™, Erector Sets™, or Tinker Toys™. (If you don't know what these are, you're not a baby boomer.) If you can put yourself in the same playful, "tinkering" frame of mind out in your garden, then drip parts may not seem so intimidating.

Start with a small system, watch the results as the plants grow, and adjust the system before expanding to more parts of the garden. Drip irrigation needn't be thought of as something you're forced to use because of the ravages of drought; it's a valuable tool to make any garden in most of the country flourish while conserving a precious resource.

No Irrigation for Ornamentals and Fruit Tress

I had great drip irrigation on my Occidental property, but the well servicing it was so poor that I almost ran out of water just before the winter rains begin (November now, but used to be October—climate change). To lessen the load on the well, I started playing with ways to reduce irrigation and still maintain good growth.

First I tried planting early in the spring, mulching very heavily, and skipping drip irrigation, but by mid- or late summer the plants looked tired or wilted. Some died; so I experimented some more.

I finally found the formula. Turns out it's fairly simple:

1. Do any soil mounding or amending in September. If you're planning to cultivate (I prefer to use no-till methods when I can), you can moisten the area to make it easier.
2. Plant only the most drought-resistant plants—native or exotic. I found that there are Mediterranean plants that are more drought-tolerant than some of the native plants in my county, for example, the *Echium* spp. shrubs that I grew until a 15°F cold snap killed them. Echium is an exotic here but a native in the Canary Islands, where the average yearly rainfall is five to 10 inches with only 22 days of rain.
3. Plant ***only in the fall*** (late October to late November in our California Mediterranean climate), before the winter rains begin. There is *no other* time you can plant successfully and avoid irrigation. This allows the roots to grow all winter during the rains. As the soil dries out in late spring, the roots have already spread to a large volume of soil, and easily adapt to the drying earth. Spring planting doesn't give the roots a chance to grow into enough soil to tolerate a summer without irrigation. This unirrigated garden technique is only for home gardeners—commercial landscapers have to plant year-round to make a living. I like it that way.
4. Water by hand until the rains begin.
5. Weed and mulch every spring thereafter.
6. Walk away and leave the plants alone as much as possible, except to admire them.

Figure 409. This is a view of that deer-resistant garden that I didn't water for 30 years. I used a lot of Mediterranean plants and many culinary herbs.

The above photo shows a section of my deer-resistant, drought-resistant garden after 30 years of no summer irrigation. This approach is very dependent on: 1) choosing the most drought-resistant plants, and 2) constant mulching. In climates where it rains less than the 58 inches per year that my garden received, you have to be even more careful—my garden's surface soil is as dry in August as any other nearby garden with less rain.

The portfolio of plants to choose from is greatly reduced by this approach. I think it can be done in any climate with the proper choice of plants. Experiment in a tiny area of your yard or garden. I pulled this off in a place where it doesn't rain from May to November. If I can do this, most climates should be manageable without irrigation, except, perhaps the "deep" southwest. (See Gary Nabhan's work in the dry southwestern climates mentioned on page __.)

Drip for Veggie Boxes

While I could get away with sparing the irrigation for ornamentals, it's a whole different story with vegetables, which usually *have* to be irrigated. It's a *very* rare gardener who can dry-farm vegetables, and only in very sandy-loam soil. For some good information on low-water gardening, see "Gardening When It Counts: Growing Food in Hard Times" by Steve Solomon (*Mother Earth News* Wiser Living Series, New Society Publishers; 2006.)

I use two or four lines of in-line drip irrigation tubing for each wooden raised bed. Here are some examples from the edible landscape I designed for the Prestons. I'll walk you through the design and then explain the drip-irrigation installation/configuration. The goal of this design was to create an informal Italian look without too much symmetry.

I call the lawn in the middle of the landscape "the green piazza." It's where the public lounges to sip wine and view the vegetable beds. The lawn faces north to the office house, but the Prestons didn't want

Figure 410. I ran the drip-irrigation pipes at Preston's up the south sides of the beds that are visible from the lawn "piazza" where the public gathers. I then masked the pipe with English lavender plants (*Lavandula angustifolia*) that are still young here, but grew to hide the plumbing within three years. I also notched the top of the box so that the fittings don't stick out.

the public to enter the south-facing French doors of the office. I installed a square brick landing for staff access, but limited the walking space with a classic Italian fountain. Most visitors took the hint, while staff could easily slip around the fountain.

As a way to keep the public out of the veggie area but still allow visitors to admire the vegetables growing, I planted a lavender hedge along the entire south side of the box-ends. This helped separate the boxes; hid some of their rigid edges; obscured the drip-irrigation tubing coming up into them, and presented a glorious bloom and fragrance each summer. (This design was installed in 1983, *long* before lavender and wineries in Sonoma County, CA became a cliché.)

Along the northern length of boxes, I made a gravel path six feet wide for wheelbarrow or garden cart access. Each box, I should mention, is only six feet long. Since the south side of the box grouping is blocked by the lavender hedge, the six-foot length allows the gardeners to toss soil or mulch easily from the northern path. (With a typical eight-foot box, the tossing would be too stressful and awkward.) Keeping full access from the north also meant that the beds could be much closer together (as little as 18 inches), a careful use of space.

Each box is watered by two lines of in-line tubing. Because the boxes are only three feet wide it's easy to reach in from the side to till and harvest, and it takes only two lines of emitters to moisten the entire volume of soil in each box. (Four-foot-wide boxes would need four lines of tubing down their length.)

The lawn uses one water valve for the sprinklers, and a different one for the lavender hedge. All of the vegetable boxes are on yet another valve and come on all at once. Each bed is equipped with a ball-valve, which allows a box to be turned off if it's empty, or to enable gardeners to cut off water to the tomato box in mid- to late summer for the best-tasting fruit. There's also a fitting that can be released; this allows the gardeners to separate the tubing harness, toss it on the

Figure 411. This is the Prestons' "piazza" landscape, where visitors to the wine room can sample vintages on an open-air lawn. The lavender hedge, in this photo about three years older than in Figure_, hides the drip irrigation hardware and the angular nature of the raised beds, which are fitted with wire bottoms to thwart gophers.

Figure 412. From the east side of the planting you can see the angular aspects of the raised beds. Visitors walking toward the office doors will see more and more lavender hiding the south end of the boxes. Culinary herbs surrounding them add color, attract beneficial insects, and soften the boxes' hard edges.

lawn, prep the soil, and retrieve/replace the tubing once the soil is ready. The illustration of the fittings for each box is from my book *Drip Irrigation for Every Landscape and All Climates*, second edition, 2009.

Off to one side, a mature orange tree provides a focal point, with a mature peach tree on the opposite side. Other perennial foods and flowers are clustered around the perimeter of the garden and under the arbor, which itself has posts on one side only, with an opening leading to the "lawn piazza".

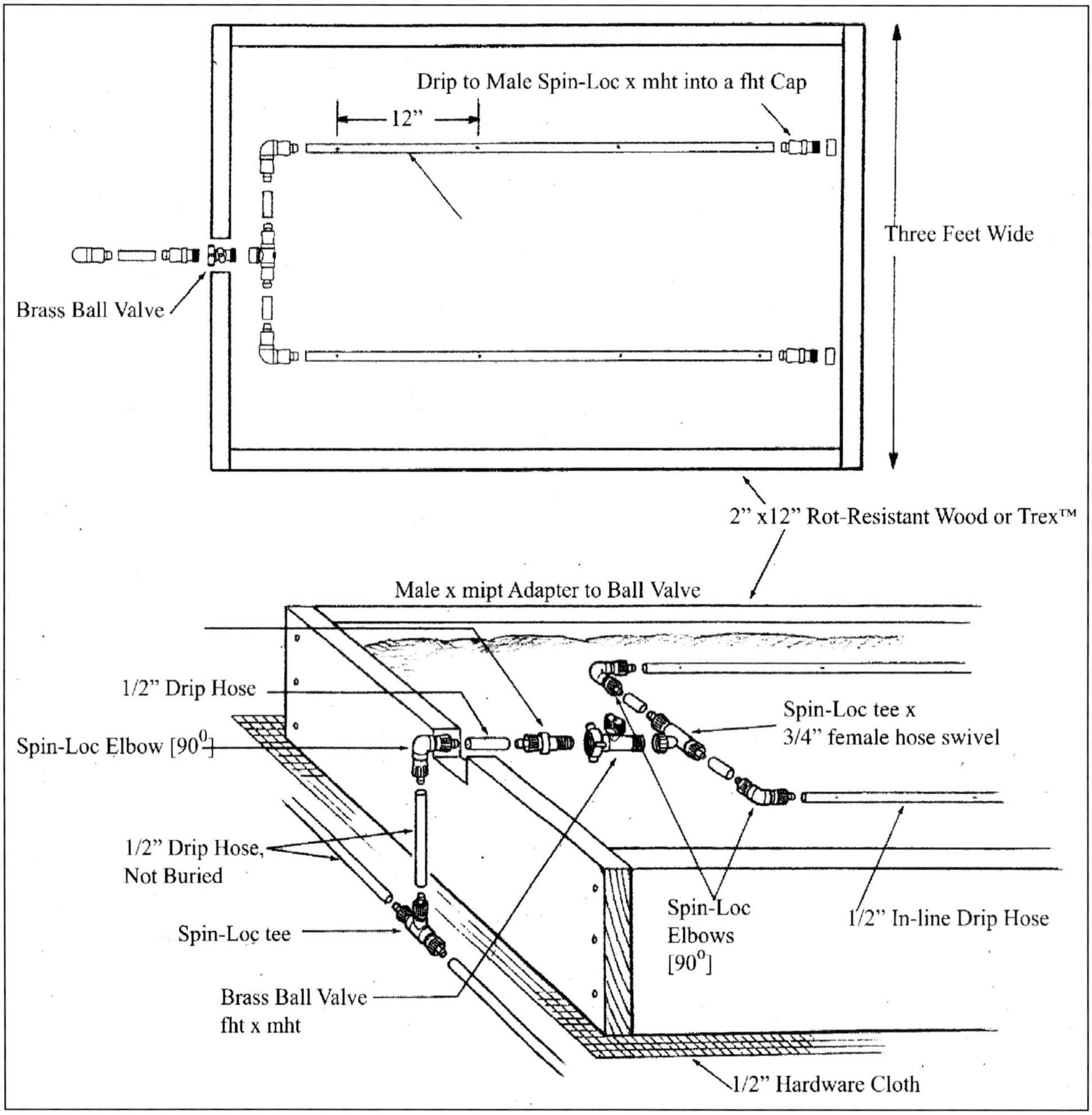

Figure 413. This shows the positioning of all the hardware that I use to irrigate a raised bed. Any landscape-supply business selling irrigation parts will have all these fittings. (If they don't, go somewhere else; a good mail-order company is DripWorks in Willets, CA.) The notch in the wood next to the Spin-Loc elbow gives the hardware a low profile. A four-foot-wide bed will need four lines of equally spaced in-line tubing. (The illustration is from my book *Drip Irrigation for Every Landscape and All Climates*, second edition, 2009.)

Figure 414. This is the edible landscape (shown in the early spring after three years of growth) that I designed for visitors to view while sampling Preston wine. The vegetable raised beds are between the arbor and the fountain. The white flowers on the arbor are the Lady Banks' roses (*Rosa banksiae*). The lawn on the mid-right in the photo is what I call the "piazza", where visitors can sip wine, wander around, view the gardens, and picnic. The plantings to the left and right of the steps to the office are a mixture of ornamentals and herbs for year-round color. If I were to do it over, I'd add a splash of hot color to the white theme. (See how Lou and Susan Preston improved my design in the next figure.)

Figure 415. In year four the Prestons added some hot-pink mallow (*Malva* spp.), which I consider a considerable improvement on my original design. The gophers loved eating the artichoke roots, but the dark-green rock-roses (*Cistus* spp.) have begun to fill in.

Figure 416. The Prestons have always made very good use of the beds, mixing vegetables with flowers each season to add a touch of flare. Here, marjoram is in the left front, with a touch of thyme in the lower right. Both soften the angles of the box corners.

Worms

If you don't have room for a cold compost pile, worms may be for you. I spent nine months living and working with Ken and Barbara Kern (of *The Owner-Built Homestead* fame) at their place near Yosemite National Park. They were two of the grandparents of the 1960s back-to-the-land movement, but *very* few people know of them now. Ken was (he died in 1986) a major proponent of worms as waste digesters. He used them to devour garden waste as well as animal and human manures, and frequently extolled the virtues of worms as decomposers and for the speed at which they consumed wastes.

I set up a simple worm bin, solely for my kitchen scraps, in an abandoned Army-surplus drum with a tight-sealing lid. I layered a foot of straw in the bottom, added six or more inches of horse manure, capped this with a six-inch layer of straw mulch, and moistened the entire bin. After the manure had started to compost the straw, I added a large coffee can's worth of red wiggler manure/worm-bin substrate (a mixture of manure worms and their castings). Shortly thereafter, I started adding kitchen scraps on a daily basis. (Raw garden scraps may have to be chopped up to work well.)

The Reality of Worm Culture

The first lesson learned was that I should've waited until the population of worms had increased, as there simply weren't enough worms in the bin at

Figure 417. These beds are three feet wide and six feet long; the gravel path provides access for mulching and harvesting, their six-foot length allowing workers to easily toss mulch while standing on the path. The gravel runs beneath the dripline of the walnut tree (here leafless in winter) to protect its roots from rot and allow space for picnic tables.

Figure 418 . This photo is from the first year at the Preston Winery. Again, the lavender is quite young, and doesn't yet cover the south-facing end of the boxes. The bricks form a "mow strip" for the wheel of a lawnmower. (As the lawn and lavender grew, the bricks became less visible.) The boxes are 12 inches tall with wire bottoms to deter gophers.

first to digest each day's waste quickly. After several months of cohabiting, however, they had obligingly increased their population enough to easily consume each day's scraps. The thoughtful wormologist feeds the new worms manure, coffee grounds, or nutmeats for a period of time to build up the population before introducing kitchen scraps. (Contrary to many guidelines, I feed my worms all the scraps left over from BBQs. No problem, except that the castings were full of bones which I had to sift out so my garden wouldn't look like it belonged to a serial killer.)

Then, tragedy struck. I had a large BBQ party and the lid was left slightly open. The extra amount of scraps, combined with the larvae of flies, caused a literal "meltdown"—the bin turned into a wet, near-slurry of castings (the discreet word for worm poop), drowned worms, and fly maggots. Enough to make you lose your lunch. I drained the surplus liquid off by puncturing small holes in the bottom of the drum, but it took many months for the worms to re-populate the bin and the fly larvae to die off.

Then another tragedy struck: my landlady cut down the large tree that shaded the drum, which the first hot spell turned into a casserole dish of baked worms. Yuck! Again, it took months for the few worm survivors to re-colonize the drum. Many months later, I was back to a healthy, happy drum-full of wiggly, hungry worms. I hope that I've been through all of Mr. Murphy's laws of worm culture!

Figure 419. California Flannel Bush (*Fremontodendron* spp.) is *very* drought resistant.

Low Maintenance
Since ground covers can lead to tedious weeding, you might consider planting shrubs. The definition of a shrub varies. One I like is: "a woody plant that is smaller than a tree and has several main stems arising at or near the ground, where it branches almost immediately."

Figure 420. Matilija Poppy (*Romneya coulteri*) can spread dozens of feet wide. It's very drought resistant. Pest free.

One of my favorite shrubs is lavender (*Lavandula* spp.), which not only looks nice and smells great, but also has another characteristic I prefer in a shrub—density. The thick foliage blocks much of the sunlight from the ground, creating a deep shade that tends to stifle weeds.

Lavender only needs one intense "haircut" (heavy pruning) a year to keep it dense. I no longer recommend it to clients, however, as they often don't prune it correctly and it gets woody and ugly in less than four or five years. They should cut the plant back heavily, to within one to three inches of the bare wood to keep it dense. I don't mind the work, as I love lavenders and it doesn't take me too long.

There are many other shrubs that need less or no pruning, so most of my pathway borders are designed with shrubs that can easily be kept to 12"-24" high (technically called subshrubs). Again, with most shrubs, the dense foliage can provide so little light that weeds are suppressed.

Don't try to grow proper clipped hedges unless you can afford a full-time gardener— it'll be job security for them. If you let shrubs grow to their natural mature size, it takes a lot fewer plants to fill the same area as a trimmed hedge. Allowing for the full-sized growth of a hedge plant means you get to skip countless hours of pruning.

One solution, if you want a lower border, is to use a smaller version. Many plants have a form often called 'compacta' or 'nana'. As an example, consider a shrub called 'mock orange.' (It's not at all related to citrus; the blossoms just happen to have a similar fragrance.) Its botanical name is *Pittosporum tobira*—and it can grow to be 33 feet tall and 10 feet wide.

One variety, however, called *Pittosporum tobira 'Compacta'* only grows 10 feet tall and wide. Another selection, *Pittosporum tobira* 'Nanum' or 'Nana', comes in a compact form that grows only two feet high and wide. Yet another selection is *Pittosporum tobira* 'Wheeler's Dwarf'—three feet tall by five feet wide. So, you do have choices. (There are even variegated selections of this plant— two feet to two and 1/2 feet tall and wide.)

Here are some other low-maintenance plants for my Northern California area, with some notations on growing them. (Check with the Master Gardeners for your area.) Most attract few pests except for deer, gophers, and rabbits. While most shrubs are not edible, many yards require at least a few shrubs or trees for form, flowers, and/or fall color:

- California Buckwheat (*Eriogonum* spp.) Remove spent flowers once per year.
- California coast Live Oak (*Quercus agrifolia*): You'll usually never have to do anything once you've established a proper canopy.
- California Flannel Bush (*Fremontodendron* spp.): If left uncut it can get rangy; with enough room it never needs pruning.
- California Fuchsia (*Epilobium* spp., formerly *Zauschneria* spp.): The low-growing ones need only one pruning, almost to the ground, each fall.
- California Myrtle (*Myrica californica*): Keeps a fairly tight form with no pruning.
- California Poppy (*Eschscholzia californica*): All you have to do is to let it self-seed.
- Ceanothus groundcovers (*Ceanothus* spp.): In the proper sunny spot, they never need pruning.
- Ceanothus shrub (*Ceanothus* 'Julia Phelps' and 'Dark Star'): This can get awkward without a small amount of shaping. Deer can damage it by rubbing their antlers on the branches to get rid of the velvet each spring.
- Coffeeberry (*Rhamnus californica*): Fairly dense, and usually requires no shaping if you give it plenty of room.
- Jerusalem Sage (*Phlomis fruticosa*): You'll need to prune spent flower-stems off early summer.
- Juniper (*Juniperus* spp.): There are lots of cultivars that never need pruning if left with plenty of room to grow or sprawl.
- Manzanitas (*Arctostaphylos* spp.): Usually no pruning once you've shaped the main scaffold (branches).
- Matilija Poppy (*Romneya coulteri*): Takes up a *lot* of room if the roots are left untrimmed. Cut the foliage to the ground each fall. No pests.
- Rosemary, (*Rosmarinus officinalis*): Only the ground-cover cultivars need a light pruning, unless you have lots of room to let the full-sized plant flop around.
- Sea Pinks *(Armeria maritima*): Needs only some moisture and a quick clip of a few spent flower stems.
- Thimbleberry (*Rubus parviflorus*): A little bit rangy but very tasty fruit.
- Wild Ginger (*Asarum caudatum*): With no care, it will ramble around via rhizomes.

Many ornamental grasses don't need pruning except to remove spent seed heads. *Stipa gigantea* is one excellent example.

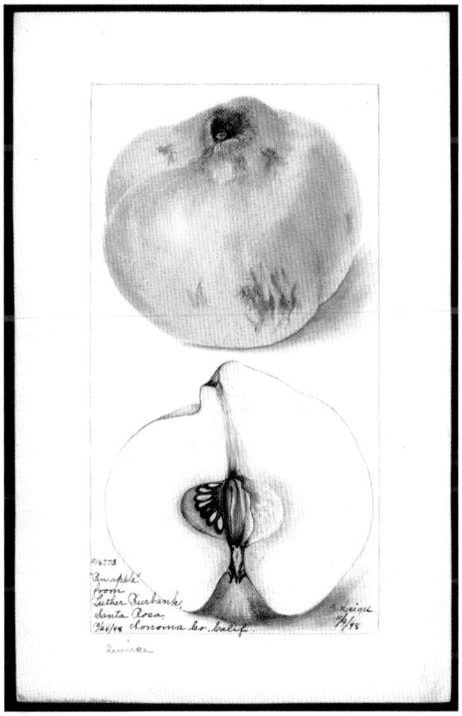

Figure 421. This pineapple quince is from Luther Burbank's farm in Sebastopol, CA, about 20 miles from my home. He provided the fruit for the USDA artists to sketch. This drawing is from 1898.

U.S. Department of Agriculture Pomologicatl Watercolor Collection. Rare and Special Collections, National Agricultural Library, Beltsville, MD 20705.

Fruitful Abundance

When considering fruit and nut trees, one should carefully choose what varieties to plant, as some require much less work than others. You might want to slightly adjust your palate to favor the more pest- and disease-free varieties of fruit and nut trees. Examples include:

- Kiwis: Except for deer, there are no major pests or diseases for the popular kiwi (*Actinidia deliciosa*) and I've found it to be drought-resistant. This is the kiwi you find in the store, but it only fruits well in USDA Zones 7-10. The cold-hardy kiwi (*Actinidia arguta*) will grow in USDA Zones 4-8, but it can be very invasive if left to its own devices, and produces smaller fruit. Both kiwis require substantial trellises, the commercial kiwi more so. Male and female vines are required for fruiting.

- Persimmons: The Asian persimmon (*Diospyros kaki*) is cold-hardy to -15°F. The American native persimmon (*Diospyros virginiana*) grows in USDA Zones 5-9 and tolerates temperatures down to -25°F Both are self-fertile. As the photo on page 21 shows, they flaunt dramatic, gorgeous fall foliage and the colorful fruits will hang on into the winter like tree ornaments. If you give the roots good drainage, there are no significant pests or diseases, and the tree will fruit plenteously without pruning. You may need to shape it once in awhile to keep heavily laden limbs from breaking.

- Pineapple Guava: I use this tasty self-fertile, evergreen fruit tree (*Acca sellowiana*; was *Feijoa sellowiana*) a lot, for hedges and shrubs, and they can also be trained as standard-shaped trees. The fruit has a wonderful tropical-like taste, and the flowers make a great addition to fruit salads. The gray-green foliage is unique to fruit trees. I have seen it grow well in a wide range of soils. Unfortunately, it's only hardy in USDA Zones 8-10. (Sorry, Missouri.) Deer eat it, but it's otherwise refreshingly carefree.

Figure 422. Pawpaws (*Asimina triloba*) are great tree crops as they are native to the midwest and tolerate a lot of shade. They're also *very* tasty.
U.S. Department of Agriculture Pomologicatl Watercolor Collection. Rare and Special Collections, National Agricultural Library, Beltsville, MD 20705.

- Quince: (*Cydonia oblonga*): is known for its jelly-making properties, but it can be used in entrée dishes like "Lamb and Quince." Just one ripe fruit will fill a room with a delightful fragrance. It's self-fertile and grows in much of the US, in USDA Zones 4-9. The foliage begins to grow on this deciduous tree in the spring, followed by the formation of flowers, thus it escapes most late-frost damage to its blossoms. Its roots prefer good drainage. Deer eat it, but carefree otherwise.

- Pawpaws (*Asimina triloba*): Another fruit for cold-winter climates, growing down to USDA Zone 5A (-15 to -20°F). This unique tree, with its tropical-flavored fruit, prefers some shade when young and tolerates shade as a mature tree (with, however, lower yields), so is adapted to more areas in a food forest than most fruit trees. It requires two different trees for pollination.

- Loquats (*Eriobotrya* spp.): This evergreen fruit tree, with its attractive shiny dark-green leaves, comes from southeast Asia, and is best grown in tropical and warm or temperate climates, tolerating temperatures to 12°F, although the blossoms, which are produced in the fall or early winter, will freeze at 26 to 27 °F. Loquats can be grown in USDA Zones 9 through 11. Temperatures above 95 degrees may cause leaf scorch and damage to the fruit.

  These are perfect gardener fruits, as they're rarely seen in markets because they bruise so easily. I consider them to be very tasty, and they're the earliest fruit of the summer. They're susceptible to fireblight, but a quick clipping usually solves the problem. (Be sure to dip the clippers in 10-percent bleach after *every* cut.) Otherwise, they harbor no diseases or pests.

- Jujubes: Often called "Chinese dates" these fruits taste a bit like a dry date but with a sweet apple flavor, and they grow on a deciduous fruit tree (*Ziziphus jujuba*), not a palm. They're not too cold-hardy (down to 10°F) but will tolerate a lot of heat. Some have thorny branches. They grow to a height of 30 feet and are self-fertile.

- Figs (*Ficus carica*): Attract no pests or diseases (well, except for gophers that seem to *really* love the roots). Fig-tree roots require very good soil drainage. They're self-fertile, hardy to USDA Zone 7-11, require lots of sun and

Figure 423. Loquats are the first fruits in early summer. They bruise too easily to ship, perfect for home gardeners. But sthey are only hardy to Zone 8.

U.S. Department of Agriculture Pomologicatl Watercolor Collection. Rare and Special Collections, National Agricultural Library, Beltsville, MD 20705.

Figure 424. Serviceberry (*Amelanchier arborea*) is native to eastern North America. Tasty with sugar.

U.S. Department of Agriculture Pomologicatl Watercolor Collection. Rare and Special Collections, National Agricultural Library, Beltsville, MD 20705.

heat to get the best flavor, and if unpruned, will grow to be at least 20 (or more) feet tall and wide.

- Serviceberry (*Amelanchier arborea*): Produces brilliant, slightly fragrant, flowers that bloom in early- to mid-spring. It tolerates a wide range of soils, but needs good drainage. The dark-purplish fruits, often called "Juneberries," are native from Maine to Iowa, and south to northern Florida and Louisiana. The fruit usually requires the addition of sugar to be enjoyable. Zones 4-9.

- Alpine strawberries (*Fragaria vesca*): These perennials are not trees, but I put them here because they flourish under trees and are pest- and disease-free. (Slugs on the fruit can occasionally be a problem, but less so than with commercial strawberries.) Even though the fruit is very small, it has an intense flavor, superior to that of many conventional strawberries. Hardy to USDA Zone 3, they will live and fruit for many years longer than typical strawberries. Protect the shallow roots with mulch.

There are also many "wilder" fruits that have not yet been selected or bred for superior fruit, and tend to be tart. Fruits with good flavor that are often sweetened include: salal (*Gaultheria shallon*), evergreen huckleberry (*Vaccinium ovatum*), and various native huckleberries (*Vaccinium* spp.).

In regard to other fruits, there are options that require somewhat less work. If your climate allows, apples are a good example. All apples form on what's called a spur, a cluster of flower buds that can resemble a tiny tree. Once a spur is formed, an apple can grow there each year for 20 years or more if it gets sufficient sunlight. Not all apple trees are the same when it comes to spur formation. Observant orchardists have long noticed that some trees form far more spurs, closer together, than others. This means more fruit per linear foot of a branch, regardless of pruning. These spur-type trees produce more fruit than conventional apple trees. When buying apple trees, look for the word "spur" or "spur-type" on the label. See the list on page 236 of the Container Gardening chapter.

Peach trees are fraught with numerous problems, the most noticeable of which is the fungus *Taphrina deformans*, known as peach-leaf curl. While trees with this fungus seldom outright die, they do become very stressed, and fruiting can be reduced or limited. Spraying to control the curl is difficult, as it has do be done before the buds open, but a rain after a spray negates the effect and the tree must be sprayed again—during mild, rainy springs, this can require repeated spraying. Many busy people often forget, and the tree gets curl.

Observant orchardists have once again come to the rescue, spotting and selecting trees resistant to the curl. One good example is the cultivar 'Frost'. Other selections and trees recently bred for curl resistance include: 'Betty', 'Salish Summer', 'Black Boy', 'Muir', 'Q-1-8', and 'Red Haven'. An old heirloom variety, 'Indian Free' peach is also curl-resistant and *very* tasty. I don't recommend any other varieties to my clients.

(Begin rant - again. Important enough to repeat.) In regard to the increasing number of books saying you should replace all your lawn with edibles (and/or native plants), I say they must have never been to St. Louis County where I grew up, and where there are miles and miles of homes with huge lawns. Our lawn, at one-half acre, was one of the smallest. The lawns I mowed to make money in high school ranged from one-half to two acres. According to the Cornell Lab of Ornithology, the average area of a lot in America is 9,801 square feet (.225 acre). Vegetable gardens take as much as 20 hours of tending per season per 100 square feet, while lawns only need one hour of labor per 100 square feet per season. To look at it another way, the average homeowner spends 150 hours per year mowing; or .4 hours per day. (Although I suspect organic care of a lawn adds extra time, especially if you get compulsive about dandelions and clover.)

The same goes for those who would have us replace all the lawns in America with grass meadows, which need a boatload of weeding to get established, and a fair amount of work forever as weeds blow in. Give me a break.

Granted, housing now is so expensive now that many houses are built on lots so small that any patch of lawn seems to be an afterthought. (End of rant.)

Disease- and Pest-"free" Trees

In a good-sized yard, not every tree needs to produce edibles. There are aesthetic reasons to plant trees, but wise homeowners will choose one that's as low-maintenance as possible.

There's a saying among old orchardists: "The best time to plant a tree is 20 years ago, the second best time is today." Do some research on your iPad or laptop, and Google up some trees. Or, call your local Master Gardeners and ask them about disease/pest-free trees—if they don't know, fire them. Of course, a reputable garden nursery will know about and stock many of these trees. You can also often see some of them as mature specimens in cemeteries.
Here's a short list to start with:

Figure 425 Ginkgo (*Ginkgo biloba*) is a hardy tree with no pests. It has, as pictured below, fantastic fall color. Its roots don't easily heave pavement.
(Photo © Saxon Holt)

Ginkgo (*Ginkgo biloba*) Zones 3-9
50'-100' H X 20'-40' W
This is my number-one choice. This living fossil— a rare deciduous conifer documented as unchanged for over 270 million years—is, as far as I know, free of all pests and diseases, and the blazing, hot-yellow fall color is spectacular. As an extra asset, the roots are much less likely to heave sidewalks than most trees. The male tree doesn't have the disgusting smell of the females' fruits, but is fairly high on the allergy scale. If you plan to live a long time, you're in luck—some trees are thought to be over 1500 years old.

Black Tupelo (*Nyssa sylvatica*) Zones 4-9
30'-50' H X 20'-30' W
Besides being relatively pest-free, this tree is very tolerant of wet soils, even flooding, and its roots are resistant to root rots (*Phytophthora* spp.). Native to the central and eastern parts of America, it's known for its spectacular fall color, and is often rated in the top ten on fall-color lists. Can attract occasional butterfly larvae.

Sweet Bay Magnolia (*Magnolia virginiana*)
Zones 5-10 10'-35' H X 10'-35'W
A very adaptable tree, it tolerates acidic soil, air pollution, moist and boggy soils, clay soils and full or part shade. Native to the eastern US, it's semi-deciduous and bears fragrant blossoms.

Katsura (*Cercidiphyllum japonicum*) Zones 4-8
40'-60' H X 25'-60' W
Likes well-drained, but moist, soils. No showy flowers, but the early-spring leaves are an attractive reddish-purple color.

Persian Parrotia (*Parrotia persica*) Zones 4-8
20'-40' H X 20'-30' W
Often found on top-ten fall-color lists. Needs well-drained soil in full sun. This deciduous tree has showy curled (exfoliating) bark that's attractive in the winter.

Paperbark Maple (*Acer griseum*) Zones 4-9
20'-30' H X 15'-25' W
This deciduous tree is very striking in the winter due to its especially beautiful exfoliating copper-orange bark. It can also have showy orange leaves in the fall. Needs some moisture in full or partial sun, and makes a great understory tree.

Other examples include: Goldenrain Tree (*Koelreuteria paniculata*); Douglas Fir (*Pseudotsuga menziesii*); Hemlock (*Tsuga* spp.); Norway Maple (*Acer platanoides*); Korean Dogwood (*Cornus kousa*); and Northern Red Oak (*Quercus rubra*).

A Summary

Polyculture is the mixture of a variety of plants, unlike a *monoculture,* which is composed of a single crop. My edible-landscape version of polyculture is contained within a given growing area, but doesn't consist of elements layered directly on top of each other. My designs create an ecosystem that's as self-sustaining as possible, but is *not* a food forest.

Here's another quick look at the interplay between aesthetics and productivity.

The American media is physically and politically urban-based, but the suburbs are apparently where it's at. Most of the country's population—and certainly most of the gardening kind—lives in suburban developments. Most suburbs, however, have a only vague cultural or written legal definition of "beauty" (usually produced by a homeowners' association), which doesn't often embrace the aesthetics of a typical "food forest." Sadly, it's still illegal in many areas to grow food in one's front yard, and where it *is* legal, aesthetics play an important role. Tacky little boxes of bare soil and scattered veggies may leave much to be desired as lawn decor, but a jumbled thicket of edible trees, shrubs, and vines can be considered equally ugly. What to do?

Figure 426. This lawn is for guests to the Preston Farm and Winery to sip wine outside the tasting room and to have a picnic.

Enter the edible landscape, which doesn't require you to plant your entire yard with vegetables. In fact, planting your entire property would probably be more work and produce more food than is compatible with a normal, busy lifestyle. Your available time and effort should determine the scale of your garden, not the lot size of your yard. Enough sunlight, a fertile soil, good drainage, and a lack of debilitating winds are the main factors to take into consideration when siting an edible landscape (although you might want to grow your herbs in soil or pots as close to the kitchen as possible).
Style is another important consideration. The Napa Valley, CA garden/edible landscape in Figure 427 is laid out on a grid of angular lines that suits the rather formal nature of the rest of the estate, and the reserved, proper nature of the owner. (It's not at all my preferred style, but the homeowner loves it.)

Figure 426 is another view of the landscape that I designed for the Preston Farm and Winery's outside "tasting room." The raised boxes are laid out in a gentle curve so as to lend a more informal and only slightly symmetrical feeling. Flower borders near and around the vegetables boxes offer year-round color to attract the eye, even when the veggie beds are looking a bit tacky between harvests. The Prestons wanted a somewhat formal, Italianate, and not too symmetrical feeling. Thus, the five boxes west of the terracotta fountain and the three to the east feel symmetrical only when you get close to the fountain. From most of the lawn "piazza" one sees the asymmetry. The lavender hedge is shaped, but not trimmed in a tight topiary fashion. A further subtle touch of asymmetry comes from plantings of lavender, grown from seed so that, unlike plants all grown with cuttings from the same plant, each shrub's foliage and flowers are slightly different, In many ways, these two gardens are polar opposites, but both would probably be considered attractive in a suburban setting.

For low maintenance, some permaculturists say that with a good design, you should be able to plant, walk away, and come back to harvest for years to come. Fat chance with most plants! But you can do this to a limited extent with terminal-bearing edible shrubs or fruit trees with spring-growth patterns that produce leaves *before* blossoms and always fruit on new growth.

I've always liked scruffy, wild-and-woolly food gardens, especially in the 1970s when I was under the influence of Edgar Anderson's "dump-heap" review. Unlike some people, I seem to be able to "see" ecological patterns, nutritional dynamics, the interdependency, the subtle use of microclimates, and the edibility factor where most people driving by would just see a jungle. The one in Figure 428.

Some gardeners can't "see" certain ecological arrangements. They just see what's there—this plant is ready to eat; this plant is dying; this veggie looks sick, etc.—but not how the plants interact environmentally. They don't register that the yellow leaves of the lemon shrub mean that its pH is not acidic enough; or that there are no nodules on the green-bean roots, so the next batch of seeds needs to be inoculated; or that the nasturtiums should be kept closer to the subtle microclimate of a warm wall so that the open air doesn't freeze the leaves in late fall.

Once again, an important distinction between a forest garden and an edible landscape is that an edible landscape is designed to be visually attractive as well as to produce tasty food. While many forest gardens are ecologically vibrant, they can also appear as an unattractive hodgepodge of plants. With a sustainable edible landscape, I believe you can have both productivity and beauty in the same place.

Figure 427. This garden in Napa, CA belongs to a woman who loves to make flower arrangements, but also grows a lot of food in the angular beds (note the bounty of carrots in the foreground). The fence accentuates the linear nature of this garden. The fruit orchard is in a separate area nearby.

Figure 428. This permaculture garden— a thicket of fruit trees, including cherry, apricot, peach, and avocado—is quite a contrast to the angular garden on the left. Vegetables fill the south-facing edge, and a squash climbs on the railing. Other low-growing plants, bird house and herbs make this a certified Wildlife Habitat.

Bonus Photos:

Figure 429. Here's an attractive design (prior to planting) by Michael Judd, showing that a permaculture landscape can actually embrace a bit of lawn, which is for events held at the adjacent restaurant. The white circle of stone receives water from the downspout and directs it toward the brown curved swales. Judd later planted an edible woodland garden here, featuring currants, pawpaw, spicebush, sweet woodruff, and fiddlehead ferns. (Photo from *Edible Landscapes with a Permaculture Twist* © Michael Judd.)

Figure 430. This is a "spine" of rocks I use to lend an air of place and history to a new landscape.

Figure 431. You can place your rocks before laying bricks so that the rocks look like they came first.

Figure 432. Susan Meyer, who lives in a mobile-home complex, prefers very intense plantings so she can pack a lot of edibles into the limited space of this permaculture garden. Ornamental flowers attract beneficial insects, and nasturtiums, herbs, and strawberries flourish below the peaches (pink blossoms) and plums (white blossoms.) Susan grows over 100 fruits and berries on her very small lot. Her paths are very close to the trees, making it easy to prune and harvest. (Photos © Susan Meyer)

Figure 433. This is Jude Hobbs' permaculture garden in Oregon as seen from a bedroom. Oregano and rosemary grow in pots in front of a beautiful dahlia, mullein, kale, and nasturtiums. In the back are an apple tree and a fig tree (not visible here). The tree in the back on the left is an ornamental maple. This area leads to a one-acre forest garden.
(Photo © Jude Hobbs/Cascadia Permaculture)

Figure 434. Here are two views of what I call a garden "throne"—a seat with an arched canopy, made of woven willow by a basket maker, who also framed the entire garden (below). This is a picturesque feature in Joy Larkcom's "Little Potager" in East Anglia, England, near Cambridge. (A potager is a vegetable plot that follows the principles of garden design to create an area both ornamental and productive.) Considering rhythm, line, color and texture, Joy introduced this focal point to bring the whole design together. In her "Little Potager," vegetables, edible flowers, herbs and sometimes fruits mingled. After gardening here for well over 25 years, she had very fertile soil, and the garden produced enormous vegetables like this king-sized cabbage and the huge fava-bean leaves on the right. Joy is the author of several books on vegetable- and salad-growing, including the ground-breaking *Oriental Vegetables,* (1991).

Figure 435. This is the deer-and drought-resistant landscape that I grew in Occidental, CA. The narrow path is accented by clay pottery, and the lavender planted closely on both sides of it allows the legs of walkers to brush the foliage—releasing wonderful aromas. The photo shows three English lavenders (*Lavandula augustifolia*) and a Spanish variety (*L. stoechas*). At lower left is the ordinary species; the white 'Alba' is just to the right of the pot; and 'Hidcote' to the right of 'Alba.' Above the pot is a green Spanish lavender ('Viridis') which, along with the blue Spanish form (not shown here), is the species I use the most for grilling.

All of these lavenders were planted in November of various years, just before the beginning of the winter rains. This timing allows the roots to grow and develop all winter. When the soil dries out in the spring, the large root system is then able to gather enough moisture for the plant to thrive. For over 30 years this garden was never irrigated, even in the dry California summers.

The flowers in the background are on the tips of pink foxgloves and yellow mullein. To allow these plants to grow naturally, they must seed themselves, that means leaving the flower stalks to dry out. Some people think this looks messy, but it's worth it to me to have the plants "walk around the garden," sprouting in different places every year and providing annual surprises.

Chapter 13
Sustainable Garden Play

Be Productive *and* Playful

The following may be the last section of the book, but it's probably the most important.

To sustain a garden you really need to play in it. Without a sense of play, your garden can become a tedious chore of *just* gardening, harvesting, and eating. (And you can, in spite of what your mother said, play with your food.) Here are some illustrations of how I play in my garden:

I've never had children, but after all my friends were getting pregnant, I was a much-desired babysitter. To amuse the kids, I replaced the innards of my old 10MB Kaypro computer with spilling-out foliage—remember in the '80s when we couldn't imagine how we would *ever* fill up a 10MB computer disk?—and put it on a low table with two small kiddie chairs. My young visitors would play for hours at a time at the keyboard, pretending they were computing or playing computer games (a precursor to cell-phone

Figure 436. I retired the damaged computer I used to write my 1986 book on edible landscaping to the garden, stripped the innards, and planted the screen and floppy-disk openings. Tiny chairs were added so the kids could play on the keyboard.

Above: Some people use this for recreation.

Figure 437. My BBQ was made from a lot of found objects, particularly round chimney flues. Notice the plates waiting to be washed by the rain—an bachelor's old trick to clean the dishes. The whole area was hosed down first to prevent a forest fire.

Figure 438. Here's the lavender chicken cooking "on the barbie." There's a layer of lavender, then the chicken, followed by another layer of lavender and a cookie sheet. Put five to ten bricks on top to squish down the lavender and steam the meat.

time, I guess). As the use of the disk-drive slot was measured in its "access-time," I planted thyme in it.

For fun, I also set up an outdoor desk as an office with another plant-filled computer and placed all manner of desk paraphernalia on top of it. It was, I told visitors, my writing study for my organic-gardening book; and when the weather eventually totally destroyed it, I told them that my writing career must be over.

BBQs, lots! When I was writing my first book (*Designing and Maintaining Your Edible Landscape—Naturally*, published in 1986), I must have averaged one BBQ meal with friends per week over the four years it took to complete it. Sometimes I had friends over three times in one week, not to mention the evenings with my lady friend(s). (I'm no longer a bachelor, and I now enjoy every meal with Paula.)

I was constantly asked to prepare my famous lavender chicken "*pressada*"—boneless chicken breasts cooked between two thick layers of Spanish lavender (*Lavandula stoechas*; not English lavender, which is too sweet, and some people think it tastes "soapy"). This recipe requires almost one five-gallon bucket of Spanish lavender cuttings, but since I collect lavender varieties, I have lots.

The recipe is simple: put a cookie sheet on the grill. Layer two inches of foliage. Add the chicken breasts. Add another two-inch layer of lavender foliage. Cover with another cookie sheet and place six or more bricks on top of it all. Cook for five to 10 minutes at the most; the heat quickly steams every cell of the meat with flavor. To get some grill marks (the steaming turns the chicken pale), take the chicken out of the sandwich of lavender and place it briefly on the grill. Serve with "lavenchello" (described below).

Here's the most-requested dish at my BBQs:

Bean-Hole Beans

For many years I've been cooking beans underground in a Dutch oven that's buried under 12 inches of sand for 20-24 hours. People whom I invite over for an evenivng of food and conviviality inevitably beg me to prepare a pot of bean-hole beans.

I start by digging a hole big enough to line with bricks and still have enough space to accommodate a Dutch oven with room to spare. My "bean-hole" is lined with bricks (including the bottom) and is about 24 inches deep.

I soak the beans for at least two days and change the water two to four times, I cheat a bit and use a pressure cooker to soften the beans a bit, but make sure they're still firm. (These steps, plus the slow cooking time, seem to make for flatulence-free beans.)

The next part of the "recipe" starts with a large fire fed by scrap wood—hardwoods work the best—to create a huge pile of hot coals and to store plenty of heat in the bricks. This takes three to four hours depending on the wood.

Next I take out half of the coals and put them into a metal wheelbarrow or bucket. I put the Dutch oven filled with the "secret" bean mixture into the hole, with a few layers of wet newspaper on the lid to keep grit from getting into the pot. Then the rest of the coals go back on top of the Dutch oven, and sand is mounded up to 12 inches above it. After the beans have slow-cooked for 20-24 hours, I dig them up with a shovel to be eaten the next evening. (Along with Lavender Chicken "*Pressada,*" of course.)

Note: I lived in a Mediterranean dry-land forest for 30 years. For fire safety when cooking beans this way, I made a 24-inch-high cone of ¼ inch hardware cloth to trap the occasional drifting ember; laid down a graveled weed-free six-foot-radius circle around the bean hole for extra protection; and hosed down all the nearby foliage.

Here, *never before revealed,* is my "secret recipe" for bean-hole beans.

Figure 439. This is the Dutch oven buried under 12 inches of sand. The coals below the sand cook the beans for 20-24 hours. The fossilized spine of a dinosaur proves this is an ancient BBQ site.

Bean-Hole Beans Recipe

2 lbs. beans, kidney, black, or white (slightly pre-cooked in a pressure cooker)
½ lb. bacon (Can be deleted by vegetarians, vegans, the overweight, or potential heart-attack victims.)
3 Tablespoons Mustard
¾ Tablespoon black pepper, freshly ground
½ cup dark molasses
2 (or more) large heads of garlic, peeled and diced into big chunks or left whole
1 cup of your favorite hot sauce
1 Tablespoon chopped rosemary leaves
1 Tablespoon or less fresh-ground French-roast coffee (The real secret ingredient.)

Combine all the ingredients and fill the Dutch oven. My Dutch oven is big enough to feed six to eight people, depending on their waist size. (You can cook the beans for eight to ten hours in a Dutch oven in an oven set to 200°F or less. But where's the fun in that!)

Enjoy!

Booze Garden—Drink Your Garden

Those of you who took up drinking—or more drinking—during the pandemic could probably use some recipes for drinks that can be made from the plants and fruit trees in your own edible landscape. These are drinks you'd seldom see at a bar, treats that only a gardener can grow and serve. Here are a few to savor. (If you don't drink, the landscape can also provide the makings for tasty herbal teas, pies, sauces, chutneys, and sumptuous fruit salads.):

First, a springtime taste-treat for the sober people:

Rhubarb (*Rheum rhabarbarum*)

English gardeners grow various strains of ornamental rhubarbs for their large attractive and colorful leaves, the culinary rhubarb can also be used as decorative foliage. The leaves are enormous, heart-shaped, and crinkled, with an umber-reddish/dark-green mix of colors, and veins laced with the reddish tint found in the stalks. The flower-stalk appears primitive, and almost archetypical, as it unfolds into grape-like clusters of small inconspicuous flowers. When focusing on the harvesting of rhubarb's edible leaf stalks, the prudent gardener cuts out the flower stalk before it gets too tall; this is to conserve the plant's nutrients for the roots and for next year's stalks.

Rhubarb-Strawberry Chutney
(Makes about 3 cups)

Although most Americans are only familiar with commercial chutneys—sweet-hot condiments typically made with mango—there are many types of chutneys, some cooked and preserved, others that can be enjoyed fresh, with no cooking.

This one is halfway between the two; it's cooked briefly but not preserved. Enjoy it with Indian curries, on grilled chicken or pork, and on grilled cheese sandwiches.

1 pound rhubarb, peeled and cut into 1-inch pieces
1 tablespoon fresh minced ginger or 2 teaspoons ground ginger
1 cup granulated sugar
½ cup apple cider vinegar
¾ cup Zante currants
1 pint fresh strawberries, chopped coarsely

In a medium-sized, nonreactive saucepan, combine the rhubarb, ginger, sugar, vinegar, and currants; set over medium heat, bring to a boil, reduce the heat to low and simmer until the rhubarb is tender and the currants are plump, about 25 minutes. Add the strawberries, stir, and taste the chutney. Add additional sugar or vinegar to correct the sweet-acid balance, simmer for five minutes and remove from the heat.

This chutney maybe be served warm or chilled, and will keep in the refrigerator for up to 10 days.

Recipe © 2018 Michele Anna Jordan, a writer based in Sonoma County, CA and the author of 24 cooking books to date. Visit her website at: michele@micheleannajordan.com.

Now, on to the drinkies:

- **Lavender Martini**

This concoction won over one rabid lavender skeptic with the first sip.

Simple, just take a sprig of English lavender (*Lavandula angustifolia*) only, and use it to stir an ice-

Figure 440. To prevent woody lavender shrubs, you must prune the plants way back—"hard." Notice how much of the stems are removed, leave just two or so inches of foliage. (The lavender cut back in this photo is the cultivar 'Grosso.' The other is 'Provence.' Neither is English lavender.)

cold martini.

- **Mojito**

Ingredients
3 ounces spiced rum
8 spearmint mint leaves
1 Tablespoon. sugar
½ ounce lime juice
2 ounces club soda water

Place mint leaves in bottom of glass and pulverize.

Add crushed ice, spiced rum, sugar, and lime juice.

Add soda water and garnish with mint leaves.

- **Mulberry Mule**

Ingredients

1/2 oz. lime juice

2 oz. vodka

6 oz. ginger beer

2 tbsp. muddled fresh mulberries

Squeeze lime juice into a copper "Moscow Mule" mug and drop in the spent rind.

Add 2 or 3 ice cubes, then pour in the vodka and fill with cold ginger beer.

Add the fresh muddled mulberries.

Serve with a stirring rod.

- **Persimmon Margarita**

Ingredients
4 ounces (1/2 cup) very-ripe Hachiya persimmon puree
3 ounces (1/2 cup) tequila
1 ounce lime juice
3 Tablespoons. simple syrup (buy it so you can easily make this drink) Sugar and ground cinnamon

Chill two margarita glasses in the freezer.

Spread a thin layer of sugar on a plate, and sprinkle it with a bit of flaky sea salt and ground cinnamon.

Add the persimmon puree, tequila, lime juice and simple syrup to a cocktail shaker.

Fill the shaker with ice and shake vigorously until cold.

Get the glasses from the freezer and press the rims into the sugar mixture.

Pour into two glasses. Garnish with round of fresh lime if desired.

Figure 441. Hachiya persimmons on an unpruned tree. Prune to keep the limbs from breaking.

- **Asian Pear Sparkler**

(Be sure to blend the pear and cucumber first.)

Ingredients

4 ½ cups 1-inch-cubed Asian pear (about 3 pears)

1 cup 1-inch-cubed peeled cucumber (1 large)

2 Tablespoons fresh lime juice (about 2 limes)

2 (12-ounce) bottles sparkling apple drink

6 chilled Asian pear slices

Place first 3 ingredients in blender; process 1 minute or until almost smooth.

Strain mixture through a sieve into a bowl, pressing with the back of a spoon; discard solids.

Cover and refrigerate 1 hour or until thoroughly chilled.

Pour 1/2 cup mixture into each of 4 glasses.

Top each serving with ¾ cup sparkling apple drink, and stir gently.

Garnish each glass with 3 slices of pear.

- **"Lavenchello"**

This spectacular drink, a victorious marriage of lemoncello liqueur and lavender, was "invented" by Suzanne Adams and Kandis Kozolanka for our monthly convivial meals. The recipe takes some effort and storage time, but is *really* worth the industry.

Ingredients
1 quart vodka
zest of 2 lemons
handful of chopped lavender leaves
¼ cup lavender flower heads
1 1/2 cups spring water
¾ cup sugar

Remove the yellow rind of the lemons with a zester, or peel and slice very thinly.

Place in a glass container that will hold at least a half-gallon. Add the lavender leaves.

Put the lavender flower heads in a blender or coffee-grinder and whir until powdery.

Add to jar and pour in the vodka. Stir well and cover.

Store in a cool dark place for at least a week.

Line a sieve with several layers of cheesecloth or usea coffee filter to strain the liquid into another large clean glass jar.

While the liquid is straining, warm the 1 1/2 cups of spring water and add the sugar. Stir well until completely dissolved to make a syrup. The strained product will be either a rosy lavender or pale lavender-blue.

Cool the syrup to room temperature.

Add the syrup to vodka mixture and stir well. Store for a few more days stirring occasionally.

Decant into clean bottles and cap.
Store in freezer. Makes one quart.

When you take it out, the lavencello will be slushy, pour into aperitif glasses. Salut!

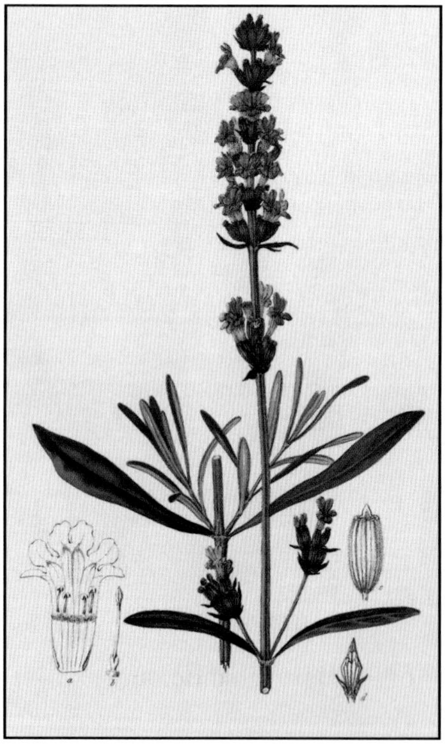

Figure 442. English lavender (*Lavandula angustifolia*)—pictured here—is the lavender you use to make lavenchello. Don't bother to remove just the flower petals; use the entire flowerheads stripped off the stem. Use the younger leaves pictured here.

And now, for your non-alcoholic pleasure—"mocktails." All the fun without the alcohol.

Mojito Mocktail
Makes 1

Ice
½ lime, cut into 4 pieces, muddled
6 to 8 spearmint leaves, muddled
6 to 8 cilantro leaves, muddled, optional
1 tablespoon simple syrup
1 ounce lime juice
Sparkling water or club soda
Small mint sprig

Fill a glass about half-full with ice. Add the muddled lime, spearmint, and cilantro, followed by the simple syrup and lime juice. Stir and top off with sparkling water. Garnish with the mint sprig and enjoy.

Variation: For a cocktail, add 3 ounces of spiced rum along with the simple syrup and lime juice.

Persimmon Margarita Mocktails
Makes 2

½ cup granulated sugar
3 teaspoons ground cinnamon
1 teaspoon kosher salt
Ice
½ cup (4 ounces) persimmon purée, from very ripe Hachiya persimmons
2 ounces freshly squeezed lime juice, plus more as needed
3 ounces freshly squeezed orange juice, preferably from Cara Cara oranges
2 tablespoons simple syrup, as needed
Small lime wedge
2 slices of lime

Put the sugar on a flat plate, add the cinnamon and salt and use a fork to mix it together. Set aside.

Put two cocktail glasses in the freezer.

Fill a cocktail shaker half full with ice. Add the persimmon purée, lime juice, and orange juice, close the shaker, and shake very well, until the outside is frosty. Taste and if it is too tart, add the simple syrup, close the container, and shake again. If the mixture is too sweet, add another ounce of lime juice.

Working quickly, remove the glasses from the freezer and coat the rims with the lime wedge. Dip the rims of both glasses into the cinnamon sugar mixture and set the glasses upright. Pour in the mocktail, garnish with a slice of lime, and enjoy.

Mulberry Mule
Makes 1 Mocktail

1 slice of fresh ginger, about the size of a quarter, muddled
1 ounce freshly squeezed lime juice
1 small wedge of lime, muddled
Ice
1 bottle best-quality ginger beer, such as Fever-Tree brand
2 tablespoons fresh muddled mulberries

Put the ginger, lime juice, and lime wedge into a Moscow Mule mug.

Add 3 or 4 ice cubes.

Pour in the ginger beer, leaving a half inch or so of space at the top.

Stir in the mulberries and enjoy right away.

Variation: For a cocktail, add 2 ounces of vodka along with the lime juice.

"Muddling" is the term barkeeps use to explain how they pound herbs, spices, and other ingredients for cocktails. If you do it frequently, you should have a muddler - an implement nearly identical to a wooden pestle – and a dedicated vessel, such as a wooden mortar. But you can muddle ingredients directly in a cocktail glass, too, provided it is a sturdy glass and you are not too vigorous. To muddle an ingredient, put it into the vessel and pound it to release its aromas and juices.

Pear Ginger Shrub
Makes About 3 pints

3 ounces fresh ginger, pounded
5 or 6 very ripe pears
1 quart apple cider vinegar or pear vinegar
Simple syrup, as needed

Put the pounded ginger into a large glass or ceramic vessel.

Add the pears and pour in as much of the vinegar as the jar will hold. Stir, cover, and refrigerate for at least 3 days and as long as 3 or 4 weeks.

Line a strainer with 3 or 4 layers of cheesecloth and set it over a deep bowl. Strain the shrub into it.

Discard the ginger. Save the fruit for another use or discard it, too.

Taste the shrub and if it is too tart or flat, add simple syrup, a tablespoon at a time. Stir and taste after each addition.

Pour the shrub into glass containers and close with a non-metallic top. It will keep for months.

Suggested Uses:

•Use in place of alcohol in simple cocktails.

•Add a splash to a glass of sparkling water or sparkling wine.

•Use to make a sweet-tart vinaigrette by adding a pinch of ground cardamon, a minced shallot, salt, pepper, and a late-harvest olive oil.

Simple Syrup
Makes about 3 cups

3 cups granulated sugar

Put the sugar into a heavy saucepan, add the sugar, and set over high heat. Do not stir.

Bring to a boil, reduce the heat, and simmer for 4 to 5 minutes, until the sugar is completely dissolved and the syrup is transparent.

Cool completely, pour into a clean bottle, and store in the refrigerator. Simple syrup will keep indefinitely.

***

For more ideas, see:
The Drunken Botanist by Amy Stewart
Cocktails, Mocktails, and Garnishes from the Garden by Katie Stryjewski
The Herbalist's Happy Hour by Susan D. Evans

The mocktails and shrub were crafted by Michele Anna Jordan, a writer based in Sonoma County, CA and the author of 24 cooking books to date. Visit her website at: michele@micheleannajordan.com.

Eccentric Gardens
I think no garden should be without the special touch of the gardener. And why not make it truly unique. Be a bit eccentric! Or, be a lot eccentric. Feed the soul as well as the tummy. Here are some illustrations of my tomfoolery.t

Figure 443. Clockwise from the top left: Clay towers by Ben Katz and bones in my garden; three clay towers lining up, facing solar south; and a chaos tornado of writing—notice the writer's mind was, in 1986, (not now) fueled by alcohol.

I Give Up!
The author's new front yard.

Finito!
The author after finishing this book.

Appendix 1
Comfrey

This comfrey text is from the **Balkan Ecology Project**. In the words of Paul Alfrey: "... (it) is a permaculture-inspired, grass-roots project based in southeastern Europe, in Shipka, Bulgaria. Our mission is to develop and promote practices that provide food and other resources while enhancing biodiversity." I really like this group because they are actively recording the harvests and time spent with polyculture gardens—the closest to the science of permaculture that I know about. Since 2014 they have been recording the inputs and outputs from their home garden annual polycultures. In 2015 they started to record data from a newly established market garden. Here's what Paul says now (2020): "In 2020 we shift focus from our 5-year study of annual herb and vegetable polycultures to perennial systems such as forest gardens and agroforestry. We are undertaking a multi-year study looking at the economics of running such a garden, and how productive it can be in terms of yields, income and biodiversity." See: https://wwwbalkep.org. Their phone number in Bulgaria is +359988342649.

Comfrey

Comfrey (a blosssom above) beds provide copious amounts of biomass rich in nitrogen, phosphorus, potassium, calcium and many other valuable plant nutrients. (The npk of dried comfrey leaves can content be 1.8 - 0.5 - 5.3, [chicken manure is 5 - 2 - 1] *RK*) The biomass is best used to feed crops in the garden via compost or a soil drench of fermented leaves.

Planting out the Comfrey Patch Spacing - The plants should be spaced 24" apart in rows and 24" apart at diagonals between rows. Plant the rows 6" from the edge of the beds.

Maintaining the Comfrey Patch Cutting - In the first year allow the plants to establish so that the roots develop well. The following year the cuttings can begin. You can scythe the beds for a quick harvest or cut each plant individually with a pair of clippers or a hand sickle. Cut to 2" -3" from ground level. The leaves are prickly so if you have sensitive hands wear gloves. You can cut the comfrey as the flowering stalks emerge, up to 4 times a year. Allow the plants to flower at least once during the season to provide bee fodder to a range of native bees as well as honey bees. Leave the last flush of leaves before the winter so that invertebrates can find winter shelter in the undergrowth.

Feeding - After you have cut the comfrey, mow the pathways with a clover cover between the beds and empty the contents to the base of the comfrey plants. Any mowings from the surrounding area can be used in a similar way. Or for extra feed and to increase yields, undiluted urine can be applied to the plant at a rate of about one quart per plant twice per growing season. Urine is an excellent feed for comfrey. (Drink plenty of Guinness Stout. *RK*)

Figure 444. This is not 'Bocking 14' comfrey. Buy the true 'Bocking 14' via the Internet. It's often sold out.

Irrigation - Comfrey will produce more biomass if irrigated, and in dry climates it's essential to irrigate. Comfrey plants wilt very fast in hot conditions and will stop photosynthesizing at this point. The beauty of biological systems is that if they are managed correctly, each year the soil in the beds will improve, resulting in less water required and higher yields.

Comfrey Beds Species Overview

***Symphytum* x *uplandicum* 'Bocking 14'**

Overview - 'Bocking 14' is a hybrid comfrey plant developed in the 1950s by an original ecotrepreneur Lawrence D. Hills specifically for its ability to produce tremendous quantities of biomass. Comfrey is adaptable to many soils but prefers moist, fertile soils. Thin soils over rock will give a poor crop, but on light sands and loams, this crop will be productive if adequate nutrients are present. Comfrey productivity is not very sensitive to soil pH, but the highest yields occur on soils with a pH of 6.0 to 7.0.

Uses – Insects The beautiful flowers are highly attractive to a wide range of bees including *Apis mellifera* - the western honey bee. Predators such as spiders, lacewings, and parasitoid wasps associate with this plant. It's best to leave some plants alone in order to sustain pest-predator relationships.

Uses - Mulch - Freshly cut comfrey leaves make good mulch because they are high in nitrogen, so they do not pull nitrogen from the soil while decomposing, as high-carbon mulches like straw and leaves do. Comfrey's high potassium content makes it especially beneficial for vegetables (such as tomatoes, peppers, and cucumbers), berries, and fruit trees. With adequate feed and watering, we have seen yields 4-6 pounds per plant per cut. Plants can be cut up to four times in a year.

Uses - Liquid Fertilizer - What we like to call "Comfert." Fill a barrel, preferably with a bottom tap and a gauze on the inside (to prevent clogging) about 3/4 full with fresh-cut comfrey and add water to fill the barrel. Cover it, and let it steep for 3 to 6 weeks. The smell from the resulting liquid is far from attractive so approach with caution. The tea may be used diluted by half or more. Don't apply before heavy rain is forecast as most of the nutrients suspended in the liquid will wash straight through the soil. For the best results apply the feed to your vegetables when they are in most need of the extra fertility. This will be different for each crop, for example, tomatoes are best fed when they are setting fruit and then any time during the fruiting period. Applying comfert before this can be counter-productive and make your plants more susceptible to pest problems. The black slurry at the bottom of the barrel can be dispersed evenly back over the comfrey patch.

Uses - Liquid fertilizer concentrate: "Comfert Plus" can also be made by packing fresh-cut comfrey

Figure 445. Here are three ways to make "Comfert" as demonstrated in the lower photograph at HDRA (Henry Doubleday Research Association) in England. Each lets the comfrey rot for days or weeks and then drain the viscous liquid off the bottom of the container. Above, Dick Barwick is draining liquid comfrey from the bottom of his small vat. A pile of tiles on top of the comfrey leaves kept them submerged. Dick is a vegan. He relys on "Comfert" to fertilize his plants. He also rotates legumes with all other crops. He uses no manures.

tops into an old bucket, weighing them down with something heavy, covering tightly, and waiting a few weeks for them to decompose into a black slurry. You can put a hole into the bottom of the bucket and collect the concentrate in another container as it drips out. Dilute this comfrey concentrate about 15 to 1 with water, and use as you would Comfert. You can seal this concentrate in plastic jugs until you are ready to use it. (Beware, it *really smells. RK)*

Biodiversity - The last growth of comfrey leaves can be left uncut before the winter. The leaves can provide habitat for nesting spiders and invertebrates. Allow the plants to flower at least once between cuts to provide bee fodder to a range of native bees as well as honey bees. Cutting back the flowering comfrey when neighboring crop plants are in flower will drive the pollinators to your crops, increasing the likelihood of successful pollination.

Propagation - Root cuttings of pieces approximately 1"-3" are the only way to propagate 'Bocking 14'. The cuttings should be grown on in small pots and planted in the spring as soon as the first leaves emerge.

Planting Material - You can plant out with crown divisions or root cuttings. A crown division is simply putting a spade through the center of a mature comfrey plant and transplanting the divided sections. Divide the crown into quarters and these establish very well in the first year. Do not cut the plants in the first year in order to allow an extensive root system to develop.

(Paul Alfrey's summary ends here. What follows is my text.)

Figure 446. A lithograph from the 1700s of a white-flowering comfrey. The roots are sideways, but that may be due to how it was excavated.

Comfrey's Deep Roots

To begin with, in the scientific literature, how deep is "deep".

In Ecosystems (2002) 5: 385-398 (Springer-Verlag. Tree Species Effects on Calcium Cycling: The Role of Calcium Uptake in Deep Soils, Feike A. Dijkstral and Mark M. Smits), there's a reference to "...deep soil ([the] rooting zone below 20 cm (eight inches) mineral soil)...." This is not especially deep by most people's assumption.

"Fine-root Dynamics in Forest Trees" by H. Persson (in Plant Root Systems and Natural Vegetation by H. Persson and I.O. Baitulin (eds): defines deep as: "... upper parts of the soil profile (LFH [top soil] - 40 cm (16 inches) mineral soil."

Some sources, as in "Nutrient Uptake as a Contributing Explanation for Deep Rooting in Arid and Semi-Arid Ecosystems," R. L. McCulley, E. G. Jobbágy, W. T. Pockman and R. B. Jackson,

Figure 447. I dug up this young comfery plant. All the roots were growing sideways, They just drooped down some after lifting it from the soil. What looks like a taproot is just a root coming directly toward the camera.

Oecologia, Vol. 141, No. 4 (Dec. 2004), pp. 620-628, say "Deep-rooted (>1 m= 3 feet) plants are ubiquitous in arid ecosystems." (Weaver and Albertson 1943; Canadell et al. 1996; Schenk and Jackson 2002)

Also, from Ecological Monographs, 72(3), 2002, pp. 311–328, 2002, "The global biogeography of roots;" H. Jochen Schenk and Robert B. Jackson: "Rooting depths were estimated by calculating the median rooting depth of deeply rooted (>2m) plant species..."

The practical applications? One needn't water as deeply as popular opinion/mythology indicates, since many plants, like comfrey, may only need decent moisture in the upper 8 to 12 inches of topsoil. In almost every case, the absorption by roots growing in the top 2 feet of soil equals about 50% of the water used in most of these agricultural studies.

The most important takeaway for me is this: it's the amount of roots, not their depth, that contributes most to moisture absorption. Deeper soils contain fewer roots to absorb both moisture and nutrients.

Figure 448. It's common for permaculturturists to recommend planting comfrey next to fruit tees. It is supposed to have the foliage chopped down and dropped to make a fertile mulch and soil. Here, in the dry west, the foliage just desiccates and does not make a fertile mulch. Many permaculturists don't plant comfrey further from the tree as the dripline expands—they are unclear on the growth of new roots. I know comfrey is a heavy feeder and I think it competes with the fruit tree for nutrients. Also, here we see a new young plant spreading away from the tree. It can really become an invasive plant if tilled.

From "Root Water Uptake and Profile Soil Water as Affected by Vertical Root Distribution" by Gui-Rui Yu, Jie Zhuang, Keiichi Nakayama and Yan Jin, Plant Ecology 189:15–30, 2007: "Rooting depth of the roots having highest length-density was more vital than the maximum rooting depth for maintaining water uptake rate under dry versus wet-soil conditions. In the wet soil, soil water was absorbed from nearly the entire field profile but with the majority from the most densely rooted layer, and occurrence of a fraction of roots in deep soil increased root water uptake. In the dry soil, the water uptake only significantly took place in the most densely rooted

(continued on page 428)

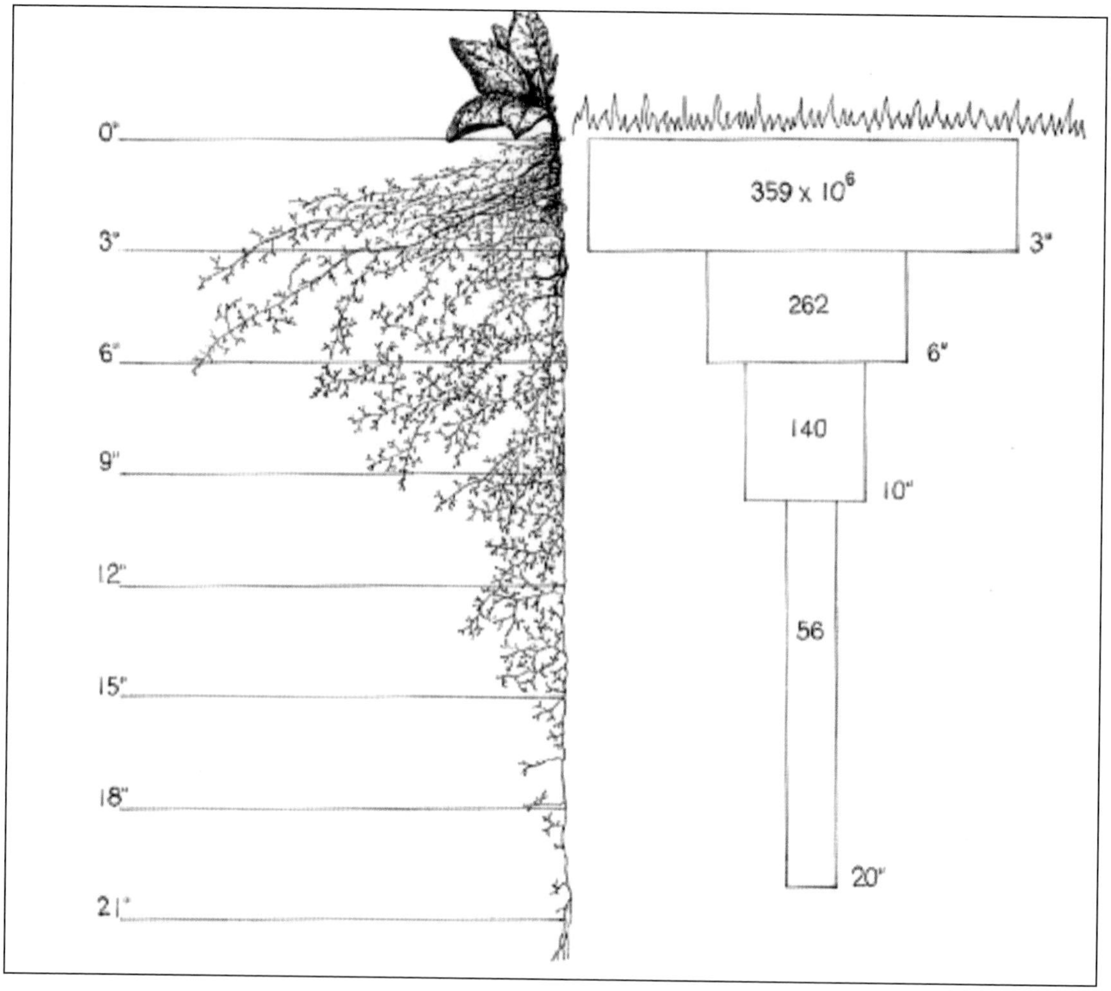

Figure 449. This important illustration shows the correlation between the beneficial bacteria and the number and placement of the roots of a spinach plant. Here the roots flourish in the top three to six inches of soil, which contain many more bacteria than are found at lower levels. (The six-inch zone contains about one-third fewer bacteria than the top three inches and is populated by fewer roots.) Again, the more roots in the aerobic zone, the greater the absorption of nutrients. (From *Understanding Roots: Discover How To Make Your Garden Flourish*; Robert Kourik, Metamorphic Press, 2015.)

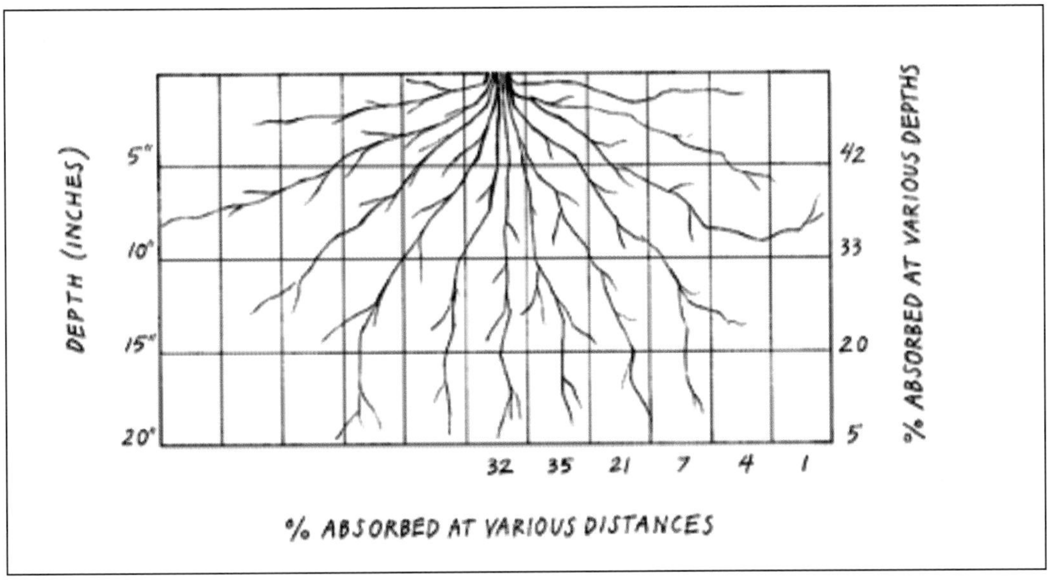

Figure 450. This chart provides an example of where the water is absorbed by plant roots. At 5 *inches* (not feet) deep, roots absorb 42 percent of the moisture, as compared to 20 percent at the 15-inch level. (Here the roots beneath the trunk gather 32 percent of the moisture, in comparison to 7 percent 3 feet away.)

Keep in mind that there are considerably fewer roots in the deeper zones, so it's still possible that upper roots gather more water, and therefore nutrition, due to a greater root mass.

In a different report we read: "The availability of important plant nutrients in the soil, with the exception of Mg and Zn, was higher in the upper 50cm (20 inches) of the soil profile where the majority of roots were found." ("On the Maximum Extent of Tree Roots" by E.L. Stone and P.J. Kalisz, Forest *Ecology and Management*, 46 59-102, 1991.)

In "*Root Location of Fruit Trees and its Agrotechnical Consequences*;" Akademiai Kiado, Budapest, 1986, J. Tamasi simply states: "The extent of water consumption is proportionate to the volume of roots."

"...[when] plants were competing with each other Nitrogen uptake was proportional to the amount of root on the plants." "Root density and competition for nutrients." (Andrews, R. E.; Newman, E. I., *Oecologia Plantarum*, Vol.5 pp.319-334, 1970.)

Keep in mind that "Roots at deeper depths function more for water uptake and support than for nutrient uptake."
(From "Variation In Sugar Maple Root Respiration With Root Diameter And Soil Depth," by Kurt S. Pregitzer, Michele J. Laskowski, Andrew J. Burton, Veronica C. Lessard and Donald R. Zak, *Tree Physiology* 18(10): 665-670; October 1998.)

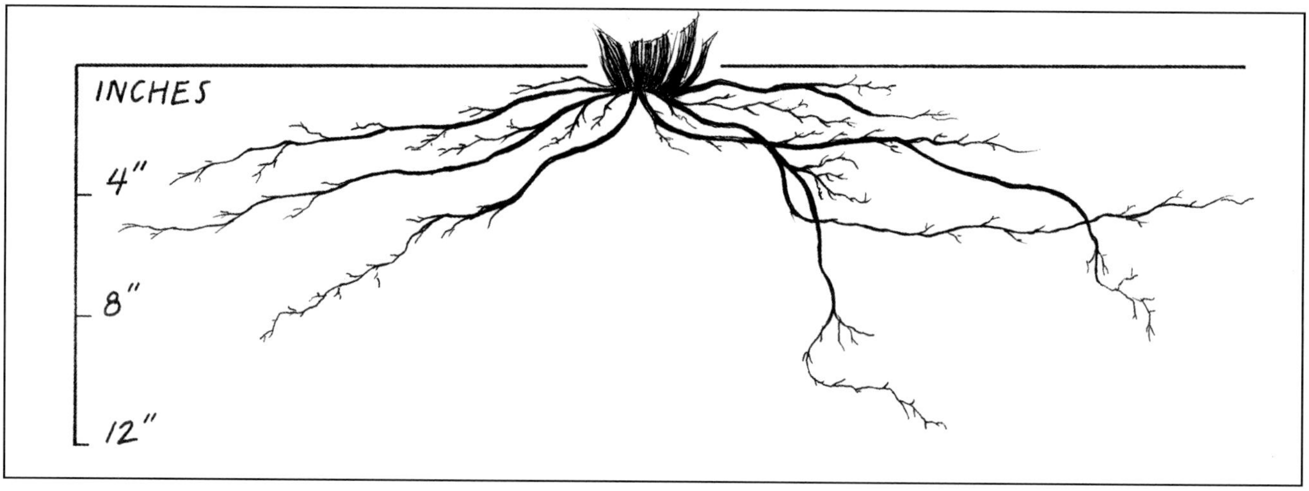

Figure 451. How deep do comfrey roots go? I've been looking for 20 years to find the source of the "10-foot-deep" references. No luck. When digging comfrey roots, you can see that the roots can definitely grow deeper than eight to 16 inches. Here is a drawing showing the depth of some young comfrey roots: From: *Understanding Roots: Discover How To Make Your Garden Flourish*; Robert Kourik, Metamorphic Press, 2015.

A more reasonable depth for comfrey roots is reported by Phytopharmaceutical Cooperation, Bonn, Germany as "12 inches deep in the earth."

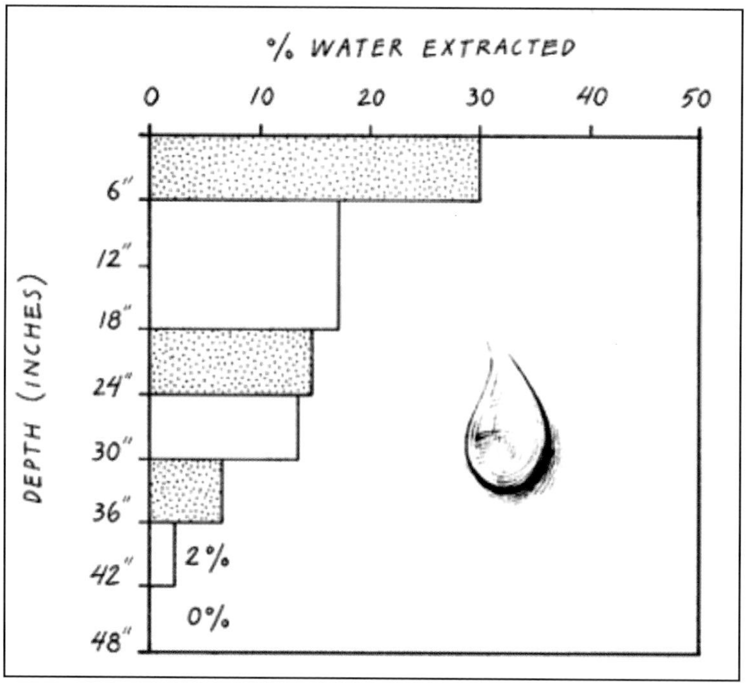

Figure 452. This chart is revealing. It helps dismantle a popular myth about watering plants. The myth is: "A deep soak on an infrequent basis is the best way to water trees and shrubs." If deep is more than two feet, you're wasting water. Thirty percent of the moisture is absorbed in just the top six *inches* of the soil.

(continued from page 425)
layer; and occurrence of the highest length density roots at soil depth of 30-cm to 40-cm (12-16 inches) where soil was less dry helped maintain a certain rate of water uptake by plants as soil progressively dehydrated..."

In many cases, plants "gather" more water near the soil surface than they do in deeper soils. This chart (from "Spatial and Temporal Distribution of Soil Water Balance for a Drip-irrigated Almond Tree" (by L. Andreu, J.W. Hopmans and L.J. Schwankl, Agricultural Water Management, 35:123-146, May 1997) shows that the roots in the top six inches absorb three times more moisture than those growing 36 inches deep.

From Dr. John Weaver (Professor of Plant Ecology at the University of Nebraska and of vegetable-root drawing fame): "The materials necessary for food manufacture were

(continued on page 430)

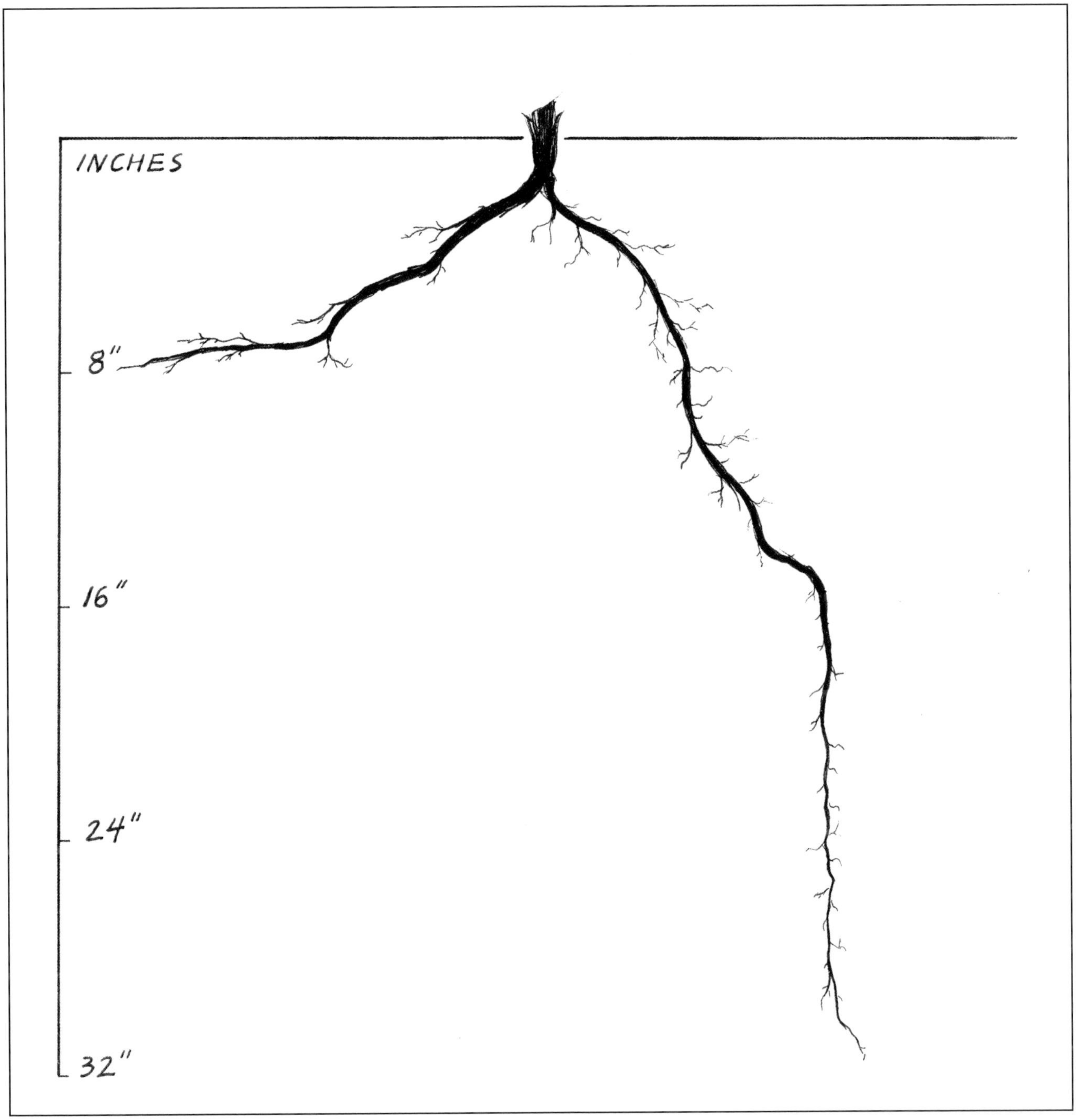

Figure 453. The comfrey root system above shows a "multi-year" plant that grew from two inches below the soil's surface. The more spindly root on the right penetrated over two feet into the soil, but should probably not be considered a taproot due to its weakened growth. However, it does seem to have more feeder roots to absorb at least moisture and thereby some nutrients. So, the jury is still out. From: *Understanding Roots: Discover How To Make Your Garden Flourish*; Robert Kourik, Metamorphic Press, 2015.

(continued from page 428)

taken from the deeper soil in considerable quantities, although to a lesser extent than from the soil nearer the surface, which the roots occupy first."

The practical applications?

One needn't water as deeply as popular opinion/mythology indicates, since many plants, like comfrey, may only need decent moisture in the upper 8 to 12 inches of topsoil. In almost every case, the absorption by roots growing in the top 2 feet of soil equals more than 50% of the water used in most of these agricultural studies.

The most important takeaway for me is this: it's the amount of roots, not their depth, that contributes most to moisture absorption. Deeper soils often contain fewer roots to absorb any moisture and nutrients present.

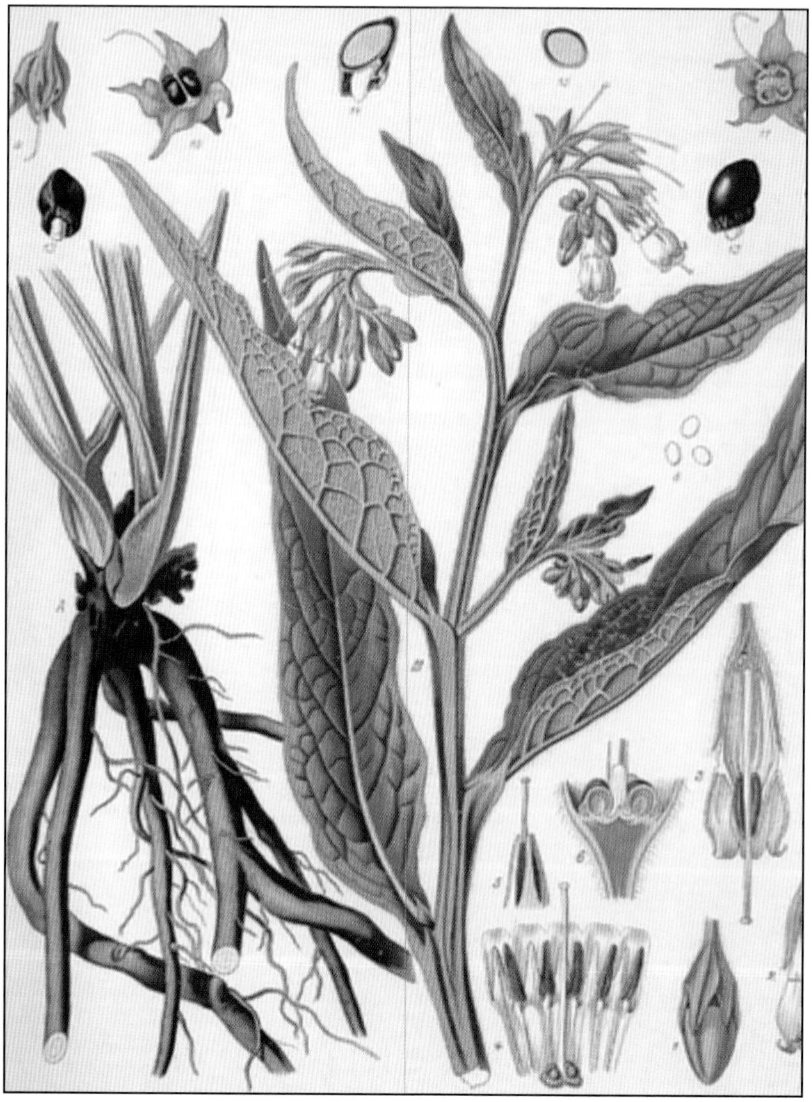

Appendix 2
Vegan Gardening

Guidelines for Vegan-Organics – the Basics

Rosa Dalzie O'Brien's (of surface cultivation fame, see page265-266) son Kenneth, was also a follower of vegan gardening (he was not a farmer). He published two comprehensive books on the subject: *The Kenneth Dalziel O'Brien Veganic Gardening Method* and *Veganic Gardening: The Alternative System for Healthier Crops*.

The largest network of vegan gardeners in the world is the Vegan Organic Network (VON, in England). What follows is their overview of vegan food gardening and farming as described on their excellent website www.veganorganic.net and www.stockfreeorganic.net, written by volunteers and vegan organic farmers from VON. I quote it because it is thorough, represents the technique with accuracy, and is more articulate than I. And, because it uses no manures, requires green manuring, rotations, and prefers surface cultivation, it is more sustainable than most gardening.

"Vegan-organics is, briefly, any system of cultivation that does not use artificial chemicals and sprays, GMOs, livestock manures and animal remains from slaughterhouses or fish processing, etc. Instead, fertility is maintained by vegetable composts, green manures, crop rotation, mulches and any other method that is sustainable, ecologically—viable and not dependent upon animal exploitation. This will ensure long-term fertility, and wholesome food for this and future generations.

While conventional cultivation relies on synthetic chemicals and animal products, traditional organic production also generally relies on animal wastes and by-products. Both can be seen to have the exploitation of living creatures in common, with conventional agriculture also leading to terrible health, social, economic, and environmental problems. Vegan-organic methods involve none of these drawbacks. Many people, some of whom are not themselves vegan or vegetarian, are coming to appreciate these benefits of animal-free growing.

Growing Without Cruelty

Organic growing involves treating the soil, the growing environment and the world environment as a resource to be husbanded for future generations, rather than exploited in the short term. The maxim of organic growing is to feed the soil and the soil will feed the plants.

Vegan-organics means doing this without any animal products at all, which is not difficult when you know how. The guidelines below do not attempt to be fully comprehensive.

Preparing the Soil

Most bacterial activity and soil organisms live in the top few inches of the soil helping to create drainage and build up fertility. Constantly digging the soil and exposing it to erosion from the elements disturbs the natural balance, resulting in the loss of availability of organic matter and the breakdown of soil structure.

When cropping you need to constantly replenish soil organic matter levels by the addition of plant-based composts, mulches and by using plants grown to improve fertility i.e. green manures. With the exception of green manures, digging is not necessary for incorporating materials as organic matter spread on the surface will soon be drawn under by worm activity and plant nutrients will be available at root level and not be buried out of reach.

[Despite surface cultivation as the goal], it is however advisable to dig heavy clays, as exposure to frost and rain can result in a more workable soil, especially over winter. Compaction, caused by standing or running the wheelbarrow over the soil when it is too wet, can be avoided by making permanent beds that are never stood on. These can be timber lined as raised beds with soil from the paths being placed on the beds to raise them. Where soil has been compacted it may be loosened by forking.

When clearing land for the first time it is important to remove all the perennial weeds such as bind weed, couch grass, ground elder and horsetail whose roots are deep and wide-spreading. Dandelions, docks and thistles have a long taproot. You remove them through a process of digging a trench a spade's depth across an area. Work backwards by pushing the soil continually forward, almost like a sieving process so that you can inspect every part of the soil and remove all the weeds. Do not stand on your new soil tilth. Finally when all the weeds are removed rake the soil level with the back of the rake to reduce hillocks and mounds. All weeds will re-grow from a small piece left in the ground so it is important that you are thorough.

After digging, the soil will soon be covered by germinating weeds blown in or brought in by birds etc. therefore it is important to manage weeds so that they do not out-compete the crop. Weeding is a constant task for the vegan-organic grower but it is always easier to hoe small weeds when they are at the white, stringy stage rather than having to uproot established weed clumps by hand.

Weeds are not all bad as they contain nutrients that have been brought to the surface level via the roots. Rather than waste this valuable resource, annual weeds (if not seeding) can be composted and perennials can be chopped and added to water to make a liquid feed, as outlined below.

Mulching and No Dig

Mulching is the method of applying organic matter to the soil surface, providing a constant supply of material to break down, suppressing weed growth, ensuring more even soil temperature and moisture. Mulch can be applied at any time except when the soil is frozen or dry.

When using no-dig methods, weed-infested sites can be cleared by firstly covering with a barrier e.g. cardboard boxes, flattened and wetted, newspapers (avoiding slick toxic-colored inks) or carpets (only those made from natural materials such as cotton and avoiding synthetic mixtures and foam backing). A sufficient quantity of organic material is then spread over the barrier. Plants can be planted into holes cut into the barrier. Any weeds growing out of the planting holes can be removed by hand. After two years of covering weed-infested land, most perennial weeds should have died off.

Organic material

Obtaining sufficient organic matter is often the most difficult aspect of vegan-organic gardening. These are some potential sources.

- Grass cuttings - rich in nutrients, mix with compost or fork into the top of the soil, or use as a mulch.
- Old hay - ideal for mulching, containing a balance of nutrients. It must be at least a year or two old so that no seeds will germinate.
- Spent hops - an excellent soil conditioner, containing some nitrogen, available from many breweries.
- Tree leaves - obtainable from your own trees or from local councils [cities]. Be careful of street trees, which may contain litter, dog feces or vehicle pollution.
- Comfrey - leaves can be cut when ready and added to water to make a liquid feed (see 'liquid feeds' below) or applied when wilted; you can grow your own comfrey for this purpose. [ed. Keep in mind that if not managed correctly it can become a noxious weed.]

- Composted garden waste - some [cities] now have recycling centers where garden waste is shredded, composted, bagged and sold as soil conditioner.

- Compost - all organic material from your own kitchen and garden or from local suppliers can be composted. Using cooked food is not recommended.
- Seaweed - use washed up seaweed, stack and allow rain to wash out salt, or buy seaweed meal.

Green Manures

There are many plants that can be grown in order to increase the fertility and humus content of the soil. Basically you sow the seed and allow the plant to grow then cut it down before it flowers. Many gardeners will then incorporate this into the soil, though it is also possible to either let it break down in situ or to remove it and compost it. [Called cover cropping in the US.] A number of green manure crops, in particular peas, beans, clovers, favas [as seen in the photograph on the first page of Appendix 2] and winter vetch, will enrich the soil with nitrogen as well as providing organic matter. Green manures have many benefits and can be grown as catch crops in land that would otherwise be empty. Species to consider include buckwheat, cereal rye, winter vetch, clovers and alfalfa. Winter vetches are good for heavy soil and lupines are good for light sandy soil types.

Liquid Feeds

When properly carried out, organic systems should not need supplementary additions as the soil provides all the necessary nutrients. However, getting enough potash for your luscious fruiting crops - tomatoes, cucumbers, aubergines and peppers - may prove difficult. It is possible to buy vegan liquid feeds or you can make your own much more cheaply. The process is simple; you fill a container, e.g. watering can with plant material such as comfrey, nettles or weeds. You then fill the container with water (and you will be surprised how much water will fit in) and leave it for a week or two. It will then be very smelly, and you dilute it by perhaps 5-1 with water and then pour it on the soil around the plants. Alternatively you can increase the dilution and spray it over the leaves of plants, preferably in the evening or on cloudy days.

There are some materials that we would not recommend:

- Peat - peat bogs are still being destroyed in order to provide peat for composts, mulching etc.
- Calcified seaweed - this material is obtained from the temperate ocean equivalent of coral reefs and is being harvested in an unsustainable way; it is being phased out of organic practice.
- Spent mushroom compost - just in case you were not aware of it, this material is usually made from animal manures, especially horse manure. It is also heavily polluted with all the chemicals they use in growing mushrooms.
- Coir (coconut fiber) - this is being used in increasing quantities as a peat replacement. Although this is an excellent soil conditioner it is imported from countries that need the fertility of their own soils improved. As an alternative to peat in composts, the best option may be to use coir.
- Slaked lime - is a by-product of the chemical industry. (Dolomite and ground limestone have to be quarried, so should be used in moderation.)

Pests and Diseases

Crops grown in a soil containing high levels of organic matter and plant nutrients with a good soil structure and texture will be healthy plants and will therefore have a greater resistance to "pests" and diseases which tend to attack weaker, sappy plants.

There are many organic sprays available for treating pests and diseases in plants. These are not recommended because whilst they are not based on artificial chemicals, they can still be poisonous (e.g. copper sulphate) to non-target insect species, many of which are very beneficial in the garden. If you want to consider the alternatives to spraying, then it is important that you adopt a holistic approach i.e. use all of the following growing suggestions to increase your likelihood of healthy crops. Think of "pests" as "competing organisms" and you will feel better towards them.

Rotations

Pests and diseases spread in monoculture so it is important that you adopt at least a four course rotation. A typical example is to split your garden into four plots:

1. Potatoes, tomatoes and cucurbits (zucchini, squash, cucumbers)
2. Legumes (peas and beans) and alliums (onions, garlic, leeks)
3. Brassicas (cabbage, cauliflower, broccoli, Brussels sprouts, turnip, radish, rutabaga, oriental salad leaves)
4. Salads, roots and miscellaneous (lettuce, chicory, beetroot, carrot, parsnip, sweet corn)

Each year the crops are rotated into different parts. So in year one potatoes are growing and they will be followed the next year by peas etc.

Providing Habitat for Predators to Live

Providing permanent predator belts, including incorporating undisturbed perennial plants, shrubs and trees for insects and beetles to over winter are all good ideas. Hedges are the best example of this. It is also possible to provide annual predator belts by leaving uncultivated strips between beds. Beneficial insects, birds and mammals will not inhabit your garden unless they have a water body in which to drink and places to shelter. Think about hedgehog shelters, log piles, bark piles (favored by ladybirds), compost heaps, lacewing hotels and bird boxes.

Planting Attractant Species for Beneficial Insects

By planting aromatic species amongst your other plants you will find that the incidence of pests and diseases will fall. Chamomile, garlic and many of the Mediterranean herbs are very useful here. Plants such as borage, limnanthes (poached egg plant), echium [Ed. invasive] and members of the umbellifer family (let several parsnips run to seed) attract predatory insects such as hover flies, parasitic wasps and other insects. Phacelia will attract bees and hoverflies and growing marigolds on the bed perimeters will help reduce greenfly. Companion planting and mixed cropping increases biodiversity and helps attract insects. [Ed. Direct intercropping and companion planting is really not necessary as described in Chapter 11 because the beneficial insects can easily migrate from nearby flower borders and hedges.]

Grow Perennial Species Where Possible

The gardening world, especially when it comes to growing food, is becoming aware of the advantages of perennial crops. This is thanks to the pioneering work of amongst others, Ken Fern (*Plants for a Future*) and Robert Hart (*Forest Gardening* techniques).
There are plenty of perennial food crops available. [A good book on the topic is: *Perennial Vegetables* by Eric Toensmeier, Chelsea Green Publishing, 2007.] These are much easier to grow and once established they will come back of their own accord year after year. Very little research has been carried out on replacing annual vegetables with perennial food crops, but if you would like more information on this then we recommend you contact Plants For A Future (http://www.pfaf.org/). This organization is vegan-organic; they have carried out extensive research into alternative food crops and other useful species."

Resources :

Organizations and groups promoting vegan, organic growing:

Vegan Organic Network, UK (They also publish the magazine "Growing Green International.")
VON enquiries to: media@veganorganic.net
www.veganorganic.net
www.stockfreeorganic.net
info@veganorganic.net

Veganic Agriculture Network for North America
Pacific Roots Magazine offers an online magazine and podcast with a strong focus on veganic agriculture. They promote plant-based farming and gardening in North America. Started by a small group of Canadians, this network is closely associated with the Vegan Organic Network in the UK. From food justice to farmer interviews, *Pacific Roots Magazine* also covers the veganic movement around the world.
www.goveganic.net
What follows are some examples of perennial foods to consider:

TWO-THIRDS LIFE SIZE *FIG.* $1 \times \frac{1}{12}$

1 JERUSALEM ARTICHOKE 1A Tuber 2 OCA 2A Tubers
3 ULLUCO 4 YSAÑO (TROPAEOLUM) 4A Tubers 179

NUTS LIFE SIZE *BRANCHES* × ¼

1 WILD HAZEL NUT 1A Flowering and fruiting branches 1B Cultivated COBNUTS
2 FILBERT NUTS 2A Fruiting branch
3 SWEET CHESTNUTS 3A Flowering and fruiting branches
4 ALMOND NUT 4A Flowering and fruiting branches

27

FRUITS

FRUITS : 1. Abricot. — 2. Orange. — 3. Mandarine. — 4. Pamplemousse. — 5. Bigarade. — 6. Bergamotte. — 7. Cédrat. — 8. Citron. — 9. Grenade. — 10. Coing du Japon. — 11. Kaki de Mazel. — 12. Kaki Tomate. — 13. Kaki elliptique. — 14. Kaki de Wiesemer. — 15. Figue violette. — 16. Figue blanche. — 17. Figue noire. — 18. Nèfle.

LÉGUMES-FRUITS : 19. Pastèque (Melon d'eau). — 20. Melon Cantaloup. — 21. Melon Sucrin brodé de Tours. — 22. Melon Noir des carmes. — 23. Melon à rames. — 24. Melon de Perse. — 25. Melon Pêche. — 26. Concombre. — 27. Chayote. — 28. Aubergine ordinaire. — 29. Aubergine ronde. — 30. Tomate ordinaire. — 31. Tomate olive. — 32. Tomate poire. — 33. Coqueret de Chine. — 34. Piment long. — 35. Piment carré.

FRUITS DES PAYS CHAUDS : 36. Feijoa. — 37. Caroube. — 38. Figue de Barbarie ou Figue d'Inde. — 39. Ananas. — 40. Anone Cœur de bœuf. — 41. Anone Pomme cannelle. — 42. Anone Corossol. — 43. Fruit de l'Arbre à pain. — 44. Poire d'Avocat. — 45. Banane blanche. — 46. Banane rouge. — 47. Banane de Chine. — 48. Noix de Coco. — 49. Goyave Pomme. — 50. Goyave Poire. — 51. Litchi. — 52. Mangoustan. — 53. Mangue blanche. — 54. Mangue rose. — 55. Grenadille (Passiflore). — 56. Papaye. — 57. Pomme Cythère. — 58. Monbin. — 59. Sapotille. — 60. Jambosa (Pomme rose). — 61. Noix d'Acajou ou de Cajou 62. Dattes.

INDEX

Note: Page numbers in *italics* refer to information in illustrations or photographs. Page numbers in **boldface** refer to information in tables and graphs.

A

B

C

D

E

G

H

M

N

O

P

T

U

V

W

X

Y

Z

Lazy-Ass Gardening

Maximize your soil; Minimize your toil

Bob has filled the book with lots of easy-to-follow advice for beginners and an astounding amount of "I did not know that" info for grizzled veterans of the garden wars.

Mike McGrath, host of "You Bet Your Garden"

The Stats:

241 pages

Size - 8.5" x 11"

78 B&W Informative Illustrations

16-page Color Insert with 68 Color Images

To Order: See www.robertkourik.com

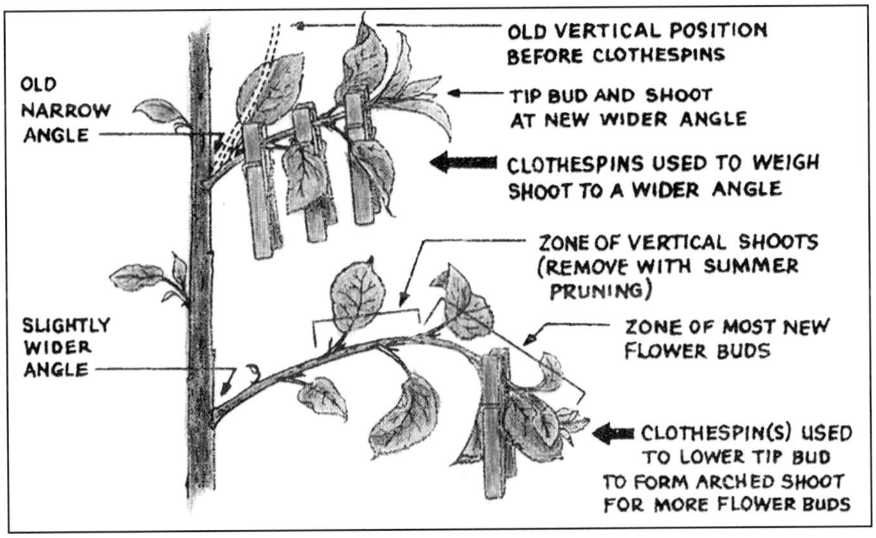

Shape highly productive apple and pear trees with ease and fun without a pruning clipper. Fully explained.

Only $20

No tax, No S&H

No Digging - For a Healthier Soil and a Sustainable Garden

Mother Nature didn't bend over much to make soil. Nature slowly builds soil from the top down—it takes a 1000 years to make an inch of topsoil. She (or he?) doesn't dig and invert soil. This handbook is an important introduction to No-Dig gardening. If you care about global climate change, (and it's not cool to deny it), you'll be amazed to learn that tillage releases a large amount of the carbon dioxide and methane. Tillage kills off many soil critters. Each one either coughs out or farts carbon dioxide and methane as they "expire." The handbook (how long does a handbook have to be before it morphs into a book?) also includes a word about surface cultivation when the top two-three inches of soil is skimmed (not inverted)—no shovels are used. This is a great stepping off handbook to an increasingly exciting way to garden. It can be easier on the backs of aging baby boomers. If nature doesn't bend over, why should you? Includes a description of permanent raised garden beds that thwart pesky gophers. (East of the Mississippi river you don't have these dreadful creatures. Lucky you.) Also includes a page of references.

$3.00 for the 10-page eBook (PDF) version. See: robertkourik.com to order.

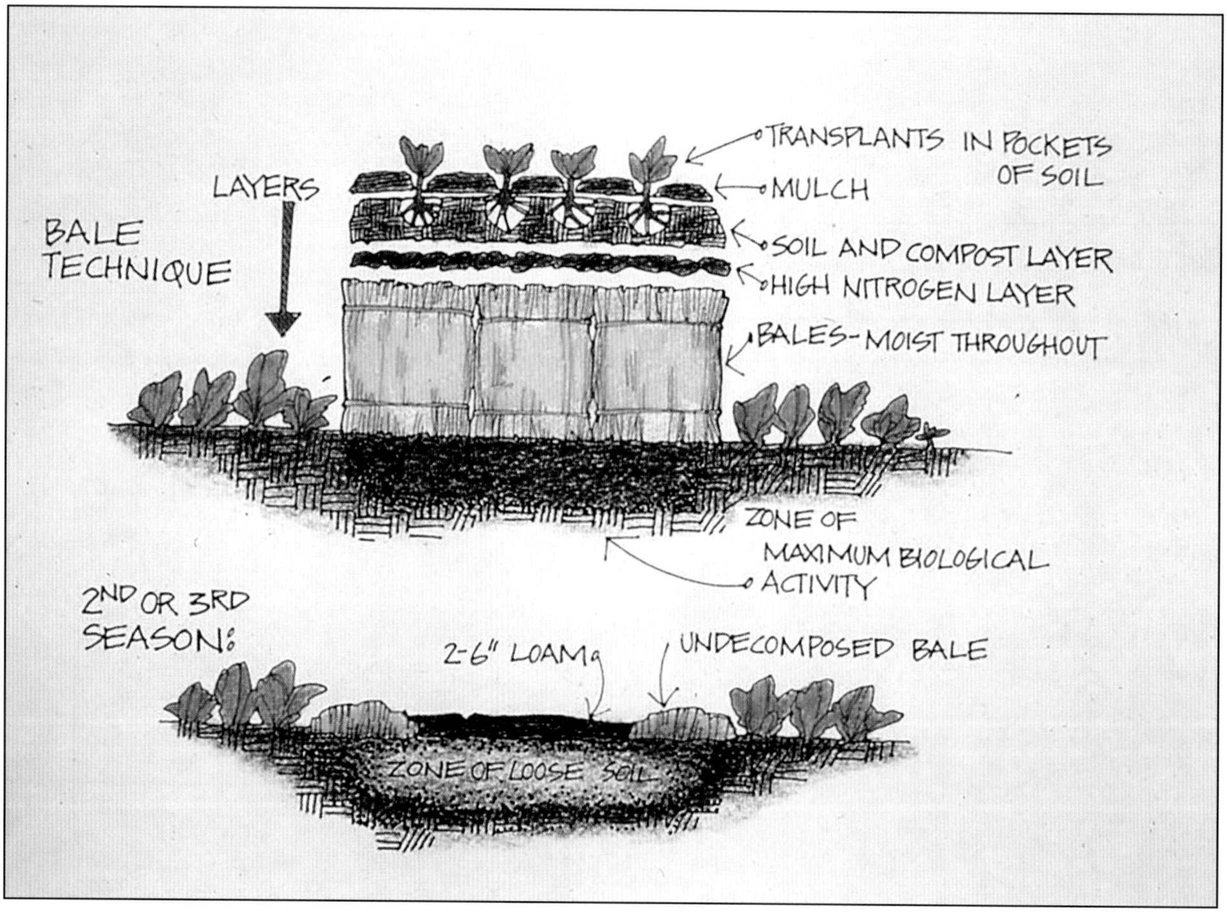

If you want an
Edible Landscape
This book is for you!

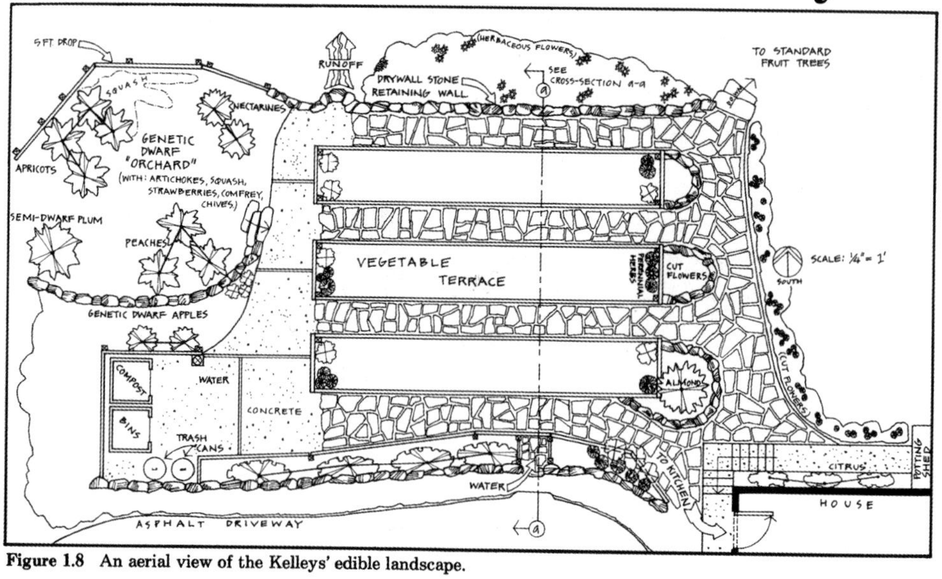

Figure 1.8 An aerial view of the Kelleys' edible landscape.

"The permaculture book before there was permaculture."

One of the most respected texts on the subject is *Designing and Maintaining Your Edible Landscape—Naturally*, by Robert Kourik. Originally published in 1986, the book continues to sell steadily as a nationally-favored gardening reference. It remains a classic text for gardeners of all skill levels. Now, nearly 40 years later, this classic book is still in print.

Part 1: The Basics of Edible Landscaping - Introduction to Edible Landscaping, Organic Gardening—Myths and Facts, Understanding Your Property.

Part 2: Designing Your Edible Landscape - Plan Before You Plant, Energy-Conserving Landscaping, Putting It All Together.

Part 3: Growing Vegetables - No-Till Gardening: Less Work, Good Yields, Tillage: A Little More Work, Greater Yields, Growing Healthier Vegetables.

Part 4: Growing Tree Crops - Choosing the Best Trees for Your Climate, Roots, Soils, ttand Rootstocks, Trees of All Sizes, Planting Trees, Pruning Fruit and Nut Trees, Caring for Your Fruit and Nut Trees.

Part 5: Biological Balance with Insects.

Part 6: Soil Is Health—Improving Soils, Free Fertilizers.

Part 7: Mixing Trees, Lawns, Flowers, Herbs, Vines, and Vegetables.

Part 8: The Edible Landscape Cookbook.

371 Pages, Soft cover - $40 USD (Was $50, includes free S&H, autograph, and no CA tax.)

To Order: See www.robertkourik.com for sample pages & more information.

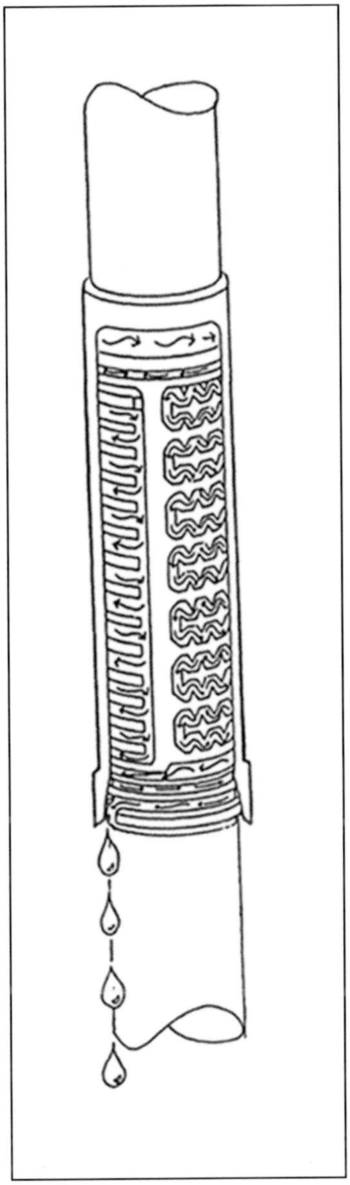

Drip Irrigation
for Every Landscape and All Climates
2nd Edition

"Fun book, great info.
Buy it and save the world's water supply."
Jim McCausland, ***Sunset Magazine***

Avant Gardener's list of the top 25 gardening books of the past 25 years rates ***Drip Irrigation for Every Landscape and All Climates*** by Robert Kourik at number seven (7)! It is simply "the best drip irrigation book on the market".

Yes, ***Drip Irrigation for Every Landscape and All Climates*** is now available in an **updated second edition**. It will give you the know-how to optimize your water use, survive a protracted drought, or work with the unpredictable intermittent rains of certain climates. Its light-hearted approach, step-by-step instructions and clearly drawn diagrams will allow you to construct and streamline your own drip-irrigation system without the clutter of superfluous widgets and gizmos.

Drip Irrigation for Every Landscape and All Climates also describes, in clear and scientific terms, how drip-irrigation systems can be designed or reconfigured to use less water while increasing yields of fruits and vegetables and improving the growth of all plants. New in this edition are descriptions of how to use two very important water sources—grey water and cisterns—as well as how to utilize satellite control or a home-based weather station to irrigate your garden with precision.

Page Count —192 (74 more pages than the 1993 edition)
Number of illustrations—89 B&W (21 new B&W illustrations)
Includes: A Bibliography, Glossary, Web sites to explore (new to this addition), and an expanded and detailed Index.

Special offer: Only $20
Save 20%, lists for $24.95
Free shipping. No tax. Autograph included.

Buy From the Source So Writers Can Keep on Writing™

Sent free via Media Mail [4th Class]. Expect delivery in one to two weeks depending on how close you are to the Northern San Franscico Bay Area.

Go to www.robertkourik.com to order via PayPal.

Understanding Roots...

discover how to make your garden flourish.

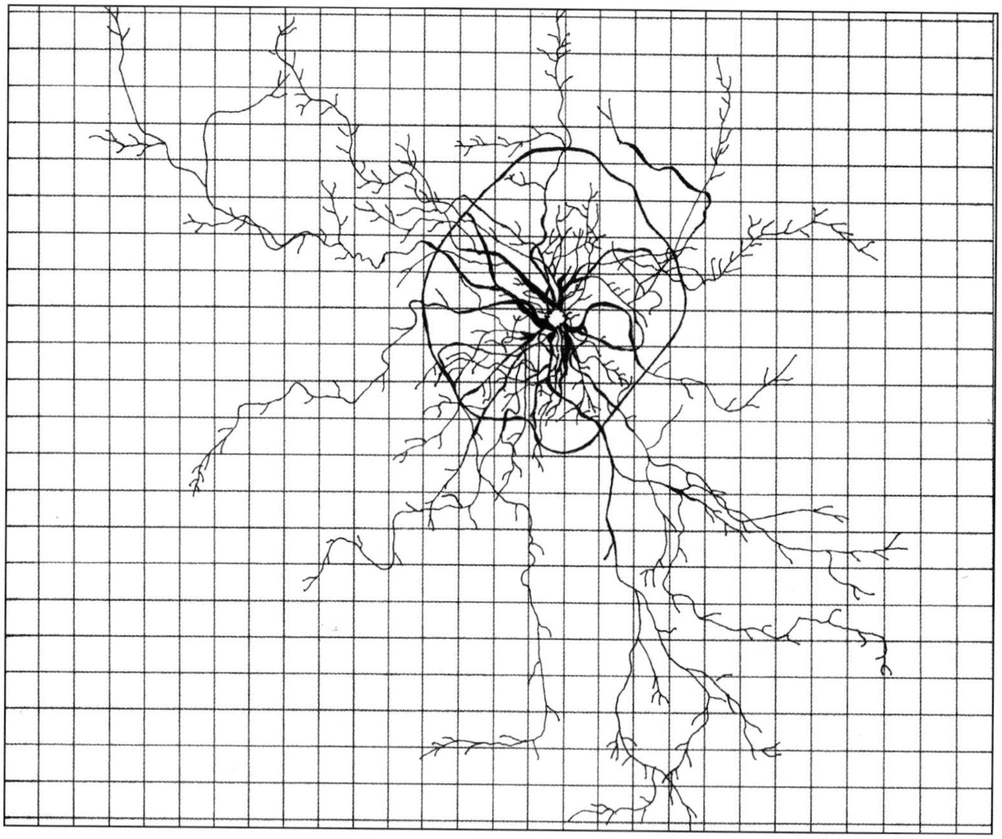

Yes, this might look like a new version of my best-selling *Roots Demystified: Change Your Garden Habits to Help Roots Thrive*, but it's oh so much more, and so different that I've given it a new title. Here're some highlights:

- Seventy-seven enchanting and revealing root drawings that most people have never seen (this is in addition to the 25 in the previous book) plus even more in-depth and up-to-date explorations of some important subjects:
- A look at drawings of roots from the Prairie, Grasslands, and the Desert, Also drawings based on excavations of Vegetable, Fruit, Nut and Ornamental tree roots.
- The mysteries of mycorrhizal association (all about the good fungus that feeds plants).
- The hidden life of soil microbes and their secret relationship to roots.
- All aspects of transplanting (the "hole" truth; a very detailed review).
- How roots work their magic to improve all the soil's nutrients.
- The real effect of cardboard mulch on apple-tree roots (the only research on this offbeat but important topic, with an additional 63 apple root drawings).
- Why grey water increases crop yields more than fresh water. (Really? Yep.)
- The science behind all the above (and citations for each scientific paper that helped me learn it).

I estimate this book contains at least eighty percent more new information and helpful guidelines on roots than the previous book. Enjoy!

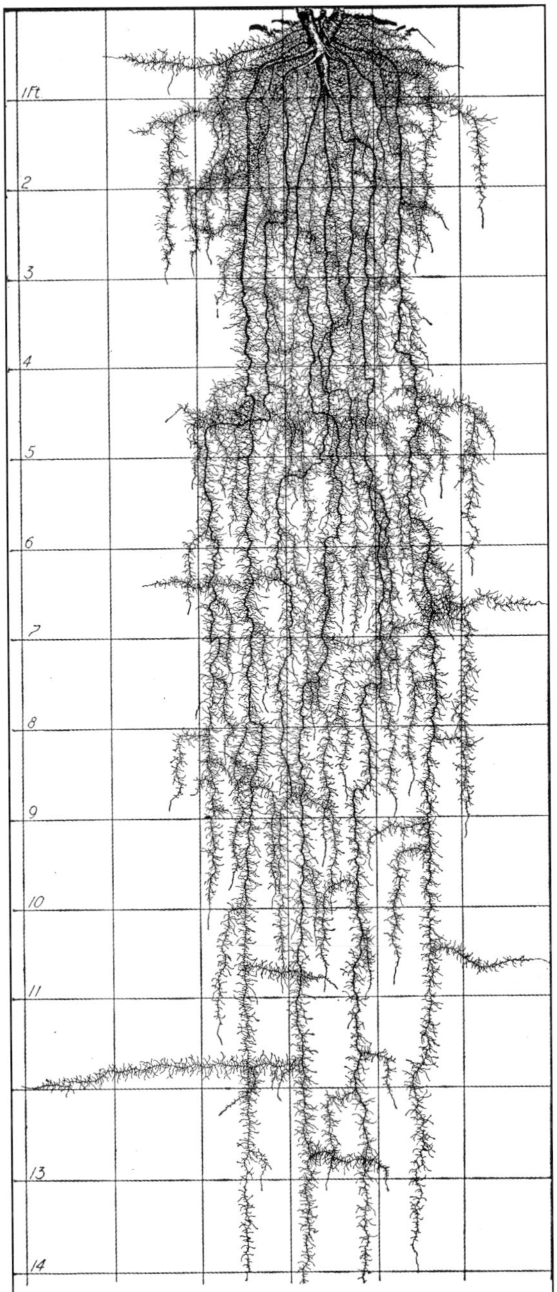

45.—Mature root system of a 10-year-old plant of horse-r

The Biggest and Best Insectary Plant List in North America

This chart provides guidelines to the blooming cycles of various insect-attracting plants—often referred to as "insectary" plants. (The term "insectary," which previously only meant a laboratory for the study of live insects, has been adopted among horticulturists as an adjective to describe plants that attract beneficial insects.)

The bloom cycle of 191 plants is presented.
13 beneficial insects are listed.
117 plants have notations of which of the 13 beneficial insects the flower attracts.
There are 30 full citations for further reading.

It will also inform you on how to choose the best plants to lure beneficial insects outside of the window of midsummer bloom. Because so many beneficial plants bloom in midsummer, that time of year is a banquet of blossom for helpful bugs. Since pests don't magically disappear in the spring and fall, this chart will help you choose the plants that will bloom in your garden during the "in-between" seasons.

Most of the bloom periods came from www.Calflora.org which is, as would be expected, Californiacentric. Some are from the Missouri Botanic Garden database. With careful observation, you will see how your garden blooms react to the seasons and "slide" the bloom-period bar on the chart toward Spring or Fall.

You will receive via email:
• A PDF document of the guidelines and citations and
• An Excel spreadsheet of the 191 plants and beneficial-insect visitations. (Use Excel version 14.7.1 or newer)
(Note: This is the product of 100-200 hours of work over three years by myself and Damien McAnany. This final set represents a tremendous amount of condensed information. The Excel spreadsheet is deemed to be the best way to present it for you.)

$5 for both files **To Order: See www.robertkourik.com**

Bloom Period of Insectary Plants

Common Name	***Botanical Name***	**Jan.**	**Feb.**	**Mar.**	**Apr.**	**May**	**June**	**July**	**Aug.**	**Sept.**	**Oct.**	**Nov.**	**Dec.**
Agastache[24, 28]	*Agastache foeniculum*												
Alfalfa[1,7,9]	*Medicago sativa*												
Alpine Cinquefoil[1]	*Potentilla concinna*												
Alyssum, Sweet[1,3,7,9,13,14,16,21]	*Lobularia maritima*												
Angelica[1,3,5,9,22,26,30]	*Angelica* spp.												
Anise Hyssop[1,9,10,22,30]	*Agastache foeniculum*												
Aquelegia[21a]	*Aquelegia x hybrida 'Blue Bird'*												
Arugula[23, 25]	*Eruca sativa*												
Asters[3, 6, 22, 30]	*Aster laterifolius, A. x frikartii* 'Monch'												

Deep Roots—No, Shallow Roots—Yes

Think deep roots get deep nutrients? Guess again.

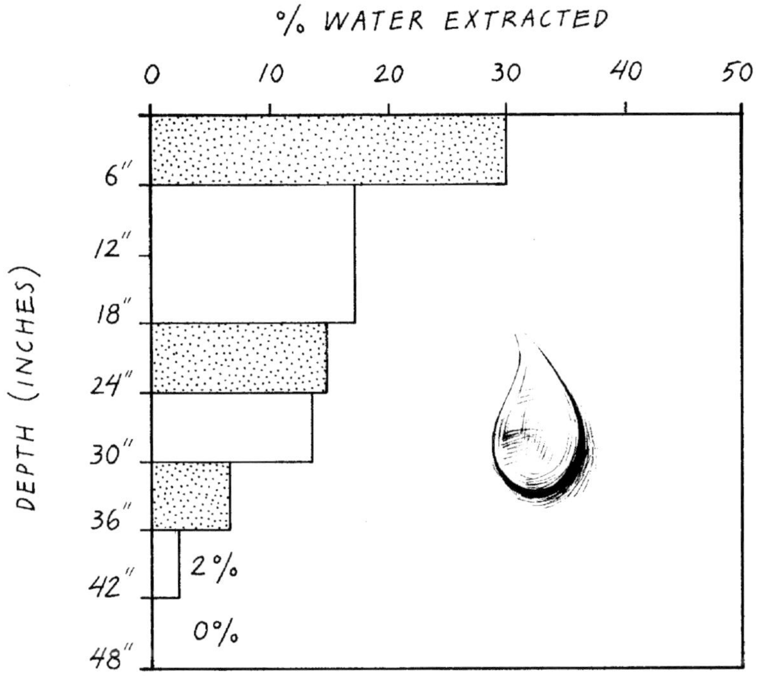

"In many cases, plants 'gather' more water near the soil surface than they do in deeper soils. This chart (from 'Spatial and Temporal Distribution of Soil Water Balance for a Drip-irrigated Almond Tree' (by L. Andreu, J.W. Hopmans and L.J. Schwankl, Agricultural Water Management, 35:123-146, May 1997) shows that the roots in the top six inches absorb three times more moisture than those growing 36 inches deep."

Greater Garden Yields with Drip Irrigation

(It's not just for droughts anymore.)

Hey dripologists, wanna see how drip irrigation increases yields? First, read how the scientific literature—boring, but still fascinating—proves that, in spite of reducing water, vegetables yields go up. Nearly 100 vegetable, nut and fruit examples in 34 citations from climates throughout the country and world. Enough examples for you? You're gonna believe it's true. It's even shown that yields can be up to 48% greater with drip irrigation than other methods of irrigation. Cool, no? You can learn why roots love drip irrigation moisture. Believe it or don't, but I think you'll learn how it works. You can also learn why so many of the soil's critters love the upper levels of the soil. Maintaining healthy bacteria, microbes and other soil life equals greater yields. So peruse this eBook on drip irrigation of your veggies. It's cheap.

78 pages of great information as a PDF.

$6, Buy now and the PDF will be sent to you via email

To Order: See www.robertkourik.com

"Once again, Robert Kourik has done it! His latest book *Greater Garden Yields With Drip Irrigation* is the perfect blend of down to earth practical advice and interesting scientifically-accurate information."

Dr. Judyth McLeod was the head of landscaping in a leading Australian university for thirty years.

Recommendations continued from back cover—

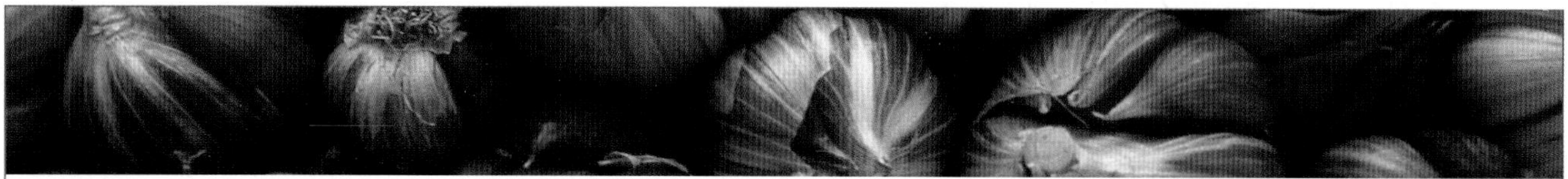

Robert Kourik's ***Sustainable Food Gardens*** is an invaluable resource for both beginning and experienced gardeners. Using copious amounts of scientific research on horticulture and agriculture, Kourik boils studies down to their essential and relevant facts for the home grower, while also dispelling long-held myths and favorite theories that have not been proven or peer-reviewed in the field. If you are looking for a guide to growing that is Earth-friendly, accurate, and does not indulge in faddish belief systems, this book is for you.

***—Kathy Jentz, speaker, author, and editor of* Washington Gardener Magazine**

If "Gardening Techniques" was a high school debate topic, I would want Robert Kourik on my team. He could argue effectively on the pros and cons of anything you are doing in your garden. In his latest book, ***Sustainable Food Gardens***, Robert looks at long-held, tried-and-true landscaping practices. He defends those techniques while at the same time pointing out their flaws: mulching (a few inches is good; more than that may cause problems); raised bed gardening (boxes are good for root development, bad for aesthetics); drip irrigation (your intentions are good, but you're probably using it wrong). And, he may be the first gardener to tell you aphids don't suck (they just pierce a plant part, and the plant juices stream into their mouths. And you're helping them get that easy meal by overfertilizing). ***Sustainable Food Gardens*** is the one-volume compilation of Robert's best thoughts and ideas from all his previous books. Keep in mind what I have learned from Robert's books and newsletters over the years: "Everything you know is wrong," and, "If it works for you, fine. But keep an open mind." After poring through all 400+ pages of this book, you may need a drink. Robert comes through again, with a unique recipe for "Lavenchello"—a concoction of vodka, lemons…and lavender.

—Fred Hoffman, producer/host, "Garden Basics with Farmer Fred" podcast

Your work is amazing. You provide so much practical, detailed info…you are unique in your ability to provide people with accurate science based useful information that is fun to read. It's a unique talent. Our younger generation is taking up gardening because they want to eat healthy food and be part of future sustainability. But if they believe the seductive pie in the sky things they read and then they fail at gardening and give up, that is a serious problem for the world. Thank you for making it possible for them to succeed.

—Sam Benowitz, Founder of Raintree Nursery

Robert Kourik's ***Sustainable Food Gardens*** is quite a book. I loved it. I am not sure I have ever read any gardening tome with more information, written so well and perfectly supplemented by charts, drawings and pictures. ***Sustainable Food Gardens*** is destined to become a classic! If you want to be a thinking and successful food gardener, just get it.

—Jeff Lowenfels "Lord of the Roots, Author of the Teaming series"

Robert Kourik's latest book, ***Sustainable Food Gardens***, is a tour de force. He calls it a sequel to his first book, ***Designing and Maintaining Your Edible Landscape***—naturally, which was published in 1986, but it is so much more than that, especially with the addition of full color photographs and graphics and all the scientific research for which Robert is so well known. Going beyond edibility deep into the realm of sustainability—so important today—it encompasses more than you imagined there was to know (from soil to roots, pests to design, forest gardening to conserving resources, and even vegan gardening), which can make anyone, beginner or experienced, a better and more knowledgeable gardener with results they can see and eat! I'm going out to change some of my container gardening practices today!

***—Cathy Wilkinson Barash, gourmet horticulturist, author of* Edible Flowers from Garden to Palate,**
past president of GardenComm (formerly Garden Writers Association